Graduate Texts in Mathematics 50

Springer
New York
Berlin
Heidelberg
Barcelona
Hong Kong
London
Milan
Paris
Singapore
Tokyo

Graduate Texts in Mathematics

(continued after index)

Harold M. Edwards

Fermat's Last Theorem

A Genetic Introduction
to Algebraic Number Theory

Work on this book was supported in part by the
James M. Vaughn, Jr., Vaughn Foundation Fund.

Springer

Harold M. Edwards
Courant Institute of Mathematical Sciences
New York University
251 Mercer Street
New York, NY 10012
USA

Mathematics Subject Classification (1991): 01A50, 01A55, 11-03, 11A15, 11A41, 11D41

Library of Congress Cataloging in Publication Data

Edwards, Harold M.
  Fermat's last theorem

  (Graduate texts in mathematics ; 50)
  Bibliography: p.
  Includes index.
  1. Algebraic number theory. 2. Fermat's theorem.
I. Title. II. Series.
QA247.E38   512'.74   77-8222
ISBN 0-387-95002-8

Printed on acid-free paper.

ISBN 0-387-95002-8 Springer-Verlag New York Berlin Heidelberg SPIN 10755990

Preface

Since it is likely that many people will open this book wanting to know what the current state of knowledge about Fermat's Last Theorem is, and since the book itself will not answer this question, perhaps the preface should contain a few indications on the subject. Fermat's Last Theorem is of course the assertion (not a theorem) that the equation $x^n + y^n = z^n$ has no solution in positive whole numbers when $n > 2$. It is elementary (see Section 1.5) to prove that $x^4 + y^4 = z^4$ is impossible. Therefore the original equation is impossible whenever n is divisible by 4. (If $n = 4k$ then $x^n + y^n = z^n$ would imply the impossible equation $X^4 + Y^4 = Z^4$ where $X = x^k$, $Y = y^k$, and $Z = z^k$.) Similarly, if $x^m + y^m = z^m$ can be proved to be impossible for any particular m, it will follow that the original equation is impossible for any n that is divisible by m. Since every $n > 2$ is divisible either by 4 or by an odd prime, *in order to prove Fermat's Last Theorem it will suffice to prove it in the cases where the exponent n is a prime.*

For the exponent 3, the theorem is not too difficult to prove (see Chapter 2). For the exponents 5 and 7 the difficulties are greater (Sections 3.3 and 3.4), but the theorem can be proved by essentially elementary methods. The main topic of this book is the powerful theory of ideal factorization which Kummer developed in the 1840s and used to prove the theorem at one stroke for all prime exponents less than 100 other than 37, 59, and 67. Specifically, Kummer's theorem states: *Let p be an odd prime. A sufficient condition for Fermat's Last Theorem to be true for the exponent p is that p not divide the numerators of the Bernoulli numbers B_2, $B_4, \ldots, B_{p-3}$.* (See Sections 5.5 and 6.19.) A prime which satisfies Kummer's sufficient condition is called "regular."

Since 1850, work on the theorem has centered on proving more and more inclusive sufficient conditions. In one sense the best known sufficient

v

conditions are now very inclusive, and in another sense they are very disappointing. The sense in which they are inclusive is that *they include all primes less than* 100,000 [W1]. The sense in which they are disappointing is that *no sufficient condition for Fermat's Last Theorem has ever been shown to include an infinite set of prime exponents*. Thus one is in the position of being able to prove Fermat's Last Theorem for virtually any prime within computational range, but one cannot rule out the possibility that the Theorem is *false* for *all* primes beyond some large bound.

The basic method of the book is, as the subtitle indicates, the genetic method. The dictionary defines the genetic method as "the explanation or evaluation of a thing or event in terms of its origin and development." In this book I have attempted to explain the basic techniques and concepts of the theory, and to make them seem natural, manageable, and effective, by tracing their origin and development in the works of some of the great masters—Fermat, Euler, Lagrange, Legendre, Gauss, Dirichlet, Kummer, and others.

It is important to distinguish the genetic method from history. The distinction lies in the fact that the genetic method primarily concerns itself with the subject—its "explanation or evaluation" in the definition above—whereas the primary concern of history is an accurate record of the men, ideas, and events which played a part in the evolution of the subject. In a history there is no place for detailed descriptions of the theory unless it is essential to an understanding of the events. In the genetic method there is no place for a careful study of the events unless it contributes to the appreciation of the subject.

This means that the genetic method tends to present the historical record from a false perspective. Questions which were never successfully resolved are ignored. Ideas which led into blind alleys are not pursued. Months of fruitless effort are passed over in silence and mountains of exploratory calculations are dispensed with. In order to get to the really fruitful ideas, one pretends that human reason moves in straight lines from problems to solutions. I want to emphasize as strongly as I can that this notion that reason moves in straight lines is an outrageous fiction which should not for a moment be taken seriously.

Samuel Johnson once said of the writing of biography that "If nothing but the bright side of characters should be shown, we should sit down in despondency, and think it utterly impossible to imitate them in anything. The sacred writers related the vicious as well as the virtuous actions of men; which had this moral effect, that it kept mankind from despair." This book does, for the most part, show only the bright side, only the ideas that work, only the guesses that are correct. You should bear in mind that this is *not* a history or biography and you should not despair.

You may well be interested less in the contrast between history and the genetic method than in the contrast between the genetic method and the more usual method of mathematical exposition. As the mathematician

Otto Toeplitz described it, the essence of the genetic method is to look to the historical origins of an idea in order to find the best way to motivate it, to study the context in which the originator of the idea was working in order to find the "burning question" which he was striving to answer [T1]. In contrast to this, the more usual method pays no attention to the questions and presents only the answers. From a logical point of view only the answers are needed, but from a psychological point of view, learning the answers without knowing the questions is so difficult that it is almost impossible. That, at any rate, is my own experience. I have found that the best way to overcome the difficulty of learning an abstract mathematical theory is to follow Toeplitz's advice and to ignore the modern treatises until I have studied the genesis in order to learn the questions.

The first three chapters of the book deal with elementary aspects of the question of Fermat's Last Theorem. They are written at a much more elementary level than the rest of the book. I hope that the reader who already has the mathematical maturity to read the later chapters will still find these first three chapters interesting and worthwhile, if easy, reading. At the same time, I hope that the less experienced reader who must work his way more gradually through the first chapters will in the course of that reading acquire enough experience to enable him, with effort, to make his way through the later chapters as well.

The next three chapters, Chapters 4–6, are devoted to the development of Kummer's theory of ideal factors and its application to prove his famous theorem, stated above, that Fermat's Last Theorem is true for regular prime exponents. This is as far as the present book takes the study of Fermat's Last Theorem. I plan to write a second volume to deal with work on Fermat's Last Theorem which goes beyond Kummer's theorem, but these later developments are difficult, and Kummer's theorem is a very natural point at which to end this volume.

The final three chapters deal with matters less directly related to Fermat's Last Theorem, namely, the theory of ideal factorization for quadratic integers, Gauss's theory of binary quadratic forms, and Dirichlet's class number formula. To study Kummer's work on Fermat's Last Theorem without studying these other aspects of number theory would be as foolish as to study the history of Germany without studying the history of France. Kummer was aware at the very outset of his work on ideal theory that it was closely related to Gauss's theory of binary quadratic forms. While the application to Fermat's Last Theorem was one of Kummer's motives for developing the theory, others (and by his own testimony more immediate ones) were the quest for the generalization of quadratic, cubic, and biquadratic reciprocity laws to higher powers, and the explication of Gauss's difficult theory of composition of forms. Moreover, Kummer's amazingly rapid development of his class number formula and his discovery of the striking relationship between Fermat's Last Theorem for the exponent p and the Bernoulli numbers mod p, were, as he says, made possible by Dirichlet's solution of the analogous problem in the quadratic case. The genetic method suggests—almost demands—that these other issues be exploited in motivating the difficult but enormously fruitful idea of "ideal prime factors" that is so essential to an understanding of Kummer's work on Fermat's Last Theorem. Moreover, the material of these final three chapters provides a necessary background for the study of the higher reciprocity laws and class field theory which, in turn, are the context of the later work on Fermat's Last Theorem to be studied in the second volume.

In this book there is a good deal of emphasis, both in the text and in the exercises, on *computation*. This is a natural concomitant of the genetic method because, as even a superficial glance at history shows, Kummer and the other great innovators in number theory did vast amounts of computation and gained much of their insight in this way. I deplore the fact that contemporary mathematical education tends to give students the idea that computation is demeaning drudgery to be avoided at all costs. If you follow the computations of the text attentively and if you regard the more computational exercises not only as time-consuming (which they will inevitably be) but also as challenging, enjoyable, and enlightening, then I believe you can come to appreciate both the power and the ultimate simplicity of the theory.

I believe that there is no such thing as a passive understanding of mathematics. It is only in actively lecturing, writing, or solving problems that one can achieve a thorough grasp of mathematical ideas. This is the reason that the book contains so many exercises and the reason that I suggest that the serious reader do as many of them as he can. Some of my colleagues have suggested that by including so many exercises I will deter readers who want to read the book merely for the fun of it. To this I reply that the exercises are offered, they are not assigned. Do with them what you will, but you might in fact find that they can be fun too.

A famous prize for the proof of Fermat's Last Theorem was established by P. Wolfskehl in 1908. One of the conditions of the prize is that the proof must appear in print, and the primary result of the offer of the prize seems to have been a plague of nonsense proofs being submitted for publication and being published privately. It was with obvious satisfaction that Mordell and other number theorists announced that the post–World War I inflation in Germany had reduced the originally munificient prize to almost nothing. However, the economic recovery of Germany after World War II has reversed this situation to some extent. The Wolfskehl prize is at the present time worth about 10,000 DM or 4,000 American dollars. In order to win the prize, a proof must be published and must be judged to be correct, no sooner than two years after publication, by the Academy of Sciences in Gottingen.

If you are inclined to try to win the prize, you have my best wishes. I would be truly delighted if the problem were solved, and especially so if the solver had found my book useful. Although it might be argued that a book full of ideas that *haven't* worked couldn't possibly be of any use to someone hoping to solve the problem, I think that the unsuccessful efforts of so many first-rate mathematicians—not to mention many not-so-first-rate ones—are enough to render a naive approach to the problem completely hopeless. The ideas in this book *do* solve the problem for all exponents less than 37, which is more than can be said of any approach to the problem which does not use Kummer's theory of ideal factorization. But before you set out to win the Wolfskehl prize there is one further fact which you should take into account: there seems to me to be no reason at all to assume that Fermat's Last Theorem is true, but the prize does not offer a single *pfennig* for a disproof of the theorem.

Note added in the fifth printing

The Annals of Mathematics for May, 1995, contains a proof of Fermat's last theorem by Andrew Wiles of Princeton University, which, in all probability, will receive the Wolfskehl prize once the requisite two years have elapsed. This great achievement, which has been celebrated by scholars and the general public all over the world, will probably enhance rather than end interest in the topic. Not surprisingly, Wiles's proof uses very sophisticated modern concepts that could not possibly have been used by Fermat. We now know that Fermat's last theorem is true, but whether Fermat himself could prove it is still unresolved. The quest for a proof that might have been accessible to Fermat will surely continue.

Acknowledgments

Merely to say that work on this book was supported by a grant from the Vaughn Foundation would give a very inadequate notion of the extent of my indebtedness to the Foundation and to Mr. James M. Vaughn, Jr. Without their invitation for proposals related to Fermat's Last Theorem, I would never have conceived of this book, let alone written it. I am profoundly grateful to Mr. Vaughn and to the Foundation for involving me in what has been an extremely stimulating, enlightening, and enjoyable project.

I would also like to thank Bruce Chandler for his friendly encouragement and for his wise counsel. Bruce's Seminar on the History of Mathematical Sciences at NYU has been a meeting place for historians of mathematics which I and many others have found to be enormously useful and inspiring.

In addition, I am grateful to many scholars for comments on the manuscript. In particular I would like to mention John Brillhart, Ed Curtis, Pierre Dugac, J. M. Gandhi, Lynette Ganim, Paul Halmos, Jean Itard, Walter Kaufmann-Bühler, Morris Kline, Carlos Moreno, Al Novikoff, Harold Shapiro, Gabriel Stolzenberg, James Vaughn, and André Weil.

Finally, I am grateful to the Courant Institute of Mathematical Sciences here at NYU for providing a good place to work, an excellent library, and the superb typing services of Helen Samoraj and her staff.

April 1977
New York City

Harold M. Edwards

Contents

with which a prime divisor divides a cyclotomic integer. The one prime divisor $(1 - \alpha)$ of λ. **4.11 The fundamental theorem.** A cyclotomic integer $g(\alpha)$ divides another $h(\alpha)$ if and only if every prime divisor which divides $g(\alpha)$ divides $h(\alpha)$ with multiplicity at least as great. **4.12 Divisors.** Definition of divisors. Notation. **4.13 Terminology.** A divisor is determined by the set of all things that it divides. "Ideals." **4.14 Conjugations and the norm of a divisor.** Conjugates of a divisor. Norm of a divisor as a divisor and as an integer. There are $N(A)$ classes of cyclotomic integers mod A. The Chinese remainder theorem. **4.15 Summary.**

Chapter 5   Fermat's Last Theorem for regular primes

5.1 Kummer's remarks on quadratic integers. The notion of equivalence of divisors. Kummer's allusion to a theory of divisors for quadratic integers $x + y\sqrt{D}$ and its connection with Gauss's theory of binary quadratic forms. **5.2 Equivalence of divisors in a special case.** Analysis of the question "Which divisors are divisors of cyclotomic integers?" in a specific case. **5.3 The class number.** Definition and basic properties of equivalence of divisors. Representative sets. Proof that the class number is finite. **5.4 Kummer's two conditions.** The types of arguments used to prove Fermat's Last Theorem for the exponents 3 and 5 motivate the singling out of the primes λ for which (A) the class number is not divisible by λ and (B) units congruent to integers mod λ are λth powers. Such primes are called "regular." **5.5 The proof for regular primes.** Kummer's deduction of Fermat's Last Theorem for regular prime exponents. For any unit $e(\alpha)$, the unit $e(\alpha)/e(\alpha^{-1})$ is of the form α^r. **5.6 Quadratic reciprocity.** Kummer's theory leads not only to a proof of the famous quadratic reciprocity law but also to a derivation of the statement of the law. Legendre symbols. The supplementary laws.

Chapter 6   Determination of the class number

6.1 Introduction. The main theorem to be proved is Kummer's theorem that λ is regular if and only if it does not divide the numerators of the Bernoulli numbers $B_2, B_4, \ldots, B_{\lambda-3}$. **6.2 The Euler product formula.** Analog of the formula for the case of cyclotomic integers. The class number formula is found by multiplying both sides by $(s - 1)$ and evaluating the limit as $s \downarrow 1$. **6.3 First steps.** Proof of the generalized Euler product formula. The Riemann zeta function. **6.4 Reformulation of the right side.** The right side is equal to $\zeta(s)L(s,\chi_1)L(s,\chi_2) \cdots L(s,\chi_{\lambda-2})$ where the χ's are the nonprincipal characters mod λ. **6.5 Dirichlet's evaluation of** $L(1,\chi)$. Summation by parts. $L(1,\chi)$ as a superposition of the series for $\log(1/(1 - \alpha^j))$, $j = 1, 2, \ldots, \lambda - 1$. Explicit formulas for $L(1,\chi)$. **6.6 The limit of the right side.** An explicit formula. **6.7 The nonvanishing of** L-**series.** Proof that $L(1,\chi) \neq 0$ for the χ's under consideration. **6.8 Reformulation of the left side.** In the limit as $s \downarrow 1$, the sum of $N(A)^{-s}$ over all divisors A in a divisor class is the same for any two classes. Program for the evaluation of their common limit. **6.9 Units: The first few cases.** Explicit derivation of all units in the cases $\lambda = 3, 5, 7$. Finite-dimensional Fourier analysis. Implicit derivation of the units in the case $\lambda = 11$. Second factor of the class number. **6.10 Units: The general case.** Method for finding, at least in principle, all units. Sum over all principal divisors written in terms of a sum over a certain set of cyclotomic integers. **6.11 Evaluation of the integral.** Solution of a problem in integral calculus. **6.12**

Comparison of the integral and the sum. In the limit to be evaluated, the sum can be replaced by the integral. **6.13 The sum over other divisor classes.** Proof that, in the limit, the sum over any two divisor classes is the same. **6.14 The class number formula.** Assembling of all the pieces of the preceding sections to give the explicit formula for the class number. **6.15 Proof that 37 is irregular.** Simplifications of the computation of the first factor of the class number. Bernoulli numbers and Bernoulli polynomials. **6.16 Divisibility of the first factor by λ.** Generalization of the techniques of the preceding section to show that λ divides the first factor of the class number if and only if it divides the numerator of one of the Bernoulli numbers $B_2, B_4, \ldots, B_{\lambda-3}$. **6.17 Divisibility of the second factor by λ.** Proof that λ divides the second factor of the class number only if it also divides the first factor. **6.18 Kummer's lemma.** (A) implies (B). **6.19 Summary.**

Chapter 7   Divisor theory for quadratic integers   245

7.1 The Prime divisors. Determination of what the prime divisors must be if there is to be a divisor theory for numbers of the form $x + y\sqrt{D}$. Modification of the definition of quadratic integers in the case $D \equiv 1 \bmod 4$. **7.2 The divisor theory.** Proof that the divisors defined in the preceding section give a divisor theory with all the expected properties. Equivalence of divisors. **7.3 The sign of the norm.** When $D > 0$ the norm assumes negative as well as positive values. In this case a divisor with norm -1 is introduced. **7.4 Quadratic integers with given divisors.** Unlike the cyclotomic case, for quadratic integers there is a simple algorithm for determining whether a given divisor is principal and, if so, of finding all quadratic integers with this divisor. It is, in essence, the cyclic method of the ancient Indians. Proof of the validity of the algorithm in the case $D < 0$. Exercises: Use of 2×2 matrices to streamline the computations of the cyclic method. **7.5 Validity of the cyclic method.** Proof in the case $D > 0$. Computation of the fundamental unit. **7.6 The divisor class group: examples.** Explicit derivation of the divisor class group for several values of D. **7.7 The divisor class group: a general theorem.** Proof that two divisors are equivalent only if application of the cyclic method to them yields the same period of reduced divisors. This simplifies the derivation of the divisor class group. **7.8 Euler's theorems.** Euler found empirically that the way in which a prime p factors in quadratic integers $x + y\sqrt{D}$ depends only on the class of p mod $4D$. He found other theorems which simplify the determination of the classes of primes mod $4D$ which split and the classes which remain prime. These theorems, unproved by Euler, imply and are implied by the law of quadratic reciprocity. **7.9 Genera.** Gauss's necessary conditions for two divisors to be equivalent. Character of a divisor class. Resulting partition of the divisor classes into genera. **7.10 Ambiguous classes.** Definition. Proof that the number of ambiguous classes is at most half the number of possible characters. **7.11 Gauss's second proof of quadratic reciprocity.** Proof that at most half of the possible characters actually occur. Gauss's deduction, from this theorem, of quadratic reciprocity.

Chapter 8   Gauss's theory of binary quadratic forms   305

8.1 Other divisor class groups. When D is not squarefree the definition of the divisor class group needs to be modified. Orders of quadratic integers.

Equivalence relative to an order. The divisor class group corresponding to an order. Exercises: Euler's convenient numbers. **8.2 Alternative view of the cyclic method.** Interpretation of it as a method for generating equivalent binary quadratic forms. Method for finding representations of given integers by given binary quadratic forms. **8.3 The correspondence between divisors and binary quadratic forms.** Proper equivalence of binary quadratic forms. The one-to-one correspondence between proper equivalence classes of properly primitive forms (positive when $D > 0$) and divisor classes for the order $\{x + y\sqrt{D} : x, y \text{ integers}\}$. **8.4 The classification of forms.** Extension of the theorem of Section 7.7 to the case where D is not squarefree. **8.5 Examples.** Derivation of the divisor class group in several cases. **8.6 Gauss's composition of forms.** How Gauss defined the product of two classes of binary quadratic forms without using divisor theory. **8.7 Equations of degree 2 in 2 variables.** Complete solution, essentially due to Lagrange, of $ax^2 + bxy + cy^2 + dx + ey + f = 0$.

9.1 The Euler product formula. Analog in the case of quadratic integers. Splitting into cases for various types of D. **9.2 First case.** The case $D < 0$, $D \not\equiv 1 \bmod 4$, D squarefree. Derivation of the class number formula. Examples. **9.3 Another case.** The case $D > 0$, $D \not\equiv 1 \bmod 4$, D squarefree. Derivation. Examples. **9.4 $D \equiv 1 \bmod 4$.** Modifications required when $D \equiv 1 \bmod 4$, D squarefree. **9.5 Evaluation of $\Sigma(\frac{D}{n})\frac{1}{n}$.** This term of the class number formula can be evaluated using the technique of Section 6.5. Fourier transform of the character $(\frac{D}{n})$ mod $4D$ is a multiple of itself. Use of this fact to reduce the formula. Exercises: Dirichlet's further reductions of the formula in the case $D < 0$, D squarefree. The sign of Gaussian sums and its relation to this formula. **9.6 Suborders.** Generalization of the class number formula to the case where D is not squarefree and, more generally, to divisor class groups corresponding to arbitrary orders of quadratic integers. **9.7 Primes in arithmetic progressions.** Dirichlet's proof that $an + b$ represents an infinity of primes when b is relatively prime to a. Use of the class number formula to prove $L(1, \chi) \neq 0$ for all real characters χ mod a.

A.1 Basic properties. Addition and multiplication. Euclidean algorithm. Congruence modulo a natural number. Chinese remainder theorem. Solution of $ax \equiv b \bmod c$. Fundamental theorem of arithmetic. Integers. **A.2 Primitive roots mod p.** Definition. Proof that every p has a primitive root.

Fermat 1

1.1 Fermat and his "Last Theorem"

When Pierre de Fermat died in 1665 he was one of the most famous mathematicians in Europe. Today Fermat's name is almost synonymous with number theory, but in his own time his work in number theory was so revolutionary and so much ahead of its time that its value was poorly understood and his fame rested much more on his contributions in other fields. These included important work in analytic geometry—which he invented independently of Descartes—in the theory of tangents, quadrature, and maxima and minima—which were the beginning of calculus —and in mathematical optics—which he enriched with the discovery that the law of refraction can be derived from the principle of least time.

There are two surprising facts about Fermat's fame as a mathematician. The first is that he was not a mathematician at all, but a jurist. Throughout his mature life he held rather important judicial positions in Toulouse, and his mathematical work was done as an avocation. The second is that he never published a single* mathematical work. His reputation grew out of his correspondence with other scholars and out of a number of treatises which circulated in manuscript form. Fermat was frequently urged to publish his work, but for unexplained reasons he refused to allow his treatises to be published and many of his discoveries—particularly his discoveries in number theory—were never put in publishable form.

This fact that Fermat refused to publish his work caused his many admirers to fear that he would soon be forgotten if an effort weren't made to collect his letters and unpublished treatises and to publish them post-humously. The task was undertaken by his son, Samuel. In addition to

*There is one slight exception. He did allow a minor work to be published in 1660 as an appendix to a book written by a colleague. However, it is an exception which proves the rule: it was published anonymously.

1

soliciting letters and treatises from his father's correspondents, Samuel de Fermat went through his father's own papers and books, and it was in this way that Fermat's famous "Last Theorem" came to be published.

Diophantus' *Arithmetic*, one of the great classics of ancient Greek mathematics which had been rediscovered and translated into Latin shortly before Fermat's time, was the book which had originally inspired Fermat's study of the theory of numbers. Samuel found that his father had made a number of notes in the margins of his copy of Bachet's translation of Diophantus, and as a first step in publishing his father's works he published a new edition of Bachet's Diophantus [D3] which included Fermat's marginal notes as an appendix. The second of these 48 "Observations on Diophantus" was written in the margin next to Diophantus' problem 8 in Book II which asks "given a number which is a square, write it as a sum of two other squares." Fermat's note states, in Latin, that "On the other hand, it is impossible for a cube to be written as a sum of two cubes or a fourth power to be written as a sum of two fourth powers or, in general, for any number which is a power greater than the second to be written as a sum of two like powers. I have a truly marvelous demonstration of this proposition which this margin is too narrow to contain." This simple statement, which can be written in symbols as "for any integer $n > 2$ the equation $x^n + y^n = z^n$ is impossible" is now known as *Fermat's Last Theorem*. If Fermat did indeed have a demonstration of it, it was truly "marvelous," because no one else has been able to find a demonstration of it in the three hundred and more years since Fermat's time. It is a problem that many great mathematicians have tried unsuccessfully to solve, although sufficient progress has been made to prove Fermat's assertion for all exponents n well up into the thousands. To this date it is unknown whether the assertion is true or false.

The origin of the name "Fermat's Last Theorem" is obscure. It is not known at what time in his life Fermat wrote this marginal note, but it is usually assumed that he wrote it during the period when he was first studying Diophantus' book, in the late 1630s, three decades before his death, and in this case it surely was not his last theorem. Very possibly the name stems from the fact that of the many unproved theorems that Fermat stated, this is the last one which remains unproved. It is perhaps worth considering that Fermat may have thought better of his "marvelous proof," especially if he did write of it in the 1630s, because his other theorems are stated and restated in letters and in challenge problems to other mathematicians—and the special cases $x^3 + y^3 \neq z^3$, $x^4 + y^4 \neq z^4$ of the Last Theorem are also stated elsewhere—whereas this theorem occurs just once, as observation number 2 on Diophantus, a sphinx to mystify posterity.

Since the *Arithmetic* of Diophantus deals exclusively with rational numbers, it goes without saying that Fermat meant that there are no *rational* numbers x, y, z such that $x^n + y^n = z^n$ $(n > 2)$. If irrational numbers were allowed then, for any pair of numbers x, y, one could obtain a solution simply by setting $z = \sqrt[n]{x^n + y^n}$. But if there were any rational solutions of $x^n + y^n = z^n$ there would be *whole number* or *integer* solutions; this is clear from the observation that if x, y, z were rational numbers satisfying $x^n + y^n = z^n$ and if d were the least common denominator of x, y, z then xd, yd, zd would be integers and $(xd)^n + (yd)^n = (x^n + y^n)d^n = (zd)^n$, so that zd would be an integer whose nth power was a sum of nth powers. Moreover, Diophantus and Fermat both dealt with *positive* numbers—negative numbers and zero were still viewed with suspicion in Fermat's time—so that the trivial case in which x or y is zero was also tacitly excluded. (For example, $2^5 + 0^5 = 2^5$ would certainly not contradict Fermat's Last Theorem.) Thus Fermat's Last Theorem amounted essentially to saying that if n is an integer greater than 2 then it is impossible to find positive whole numbers x, y, z such that $x^n + y^n = z^n$. This is the form in which the theorem is usually stated.

In the three centuries since Fermat's death, his work in areas other than number theory has tended to fall into obscurity, not because it was anything less than first-rate but because it consisted of important first steps in the development of important theories which now are much more clearly understood and can be explained more simply using language and symbolism which did not exist in Fermat's time. On the other hand, Fermat's work in the theory of numbers is of lasting renown, not only his Last Theorem but his many other discoveries and ideas, some of which are discussed later in this chapter. It seems altogether fitting that this should be the case, because, as is clear from Fermat's correspondence, however important he might have felt the other work was to the development of mathematics, his real love was number theory, the study of properties of positive whole numbers, which seemed to him the greatest challenge to the power of pure mathematical reasoning and the greatest treasury of pure mathematical truths.

1.2 Pythagorean triangles

The proposition in Diophantus' *Arithmetic* which inspired Fermat's Last Theorem deals with one of the oldest problems in mathematics, "to write a square as the sum of two squares." One solution of this problem derives from the equation $3^2 + 4^2 = 5^2$, which implies that for any square a^2, $a^2 = (3a/5)^2 + (4a/5)^2$. In a similar way, any triple of positive integers x, y, z such that $x^2 + y^2 = z^2$ gives a solution $a^2 = (xa/z)^2 + (ya/z)^2$ and, as is easy to see, every solution arises in this way. In short, Diophantus' problem amounts to the problem of finding triples of positive whole numbers that satisfy $x^2 + y^2 = z^2$.

When the problem is stated in this way its relation to the Pythagorean theorem becomes obvious. The equation $3^2 + 4^2 = 5^2$ implies, by the

Pythagorean theorem,* that a triangle whose sides are in the ratio 3:4:5 is a right triangle. More generally, any triple of positive integers x, y, z which satisfies $x^2 + y^2 = z^2$ determines a set of ratios $x:y:z$ such that a triangle whose sides are in this ratio is a right triangle. This means that Diophantus' problem can be expressed geometrically as the problem "find right triangles in which the lengths of the sides are *commensurable*,† that is, in which the ratios of the lengths of the sides can be expressed in terms of ratios of whole numbers." Because of this geometrical interpretation of the problem, a triple of positive integers that satisfies $x^2 + y^2 = z^2$ is called a *Pythagorean triple*.

The Pythagorean triple $3^2 + 4^2 = 5^2$ is the simplest and best-known example. Another example is $5^2 + 12^2 = 13^2$. Of course it is only the *ratios* that are important, and the triple 6, 8, 10, which is in the same ratio as 3, 4, 5, is also a Pythagorean triple. Similarly 9, 12, 15 and 10, 24, 26, are Pythagorean triples which do not differ essentially from those above. But the triple 7, 24, 25 does differ essentially. Diophantus' problem is tò *find Pythagorean triples*. This problem was treated several centuries before Diophantus (circa 250 A.D.?) by Euclid (circa 300 B.C.) in the *Elements*, Book X, Lemma 1 between Propositions 28 and 29. However, according to one of the most surprising discoveries of twentieth century archaeology, the problem was studied over a thousand years before Euclid by the ancient Babylonians.

The archaeological collection of Columbia University includes a cuneiform tablet dating from about 1500 B.C. which turns out upon study (see Neugebauer [N1]) to contain a list of Pythagorean triples. The triple 3, 4, 5 is one of the triples in the list, but so is 4961, 6480, 8161. (You might find it interesting to carry out the arithmetic needed to verify that this is indeed a Pythagorean triple.) This triple certainly shows that the list was compiled by some method other than trial and error, and shows that the ancient Babylonians were in possession of some method of solution of Diophantus' problem. We do not know what this method was, nor do we know what reasons the Babylonians may have had for investigating Pythagorean triples, but in Neugebauer's opinion it is probable that the geometrical significance of Pythagorean triples was known to them—in other words the Babylonians knew the Pythagorean theorem a thousand years before Pythagoras lived—and that they obtained the triples by some method like

---

*If the Pythagorean theorem is stated in the usual form "in a right triangle the square on the hypotenuse is equal to the sum of the squares on the sides" then what is needed here is the converse of the Pythagorean theorem. However, the stronger form of the theorem "if a, b, c are the sides of a triangle then the angle opposite c is acute, right, or obtuse according as $a^2 + b^2$ is greater, equal, or less than c^2" implies the converse "if $a^2 + b^2 = c^2$ then the triangle is right."

†The concept of commensurability is a central one in Greek mathematics and much of Euclid's *Elements* deals with it. A basic inspiration of Greek mathematics was the Pythagorean discovery that the sides of an isosceles right triangle are incommensurable or, what is the same, that $\sqrt{2}$ is irrational.

the one found in Diophantus and, with less exactitude, in Euclid. This method is explained in the next section.

It is interesting and perhaps not entirely coincidental that this problem from the prehistory of mathematics should have provided the inspiration for one of the most celebrated mathematical problems of modern times.

1.3 How to find Pythagorean triples

Although the ideas in the following solution of the problem of finding Pythagorean triples are essentially identical to the ideas in Diophantus' solution,* the notation, the vocabulary, and the point of view of the presentation are modern. It is certainly worthwhile for the serious student to read Diophantus' original presentation (there is a fine English translation by Heath [H1]), but it requires a certain amount of effort to reconstruct Diophantus' notation and point of view, and since the concern of this book is with more modern mathematics, no attempt at such a reconstruction will be made here. The argument is a very important one and its central idea occurs over and over in the study of Fermat's Last Theorem.

The method of the solution that follows is what was known in classical Greek mathematics as the *analytic*† method: it is assumed at the outset that a solution $x^2 + y^2 = z^2$ (x, y, z positive integers) is given and the properties of the given solution are *analyzed* or broken down to find characteristics of such solutions which make it possible to construct them.

Note first that if d is any number which divides all three numbers x, y, z then d^2 can be cancelled from the equation $x^2 + y^2 = z^2$ and the integers $x/d, y/d, z/d$ also form a Pythagorean triple. If d is the greatest common divisor of x, y, z then $x/d, y/d, z/d$ have no common divisor other than 1 and form what is known as a *primitive* Pythagorean triple, that is, a Pythagorean triple in which the numbers have no common divisor other than 1. In this way every Pythagorean triple can be reduced—by dividing by the greatest common divisor—to a primitive one. Conversely, given a primitive Pythagorean triple, say $a^2 + b^2 = c^2$, any Pythagorean triple which reduces to it can be obtained by choosing the appropriate integer d and setting $x = ad$, $y = bd$, $z = cd$. Therefore it will suffice to be able to construct primitive Pythagorean triples and it can be assumed at the outset that the given triple x, y, z is primitive.

This assumption implies that no *two* of the three numbers x, y, z have a common divisor greater than 1. For example, if d were a divisor of x and y then, by $z^2 = x^2 + y^2$, it would follow that d^2 divided z^2 and hence‡ that d

---

*Oddly enough this solution does not appear as Diophantus' solution to the problem "write a square as the sum of two squares" in Book II but in his solutions of problems dealing with right triangles. See p. 93 of Heath's edition of Diophantus [H1].

†See Pappus' *Mathematical Collection*.

‡See Exercise 3 for a proof of this fact. Very briefly, the square of a fraction in lowest terms is itself in lowest terms; therefore the only way that $(z/d)^2$ can be an integer is for z/d to be an integer.

divided z. Therefore d would divide all three and, because the triple is primitive, d would have to be 1. Similarly the only common divisor of x and z or of y and z is 1.

In particular, no two of the three numbers x, y, z can be even (have the divisor 2). Therefore at least two are odd. But obviously all three cannot be odd because then $x^2 + y^2 = z^2$ would give odd + odd = odd which is impossible. Therefore exactly one of them is even. By considering congruences modulo 4, it is easy to see that z is not the even one as follows: The square of an odd number $2n + 1$ is one more than a multiple of 4, namely $4n^2 + 4n + 1$. The square of an even number $2n$ is a multiple of 4, namely $4n^2$. Thus if x, y were odd and z even, then $x^2 + y^2 = z^2$ would give a multiple of 4 equal to the sum of two numbers each of which was one more than a multiple of 4, which is obviously impossible. Therefore z is odd and x, y are of opposite parities—one odd and one even. By interchanging x and y if necessary, it can be assumed that in the given primitive Pythagorean triple x is even while y and z are odd.

Now rewrite the equation $x^2 + y^2 = z^2$ in the form $x^2 = z^2 - y^2$ and factor the right side to find $x^2 = (z + y)(z - y)$. Because $x, z + y$, and $z - y$ are all even numbers there are positive integers u, v, w such that $x = 2u$, $z + y = 2v$, $z - y = 2w$. Then $(2u)^2 = (2v)(2w)$ or $u^2 = vw$. Moreover, the greatest common divisor of v and w is one, because any number that divided them both would also divide $v + w = \frac{1}{2}(z + y) + \frac{1}{2}(z - y) = \frac{1}{2}(2z)$ $= z$ and $v - w = \frac{1}{2}(z + y) - \frac{1}{2}(z - y) = y$ and could therefore only be 1. In other words, v and w are relatively prime.

The main step in the argument is the following one. *The only way that it is possible for the product of two relatively prime numbers v and w to be a square $vw = u^2$ is for the numbers v and w themselves to be squares.* This is obvious if you think of v and w in terms of their prime factorizations; since they are relatively prime, no prime occurs in both v and w, so the prime factorization of vw is just the two prime factorizations set side-by-side; if vw is a square then all primes in the factorization of vw must occur to even powers, but by the above observation this implies that all primes in the factorization of v and of w occur to even powers, and hence that v and w are squares. A complete proof of this proposition is given in the next section.

Therefore there are positive integers p, q such that $v = p^2$, $w = q^2$. Moreover, p and q are relatively prime because v and w are. Then

$$z = v + w = p^2 + q^2,$$

$$y = v - w = p^2 - q^2.$$

This shows that p must be greater than q (y is positive) and that p and q must be of opposite parity (because z and y are odd). Moreover, x can easily be expressed in terms of p and q by writing

$$x^2 = z^2 - y^2 = p^4 + 2p^2q^2 + q^4 - p^4 + 2p^2q^2 - q^4$$
$$= 4p^2q^2 = (2pq)^2$$
$$x = 2pq.$$

That is, given any primitive Pythagorean triple there exist relatively prime positive integers p, q such that $p > q$, such that p and q have opposite parity, and such that the triple consists of the numbers $2pq$, $p^2 - q^2$, and $p^2 + q^2$.

This completes the analysis because it is easy to show that given any pair p, q such that (1) p and q are relatively prime, (2) $p > q$ and (3) p, q are of opposite parity, the numbers $2pq$, $p^2 - q^2$, $p^2 + q^2$ form a primitive Pythagorean triple. This follows from the observation that

$$(2pq)^2 + (p^2 - q^2)^2 = (p^2 + q^2)^2$$

is an algebraic identity true for all p, q and the observation that the conditions on p, q imply that $2pq$ and $p^2 - q^2$ are relatively prime (and hence the triple is primitive) because if they had a common divisor greater than 1 they would have a prime common divisor, say P; since $p^2 - q^2$ is odd, P could not be 2, so that P would divide p or q (because it divides $2pq$) but not both (because p and q are relatively prime) and since this would contradict the assumption that P divides $p^2 - q^2$ it is impossible.

This completely solves the problem of constructing Pythagorean triples. The Pythagorean triples corresponding to pairs p, q in which $p \leqslant 8$ are shown in Table 1.1. Note that this table includes the standard examples 3,

Table 1.1 Pythagorean Triples.

p	q	x	y	z
2	1	4	3	5
3	2	12	5	13
4	1	8	15	17
4	3	24	7	25
5	2	20	21	29
5	4	40	9	41
6	1	12	35	37
6	5	60	11	61
7	2	28	45	53
7	4	56	33	65
7	6	84	13	85
8	1	16	63	65
8	3	48	55	73
8	5	80	39	89
8	7	112	15	113

4, 5; 5, 12, 13; and 7, 24, 25. Note also that it is very easy to extend the table to larger values of p, including just those values of q less than p that are relatively prime to p and of opposite parity.

EXERCISES

1. Find the values of p and q which correspond to the Pythagorean triple of the Babylonian tablet of Section 1.2.

2. Extend Table 1.1 to $p = 12$.

3. It is to be proved in the next section that if $vw = u^2$ and v and w are relatively prime then v and w are squares. Use this to prove that if d^2 divides z^2 then d divides z. [Let $d = cD$, $z = cZ$ where c is the greatest common divisor of d and z. Then $Z^2 = kD^2$ where k and D^2 are relatively prime.]

1.4 The method of infinite descent

Fermat invented the method of infinite descent and it was an invention of which he was extremely proud. In a long letter written toward the end of his life he summarized his discoveries in number theory and he stated very definitely that all of his proofs used this method. Briefly put, the method proves that certain properties or relations are impossible for whole numbers by proving that if they held for any numbers they would hold for some smaller numbers; then, by the same argument, they would hold for some numbers that were smaller still, and so forth *ad infinitum*, which is impossible because a sequence of positive whole numbers cannot decrease indefinitely.

For example, consider the proposition that was used in the preceding section, namely, that *if v and w are relatively prime and if vw is a square then v and w must both be squares*. As Fermat himself emphasized, the method of infinite descent is a method for proving that things are *impossible*. In the case at hand, what is to be shown is that it is impossible for there to be numbers v and w such that (1) v and w are relatively prime, (2) vw is a square, and (3) v and w are not both squares.

Assume such v and w can be found. By interchanging v and w if necessary, one can assume that v is not a square. In particular v is not 1. Therefore v is divisible by at least one prime number. Let P be a prime number which divides v, say $v = Pk$. Then P also divides vw which is a square, say $vw = u^2$. By the basic property of prime numbers (see Appendix A.1) if P divides $u \cdot u$ then P must divide u or u, that is, P must divide u, say $u = Pm$. Then $vw = u^2$ can be rewritten $Pkw = (Pm)^2 = P^2m^2$ which implies $kw = Pm^2$. Since P divides the right side it must divide the left. Therefore by the basic property of prime numbers P must divide either k or w. But P does not divide w because it divides v, and v and w are relatively prime. Therefore P divides k, say $k = Pv'$. Then $kw = Pm^2$

becomes $Pv'w = Pm^2$ which gives $v'w = m^2$. Since $v = Pk = P^2v'$, any divisor of v' is a divisor of v, and therefore v' and w can have no common divisor greater than 1. Moreover, if v' were a square then $v = P^2v'$ would be a square, which it is not; therefore v' is not a square. Thus the numbers v', w have the properties (1), (2), (3) listed above and $v' < v$. The same argument then shows that there is another positive integer $v'' < v'$ such that v'', w have these same three properties. Repeating the argument indefinitely would then give a sequence of positive integers $v > v' > v'' > v''' > \cdots$ which decreased indefinitely. Since this is impossible (the number v itself gives an upper bound on the number of times it can be decreased) it is impossible for two numbers v and w to have the three stated properties. This proves the proposition.

In summary, the method of infinite descent rests on the following principle: *Suppose that the assumption that a given positive integer has a given set of properties implies that there is a smaller positive integer with the same set of properties. Then no positive integer can have this set of properties.*

1.5   The case $n = 4$ of the Last Theorem

To prove the case $n = 4$ of Fermat's Last Theorem, it suffices to combine the techniques of the preceding two sections—the method of infinite descent and the method of constructing Pythagorean triples—and to follow one's nose.

Suppose x, y, z are given such that $x^4 + y^4 = z^4$. As in the case of Pythagorean triples, it can be assumed at the outset that x, y, z have no common divisor greater than 1, and even that no two of them have a common divisor greater than 1, because then by $x^4 + y^4 = z^4$ the third one would have the same divisor and its fourth power could be divided out of the equation. Therefore x^2, y^2, z^2 are a primitive Pythagorean triple and, interchanging x and y if necessary, one can write

$$x^2 = 2pq$$
$$y^2 = p^2 - q^2$$
$$z^2 = p^2 + q^2$$

where p, q are relatively prime of opposite parity and $p > q > 0$. The second of these three equations can be written $y^2 + q^2 = p^2$ and it follows, since p and q are relatively prime, that y, q, p is a primitive Pythagorean triple. Therefore p is odd and, since p and q have opposite parities, q is even. Hence

$$q = 2ab$$
$$y = a^2 - b^2$$
$$p = a^2 + b^2$$

9

where a, b are relatively prime of opposite parity and $a > b > 0$. Thus

$$x^2 = 2pq = 4ab(a^2 + b^2).$$

This shows that $ab(a^2 + b^2)$ is a square, namely, the square of half the even number x. But ab and $a^2 + b^2$ are relatively prime because any prime P that divided ab would have to divide a or b (by the basic property of prime numbers) but not both (because a and b are relatively prime) and could not, therefore, divide $a^2 + b^2$. Therefore ab and $a^2 + b^2$ must both be squares. But then, since ab is a square and a and b are relatively prime, a and b must both be squares, say $a = X^2$, $b = Y^2$. Therefore $X^4 + Y^4 = a^2 + b^2$ is a square. But this suffices to set the infinite descent in motion once it is noticed that the only fact about the original assumption $x^4 + y^4 = z^4$ that was used was that z^4 was a *square*, not that it was a fourth power. In other words, if x and y are positive integers such that $x^4 + y^4$ is a square then the above sequence of steps gives a new pair of positive integers X, Y such that $X^4 + Y^4$ is a square. Moreover, $X^4 + Y^4 = a^2 + b^2 = p < p^2 + q^2 = z^2 < z^4 = x^4 + y^4$. Therefore there is established an infinite, descending sequence of positive integers, which is impossible. Therefore the sum of two fourth powers can't even be a square, much less a fourth power. This proves Fermat's Last Theorem for fourth powers.

It obviously follows that $x^{4m} + y^{4m} = z^{4m}$ is impossible when m is any positive integer, since otherwise $X = x^m$, $Y = y^m$, $Z = z^m$ would be a solution of $X^4 + Y^4 = Z^4$. Thus Fermat's Last Theorem is true for all exponents n that are divisible by 4. An exponent $n > 2$ which is not divisible by 4 is not a power of 2 and must, therefore, be divisible by some prime $p \neq 2$, say $n = pm$. In order to prove that $x^n + y^n = z^n$ is impossible it will obviously, for the same reason as above, suffice to prove that $x^p + y^p = z^p$ is impossible. Thus, *once Fermat's Last Theorem has been proved in the case $n = 4$ the proof of the general case reduces to the proof of the case in which $n > 2$ is prime.* For this reason, in the remainder of the book only the cases of Fermat's Last Theorem in which n is a prime, $n \neq 2$, will be considered.

1.6 Fermat's one proof

It seems that only one proof is to be found in all of Fermat's surviving work on number theory. It is the proof of a particular proposition which Fermat stated a number of times* in his correspondence but which, characteristically, he did not prove in his correspondence, leaving it to his correspondents to try to solve the problem for themselves. The proof, like the statement of the Last Theorem, was found by his son Samuel in the margin of his copy of Diophantus and was included in the posthumously published works as Observation 45 on Diophantus.

*See Dickson [D2], vol. 2, p. 616.

Fermat often states in his letters that he has no wish to conceal his work, that he believes that the advancement of knowledge depends on the efforts of many—presumably cooperative—scholars, and that he would be happy to describe his methods to anyone who asked. Actions however, speak louder than words. The fact that none of the many letters of Fermat which survive gives any real indication of his methods surely means that, consciously or unconsciously, he was very jealous, secretive, and competitive about his work, as were all of his contemporaries.

The proposition which Fermat proves states that *the area of a right triangle cannot be a square*, meaning, of course that the area of a *rational* right triangle cannot be equal to a *rational* square. By the numerical operation of multiplying all lengths by their least common denominator or, what is the same, by the geometrical operation of selecting a new unit of length in which the sides of the triangle and of the square are whole numbers (going back to the Greek idea of commensurability) this proposition can be reformulated as the statement that *there is no Pythagorean triple $x^2 + y^2 = z^2$ such that $\frac{1}{2} xy$ is a square*. (Note that x, y cannot both be odd, and that $\frac{1}{2} xy$ is therefore necessarily an integer.) This is the form of the proposition which Fermat proves.

As is usual with Fermat's problems, this problem does not come from thin air but is based on earlier literature. Book VI of Diophantus' *Arithmetic* contains several problems dealing with finding Pythagorean triangles which satisfy various conditions on their areas—including the problems of finding Pythagorean triangles whose areas plus or minus given numbers are squares—but Diophantus is satisfied, as always, to give solutions of particular cases of his problems and he does not examine the conditions under which the given problem does or does not have a solution. The edition of Diophantus which Fermat used was the edition of Bachet (1581–1638), who not only translated Diophantus from the original Greek into Latin, but who also added many comments and supplements to the text. In his supplement to Book VI, Bachet gave a necessary and sufficient condition for a number A to be the area of a Pythagorean triangle, namely, the condition that there be a number K such that $(2A)^2 + K^4$ is a square.* Thus Fermat was pursuing a line of investigation very explicitly initiated by Bachet—who in turn was inspired by Diophantus—when he asked whether it was possible for a triangle to have an area equal to a square.

Fermat's proof of this proposition is more subtle than the proof of the Last Theorem for $n = 4$ given in the preceding section. It reads as follows.

"If the area of a right-angled triangle were a square, there would exist two biquadrates the difference of which would be a square number. Consequently there would exist two square numbers the sum and difference of which would both be squares. Therefore we should have a

*See Dickson [D2], the beginning of Chapter XXII of vol. 2.

square number which would be equal to the sum of a square and the double of another square, while the squares of which this sum is made up would themselves have a square number for their sum. But if a square is made up of a square and the double of another square, its side, as I can very easily prove, is also similarly made up of a square and the double of another square. From this we conclude that the said side is the sum of the sides about the right angle in a right-angled triangle, and that the simple square contained in the sum is the base and the double of the other square is the perpendicular.

"This right-angled triangle will thus be formed from two squares, the sum and differences of which will be squares. But both these squares can be shown to be smaller than the squares originally assumed to be such that both their sum and difference are squares. Thus if there exist two squares such that their sum and difference are both squares, there will also exist two other integer squares which have the same property but have a smaller sum. By the same reasoning we find a sum still smaller than that last found, and we can go on ad infinitum finding integer square numbers smaller and smaller which have the same property. This is, however, impossible because there cannot be an infinite series of numbers smaller than any given integer we please. The margin is too small to enable me to give the proof completely and with all detail." (Translation from Fermat's Latin by Heath, [H1, p. 293].)

This proof is, as you will find if you follow it step by step, quite obscure at two important points.

The first sentence is easy enough to follow. Since the most general Pythagorean triple is $x = (2pq)d$, $y = (p^2 - q^2)d$, $z = (p^2 + q^2)d$ (p, q relatively prime positive integers of opposite parity with $p > q$, d a positive integer) the problem is to make $\frac{1}{2}xy = pq(p^2 - q^2)d^2$ a square. This will be true if and only if $pq(p^2 - q^2)$ is a square. (If Ad^2 is a square then A must be a square. See Exercise 1.) Since p, q are relatively prime, both of them must be relatively prime to $p^2 - q^2$. Therefore $pq(p^2 - q^2)$ can be a square only if p, q, and $p^2 - q^2$ are all squares. In other words, a triangle with area a square leads to a pair of relatively prime integers p, q of opposite parity such that p, q, and $p^2 - q^2$ are all squares. Since p, q are squares, $p^2 - q^2$ is a difference of fourth powers (biquadrates); this is Fermat's "two biquadrates whose difference is a square." Moreover, $p^2 - q^2 = (p - q)(p + q)$ is a decomposition into relatively prime factors because any factor which $p - q$ and $p + q$ had in common would also be a factor of $(p - q) + (p + q) = 2p$ and $(p + q) - (p - q) = 2q$ and therefore, since p and q are relatively prime, could only be 2 or 1; but p and q have opposite parity, so $p - q$ and $p + q$ are both odd and they have no common factor other than 1. Therefore the assumption that $p^2 - q^2$ is a square implies that $p - q$ and $p + q$ are both squares. This is Fermat's "two squares whose sum and difference are squares." Fermat's third sentence refers to the equations

$$(p - q) + 2q = p + q, \qquad (p - q) + q = p$$

in which p, q, $p - q$, and $p + q$ are all squares.

It is the next two sentences that are the difficult ones to follow. Let $p + q = r^2$, $p - q = s^2$. The first of these two sentences says that r can be written in the form $r = u + v$ where one of the numbers u, v is a square and the other is the double of a square; the second says that u, v are the sides of a right triangle, that is, $u^2 + v^2$ is a square. Fermat does not say how to prove the first statement* other than to say that he can prove it "easily" and, on the face of it at least, there is no way to "conclude" from it that the second statement is true. The interpretation of these sentences is therefore necessarily conjectural. The interpretation which follows (due essentially to Dickson, [D2], vol. 2, pp. 615–616) may or may not be the one that Fermat intended. In any event it does prove the two assertions in question very simply.

Since p and q have opposite parity, $p - q = s^2$, $p + q = r^2$ are both odd, and therefore r and s are both odd. Moreover, r and s are relatively prime because, as was shown above, $p - q$ and $p + q$ are relatively prime. Now define positive integers u, v by

$$u = \frac{r - s}{2}, \qquad v = \frac{r + s}{2}.$$

Then u and v are relatively prime because any common factor would be a common factor of the sum and difference $r = u + v$, $s = v - u$ which are relatively prime. Moreover

$$uv = \frac{r^2 - s^2}{4} = \frac{(p + q) - (p - q)}{4} = \frac{q}{2}.$$

Since q is a square, $\frac{1}{2}q$ can be an integer only if it is an even integer. Therefore $\frac{1}{2}uv = \frac{1}{4}q$ is an integer and, because it is a quotient of squares, it is a square. Either u or v must be even (because $\frac{1}{2}uv$ is an integer) but not both (because they are relatively prime). Now half the even one is relatively prime to the odd one and their product $\frac{1}{2}uv$ is a square, therefore the factors are squares, and therefore the even one is twice a square and the odd one is a square. Thus $r = u + v$ expresses r as the sum of a square and twice a square as desired. Moreover,

$$u^2 + v^2 = \frac{r^2 - 2rs + s^2}{4} + \frac{r^2 + 2rs + s^2}{4}$$
$$= \frac{r^2 + s^2}{2} = \frac{(p + q) + (p - q)}{2} = p.$$

Thus $u^2 + v^2$ is a square and Fermat's two statements are proved.

The remainder of the proof is easy to follow. The Pythagorean triple with "sides" u, v is primitive because u, v are relatively prime, and therefore it is of the form $2PQ$, $P^2 - Q^2$, $P^2 + Q^2$ where P, Q are relatively prime of opposite parity and $P > Q$. Since $\frac{1}{2}uv = PQ(P^2 - Q^2)$

---

*Fermat's actual statement is inaccurate because $15^2 = 5^2 + 2 \cdot 10^2$ but 15 is not a square plus the double of a square. His conclusion is correct in the case at hand because $r^2 = (p - q) + 2q$ where $p - q$ and q are *relatively prime* squares.

is a square, it follows as before that P, Q, $P - Q$, and $P + Q$ must all be squares. But $P + Q \leqslant (P + Q)PQ(P - Q) = \frac{1}{2}uv = q/4 < q < p + q$ and the process can be repeated to find two more squares P', Q' such that $P' - Q'$ and $P' + Q'$ are squares and such that $P' + Q' < P + Q$. This can be prolonged indefinitely to give an infinitely descending sequence $p + q > P + Q > P' + Q' > \cdots$. Since such an infinite descent is impossible, it must also be impossible for a Pythagorean triangle to have area equal to a square.

Interestingly enough, this proof of Fermat, the only one we have, also proves the case $n = 4$ of the Last Theorem, because it shows that $z^4 - x^4$ can not be a square, much less a fourth power (see Exercise 2). However, the fact that $x^4 + y^4 = z^4$ is impossible can be proved more simply by the method of the preceding section and there is every reason to suppose that Fermat was aware of this simpler proof.

EXERCISES

1. Prove that if Ad^2 is a square then A is a square.

2. Prove that $x^4 - y^4 = z^2$ is impossible in nonzero integers.

1.7 Sums of two squares and related topics

One of the first topics in number theory that Fermat studied, and one that led him to many other important questions, was the problem of representing numbers as sums of two squares. As in so many other instances, Fermat's interest in this subject stemmed from Diophantus' *Arithmetic*.

There are at least three passages in Diophantus that relate to representations as sums of two squares and that show that his knowledge of the subject must have been substantial. In one place (III, 19) he notes that 65 can be written in two ways as the sum of two squares $65 = 1^2 + 8^2 = 4^2 + 7^2$ and says that this "is due to the fact that 65 is the product of 13 and 5, each of which numbers is the sum of two squares." In a second place (the "necessary condition" of V, 9) he states what amounts to a necessary condition for a number to be representable as a sum of two squares, but, in the words of the translator and editor Heath "unfortunately the text of the added condition is uncertain." Finally, in a third place (VI, 14) Diophantus remarks in passing that 15 is not the sum of two (rational) squares.

A basic fact in the study of numbers that are sums of two squares is the formula

$$(a^2 + b^2)(c^2 + d^2) = (ac - bd)^2 + (ad + bc)^2 \tag{1}$$

which shows that if two numbers are sums of two squares then their product is also a sum of two squares. Formula (1) is a simple consequence of the commutative, associative, and distributive laws, which show that both sides are equal to $a^2c^2 + a^2d^2 + b^2c^2 + b^2d^2$. Applied in two different ways to $5 = 2^2 + 1^2$ and $13 = 3^2 + 2^2$, it shows that $65 = 5 \cdot 13$ can, as Diophantus said, be written in two ways as a sum of two squares, namely, $65 = 5 \cdot 13 = (2^2 + 1^2)(3^2 + 2^2) = (2 \cdot 3 - 1 \cdot 2)^2 + (2 \cdot 2 + 1 \cdot 3)^2 = 4^2 + 7^2$ and $65 = 5 \cdot 13 = (2^2 + 1^2)(2^2 + 3^2) = (2 \cdot 2 - 1 \cdot 3)^2 + (2 \cdot 3 + 1 \cdot 2)^2 = 1^2 + 8^2$. This formula was known to Leonardo of Pisa (Fibonacci) in the thirteenth century, is implicit in the algebra of ancient India (see Section 1.9 below), and, in some form or other, was surely known to Diophantus.

Fermat was not the first scholar to attempt to clarify the passages in Diophantus on sums of two squares. Bachet, in his comments on Diophantus, tried, with limited success, to do so. Another who did was François Viete (1540–1603), one of the founding fathers of modern algebra. A third was Albert Girard (1595–1632) who succeeded in giving necessary and sufficient conditions for a number to be representable as a sum of two squares some years before Fermat's earliest known writings on the subject (see Heath [H1], p. 106). Girard evidently included $0^2 = 0$ as a square, and his conditions were that a number could be written as a sum of two squares if and only if it was (1) a square or (2) a prime number which is 1 more than a multiple of 4 or (3) the number 2 or (4) any product of such numbers. The validity of these conditions is borne out by Table 1.2, which shows all numbers less than 250 that can be written as sums of two squares. Whether Girard based his statement on empirical evidence such as this or whether he had proofs of them is unclear. He does not seem to claim to be able to prove that his conditions are necessary and sufficient.

Fermat, on the other hand, did claim to be able to prove the necessity and sufficiency of Girard's conditions.* The difficult half of this theorem is

Table 1.2. All numbers less than 256 that are sums of two squares. The primes are in boldface.

0	1	4	9	16	25	36	49	64	81	100	121	144	169	196	225
	2	**5**	10	**17**	26	**37**	50	65	82	**101**	122	145	170	**197**	226
		8	**13**	20	**29**	40	**53**	68	85	104	125	148	**173**	200	**229**
			18	25	34	45	58	**73**	90	**109**	130	153	178	205	234
				32	**41**	52	65	80	**97**	116	**137**	160	185	212	**241**
					50	**61**	74	**89**	106	125	146	169	194	221	250
						72	85	100	117	136	**157**	180	205	232	
							98	**113**	130	**149**	170	**193**	218	245	
								128	145	164	185	208	**233**		
									162	**181**	202	225	250		
										200	221	244			
										242					

*There is no evidence that Fermat knew of Girard's work. He stated the conditions independently and in a slightly different way.

to prove that the conditions are *sufficient*. Since a square a^2 is trivially a sum of two squares $a^2 + 0^2$, since $2 = 1^2 + 1^2$, and since formula (1) shows that products of sums of two squares are themselves sums of two squares, this amounts to proving that *every prime number of the form* $4n + 1$ *can be written as a sum of two squares*. Fermat stated this theorem many times and stated very definitely that he could prove it rigorously, although, as usual, he is not known ever to have put the proof in writing. Fermat also went beyond Girard in stating that he could prove that there is *only one* representation of such a prime as a sum of two squares and that he had a general *method* of finding this representation without resorting to trial and error. Proofs of all these theorems, which may or may not be the proofs that Fermat had in mind, are given in Sections 2.4 and 2.6.

The *necessity* of Girard's conditions can be reformulated as the statement that *if the quotient of a number by the largest square it contains is divisible by a prime of the form* $4n + 3$ *then the number cannot be written as a sum of two squares*. This too is one of Fermat's theorems. It is much less difficult than the other half of the theorem, but it is by no means trivial. (See Exercise 2 of Section 1.8 for a proof.)

Another problem which Fermat considered in detail was that of finding the *number* of representations of a given number as a sum of two squares. This is not a problem which is relevant to this book, and it will not be considered here, but the essence of the solution is contained in Section 2.5.

Fermat discovered that rules similar to those which govern representations of numbers as sums of two squares also apply to representations of numbers in either of the forms $x^2 + 2y^2$ or $x^2 + 3y^2$. Representations in the form $x^2 + 2y^2$ will not be of any importance in the study of Fermat's Last Theorem, and just a brief summary of the facts will be given here. Table 1.3 shows all numbers less than 256 that are of the form $x^2 + 2y^2$. A check of this table shows that, as in the previous case, the numbers in the table

Table 1.3. All numbers less than 256 that can be written in the form $x^2 + 2y^2$. The primes are in boldface.

0	1	4	9	16	25	36	49	64	81	100	121	144	169	196	225
2	**3**	6	**11**	18	27	38	51	66	**83**	102	123	146	171	198	**227**
8	9	12	**17**	24	33	44	57	72	**89**	108	129	152	177	204	**233**
18	**19**	22	27	34	**43**	54	**67**	82	99	118	**139**	162	187	214	243
32	33	36	**41**	48	57	68	81	96	**113**	132	153	176	201	228	
50	51	54	**59**	66	75	86	99	114	**131**	150	171	194	219	246	
72	**73**	76	81	88	**97**	108	121	136	153	172	**193**	216	**241**		
98	99	102	**107**	114	123	134	147	162	**179**	198	219	242			
128	129	132	**137**	144	153	164	177	192	209	228	249				
162	**163**	166	171	178	187	198	**211**	226	243						
200	201	204	209	216	225	236	249								
242	243	246	**251**												

can be described as (1) squares or (2) primes that appear in the table or (3) any product of such numbers. To write a given number in the form $x^2 + 2y^2$ the procedure, then, would be to write it as a square times a product of primes to the first power and to check whether the primes which occur are themselves of the form $x^2 + 2y^2$. If they are, then an analog of formula (1), namely,

$$(a^2 + 2b^2)(c^2 + 2d^2) = (ac - 2bd)^2 + 2(ad + bc)^2$$

(both sides are $a^2c^2 + 2b^2c^2 + 2a^2d^2 + 4b^2d^2$) shows how the original number itself can be written in this form. If any one of them is not, then, by analogy with the case $x^2 + y^2$ and by inspection of the table, the expectation is that the original number itself is not of the form $x^2 + 2y^2$. To complete the analogy with the case $x^2 + y^2$ it is necessary to prove this theorem (if the number divided by the largest square it contains has a prime factor that is not of the form $x^2 + 2y^2$ then the number itself is not of this form) and to characterize the primes which are of the form $x^2 + 2y^2$. Fermat stated, and explicitly claimed he had a rigorous proof, that *an odd prime is of the form $x^2 + 2y^2$ if and only if it is of the form $8n + 1$ or the form $8n + 3$*. To prove that odd primes *not* of the form $8n + 1$ or $8n + 3$ are *not* of the form $x^2 + 2y^2$ is the simple half of this theorem (see Exercise 3); the converse half is not at all simple (see Exercises 6 and 7 of Section 2.4).

Unlike representation in the form $x^2 + 2y^2$, representations in the form $x^2 + 3y^2$ play an important role in the study of Fermat's Last Theorem —specifically in Euler's proof of the case $n = 3$—and the analogous theorems in this case will be proved in detail in the next chapter. Table 1.4 shows all numbers of the form $x^2 + 3y^2$ less than 256. Examination of this table indicates that once again it is true that *a number can be represented in the form $x^2 + 3y^2$ if and only if it is (1) a square or (2) a prime number of this form or (3) a product of such numbers*. Moreover, an examination of the primes which occur in the table shows that, apart from the prime $3 = 0^2 + 3 \cdot 1^2$ which is clearly exceptional in this case, they differ among themselves

Table 1.4. All numbers less than 256 that can be written in the form $x^2 + 3y^2$. The primes are in boldface.

0	1	4	9	16	25	36	49	64	81	100	121	144	169	196	225
3	4	**7**	12	**19**	28	39	52	**67**	84	**103**	124	147	172	**199**	228
12	**13**	16	21	28	**37**	48	**61**	76	93	112	133	156	**181**	208	237
27	28	**31**	36	**43**	52	63	76	91	108	**127**	148	171	196	**223**	252
48	49	52	57	64	**73**	84	**97**	112	129	148	169	192	217	244	
75	76	**79**	84	91	100	111	124	**139**	156	175	196	219	244		
108	**109**	112	117	124	133	144	**157**	172	189	208	**229**	252			
147	148	**151**	156	**163**	172	183	196	**211**	228	**247**					
192	**193**	196	201	208	217	228	**241**								

by multiples of 6 and are in fact all one greater than a multiple of 6. Since *all* primes of the form $6n + 1$ are in the table for as far it goes (but not all *numbers* of this form because 55 is absent), it is natural to guess that *a prime other than 3 can be represented in the form* $x^2 + 3y^2$ *if and only if it is of the form* $6n + 1$. Once again it is easy (see Exercise 2) to prove the "only if" half of this theorem. There is an analog of formula (1), namely,

$$(a^2 + 3b^2)(c^2 + 3d^2) = (ac - 3bd)^2 + 3(ad + bc)^2$$

which makes the "if" half of the first theorem obvious. Moreover, every prime other than 3 is of one of the forms $3n + 1$ or $3n + 2$, and those of the form $3n + 1$ must be of the form $6n + 1$, because if n is odd then $3n + 1$ is even and therefore not prime. These observations reduce the two theorems stated above to the statements that *if a number divided by the greatest square it contains has a prime factor of the form* $3n + 2$ *then it is not of the form* $x^2 + 3y^2$ and *every prime of the form* $3n + 1$ *can be written in the form* $x^2 + 3y^2$. These theorems are proved in the next chapter.

These facts about representations of numbers in the forms $x^2 + y^2$, $x^2 + 2y^2$, $x^2 + 3y^2$ are thrown into high relief by consideration of the next case. Since representations in the form $x^2 + 4y^2$ are subsumed in representations as sums of two squares, the next case is not $x^2 + 4y^2$ but $x^2 + 5y^2$. Table 1.5 shows all numbers less than 100 which are of the form $x^2 + 5y^2$. Note that 21 occurs *twice*, as $1^2 + 5 \cdot 2^2$ and as $4^2 + 5 \cdot 1^2$, but that its prime factors 3 and 7 do not occur at all. This shows that nothing like the above theorems apply to this case. Fermat made a very penetrating conjecture about numbers of the form $x^2 + 5y^2$ which shows very definitely that he had considered the problem and that he realized that it was entirely different from the previous cases. He conjectured that if two primes p_1 and p_2 both are of the form $4n + 3$ and both have 3 or 7 as their last digit then $p_1 p_2$ is of the form $x^2 + 5y^2$. (The primes 3, 7, 23, 43, 47, 67, ... are 3 mod 4 and end in a 3 or 7. The conjecture is that the product of any two of these numbers is of the form $x^2 + 5y^2$. For example, $3 \cdot 3 = 2^2 + 5 \cdot 1^2$, $3 \cdot 7 = 4^2 + 5 \cdot 1^2$, $7 \cdot 7 = 2^2 + 5 \cdot 3^2$, $3 \cdot 23 = 8^2 + 5 \cdot 1^2$, $3 \cdot 43 = 2^2 + 5 \cdot 5^2$, $7 \cdot 23 = 9^2 + 5 \cdot 4^2$, etc.) This conjecture is not only correct

Table 1.5. All numbers less than 100 that can be written in the form $x^2 + 5y^2$. The primes are in boldface.

0	1	4	9	16	25	36	49	64	81
5	6	9	14	21	30	**41**	54	69	86
20	21	24	**29**	36	45	56	69	84	
45	46	49	54	**61**	70	81	94		
80	81	84	**89**	96					

but it is the central fact about numbers of the form $x^2 + 5y^2$. (See Exercise 1 of Section 8.6.) It is one more example of Fermat's great genius as a number theorist.

EXERCISES

1. Prove that if $x^2 + y^2$ is an odd prime then it is of the form $4n + 1$.

2. Prove that if $x^2 + 3y^2$ is a prime other than 3 then it is of the form $6n + 1$.

3. Prove that if $x^2 + 2y^2$ is an odd prime then it is either of the form $8n + 1$ or of the form $8n + 3$.

4. Prove that the necessity of Girard's condition can, as is stated in the text, be reformulated as the statement that if the quotient of a number by the largest square it contains is divisible by a prime of the form $4n + 3$ then the number cannot be written as a sum of two squares.

5. Show that Girard's theorem implies Diophantus' statement that it is impossible to find *rational* numbers x and y such that $x^2 + y^2 = 15$.

6. Prove that a product of two numbers of the form $x^2 + 5y^2$ is also of this form.

1.8 Perfect numbers and Fermat's theorem

The study of "perfect numbers" goes back to the prehistory of number theory, back to its origins in the mysticism of numerology. A number is said to be "perfect" if it is the sum of its proper divisors. For example, 6 is perfect because $6 = 1 + 2 + 3$. Although this is not a very compelling concept for the modern mathematician and there doesn't appear to be much interest in the subject today, it is one which held great fascination for scholars over many centuries. The discussion of perfect numbers ranged from such mystical things as the statement by Alcuin (735–804) of York and Tours that the fact that 6 is a perfect number explains the fact that the universe was created in 6 days, to more scientific attempts to identify perfect numbers. Interest in perfect numbers was still quite lively in Fermat's time and there was considerable correspondence among Fermat, Mersenne, Descartes, Frenicle and others in the late 1630s concerning perfect numbers and related topics.

Euclid showed in the *Elements* (Book IX, Proposition 36) that if the number $1 + 2 + 4 + \ldots + 2^{n-1} = 2^n - 1$ is prime then the number $2^{n-1} \times (2^n - 1)$ is perfect. For example, $3 = 1 + 2$ is prime so $2 \cdot 3 = 6$ is perfect, and $7 = 1 + 2 + 4$ is prime so $2^2 \cdot 7 = 28$ is perfect. In modern notation this is very simply proved by noting that if $p = 2^n - 1$ is prime then the proper divisors of $2^{n-1}p$ are $1, 2, 4, \ldots, 2^{n-1}, p, 2p, 4p, \ldots, 2^{n-2}p$ and the sum of these divisors is $1 + 2 + 4 + \cdots + 2^{n-1} + p(1 + 2 + 4 + \cdots + 2^{n-2})$ $= p + p(2^{n-1} - 1) = 2^{n-1}p$ as was to be shown. Euclid's condition is a *sufficient* condition for a number to be perfect. No examples of any other

19

perfect numbers are known. It was stated by Descartes and proved by Euler that a perfect number is of Euclid's form if—and of course only if—it is even. Whether there exist any odd perfect numbers is a famous unsolved problem. Fortunately, it is not one that there is any need to pursue here.

Euclid's condition implies that in order to find perfect numbers it suffices* to find prime numbers in the sequence 3, 7, 15, 31, 63, 127, 255, 511, 1023, 2047, 4095, 8191, ... and this is the problem which is of interest here. In short, *for which values of n is $2^n - 1$ prime*? It was in studying this problem that Fermat discovered the basic fact that is now known as Fermat's theorem.

In the first place 15, 63, 255, 1023($= 3 \cdot 341$), 4095 are obviously not prime. More generally, if n is even and greater than 2 then $2^n - 1 = 2^{2k} - 1 = (2^k - 1)(2^k + 1)$ is not prime. The odd values $n = 3, 5, 7$ all lead to primes 7, 31, 127, but the odd value $n = 9$ leads to 511, which is quickly seen to be divisible by 7. This leads to the conjecture that if n is not prime then $2^n - 1$ is not prime, a conjecture that is easily verified by noting that $2^{km} - 1 = (2^k - 1)(2^{k(m-1)} + 2^{k(m-2)} + \cdots + 2^k + 1)$. This observation reduces the question to the question *for which primes p is $2^p - 1$ prime*? Primes of this form are called *Mersenne primes* in honor of Fermat's contemporary and frequent correspondent, Father Marin Mersenne (1588–1648).

The primes 2, 3, 5, 7 correspond to the Mersenne primes 3, 7, 31, 127 (and hence to the perfect numbers 6, 28, 496, 8128) but it remains to be seen whether $2^{11} - 1$ is prime. This question can be resolved very simply by finding $2^{11} - 1 = 2047$ explicitly and dividing by all primes less than $\sqrt{2047}$ to see if any of them divide it evenly. It is more instructive, however, to approach the problem in a different manner which can be used for testing $2^p - 1$ when p is larger than 11.

Consider the question of whether 7 divides $2^{11} - 1$. Imagine the powers of 2 written on one line and their remainders after division by 7 written underneath them.

powers of 2	1	2	4	8	16	32	64	128	256	512	...
remainders	1	2	4	1	2	4	1	2	4	1	...

The pattern of the remainders is obvious and it is obvious that 2^n divided by 7 will leave a remainder of 1 if and only if n is a multiple of 3, that is, 7 *divides* $2^n - 1$ *if and only if* 3 *divides n*. Therefore 7 does not divide $2^{11} - 1$. The same method can be used for other primes. Table 1.6 contains some results. For each prime the remainders show a cyclic pattern and *there is an integer d such that p divides $2^n - 1$ if and only if d divides n*. Once this fact has been noted it is easily proved. Since there are only $p - 1$ possible

*The confusion of necessary and sufficient conditions seems to be a basic failing of the human intellect. Even Fermat made the statement that there are no perfect numbers with 20 or 21 digits when what he meant was that there were none of *Euclid's type*.

remainders, there must be at least one repetition, say 2^n and 2^{n+m} leave the same remainder. Then p divides their difference $2^{n+m} - 2^n = 2^n(2^m - 1)$, and it follows, since p is prime and does not divide 2, that p divides $2^m - 1$ and therefore that 2^m leaves a remainder of 1 when divided by p. Therefore 1 occurs as a remainder. Let 2^d be the least power of 2 which leaves the remainder 1. Then 2^{md} also leaves a remainder of 1 when divided by p because $2^{md} - 1 = (2^d - 1)(2^{(m-1)d} + \cdots + 2^d + 1)$ is divisible by p. Conversely, the only powers of 2 that leave the remainder 1 are those corresponding to multiples of d because if 2^m leaves a remainder of 1 and if $m = qd + r$ ($q \geqslant 0, 0 \leqslant r < d$) then both $2^m = 2^{qd}2^r$ and 2^{qd} leave remainders of 1 and their difference $2^{qd}(2^r - 1)$ is divisible by p. Since 2^{qd} is not divisible by p this contradicts the definition of d (recall that $0 \leqslant r < d$) unless $r = 0$ and m is a multiple of d, as was to be shown. Moreover, it was noted above that 2^{n+m} and 2^n leave the same remainders only if 2^m leaves the remainder 1, that is, repeated remainders occur only at intervals which are multiples of d. It follows that there are exactly d different remainders and that they occur in a cyclic pattern as in the examples.

This observation about the remainders implies that in order to determine whether p divides $2^{11} - 1$ it suffices to determine the corresponding d and see whether it divides 11. Since 11 is prime this is the same as asking whether $d = 11$. The answer to this question is in the negative for all the primes considered so far.

There is an easier way to find the sequence of remainders than the straightforward method of finding 2^n and dividing by p for $n = 1, 2, 3, \ldots$. For example, once it is known that 128 divided by 13 leaves a remainder of 11 it is easy to find the remainder of 256 divided by 13 simply by doubling 11 and dividing by 13 to find $22 - 13 = 9$ as the next remainder. The following remainder is twice 9 minus 13 or 5, and the one following that is twice 5. In general, each remainder is either twice the preceding one or twice the preceding one minus p, whichever lies in the range $1 \leqslant r \leqslant p - 1$ of the remainders. This follows immediately from the fact that if $2^n - r$ is divisible by p then $2^{n+1} - 2r$ is divisible by p, which

Table 1.6

$p = 3$	1	2	4	8	16	32	64	$\ldots$						
	1	2	1	2	1	2	1	$\ldots$					$d = 2$	
$p = 5$	1	2	4	8	16	32	64							
	1	2	4	3	1	2	4	$\ldots$					$d = 4$	
$p = 11$	1	2	4	8	16	32	64	128	256	512	1024	2048	$\ldots$	
	1	2	4	8	5	10	9	7	3	6	1	2	$\ldots$	$d = 10$
$p = 13$	1	2	4	8	16	32	64	128	256	512	1024	2048	4096	8192	$\ldots$
	1	2	4	8	3	6	12	11	9	5	10	7	1	2	$\ldots$ $d = 12$
$p = 17$	1	2	4	8	16	32	64	128	256	512	1024	2048	$\ldots$	
	1	2	4	8	16	15	13	9	1	2	4	8	$\ldots$	$d = 8$

implies that 2^{n+1} and $2r$ leave the same remainder when divided by p. Thus for $p = 19$ the remainders are 1, 2, 4, 8, 16, 13, 7, 14, 9, 18, 17, 15, 11, 3, 6, 12, 5, 10, 1, 2, 4, . . . , from which it follows that $d = 18 \neq 11$ and therefore that 19 does not divide $2^{11} - 1$. For the next prime $p = 23$ the remainders are 1, 2, 4, 8, 16, 9, 18, 13, 3, 6, 12, 1, 2, 4, . . . , from which it follows that $d = 11$ and 23 *does* divide $2^{11} - 1$. Thus $2^{11} - 1$ is not prime and in fact $2^{11} - 1 = 2047 = 23 \cdot 89$.

The problem of determining whether $2^{13} - 1$ is prime (and consequently whether $2^{12}(2^{13} - 1)$ is perfect) can be approached in the same way. The problem is to determine whether there is a prime p for which the corresponding d is 13. All the primes considered so far are therefore eliminated (because their d's did not turn out to be 13) and it is now a question of testing whether d is 13 for $p = 29, 31, 37, . . . ,$ up to the least prime less than $\sqrt{2^{13} - 1} = \sqrt{8191} < 91$. Fermat made a simple observation about the value of d which immediately eliminates from consideration all but a very few of these primes. See if you can make the same observation. The values found so far, plus a few more, are

p	3	5	7	11	13	17	19	23	29	31	37
d	2	4	3	10	12	8	18	11	28	5	36

Of course, since there are d distinct remainders, d is at most $p - 1$. What Fermat observed is that d actually divides $p - 1$. This implies that d can be 13 only if 13 divides $p - 1$. Thus the possible value of p are $13 + 1$, $26 + 1$, $39 + 1$, $52 + 1$, ... Of these only the odd ones, 27, $27 + 26 = 53$, $53 + 26 = 79$, . . . can be prime. Since 27 is not prime and since $79 + 26$ is greater than $\sqrt{2^{13} - 1}$ this means that *only two* primes need to be tried, namely, 53 and 79. The corresponding d's, determined by the method used above, are 52 and 39. Consequently, provided one is convinced by the empirical fact that d divides $p - 1$, it follows from these brief computations that $2^{13} - 1 = 8191$ *is prime*.

In view of the fact that d divides m if and only if $2^m - 1$ is divisible by p (see above), another way of saying that d divides $p - 1$ is to say that p divides $2^{p-1} - 1$, or that p divides $2^p - 2$. Fermat states his theorem in this form as well as in the form* $d|(p - 1)$. As usual, Fermat asserts that he has a proof of the theorem but omits it. Perhaps his proof was the following.

Virtually by definition there are exactly d of the possible remainders 1, 2, 3, . . . , $p - 1$ which actually occur as remainders of powers of 2 after division by p. If all $p - 1$ remainders occur, then $d = p - 1$ and the desired conclusion $d|(p - 1)$ holds. Otherwise there is at least one remainder k which does not occur. Consider the possible remainders 1, 2, 3, . . . , $p - 1$ which actually occur as remainders of numbers of the form $k, 2k, 4k, 8k, . . . , 2^n k, . . .$ after division by p. There are exactly d such

---

*The vertical bar means "divides" and $d|(p - 1)$ is read "d divides $p - 1$". The vertical bar with crossbar $\nmid$ means "does not divide."

remainders and none of them is included in the original set of d remainders; the first of these two facts follows from the observation that $2^{n+m}k$ and $2^{n}k$ leave the same remainder if and only if $2^{n}k(2^{m} - 1)$ is divisible by p, which is true if and only if m is divisible by d, and the second follows from the observations that if 2^{n} and $2^{m}k$ were to leave the same remainder then so would 2^{n+1} and $2^{m+1}k$ (because $2^{n} - 2^{m}k$ is divisible by p if and only if $2(2^{n} - 2^{m}k)$ is divisible by p) and so would 2^{n+2} and $2^{m+2}k$ and so forth which, when $m + j$ is made divisible by d, shows that 2^{n+j} and k leave the same remainder when divided by p, contrary to the choice of k. If these two sets of remainders exhaust all $p - 1$ possible remainders then $p - 1 = 2d$ and $d|(p - 1)$ as was to be shown. Otherwise there is another possible remainder k' which is not in either set. Then, in the same way as before, the remainders of $k', 2k', 4k', \ldots, 2^{n}k', \ldots$ give a set of d more distinct remainders none of which is in either of the two sets already found. When this process is continued, it partitions the $p - 1$ possible remainders into sets of d each, which clearly implies that d must divide $p - 1$, as was to be shown.

For example, when $p = 31$ the powers of 2 leave the remainders 1, 2, 4, 8, 16. This list does not include 3 and the numbers $3, 2 \cdot 3, 4 \cdot 3, 8 \cdot 3, \ldots$ leave the remainders 3, 6, 12, 24, 17. Neither of these lists includes 5 and the numbers $5, 2 \cdot 5, 4 \cdot 5, 8 \cdot 5, \ldots$ leave the remainders 5, 10, 20, 9, 18. Continuing in this way groups the remainders $1, 2, \ldots, 30$ into six sets of five, the three sets above and the sets (7, 14, 28, 25, 19), (11, 22, 13, 26, 21), (15, 30, 29, 27, 23). The fact that the $p - 1$ remainders are always divided up into sets of d in this way guarantees that d must always divide $p - 1$.

None of these arguments depends on any special properties of the number 2—which was chosen only because it arose in connection with the search for perfect numbers—and for any other positive integer a the same arguments prove: If p is a prime which does not divide a then there is an integer d such that p divides $a^{m} - 1$ if and only if d divides m. Fermat's theorem states that d divides $p - 1$. By the defining property of d, this amounts to saying that p divides $a^{p-1} - 1$ or, what is the same, that p divides $a^{p} - a$. The last statement is the most succinct one because it does not involve d at all and because it is true even if p divides a. It is the usual statement of *Fermat's theorem*: *If p is any prime and if a is any integer then p divides $a^{p} - a$.*

Fermat's theorem is one of the most fundamental properties of the arithmetic of whole numbers and it will be essential in later chapters of this book. By contrast, the remainder of this section is devoted to a topic which is of little importance except as a footnote in history. This is the topic of the so-called *Fermat numbers* $2^{1} + 1, 2^{2} + 1, 2^{4} + 1, 2^{8} + 1, 2^{16} + 1, 2^{32} + 1, \ldots$. Fermat in his correspondence repeatedly expressed his conviction that these numbers are all *prime*. (Note that $2^{n} + 1$ is *not* prime if n is *not* a power of two because if n has an odd factor k, say $n = km$, then

$2^n + 1 = (2^m + 1)(2^{m(k-1)} - 2^{m(k-2)} + \cdots + 2^2 - 2 + 1).$) Thus he believed he had solved the ancient problem of finding a formula which yields arbitrarily large prime numbers. He even went so far as to say,* late in his life, that he could *prove* that these numbers are all prime.

The first few Fermat numbers are prime: $2^1 + 1 = 3$, $2^2 + 1 = 5$, and $2^4 + 1 = 17$ are of course prime. $2^8 + 1 = 257$ can be shown to be prime as follows. If p divides $2^8 + 1$ then it divides $(2^8 + 1)(2^8 - 1) = 2^{16} - 1$. Therefore the d corresponding to this p (for $a = 2$) must divide 16. But the only divisors of 16 are 1, 2, 4, 8, 16 and d cannot be 1, 2, 4, or 8 because p would then divide $2^8 - 1$, which would contradict the assumption that p divides $2^8 + 1$. Therefore $d = 16$ and by Fermat's theorem $p = 16n + 1$ for some integer n. But the smallest such prime $p = 17$ is already greater than $\sqrt{2^8 + 1}$ and $2^8 + 1$ has no proper divisors, as was to be shown.

Similarly the only prime divisors $2^{16} + 1$ could have would be primes of the form $p = 32n + 1$. Since $\sqrt{2^{16} + 1}$ is just slightly larger than $2^8 = 256$ the only ones that need to be tested are the primes in the list 33, 65, 97, 129, 161, 193, 225 which includes just two primes 97, 193. Neither of these divides $2^{16} + 1$ because the remainders left by the powers 1, 2, 4, 8, 16, . . . , 2^{16} after division by 97 are 1, 2, 4, 8, 16, 32, 64, 31, 62, 27, 54, 11, 22, 44, 88, 79, 61 so that $2^{16} + 1$ divided by 97 leaves a remainder 62 and the remainders after division by 193 are 1, 2, 4, 8, 16, 32, 64, 128, 63, 126, 59, 118, 43, 86, 172, 151, 109 so that $2^{16} + 1$ divided by 193 leaves a remainder of 110. Therefore $2^{16} + 1$ is prime.

The analogous proof for $2^{32} + 1$ is much longer and Fermat must not have undertaken it seriously or he must have made a computational error. The only primes which could divide $2^{32} + 1$ would be, by the same argument as before, primes of the form $p = 64n + 1$. The smallest prime factor, if $2^{32} + 1$ is not prime, could be no larger than 2^{16}. Since the numbers $64n + 1$ are at intervals of $64 = 2^6$ there are roughly $2^{10} = 1024$ numbers to be tried, but every third one of these is divisible by 3, every fifth one divisible by 5, and so forth, so that the number of *primes* $p = 64n + 1$ in the critical range is only of the order of magnitude of 500 or so. To prove in this way that $2^{32} + 1$ is prime is a quite long job, but surely not more than a few days at most However $2^{32} + 1$ is *not* prime; it has the factor 641. If the procedure outlined above is followed, the primes to be tried are 193, 257, 449, 577, and then 641. Thus 641 is only the *fifth*[†] prime that must be tried. The factor 641 of $2^{32} + 1$ was discovered by Euler.

This incident seems to be the one serious smirch on Fermat's reputation as a number theorist. To make matters even worse, it is now known that the next few Fermat numbers $2^{64} + 1$, $2^{128} + 1$, $2^{256} + 1$ and several others are *all* composite. *No* prime Fermat numbers have been found beyond $2^{16} + 1$. However, even here there is a redeeming aspect, a confirmation of

*E. T. Bell insists ([B1], p. 256) that he says no such thing. I see no other interpretation of his letter to Carcavi [F5].

[†]For an improvement on this argument see Exercise 10, Section 2.4.

the soundness of Fermat's instinct in choosing the problem. A century and a half after Fermat made his conjecture, the youthful Gauss showed that Euclid's ruler and compass constructions of the pentagon is intimately connected with the fact that $5 = 2^2 + 1$ is a Fermat number. More generally, Gauss showed that ruler and compass construction of a regular n-gon is possible whenever n is a *prime* Fermat number. Conversely, as Gauss stated and Wantzel proved (see [K1a]), the only regular n-gons that can be constructed with ruler and compass are those in which $n = 2^k p_1 p_2 \cdots p_m$ where the p's are distinct prime Fermat numbers and $k \geqslant 0$.

EXERCISES

1. Prove that $2^{37} - 1$ is not prime. [In testing a given prime it is not necessary to find d but only to determine whether the prime divides $2^{37} - 1$. To do this you need not find the remainders left by all the numbers 1, 2, 4, 8, 16, ... since it suffices to find the remainders left by 2, 2^2, 2^4, 2^8, 2^{16}, 2^{32}, 2^{32+4}, 2^{37}.]

2. Prove that a number which does not satisfy Girard's conditions cannot be written as a sum of two squares. More specifically, prove that if p divides $x^2 + y^2$ and p is a prime of the form $4n + 3$ then p divides both x and y. [If p divides $x^2 + y^2$ then it divides $(x^2)^{2n+1} + (y^2)^{2n+1}$. Unless p divides both x and y this number is either 1 or 2 greater than a multiple of p and therefore not divisible by p.]

3. Show how a regular 3-gon can be constructed by ruler and compass. Look up Euclid's construction of the regular 5-gon (Elements, Book IV, Proposition 11). Show how the two can be combined to construct a 15-gon. Using the theorem of Gauss and Wantzel find all values of $n \leqslant 30$ such that the regular n-gon can be constructed by ruler and compass.

1.9 Pell's equation

Rather late in his career, in 1657, Fermat issued a challenge problem addressed to other mathematicians and in particular to the mathematicians of England, among whom Fermat hoped to find someone to share his interest in the arithmetic of whole numbers.

"There is hardly anyone who propounds arithmetical questions, hardly anyone who understands them. Is this due to the fact that up to now arithmetic has been treated geometrically rather than arithmetically? This has indeed generally been the case both in ancient and modern works; even Diophantus is an instance. For, although he has freed himself from geometry a little more than others in that he confines his analysis to the consideration of rational numbers, yet even there geometry is not entirely absent, as is sufficiently proved by the *Zetetica* of Viete, where the method of Diophantus is extended to continuous magnitude and therefore to geometry.

"Now arithmetic has, so to speak, a special domain of its own, the theory of integral numbers. This was only lightly touched upon by Euclid in his *Elements*, and was not sufficiently studied by those who followed him (unless, perchance, it is contained in those Books of Diophantus of which the ravages of time have robbed us); arithmeticians have therefore now to develop it or restore it.

"To arithmeticians therefore, by way of lighting up the road to be followed, I propose the following theorem to be proved or problem to be solved. If they succeed in finding the proof or solution, they will admit that questions of this kind are not inferior to the more celebrated questions in geometry in respect of beauty, difficulty, or method of proof.

"*Given any number whatever which is not a square, there are also given an infinite number of squares such that, if the square is multiplied into the given number and unity is added to the product, the result is a square.*

"Example. Let 3, which is not a square, be the given number; when it is multiplied into the square 1, and 1 is added to the product, the result is 4, being a square.

"The same 3 multiplied by the square 16 gives a product which, if increased by 1, becomes 49, a square.

"And an infinite number of squares besides 1 and 16 can be found which have the same property.

"But I ask for a general rule of solution when any number not a square is given.

"E.g. let it be required to find a square such that, if the product of the square and the number 149, or 109, or 433 etc. be increased by 1 the result is a square." (Translation from Fermat's Latin by Heath, [H1], pp. 285–286.)

Fermat's preamble to the statement of the problem shows clearly that he makes a sharp distinction between the Diophantine tradition of solution in rational numbers and the tradition, with which he now associates himself, of solutions in *integers*. (It is ironic that in contemporary terminology "Diophantine" means "integers," because Diophantus never dealt with solutions in integers in the portions of his work that survive.) Strangely enough, the preamble was omitted by one of the intermediaries in the copy of the problem that was ultimately transmitted to the English, with the result that they thought the problem was a foolish one with the trivial Diophantine solution

$$Ax^2 + 1 = y^2 \quad \text{(given } A, \text{ find } x, y),$$

$$y = 1 + \frac{m}{n} x \quad \text{(say),}$$

$$Ax^2 + 1 = 1 + \frac{2m}{n} x + \frac{m^2}{n^2} x^2,$$

$$An^2x^2 - m^2x^2 = 2mnx,$$

$$x = \frac{2mn}{An^2 - m^2}, \qquad y = \frac{An^2 + m^2}{An^2 - m^2},$$

which gives infinitely many rational solutions. When the stipulation that x and y be whole numbers was added, the English found that this "solution" was of no value at all and they complained that Fermat had changed the problem. Surely they were justified in their complaint by the strong

Diophantine tradition, but, as Fermat pointed out, it was rather ridiculous of them to imagine that he had posed a problem which was so trivial.

Of course the study of the properties of whole numbers was not new with Fermat. The books of Euclid's *Elements* which deal with arithmetic deal with whole numbers exclusively, and Plato in various places expresses a philosophical preference for the study of whole numbers. But this ancient tradition had had a rebirth with Fermat and he was, by means of his challenge, trying to awaken the same interests in others. In fact, having failed to interest his countrymen—notably Pascal—in his studies of the arithmetic of whole numbers, he was now prepared to enter into an international correspondence in the attempt to find other scholars to share his interests.

Because the motivation of this problem "given a positive integer A not a square, show that there are infinitely many integers x for which $Ax^2 + 1$ is a square" lies in Fermat's own work and because Fermat was so secretive about his work, it is impossible to reconstruct the way in which Fermat was lead to this problem. To some extent the problem is implicit in portions of Diophantus (see Dickson [D2], vol. 2, pp. 345–346) and to some extent it is clear that some of Diophantus' problems and Bachet's commentaries on them led Fermat to problems which involved finding whole number solutions of quadratic equations in two variables, but the details are obscure. What is certain is that this problem was not chosen at random. As later scholars were to find, this special quadratic equation in integers $y^2 - Ax^2 = 1$ is of fundamental importance in the solution of any quadratic equation in two variables in integers.

Fermat was certainly not the first person to recognize the importance of this problem. There are indications of interest in the problem in ancient Greek mathematics—for example in the Pythagorean solution* of the case $A = 2$, which is related to the irrationality of $\sqrt{2}$, and in Archimedes' offhand statement that $1351/780 > \sqrt{3}$ which shows an awareness of the solution $3 \cdot 780^2 + 1 = 1351^2$ of Fermat's problem in the case $A = 3$—and many historians have conjectured that the Greeks had considerable knowledge of this subject which has not survived. The evidence of interest in the problem in ancient India is more fully documented. (See [D2, vol. 2, pp. 346–350], [C3], or [H1 pp. 281–285].) The solution $2 \cdot 408^2 + 1 = 577^2$ was known in very ancient India (a few centuries B.C.) and the solution $92 \cdot 120^2 + 1 = 1151^2$, the smallest solution in the case $A = 92$, together with a sophisticated technique of deriving it, was given by Brahmagupta (born 598 A.D.). A general technique of solution called the "cyclic method"

---

*This solution can be derived from the formula $(a^2 - 2b^2)(c^2 - 2d^2) = (ac + 2bd)^2 - 2(ad + bc)^2$ and the equation $1^2 - 2 \cdot 1^2 = -1$ by $1 = (-1)(-1) = (1^2 - 2 \cdot 1^2)(1^2 - 2 \cdot 1^2) = 3^2 - 2 \cdot 2^2$, $-1 = (-1) \cdot 1 = (1^2 - 2 \cdot 1^2)(3^2 - 2 \cdot 2^2) = 7^2 - 2 \cdot 5^2$, $1 = (1^2 - 2 \cdot 1^2)(7^2 - 2 \cdot 5^2) = 17^2 - 2 \cdot 12^2$, and so forth. The nth equation of this process gives $d_n^2 - 2 \cdot s_n^2 = (-1)^n$ where $d_n = d_{n-1} + 2s_{n-1}$ and $s_n = d_{n-1} + s_{n-1}$. The even-numbered equations give solutions s_{2k} of Fermat's problem "$2 \cdot s_{2k}^2 + 1$ is a square." See Dickson, vol. 2, p. 341.

was given by Bháscara Achárya (born 1114 A.D.). The essence of this method can be described as follows.

Suppose $A = 67$ so that the problem is to find an integer x such that $67x^2 + 1$ is a square, or, what is the same, to find integers x and y such that $y^2 - 67x^2 = 1$. As a first approximation to such integers x, y one might choose 1, 8 because $8^2 - 67 \cdot 1^2 = -3$ is fairly close to 1. Consider now the analog of formula (1) of Section 1.7 for this case, namely the formula

$$(a^2 - 67b^2)(c^2 - 67d^2) = (ac + 67bd)^2 - 67(ad + bc)^2.$$

Let this be applied to the equations $8^2 - 67 \cdot 1^2 = -3$ and $r^2 - 67 \cdot 1^2 = s$ (where r and hence s are to be determined later) to find

$$(8r + 67)^2 - 67(r + 8)^2 = -3s.$$

To attempt to make this number as small as possible simply by making s as small as possible would lead to the choice of $r = 8, s = -3$ again and would give $131^2 - 67 \cdot 16^2 = 9$. This clearly leads nowhere. The "cyclic method" is to *choose r in such a way that $r + 8$ is divisible by* 3, at the same time making s as small as possible. When this is done, the equation shows that $8r + 67$ must then be divisible by 3 as well and, consequently, the left side must be divisible by 3^2; therefore s must be divisible by 3 and the entire equation must be divisible by 9. This gives an essentially new case in which $y^2 - 67x^2$ is a small number.

Explicitly, for $r + 8$ to be divisible by 3, r must have one of the values 1, 4, 7, 10, 13, ... The choice $r = 7, s = -18$ gives the smallest (in the sense of absolute value) s and gives $123^2 - 67 \cdot 15^2 = 54$ which after division by 9 becomes

$$41^2 - 67 \cdot 5^2 = 6.$$

The process can now be repeated. Multiplication of this equation by $r^2 - 67 \cdot 1^2 = s$ gives $(41r + 67 \cdot 5)^2 - 67(5r + 41)^2 = 6s$. As before, if r is choosen so that $5r + 41$ is divisible by 6 then it will be possible to cancel 6^2 from the equation. For $5r + 41$ to be divisible by 6 it is necessary and sufficient for $r + 1 = 6(r + 7) - (5r + 41)$ to be divisible by 6, which is true for $r = 5, 11, 17, 23, \ldots$. The choice $r = 5$ gives the smallest value of s and gives $540^2 - 67 \cdot 66^2 = 6(-42)$, which after division by 6^2 becomes

$$90^2 - 67 \cdot 11^2 = -7.$$

It is certainly not clear that this has brought the problem any nearer to a solution, but is is at least conceivable that continuation of the process may eventually lead to an equation in which the right side is 1, as desired. This is in fact the case with the present problem: the continuation of the process of the "cyclic method" yields in this case

$$
\begin{aligned}
1^2 - 67 \cdot 0^2 &= 1 \\
8^2 - 67 \cdot 1^2 &= -3 & r &= 8 \\
41^2 - 67 \cdot 5^2 &= 6 & r &= 7 \\
90^2 - 67 \cdot 11^2 &= -7 & r &= 5 \\
221^2 - 67 \cdot 27^2 &= -2 & r &= 9 \\
1899^2 - 67 \cdot 232^2 &= -7 & r &= 9 \\
3577^2 - 67 \cdot 437^2 &= 6 & r &= 5 \\
9053^2 - 67 \cdot 1106^2 &= -3 & r &= 7 \\
48842^2 - 67 \cdot 5967^2 &= 1 & r &= 8
\end{aligned}
$$

which solves the problem. (The actual Indian solution of this problem uses a shortcut. The formula $221^2 - 67 \cdot 27^2 = -2$ is squared to find $(221^2 + 67 \cdot 27^2)^2 - 67(2 \cdot 27 \cdot 221)^2 = (-2)(-2)$ and the final solution $48842^2 - 67 \cdot 5967^2$ is obtained by cancelling 2^2 from this equation. Note that this is the Diophantine solution above with $m = 221$, $n = 27$, $An^2 - m^2 = 2$.) In short, if $A = 67$ is the given number, then $x = 5967$ has the property that $Ax^2 + 1$ is a square. The "infinite number" of solutions which Fermat requires can be obtained either by continuing the process to find more cases in which the right side is 1 (it will be found that if the process is continued, the numbers on the right side follow a cyclic pattern $1, -3, 6, -7, -2, -7, 6, -3, 1, -3, 6, \ldots$ and this may be the reason that it is called the "cyclic method") or by using the solution already derived to find others by squaring

$$
1 = (48842^2 - 67 \cdot 5967^2)^2
$$

$$
= (48842^2 + 67 \cdot 5967^2)^2 - 67(2 \cdot 5967 \cdot 48842)^2
$$

cubing

$$
1 = \left[(48842^2 + 67 \cdot 5967^2)^2 - 67(2 \cdot 5967 \cdot 48842)^2\right](48842^2 - 67 \cdot 5967^2)
$$

$$
= \left[48842(48842^2 + 67 \cdot 5967^2) + 67 \cdot 2 \cdot 48842 \cdot 5967^2\right]^2
$$

$$
- 67\left[5967(48842^2 + 67 \cdot 5967^2) + 48842^2 \cdot 2 \cdot 5967\right]^2
$$

$$
= (48842^3 + 3 \cdot 67 \cdot 48842 \cdot 5967^2)^2 - 67(3 \cdot 48842^2 \cdot 5967 + 67 \cdot 5967^3)^2,
$$

raising to the fourth power, and so forth. This solves Fermat's problem in the particular case $A = 67$.

The cyclic method can be applied in the same way to give the solution of Fermat's problem for any A not a square. In summary, the procedure is the following. At the first stage, take the equation $1^2 - A \cdot 0^2 = 1$. Suppose that at the nth stage the equation is $p^2 - Aq^2 = k$. To reach stage $n + 1$ multiply this equation by $r^2 - A = s$ to find $(pr + qA)^2 - A(p + qr)^2 = ks$, where r, and hence s, are to be determined. Choose r to be a positive integer for which $p + qr$ is divisible by k and for which the corresponding s is as small as possible. Then $pr + qA$ is divisible* by k and division of $(pr + qA)^2 - A(p + qr)^2 = ks$ by k^2 gives the next stage $P^2 - AQ^2 = K$ where $P = (pr + qA)/|k|$, $Q = (p + qr)/|k|$, and $K = s/k$. The process is continued until a stage is reached at which the equation has the desired form $p^2 - Aq^2 = 1$. Then $x = q$ is a solution of Fermat's problem because $Aq^2 + 1 = p^2$ is a square. An infinite number of solutions can then be found either by continuing the cyclic method to find other equations $p^2 - Aq^2 = k$ in which $k = 1$ or by raising the first solution to powers as above.

This method in fact succeeds in producing solutions of Fermat's problem for all values of A not a square. Table 1.7 gives a list of the smallest solutions x of the equation "$Ax^2 + 1 = $ square" for $A = 2, 3, 5, 6, \ldots$. The table shows that the particular cases $A = 149$, $A = 109$ which Fermat proposes are cases in which the solution is extremely difficult. The case $A = 61$, which is by far the hardest one for $A < 100$, was also singled out[†] by Fermat when he posed the same problem to Frenicle at about the same time. In case there was any doubt, this certainly shows that Fermat was in possession of a procedure for solving the problem which enabled him to find the most difficult cases. (However, $A = 433$ does not seem to be especially difficult; the solution has 19 digits, but the solution when $A = 421$ has 33 digits.)

It is to the credit of the English that they were able to find not only the particular solutions in the three cases Fermat proposed but also a general procedure for deriving the solution for any value of A. The author of this procedure is somewhat in doubt because, although it is John Wallis who wrote up the description of the procedure and the derivation of the three particular solutions, Wallis ascribes the method to Viscount William Brouncker. There is no evidence in Wallis's published correspondence that Brouncker ever communicated the method to him except in the form of a very simple set of observations which may have been the germ of the idea which Wallis later developed. Quite possibly Wallis was being excessively

*It is clear that k divides $(pr + qA)^2$. For most values of k this implies that k divides $pr + qA$, namely, for all values of k that are not divisible by the square of a prime. What is being asserted here is that in the application of the cyclic method, even if a value of k which is divisible by the square of a prime does occur, k always divides $pr + qA$. This, like the fact that a stage must be reached at which $k = 1$, must be proved in order to prove that the cyclic method always succeeds.

†The case $A = 61$ was also singled out by Bháscara Achárya, who gave the correct solution $x = 226153980$ five centuries before Fermat.

Table 1.7 Smallest solutions x of $Ax^2 + 1 = $ square.

A	x	A	x	A	x
1	—	51	7	101	20
2	2	52	90	102	10
3	1	53	9100	103	22419
4	—	54	66	104	5
5	4	55	12	105	4
6	2	56	2	106	3115890
7	3	57	20	107	93
8	1	58	2574	108	130
9	—	59	69	109	15140424455100
10	6	60	4	110	2
11	3	61	226153980	111	28
12	2	62	8	112	12
13	180	63	1	113	113296
14	4	64	—	114	96
15	1	65	16	115	105
16	—	66	8	116	910
17	8	67	5967	117	60
18	4	68	4	118	28254
19	39	69	936	119	11
20	2	70	30	120	1
21	12	71	413	121	—
22	42	72	2	122	22
23	5	73	267000	123	11
24	1	74	430	124	414960
25	—	75	3	125	83204
26	10	76	6630	126	40
27	5	77	40	127	419775
28	24	78	6	128	51
29	1820	79	9	129	1484
30	2	80	1	130	570
31	273	81	—	131	927
32	3	82	18	132	2
33	4	83	9	133	224460
34	6	84	6	134	12606
35	1	85	30996	135	21
36	—	86	1122	136	3
37	12	87	3	137	519712
38	6	88	21	138	4
39	4	89	53000	139	6578829
40	3	90	2	140	6
41	320	91	165	141	8
42	2	92	120	142	12
43	531	93	1260	143	1
44	30	94	221064	144	—
45	24	95	4	145	24
46	3588	96	5	146	12
47	7	97	6377352	147	8
48	1	98	10	148	6
49	—	99	1	149	2113761020
50	14	100	—	150	4

deferential in an effort to win Brouncker's favor and his patronage when he called the method Brouncker's; Brouncker was not only a nobleman, he was also the first President of the Royal Society.*

The English method is very different in conception from the "cyclic method" described above, but the resulting computations are very similar. In particular, both methods have the property that they can be applied to particular cases to find a solution without it being certain at the outset that the method will succeed. For example, the case $A = 67$ was solved above when an equation was reached in which the right-hand side was 1, and in the same way it will be found that when the cyclic method is applied to any particular case a right-hand side of 1 will eventually be reached and a solution of the problem eventually produced. However, there is no obvious reason why a right-hand side of 1 must necessarily occur in all cases and, in an analogous way, there is no obvious reason why the English method must always succeed.

Thus the English actually didn't solve Fermat's problem—which was to prove that "given any A not a square there exist infinitely many x such that $Ax^2 + 1$ is a square"—even though they did give a procedure for finding an x when A is given. What is missing, of course, is the proof that the procedure always succeeds. The English gave no such proof and did not appear to be aware that one was needed. However, the point is certainly not a minor one, because even Euler failed to find a proof that the English method always succeeds. The first proof was given by Lagrange about 110 years after Wallis produced his answer to Fermat's challenge. Lagrange's proof is outlined in the exercises which follow this section.

It is not altogether clear that Fermat was aware of the defect in Wallis's solution and it is even less certain that Fermat's own method of solution was free of the same defect. Fermat did write a letter in which he acknowledged that the English had finally succeeded in solving his problem and in which he showed no dissatisfaction with their method despite the lack of a proof that a solution is always reached. The main point of this letter, however, is that Fermat is urging the English to concede that the problem had been an interesting one that was worthy of their attention. In short, Fermat was still hopeful of awakening in Wallis and his associates an interest in the study of whole numbers, and he may have been willing to overlook their omission in order to encourage them in these pursuits.

A few years later, when Fermat was summarizing some of his discoveries in number theory in a letter to Carcavi, he pointed out that the

---

*At least two noted scholars have told me that they disagree with this opinion, one on the grounds that there is good evidence that Lord Brouncker was a very capable mathematician, and the other on the grounds that Wallis's personality was such that though he might claim credit himself for something someone else had done he could surely not give credit to someone else for something he had done. Limited support of my view comes from H. J. S. Smith's classic *Report on the Theory of Numbers* [S3, Art. 96, p. 193] which states that Wallis gave the method "attributing it to Lord Brouncker, though he seems himself to have had some share in its invention."

English had succeeded in solving his problem $Ax^2 + 1 =$ square only in certain particular cases and that they had failed to give a "general proof." The obvious interpretation of this comment is that Fermat saw the failure to prove that the process always arrives at a solution, but it might also be interpreted as the less profound criticism that the process was not described in sufficiently general terms. Fermat said that he could supply the needed "general proof" by means of the method of infinite descent "duly and properly applied." It is difficult to see how infinite descent can be used to prove that the process—either the Wallis method or the closely related Indian cyclic method—always arrives at a solution, and for this reason Fermat's statement cannot be taken as unambiguous evidence that he was in possession of a thoroughly satisfactory solution of his problem.

This problem of Fermat is now known as "Pell's equation" as a result of a mistake on the part of Euler. In some way, perhaps from a confused recollection of Wallis's *Algebra*, Euler gained the mistaken impression that Wallis attributed the method of solving the problem not to Brouncker but to Pell, a contemporary of Wallis who is frequently mentioned in Wallis's works but who appears to have had nothing to do with the solution of Fermat's problem. Euler mentions this mistaken impression as early as 1730, when he was only 23 years old, and it is included in his definitive *Introduction to Algebra* [E9] written around 1770. Euler was the most widely read mathematical writer of his time, and the method from that time on has been associated with the name of Pell and the problem that it solved—that of finding all integer solutions of $y^2 - Ax^2 = 1$ when A is a given number not a square—has been known ever since as "Pell's equation," despite the fact that it was Fermat who first indicated the importance of the problem and despite the fact that Pell had nothing whatever to do with it.

EXERCISES

1. Derive several of the entries in Table 1.7 using the cyclic method.

2. The English method, that is, the method of Wallis and Brouncker, differs from the cyclic method chiefly in that instead of making $r^2 - A$ as small as possible in absolute value it makes $r^2 - A$ *negative* and makes r as large as possible subject to $r^2 < A$. In this way the sign of the right-hand side of the equation alternates. Use this method to solve the cases $A = 13$ and $A = 67$, and compare the resulting equations to those which result from the cyclic method. (For $A = 13$ the right-hand sides are $1, -4, 3, -3, 4, -1, 4, -3, 3, -4, 1$ and for $A = 67$ they are $1, -3, 6, -7, \ldots$.)

3. Show that if $p^2 - Aq^2 = k$, and $P^2 - AQ^2 = K$ are successive lines in the process of either the cyclic method or the English method then r drops out of $pQ - Pq$ to give $pQ - Pq = \pm 1$. Conclude that P and Q are relatively prime. Next conclude that Q and K are relatively prime so that the congruence "$QR + P$ is divisible by K," which determines the next r, has solutions. Finally,

33

conclude that any solution R of this congruence also satisfies "$QA + PR$ is divisible by K" so that the next step of the cyclic method can be carried out.

4. Show that if r and R are the values of r which precede and follow the line $P^2 - AQ^2 = K$ then $r + R$ is divisible by K. [$P - rQ$ is divisible by K.] Note that this vastly simplifies the determination of R at each step.

5. Note that the simplication of Exercise 4 makes it possible to compute the sequence of k's and r's without ever computing any p's or q's. For example, show that the cyclic method in Fermat's case $A = 149$ in which lines 1 and 2 are $1^2 - 149 \cdot 0^2 = 1$ and $12^2 - 149 \cdot 1 = -5$, respectively, will arrive at a solution $p^2 - 149q^2 = 1$ on line 15. (Do not compute the solution.) Show that the English method will arrive at a solution on line 19.

6. Show that, when the r's and k's are known, the computation of the p's and q's can be simplified as follows. Let $p^2 - Aq^2 = k$, $P^2 - AQ^2 = K$, $\mathcal{P}^2 - A\mathcal{Q}^2 = \mathcal{K}$ be three successive lines and let r and R be the intervening values of r. Then

$$\mathcal{P} = nP \pm p \qquad \mathcal{Q} = nQ \pm q$$

where the signs are both plus if $Kk < 0$ and both minus if $Kk > 0$, and where n is the integer such that $r + R = n|K|$. In terms of matrix multiplication these formulas can be written

$$\begin{bmatrix} \mathcal{P} & P \\ \mathcal{Q} & Q \end{bmatrix} = \begin{bmatrix} P & p \\ Q & q \end{bmatrix} \begin{bmatrix} n & 1 \\ \pm 1 & 0 \end{bmatrix}.$$

Since the first two lines of the table are simple to find, this gives a means of computing the p's and q's. For example, the p and q on the 9th line of the computation for $A = 67$, are the entries in the first column of the matrix

$$\begin{bmatrix} 8 & 1 \\ 1 & 0 \end{bmatrix} \begin{bmatrix} 5 & 1 \\ 1 & 0 \end{bmatrix} \begin{bmatrix} 2 & 1 \\ 1 & 0 \end{bmatrix} \begin{bmatrix} 2 & 1 \\ 1 & 0 \end{bmatrix} \begin{bmatrix} 9 & 1 \\ -1 & 0 \end{bmatrix} \begin{bmatrix} 2 & 1 \\ -1 & 0 \end{bmatrix} \begin{bmatrix} 2 & 1 \\ 1 & 0 \end{bmatrix} \begin{bmatrix} 5 & 1 \\ 1 & 0 \end{bmatrix}$$

which greatly simplifies the computation. Using the r's and k's found in Exercise 5, write the solution of $p^2 - 149q^2 = 1$ on the 15th line of the cyclic method as the first column of a product of 14 matrices. Multiply the matrices and find the solution $x = 2113761020$ of Pell's equation when $A = 149$.

7. Show that if two successive values of k have the same absolute value $K = \pm k$ then $pP + AqQ$ and $pQ + qP$ are both divisible by k. [Use the formula for $P - rQ$.] This implies that the two lines of the table can be multiplied and k^2 can be cancelled from the result to give a solution of $x^2 - Ay^2 = \pm 1$. In the case $A = 149$ the 4th and 5th lines can be used in this way to find the 8th line. The 8th line can then be squared to give the 15th line. Carry out these computations.

8. Find the solution $x = 15140424455100$ in Fermat's case $A = 109$.

The remaining exercises are devoted to showing that the cyclic method must arrive at a solution of Pell's equation and all solutions of Pell's equations are obtained.

9. The English method (see Exercise 2) is less efficient than the cyclic method, but it is simpler in that the signs follow a regular pattern (the signs in Exercise 6 are always plus). Consequently it is easier to prove that the English method always yields a solution of Pell's equation, and in fact yields them all, than it is to prove the same of the cyclic method. Prove that if the line $p^2 - Aq^2 = k$ occurs in both the English method and the cyclic method and if the line following $p^2 - Aq^2 = k$ is different in the two methods—say $P^2 - AQ^2 = K$ in the English method and $\mathcal{P}^2 - A\mathcal{Q}^2 = \mathcal{K}$ in the cyclic method—then $K \neq 1$ and the line following $P^2 - AQ^2 = K$ in the English method is $\mathcal{P}^2 - A\mathcal{Q}^2 = \mathcal{K}$. In short, the only difference is that the cyclic method may omit some lines of the English method, but the omitted lines are never solutions of Pell's equation. Therefore in order to prove that the cyclic method produces all solutions of Pell's equation it suffices to prove that the English method does. [Let r and $\mathcal{R}$ be the values of r which follow $p^2 - Aq^2 = k$ in the English and the cyclic methods respectively, and let R be the value which follows $P^2 - AQ^2 = K$ in the English method. The inequality $A - r^2 > (r + |k|)^2 - A > 0$ implied by the fact that the two methods lead to different choices gives $K > -K + 2r - k$ if $k < 0$, $-K > K + 2r + k$ if $k > 0$. In the first case $(-r + nK)^2 - A$ is negative if $n = 1$ and positive if $n = 2$, which implies $R = -r + K$. In the second case $R = -r - K$. Thus $R = -r + |K|$. From this it follows that the next k in the English method is $(R^2 - A)/K = [(r + |k|)^2 - A]/k = \mathcal{K}$. Similarly the next q in the English method is $|K|^{-1}[P + Q(-r + |K|)] = |k|^{-1}[p + q(r + |k|)] = \mathcal{Q}$ and it follows that the next p in the English method is $\mathcal{P}$. It remains to show that $|K| \neq 1$. For this it suffices to prove that in the English method the inequality $r > 0$ always holds, a fact that will be proved in the following Exercise. Note that it has been assumed here that in case of a tie $A - r^2 = (r + |k|)^2 - A$ the cyclic method agrees with the English method, that is, that it chooses the smaller of the two possible values of r. The above proof actually shows that if instead the larger value is chosen then a line in which $K \neq 1$ is skipped.]

10. The proper definition of the English method requires a *proof* that there exist values of r satisfying the congruence "$qr + p$ divisible by k" for which $r^2 - A < 0$. This of course depends on the fact that large values of k do not arise. In practice one finds that values of r always exist and that they are always positive. Moreover, one finds that the cycles of k's and r's which occur in practice are always *palindromes*, that is, they are the same when they are read backwards. Now if k is positive then $(r + k)^2 - A$ is positive and consequently $k + 2r + K$ is positive, but if k is negative then $(r - k)^2 - A$ is positive and $k - 2r + K$ is negative. The palindrome property implies a symmetry between k and K and these observations lead one to expect that each set of two successive k's with an intervening r, say k, r, K, satisfies the inequalities $k + 2r + K > 0$, $k - 2r + K < 0$. These imply $r > 0$. Show that they hold by proving that they hold for the first step where $k = 1$, $r = [\sqrt{A}]$ = largest integer less than $\sqrt{A}$, $K = [\sqrt{A}]^2 - A$, and by proving that if they hold at any given step then they hold at the following step.

11. Exercise 10 shows that the English method defines an infinite sequence of k's. Show that if $k, K, \mathcal{K}$ are three consecutive values of k and if the values $K, \mathcal{K}$

occur consecutively later in the sequence of k's then they are preceded by k. Show that there are only finitely many possibilities for k and conclude that *the English method arrives at a solution of Pell's equation*. [The palindrome property mentioned in Exercise 10 suggests that if k, K, $\mathcal{K}$ are consecutive values of k and r, R the intervening values of r then r, k can be found from $\mathcal{K}$, R, K by the same rules as R, $\mathcal{K}$ can be found from k, r, K.]

12. Prove that the cycle of k's and r's produced by the English method is always a palindrome and always contains an even number of steps.

13. Prove that the English method produces *all* solutions of Pell's equation. [One can proceed as follows. Let $x^2 - Ay^2 = 1$. Assume without loss of generality that x and y are positive. Set $x = x_0, y = y_0$, and apply the English method to $x_0^2 - Ay_0^2 = 1$ to find $x_1^2 - Ay_1^2 = k_1, x_2^2 - Ay_2^2 = k_2, \ldots$. Then, by Exercise 6, $x_{i+1} = n_i x_i + x_{i-1}, y_{i+1} = n_i y_i + y_{i-1}$ where the sequence of positive integers $n_1, n_2, n_3, \ldots$ is periodic. By periodicity, n_i can be defined for all integers i, not just positive integers. Then x_i, y_i can be defined for all integers i. Prove that $x_i^2 - Ay_i^2 = k_i$ where k_i is defined for negative i by periodicity. Obviously if x_i and y_i are both positive for all $i \geqslant j$ (as they are in the case $j = 0$) then $|x_{j+2}| > |x_j|$. By infinite descent there must be a j such that x_i and y_i are both positive for $i > j$ but not for $i = j$. Consider the matrix equation

$$\begin{bmatrix} x_1 & x_0 \\ y_1 & y_0 \end{bmatrix} = \begin{bmatrix} n_0 & 1 \\ 1 & 0 \end{bmatrix}\begin{bmatrix} n_1 & 1 \\ 1 & 0 \end{bmatrix} \cdots \begin{bmatrix} n_{j+1} & 1 \\ 1 & 0 \end{bmatrix}\begin{bmatrix} x_{j+1} & x_j \\ y_{j+1} & y_j \end{bmatrix}.$$

Since $x_{j+1}y_j - y_{j+1}x_j = \pm 1$ and since x_{j+1} and y_{j+1} are positive, x_j and y_j can not have opposite signs. Since $x_j^2 - Ay_j^2 = k_j$ and $|k_j| < A$, $x_j \neq 0$. Therefore $y_j = 0$. Then $x_j = \pm 1$ and $k_j = 1$. Let

$$\begin{bmatrix} P & p \\ Q & q \end{bmatrix} = \begin{bmatrix} n_0 & 1 \\ 1 & 0 \end{bmatrix}\begin{bmatrix} n_1 & 1 \\ 1 & 0 \end{bmatrix} \cdots \begin{bmatrix} n_{j+1} & 1 \\ 1 & 0 \end{bmatrix}.$$

[Then $x_0 = x_j P$ so $x_j = 1$. Therefore $x_0 = P$ and $y_0 = Q$. One need only show that the English method gives the solution $P^2 - AQ^2 = 1$ of Pell's equation, something which is clear from Exercise 6.]

1.10 Other number-theoretic discoveries of Fermat

Fermat's many unproved assertions stood as a challenge to his successors, and although he failed to engage the interest of Wallis and others in the generation which immediately followed him, later generations, beginning especially with Euler, were immensely stimulated by the attempt to prove or disprove these assertions. With the exception of the assertion that the numbers $2^{32} + 1$, $2^{64} + 1$, $2^{128} + 1, \ldots$ are prime, and with the possible exception of Fermat's Last Theorem $x^n + y^n \neq z^n$ $(n > 2)$, all of them have been found to be true. These assertions—other than the ones already discussed above—do not bear directly on the subsequent history of Fermat's Last Theorem and they will not be considered in detail here, but it is of interest to mention some of them, if only to give an idea of the scope of Fermat's work.

One of these assertions is the statement, implicit in Diophantus and explicit in Bachet's commentary on Diophantus, that *every number can be written as a sum of four squares*. Fermat claimed that he could prove not only this statement but also the generalization that *every number can be written as a sum of three triangular numbers or four squares or five pentagonal or six hexagonal numbers*, and so forth *ad infinitum*. Here the triangular numbers are $0, 0 + 1 = 1, 1 + 2 = 3, 3 + 3 = 6, 6 + 4 = 10, 10 + 5 = 15$, and so forth, the square numbers are $0, 0 + 1 = 1, 1 + 3 = 4, 4 + 5 = 9, 9 + 7 = 16, 16 + 9 = 25$, and so forth, the pentagonal numbers are $0, 0 + 1 = 1, 1 + 4 = 5, 5 + 7 = 12, 12 + 10 = 22, 22 + 13 = 35$ and so forth. In general the n-gonal numbers $0, 1, a_2, a_3, a_4, \ldots$ are obtained by successively adding the terms of the arithmetic progression $1, n - 1, 2n - 3, 3n - 5, \ldots$. Thus $a_0 = 0, a_1 = 1, a_2 = n, a_3 = 3n - 3, a_4 = 6n - 8, a_5 = 10n - 15, \ldots, a_j = \frac{1}{2}j(j - 1)n - j(j - 2)$. (The polygonal numbers are the subject of various ancient treatises, including a treatise by Diophantus of which only a fragment survives.) This beautiful theorem later engaged the interest of some of the greatest names in the history of mathematics; it was Lagrange who first proved the statement for squares, Gauss who first proved it for triangular numbers, and Cauchy who first proved the general case.

Another of Fermat's unproved statements is the assertion that $25 + 2 = 27$ is the only solution in whole numbers of $x^2 + 2 = y^3$ and, similarly, that $4 + 4 = 8$ and $121 + 4 = 125$ are the only solutions of $x^2 + 4 = y^3$. These simple statements seem to come from nowhere and there seems to be no natural way to go about trying to prove them.* Another statement of this type is that $1 + 1 + 1 + 1 = 4$ and $1 + 7 + 49 + 343 = 400$ are the only solutions of $1 + x + x^2 + x^3 = y^2$ (Exercise 1).

The challenge to Wallis and the other English mathematicians to solve Pell's equation was actually Fermat's *second* challenge. The first challenge was even harder and appears to have gone over their heads completely. It consisted of two problems. (1) Find a cube which, added to its proper divisors, gives a square. For example, 343 added to its proper divisors gives $343 + 49 + 7 + 1 = 400$, a square. Find another cube with the same property. (2) Find a square which, added to its proper divisors, gives a cube. For solutions of these problems, see Dickson [D2, vol. 1, pp. 54–58].

Yet another problem of Fermat's, one which inspired Euler, is: given a number which is a sum of two cubes, write it in another way as a sum of two cubes. Here the cubes are rational, but the denominators can be cleared in the usual way and the problem is readily converted to the problem of finding all solutions of $x^3 + y^3 = u^3 + v^3$ in positive integers. Frenicle, to whom Fermat posed the problem, found several solutions such as $1729 = 9^3 + 10^3 = 1^3 + 12^3$ and $40033 = 16^3 + 33^3 = 9^3 + 34^3$, perhaps by educated trial and error.

*Bachet had, before Fermat, studied *rational* solutions x, y of $x^2 + 2 = y^3$ (see Dickson [D2], vol. 2, p. 533). For proofs of these statement of Fermat see the Exercises of Section 2.5.

One final theorem, simple to state but not at all simple to prove: *no triangular number greater than* 1 *is also a fourth* power. In other words, for $x > 2$ the equation $\frac{1}{2}x(x - 1) = y^4$ has no solution in integers. The first published proof of this fact was in Legendre's *Theorie des Nombres*, roughly a hundred and fifty years later.

Fermat concluded his summary to Carcavi of his favorite discoveries in number theory with the thought, "Perhaps posterity will be grateful to me for having shown that the Ancients didn't know everything." Nothing could show more dramatically the difference in attitude between Fermat's time and our own. It is inconceivable that a twentieth century mathematician could think that the Ancients knew everything. On the contrary, in our times the general attitude is that, at least where mathematics is concerned, the Ancients knew nothing at all. We might instead be grateful to Fermat for showing us how much stimulation and insight can be gained from a study of the works of great men of the past.

EXERCISES

1. Prove Fermat's statement that the only natural numbers x for which $1 + x + x^2 + x^3$ is a square are $x = 1$ and $x = 7$. The only integers with this property are these and $x = 0, -1$. [This is an excellent but very difficult problem. Use the fact that $x^4 + y^4$ can never be a square (Section 1.5) and that $x^4 - y^4 = z^2$ only in the trivial cases $y = 0$ or $z = 0$ (Section 1.6, Exercise 2).]

Euler 2

2.1 Euler and the case $n = 3$

Leonhard Euler (1707–1783) was without a doubt the greatest mathematician of his time. His contributions were in every imaginable field, from applied mathematics to algebraic topology and number theory, and they took the form not only of the discovery of new theorems and new techniques but also of a whole series of textbooks in algebra, calculus, mathematical physics, and other fields, which provided the core of the education of several later generations of mathematicians.

It is a measure of Euler's greatness that when one is studying number theory one has the impression that Euler was primarily interested in number theory, but when one studies divergent series one feels that divergent series were his main interest, when one studies differential equations one imagines that actually differential equations was his favorite subject, and so forth. Of course the major interest in this book is in Euler's contributions to number theory and in particular to Fermat's Last Theorem. Whether or not number theory was a favorite subject of Euler's, it is one in which he showed a lifelong interest and his contributions to number theory alone would suffice to establish a lasting reputation in the annals of mathematics.

There is conflicting testimony in the histories of Fermat's Last Theorem as to whether Euler proved the case $n = 3$. The most common statement is that Euler did give a proof of the case $n = 3$ of Fermat's Last Theorem but that his proof was "incomplete" in an important respect. This is as close as one can come to the truth of the matter in a few words. The full story is more complex. The proof which he gave contained a basic fallacy which he apparently did not recognize. To correct his proof by the most direct method—that of supplying an alternative proof of the statement for which

39

Euler's proof is fallacious—is not at all simple. However, as will be shown in Section 2.5, the proof can be corrected in a less direct way by bringing in arguments which Euler used to prove other propositions of Fermat. This method leaves in doubt the question of whether Euler's original argument can be patched up—which because of its elegance and generality seems highly desirable—but it does prove that $x^3 + y^3 = z^3$ is impossible for x, y, z positive integers.

2.2 Euler's proof of the case $n = 3$

The basic method of Euler's proof of the case $n = 3$ of Fermat's Last Theorem is Fermat's method of *infinite descent*. He shows that if positive whole numbers x, y, z could be found for which $x^3 + y^3 = z^3$ then smaller positive whole numbers could be found with the same property; thus it would be possible to find a sequence of such triples of positive integers which continually decreased and never terminated, which is manifestly impossible. Therefore no such x, y, z can be found.

Assume therefore that $x^3 + y^3 = z^3$. Any factor which divided two of the numbers x, y, z would, by virtue of this equation, also divide the third. Therefore all common factors can be removed and one can assume at the outset that the numbers x, y, z are *pairwise relatively prime*, that is, the greatest common divisor of x, y or of x, z or of y, z is 1. In particular, then, at most one of the three numbers x, y, z is even. On the other hand, at least one is even because if x, y are both odd then z is even. Therefore exactly one is even.

Assume first that x, y are odd and z is even. Then $x + y$ and $x - y$ are both even, say $2p$ and $2q$ respectively, and $x = \frac{1}{2}(2p + 2q) = p + q$, $y = \frac{1}{2}(2p - 2q) = p - q$. When $x^3 + y^3 = (x + y)(x^2 - xy + y^2)$ is expressed in terms of p and q it becomes

$$2p\left[(p + q)^2 - (p + q)(p - q) + (p - q)^2\right] = 2p(p^2 + 3q^2).$$

Here p and q are of opposite parities—because $p + q$ and $p - q$ are odd—and they are relatively prime because any factor they had in common would divide both $x = p + q$ and $y = p - q$, and therefore could only be 1. Moreover, p and q can both be assumed to be positive. (If $x < y$ then x and y can be interchanged to give $q > 0$. The case $x = y$ is impossible because then $x = y = 1$, $z^3 = 2$.) Therefore the assumption that $x^3 + y^3 = z^3$ is possible with x, y both odd implies that there exist relatively prime positive integers p, q of opposite parity such that

$$2p(p^2 + 3q^2) = \text{cube}.$$

The same conclusion can be reached if z is odd and x or y is even. In this case the odd one, say y^3, can be moved to the right side

$$x^3 = z^3 - y^3 = (z - y)(z^2 + zy + y^2).$$

Then $z - y = 2p$, $z + y = 2q$, $z = q + p$, $y = q - p$, and

$$x^3 = 2p\left[(q + p)^2 + (q + p)(q - p) + (q - p)^2\right]$$

which leads to the same conclusion

$$2p(p^2 + 3q^2) = \text{cube}$$

where p, q are relatively prime positive integers of opposite parity.

The next step in the argument is roughly to say that $2p$ and $p^2 + 3q^2$ are relatively prime and to conclude that the only way that their product can be a cube is for each of them separately to be a cube. Note the analogy here to the method of Section 1.3 whereby the solutions of $x^2 + y^2 = z^2$ were analyzed. However, the statement that $2p$ and $p^2 + 3q^2$ are relatively prime is not quite justified. Since p and q have opposite parity, $p^2 + 3q^2$ is odd and any common factor of $2p$, $p^2 + 3q^2$ would be a common factor of p, $p^2 + 3q^2$ and therefore a common factor of p, $3q^2$. Since p and q are relatively prime, this implies that the only possible common factor is 3. But if 3 does divide p then clearly it also divides $p^2 + 3q^2$ and $2p$, $p^2 + 3q^2$ are not relatively prime. The proof therefore splits into two cases, the one in which 3 does not divide p and consequently $2p$, $p^2 + 3q^2$ are relatively prime, and the other in which 3 does divide p. The first of these two cases will be considered first and the second will then be treated as a simple modification of the first.

Assume therefore that 3 does not divide p and that $2p$ and $p^2 + 3q^2$ are consequently both cubes. Using the formula

$$(a^2 + 3b^2)(c^2 + 3d^2) = (ac - 3bd)^2 + 3(ad + bc)^2$$

of Section 1.7, one can find cubes of the form $p^2 + 3q^2$ by writing

$$(a^2 + 3b^2)^3 = (a^2 + 3b^2)\left[(a^2 - 3b^2)^2 + 3(2ab)^2\right]$$

$$= \left[a(a^2 - 3b^2) - 3b(2ab)\right]^2 + 3\left[a(2ab) + b(a^2 - 3b^2)\right]^2$$

$$= (a^3 - 9ab^2)^2 + 3(3a^2b - 3b^3)^2.$$

That is, one way to find cubes of the form $p^2 + 3q^2$ is to choose a, b at random and to set

$$p = a^3 - 9ab^2 \qquad q = 3a^2b - 3b^3$$

so that $p^2 + 3q^2 = (a^2 + 3b^2)^3$. The major gap to be filled in Euler's proof is the proof that this is the *only* way that $p^2 + 3q^2$ can be a cube; that is, if $p^2 + 3q^2$ is a cube then there must be a, b such that p and q are given by the above equations. Euler bases this conclusion on the fallacious argument described in the next section, but he could also have based it on the argument of Section 2.5, which is essentially his. In any case, once this conclusion is granted the remainder of the proof follows relatively easily.

The expressions for p and q can be factored

$$p = a(a - 3b)(a + 3b) \qquad q = 3b(a - b)(a + b).$$

Of course a and b are relatively prime because any factor they had in common would also divide both p and q contrary to assumption. Moreover,

$$2p = 2a(a - 3b)(a + 3b) = \text{cube}.$$

The parities of a and b must be opposite because otherwise p and q would both be even. Therefore $a - 3b$, $a + 3b$ are both odd and the only possible common factor of $2a$, $a \pm 3b$ would be common factors of a, $a \pm 3b$ and therefore of a, $\pm 3b$. Similarly, any common factor of $a + 3b$ and $a - 3b$ would be a factor of a and of $3b$. In short, the only possible common factor is 3. But 3 does not divide a because if it did it would divide p, contrary to assumption. Therefore $2a$, $a - 3b$, $a + 3b$ are relatively prime and all three of them must be cubes, say $2a = \alpha^3$, $a - 3b = \beta^3$, $a + 3b = \gamma^3$. Then $\beta^3 + \gamma^3 = 2a = \alpha^3$ and this gives a solution of $x^3 + y^3 = z^3$ in smaller numbers than the original solution.

More specifically, $\alpha^3 \beta^3 \gamma^3 = 2a(a - 3b)(a + 3b) = 2p$, which is positive and a divisor of z^3 if z is even and a divisor of x^3 if x is even. In any case, then, $\alpha^3 \beta^3 \gamma^3$ is less than z^3. There is nothing to prevent α, β, or γ from being negative, but since $(-\alpha)^3 = -\alpha^3$, negative cubes can be moved to the opposite side of the equation to become positive cubes and the resulting equation is of the form $X^3 + Y^3 = Z^3$ in which X, Y, Z are all positive and $Z^3 < z^3$. Therefore the descent has been accomplished in the case where 3 does not divide p.

Consider finally the case $3 \mid p$. Then $p = 3s$, say, and 3 does not divide q. Then $2p(p^2 + 3q^2) = 3^2 \cdot 2s(3s^2 + q^2)$. The numbers $3^2 \cdot 2s$ and $3s^2 + q^2$ are easily seen to be relatively prime, and therefore both of them are cubes. By the lemma to be proved later on, $3s^2 + q^2$ can be a cube only if

$$q = a(a - 3b)(a + 3b) \qquad s = 3b(a - b)(a + b)$$

for some integers a, b. Since $3^2 \cdot 2s$ is a cube, $3^3 \cdot 2b(a - b)(a + b)$ is a cube and therefore $2b(a - b)(a + b)$ is a cube. The factors are easily seen to be relatively prime, $2b = \alpha^3$, $a - b = \beta^3$, $a + b = \gamma^3$, $\alpha^3 = 2b = \gamma^3 - \beta^3$ and an equation of the form $X^3 + Y^3 = Z^3$ with $Z^3 < z^3$ can be derived, all exactly as before.

In any case, then, the existence of a cube which was the sum of two cubes would imply the existence of a smaller cube of the same type and is therefore impossible. All that remains to be done in order to complete this proof is to show that if p and q are relatively prime integers such that $p^2 + 3q^2$ is a cube then there must be integers a and b such that $p = a^3 - 9ab^2$ and $q = 3a^2b - 3b^3$. This is done in Section 2.5.

2.3 Arithmetic of surds

The technique by which Euler tried to establish the needed conclusion about cubes of the form $p^2 + 3q^2$ is based on the bold idea of extending the arithmetic of whole numbers by applying addition, subtraction, and multiplication to "numbers" of the form $a + b\sqrt{-3}$ where a and b are integers. It is clear how to add, subtract, and multiply such "numbers" —only the rule for multiplication $(a + b\sqrt{-3})(c + d\sqrt{-3})$ $= ac + ad\sqrt{-3} + bc\sqrt{-3} + bd(-3) = (ac - 3bd) + (ad + bc)\sqrt{-3}$ is worthy of any special mention—and the usual laws of commutativity, associativity, and distributivity apply, as well as

$$1 \cdot (a + b\sqrt{-3}) = a + b\sqrt{-3}.$$

In the language of modern algebra, the "numbers" $a + b\sqrt{-3}$ form a commutative ring with unit.

The idea of computing with such surds $a + b\sqrt{-3}$ simplifies the derivation of the sufficient condition for $p^2 + 3q^2$ to be a cube. Instead of using the formula

$$(a^2 + 3b^2)(c^2 + 3d^2) = (ac - 3bd)^2 + 3(ad + bc)^2$$

one can argue as follows. Factor $p^2 + 3q^2$ as $(p + q\sqrt{-3})(p - q\sqrt{-3})$. If one of these factors is a cube, say $p + q\sqrt{-3} = (a + b\sqrt{-3})^3$, then it is easily checked that its conjugate, found by replacing $\sqrt{-3}$ by $-\sqrt{-3}$, is the cube of the conjugate of $a + b\sqrt{-3}$; that is $p - q\sqrt{-3}$ $= (a - b\sqrt{-3})^3$; therefore, by the commutative law of multiplication, $(p + q\sqrt{-3})(p - q\sqrt{-3}) = [(a + b\sqrt{-3})(a - b\sqrt{-3})]^3$, that is, $p^2 + 3q^2 = (a^2 + 3b^2)^3$. In other words, in order to find a cube of the form $p^2 + 3q^2$ it suffices to set $p + q\sqrt{-3} = (a + b\sqrt{-3})^3$. When $(a + b\sqrt{-3})^3$ is expanded using the binomial theorem

$$p + q\sqrt{-3} = a^3 + 3a^2b\sqrt{-3} + 3ab^2(-3) + b^3(-3)\sqrt{-3}$$

it follows that in order to write $p^2 + 3q^2$ as a cube it suffices to find integers a and b such that $p = a^3 - 9ab^2$, $q = 3a^2b - 3b^3$. This is the sufficient condition found in the preceding section.

Euler very seriously confuses necessary and sufficient conditions in this part of his *Algebra* [E9] and it is very difficult to determine what he meant to say. In his examples he seems for the most part to be dealing with sufficient conditions, starting with a, b and finding p, q. But he does make some quite erroneous statements. For example, he says, "When, for example, $x^2 + cy^2$ is to be a cube, one can certainly conclude that both of its irrational factors, namely, $x + y\sqrt{-c}$ and $x - y\sqrt{-c}$, must be cubes, because they are relatively prime in that x and y have no common divisors," although he gives no proof that $x + y\sqrt{-c}$ and $x - y\sqrt{-c}$

must be cubes. At the conclusion of the same paragraph (§191 of the last part) he says very unequivocally, "When $ax^2 + cy^2$ cannot be factored into two rational factors then there are no other solutions than the ones given here," that is, there is no other way for $ax^2 + cy^2$ to be a cube than for there to be integers p and q such that $x\sqrt{a} + y\sqrt{-c} = (p\sqrt{a} + q\sqrt{-c})^3$. The only hint of a proof of this fact is the implied argument by analogy given above, namely, *if AB is a cube and if A and B are relatively prime then A and B must be cubes.* This is a theorem which can be proved* for *integers A and B*, but if A and B are numbers of the form $p + q\sqrt{-3}$ or of the form $x\sqrt{a} + y\sqrt{-c}$ then the proof no longer applies. Euler's statement is in fact true in the special case $a = 1$, $c = 3$—though the proof is quite different from the proof for integers—but there are other values of a and c for which it is false.

Euler's discussion of the case "$x^2 + cy^2$ = cube" follows a rather more complete discussion of the case "$x^2 + cy^2$ = square." There he states that "if the product of two numbers is to be a square, as for example, pq, then necessarily either $p = r^2$ and $q = s^2$, that is, each factor is itself a square, or $p = mr^2$ and $q = ms^2$, that is, the factors are squares multiplied by one and the same number." Once again, this is true if by "numbers" one means "integers" and this fact was proved in Section 1.4 above, but Euler immediately applies the same principle to numbers which are not integers but are of the form $x + y\sqrt{-c}$. The discussion of squares is more complete than the discussion of cubes in that Euler gives what appears to be an alternative proof that if $x^2 + cy^2$ is a square and if x and y are relatively prime integers then $x + y\sqrt{-c} = (a + b\sqrt{-c})^2$ for some integers a and b. This "proof" follows the standard Diophantine technique of writing the square root of $x^2 + cy^2$ in the form $x + (p/q)y$, where p and q are integers so defined, and then simplifying

$$\left(x + \frac{p}{q} y\right)^2 = x^2 + cy^2$$

$$\frac{2p}{q} xy + \frac{p^2}{q^2} y^2 = cy^2$$

$$\frac{2p}{q} \frac{x}{y} = \frac{cq^2 - p^2}{q^2}$$

$$\frac{x}{y} = \frac{cq^2 - p^2}{2pq}.$$

"But x and y are to be relatively prime, as are p and q, and therefore

---

*The proof of the analogous statement for squares, given in Section 1.4, generalizes immediately to cubes.

$x = cq^2 - p^2$ and $y = 2pq$" so that* $-x + y\sqrt{-c} = (p + q\sqrt{-c})^2$. Euler says that this alternative derivation "confirms[†] the correctness of the method" but the natural assumption that p and q are relatively prime does not imply that $cq^2 - p^2$ and $2pq$ are relatively prime and hence does not imply Euler's conclusion $x = cq^2 - p^2$, $y = 2pq$ at all. In fact, the example $49 = 2^2 + 5 \cdot 3^2$ shows not only that the argument is inadequate but also that the conclusion is incorrect, because the equations $2 = 5q^2 - p^2$, $3 = 2pq$ have no rational solutions p, q, much less integral solutions p, q. (These hyberbolas intersect at $p = \sqrt{5/2}$, $q = \sqrt{9/10}$ and at $p = -\sqrt{5/2}$, $q = -\sqrt{9/10}$.)

It has been pointed out (Dickson [D2], vol. 2, Chap. XX, and also p. xiv) that Euler himself notes a failure in his method when (§195) he notes that to find a solution of $2x^2 - 5 =$ cube his method would call for setting $x\sqrt{2} + \sqrt{5} = (a\sqrt{2} + b\sqrt{5})^3 = a^3 2\sqrt{2} + 3a^2 b 2\sqrt{5} + 3ab^2 5\sqrt{2} + b^3 5\sqrt{5}$ and consequently $x = 2a^3 + 15ab^2$, $1 = 6a^2 b + 5b^3$. The last equation implies $b = \pm 1$ and $6a^2 + 5b^2 = \pm 1$, which is clearly impossible. Therefore the method indicates that $2x^2 - 5$ cannot be a cube even though it is a cube in the case $x = 4$. Euler's initial response to this contradiction is to say that "it is of the greatest importance to study the basis of it." It is natural to imagine that Euler was acknowledging the fundamental defectiveness of his method at this point and was suggesting that the defect be studied. However, the next two sections (§196, §197) make very clear that Euler was convinced that the source of the difficulty lay in the minus sign in $2x^2 - 5y^2$ and the related fact that Pell's equation $x^2 - 10y^2$ has solutions other than the trivial solution $x = \pm 1$, $y = 0$. (See Exercise 2.) Without going into details, suffice it to say that Euler appears to be sure that in cases where the sign is plus, the same type of difficulties cannot arise even though, as was noted above, the method fails in the case of $49 = 2^2 + 5 \cdot 3^2$ and in this case the sign is plus and the related equation $x^2 + 5y^2 = 1$ has only the trivial solution.

Euler stated in 1753 that he could prove the case $n = 3$ of Fermat's Last Theorem (letter of Aug. 4, 1753 to Goldbach in Fuss [F6]) but the only proof he published was this one in the *Algebra* (1770). A reasonable speculation about this erroneous proof is that his original method used a less imaginative argument to prove that $x^2 + 3y^2 =$ cube implies $x = a^3 - 9ab^2$, $y = 3a^2 b - 3b^3$, and that it was only later that he had the elegant—but wrong—idea of proving this by using the "fact" that if $(x + y\sqrt{-3})(x - y\sqrt{-3})$ is a cube and if the factors $x + y\sqrt{-3}$ and $x - y\sqrt{-3}$ are relatively prime then the factors themselves must be cubes. Whether this hypothesis is correct or not, it is true that ideas which Euler had used in his

---

*The slight discrepancy in sign can be overcome by setting $-x = cq^2 - p^2$ and $-y = 2pq$ so that $x + y\sqrt{-c} = (p - q\sqrt{-c})^2$.

[†]Note the implied uncertainty about the method.

earlier work are sufficient to prove the necessary lemma about cubes of the form $x^2 + 3y^2$.

EXERCISES

1. A "number" $x + y\sqrt{-c}$ is said to be a *unit* if it is a divisor of 1, that is, if there is another "number" of the same form whose product with it is 1. Show that the units correspond one-to-one to solutions of the equation $x^2 + cy^2 = \pm 1$. Find all units of the form $x + y\sqrt{2}$. (Give a way of finding infinitely many, and try to include them all. You need not prove that you have found them all.) Find all units of the form $x + y\sqrt{-41}$; of the form $x + y\sqrt{-7}$; of the form $x + y\sqrt{7}$; of the form $x + y\sqrt{-1}$.

2. Prove that if $f^2 - 10g^2 = 1$ and if $x\sqrt{2} + y\sqrt{5} = (f + g\sqrt{10})(a\sqrt{2} + b\sqrt{5})^3$ then $2x^2 - 5y^2$ is a cube even though $x\sqrt{2} + y\sqrt{5}$ may not be a cube of the form $(u\sqrt{2} + v\sqrt{5})^3$. Euler believed that this phenomenon explained the failure of his method to find the solution $x = 4$ of $2x^2 - 5 = $ cube. Show that $4\sqrt{2} + \sqrt{5}$ is of this form with $a = 2$, $b = -1$.

2.4 Euler on sums of two squares

Euler succeeded in 1747, when he was forty, in proving Fermat's theorem that every prime of the form $4n + 1$ is a sum of two squares. In the letter to Goldbach dated 6 May, 1747 [F6] in which he announces and proves this theorem, Euler makes clear that his primary objective was to prove another theorem of Fermat, namely, that every number can be written as a sum of four or fewer squares. However, this latter theorem eluded Euler until it was proved by Lagrange in 1770 (after which Euler succeeded in greatly simplifying the proof) and the theorems on sums of two squares which he did prove are of considerable importance. In particular, the techniques Euler developed in order to deal with this problem also enabled him to prove basic facts about numbers of the form $x^2 + 3y^2$ and, as will be shown in the next section, the same techniques can be used to prove the facts about cubes of the form $x^2 + 3y^2$ needed to prove Fermat's Last Theorem in the case $n = 3$.

Euler's proof that every prime of the form $4n + 1$ is a sum of two squares does not require much space and is quite elementary. (See [E8] as well as the letters to Goldbach.)

(1) *The product of two numbers, each of which is a sum of two squares, is itself a sum of two squares.* This follows immediately from the formula $(a^2 + b^2)(c^2 + d^2) = (ac - bd)^2 + (ad + bc)^2$.

(2) *If a number which is a sum of two squares is divisible by a prime which is a sum of two squares then the quotient is a sum of two squares.* Suppose for example that $a^2 + b^2$ is divisible by $p^2 + q^2$ and that $p^2 + q^2$ is prime. Then $p^2 + q^2$ divides $(pb - aq)(pb + aq) = p^2b^2 - a^2q^2 = p^2b^2 + p^2a^2 - p^2a^2 - a^2q^2 = p^2(a^2 + b^2) - a^2(p^2 + q^2)$. Since it is prime it must

therefore divide either $pb - aq$ or $pb + aq$. Suppose first that $p^2 + q^2$ divides $pb + aq$. Then from $(a^2 + b^2)(p^2 + q^2) = (ap - bq)^2 + (aq + bp)^2$ it follows that $p^2 + q^2$ must also divide $(ap - bq)^2$. Therefore the equation can be divided by the *square* of $p^2 + q^2$ and the result is an expression of $(a^2 + b^2)/(p^2 + q^2)$ as a sum of two squares as required. The second case, in which $p^2 + q^2$ divides $pb - aq$, can be handled in the same way using $(a^2 + b^2)(q^2 + p^2) = (aq - bp)^2 + (ap + bq)^2$.

(3) *If a number which can be written as a sum of two squares is divisible by a number which is not a sum of two squares then the quotient has a factor which is not a sum of two squares.* This is essentially just the contrapositive of (2). Suppose x divides $a^2 + b^2$ and that the quotient, factored into its prime factors, is $p_1 p_2 \ldots p_n$. Then $a^2 + b^2 = x p_1 p_2 \ldots p_n$. If all of the factors $p_1, p_2, \ldots, p_n$ could be expressed as sums of two squares, then $a^2 + b^2$ could be divided successively by $p_1, p_2, \ldots, p_n$ and (2) would imply that each quotient, concluding with x, was a sum of two squares. Therefore if x is not the sum of two squares, one of the primes $p_1, p_2, \ldots, p_n$ must not be the sum of two squares.

(4) *If a and b are relatively prime then every factor of $a^2 + b^2$ is a sum of two squares.* Let x be a factor of $a^2 + b^2$. By division $a = mx \pm c$, $b = nx \pm d$ where c and d are at most $\frac{1}{2} x$ in absolute value. Since $a^2 + b^2 = m^2 x^2 \pm 2mxc + c^2 + n^2 x^2 \pm 2nxd + d^2 = Ax + (c^2 + d^2)$ is divisible by x, $c^2 + d^2$ must be divisible by x, say $c^2 + d^2 = yx$. If c and d have any common divisor greater than one, this divisor cannot divide x because then it would divide both a and b contrary to assumption. Therefore the equation $c^2 + d^2 = yx$ can be divided by the square of the greatest common divisor of c and d to give an equation of the form $e^2 + f^2 = zx$. Moreover $z \leqslant \frac{1}{2} x$ because $zx = e^2 + f^2 \leqslant c^2 + d^2 \leqslant (\frac{1}{2} x)^2 + (\frac{1}{2} x)^2 = \frac{1}{2} x^2$. If x were not the sum of two squares then by (3) there would be a factor of z, call it w, which could not be written as a sum of two squares. But this would lead to an infinite descent, going from a number x not a sum of two squares but a factor of the sum of two relatively prime squares to another smaller number w with the same properties. Therefore x must be a sum of two squares.

(5) *Every prime of the form $4n + 1$ is a sum of two squares.* Euler first communicated the following elegant proof of this fact to Goldbach in 1749, two years after his original proof which was rather vague on this point. If $p = 4n + 1$ is prime, then by Fermat's theorem each of the numbers $1, 2^{4n}, 3^{4n}, 4^{4n}, \ldots, (4n)^{4n}$ is one more than a multiple of p. (The next term p^{4n} is divisible by p and all the succeeding terms $(p + 1)^{4n}, \ldots$ up to $(2p - 1)^{4n}$ are one more than a multiple of p, and so forth.) Therefore the differences $2^{4n} - 1, 3^{4n} - 2^{4n}, \ldots, (4n)^{4n} - (4n - 1)^{4n}$ are all divisible by p. Each of these differences can be factored $a^{4n} - b^{4n} = (a^{2n} + b^{2n})(a^{2n} - b^{2n})$ and p, being prime, must divide one of the factors. If in any of the $4n - 1$ cases it divides the first factor then, by (4) and the fact that a number and its successor are always relatively prime, it follows that p is a

47

sum of two squares as desired. Therefore it suffices to show that p cannot divide all $4n - 1$ numbers $2^{2n} - 1, 3^{2n} - 2^{2n}, \ldots, (4n)^{2n} - (4n - 1)^{2n}$. This is easily done as follows. If p did divide all these numbers then it would divide all $4n - 2$ differences of successive numbers, all $4n - 3$ differences of the differences, and so forth. But it is elementary algebra (see Exercise 2) to prove that the kth differences of the sequence $1^k, 2^k, 3^k, 4^k, \ldots$ are constant and are equal to $k!$ (see Table 2.4.1). Thus the $2n$th differences of the sequence $1, 2^{2n}, 3^{2n}, 4^{2n}, \ldots$ are all $(2n)!$ and are therefore not divisible by $p = 4n + 1$. If p divided the first $4n - 1$ of the first differences $2^{2n} - 1, 3^{2n} - 2^{2n}, \ldots, (4n)^{2n} - (4n - 1)^{2n}$, then it would divide the first $4n - 2n$ of the $2n$th differences, which it does not. Therefore p must fail to divide at least one of these $4n - 1$ first differences, as was to be shown.

In an important 1658 letter to Digby which was published in Wallis's collected works and was therefore probably known to Euler, Fermat [F4] said that he had irrefutable proofs (*firmissimis demonstrationibus*) that (a) every prime of the form $4n + 1$ is a sum of two squares, (b) every prime of the form $3n + 1$ is of the form $a^2 + 3b^2$, (c) every prime of the form $8n + 1$ or $8n + 3$ is of the form $a^2 + 2b^2$, and (d) every number is the sum of three or fewer triangular numbers, four or fewer squares, five or fewer pentagonal numbers, and so forth. Euler was very eager to prove the last of these statements—particularly the fact that every number is a sum of four or fewer squares—so it was natural for him to try to use the techniques by which he had proved the first one to prove the others. He found that (c)

Table 2.4.1 Differences of x^k.

$k = 1$	1	2	3	4	5	6	7		
1st differences		1	1	1	1	1	1	$\ldots$	
2nd differences			0	0	0	0	0	0	
$k = 2$	1	4	9	16	25	36	49	$\ldots$	
1st differences		3	5	7	9	11	13	$\ldots$	
2nd differences			2	2	2	2	2	$\ldots$	
3rd differences				0	0	0	0	$\ldots$	
$k = 3$	1	8	27	64	125	216	343	512	$\ldots$
1st differences		7	19	37	61	91	127	169	$\ldots$
2nd differences			12	18	24	30	36	42	$\ldots$
3rd differences				6	6	6	6	6	$\ldots$
4th differences					0	0	0	0	$\ldots$
$k = 4$	1	16	81	256	625	1296	2401	4096	$\ldots$
1st differences		15	65	175	369	671	1105	1695	$\ldots$
2nd differences			50	110	194	302	434	590	$\ldots$
3rd differences				60	84	108	132	156	$\ldots$
4th differences					24	24	24	24	$\ldots$
5th differences						0	0	0	$\ldots$

and (d) were still beyond his reach but that (b) could be proved by almost the same arguments as (a).

The main difference between the proof of (b) and the proof of (a) is in the treatment of the prime 2. Note that statement (4) above is false for representations of numbers in the form $a^2 + 3b^2$ because $1^2 + 3 \cdot 1^2$ is divisible by 2 even though 2 is not of the form $a^2 + 3b^2$. Note also that if the proof of (4) is applied to representations in the form $a^2 + 3b^2$ then the inequality $zx \leqslant (\frac{1}{2}x)^2 + (\frac{1}{2}x)^2 = \frac{1}{2}x$ is replaced by $zx \leqslant (\frac{1}{2}x)^2 + 3(\frac{1}{2}x)^2 = x^2$ so that $z \leqslant x$ and the strict inequality needed to accomplish the descent is lacking. However, if x is *odd* then the inequalities $|c| \leqslant \frac{1}{2}x$, $|d| \leqslant \frac{1}{2}x$ are *strict* inequalities $|c| < \frac{1}{2}x$, $|d| < \frac{1}{2}x$ and the needed inequality $z < x$ will follow. The proof of the analog of statement (4) for representations in the form $a^2 + 3b^2$ can be given as follows.

($1'$) *The product of two numbers, each of which can be written in the form* $a^2 + 3b^2$, *can itself be written in this form.* This follows from $(a^2 + 3b^2)(c^2 + 3d^2) = (ac - 3bd)^2 + 3(ad + bc)^2$.

($2'$) *If a number of the form* $a^2 + 3b^2$ *is divisible by 2 then it must be divisible by 4, and its quotient by 4 must itself be of the form* $c^2 + 3d^2$. If a and b have opposite parities then $a^2 + 3b^2$ is not divisible by 2. If a and b are both even, then $a^2 + 3b^2$ is divisible by 2^2 and the quotient is of the form $c^2 + 3d^2$ with $c = \frac{1}{2}a$, $d = \frac{1}{2}b$. Finally, consider the case where a and b are both odd. Then $a = 4m \pm 1$ and $b = 4n \pm 1$ when m and n and the signs are properly chosen. Therefore either $a + b$ or $a - b$ is divisible by 4. If $a + b$ is divisible by 4 then $4(a^2 + 3b^2) = (1^2 + 3 \cdot 1^2)(a^2 + 3b^2) = (a - 3b)^2 + 3(a + b)^2$ is divisible by 4^2 because $a - 3b = (a + b) - 4b$, and it follows that $(a^2 + 3b^2)/4$ is of the form $c^2 + 3d^2$. If $a - b$ is divisible by 4 then the same conclusion can be reached by using $4 = (-1)^2 + 3 \cdot 1^2$ in place of $4 = 1^2 + 3 \cdot 1^2$.

($3'$) *If a number of the form* $a^2 + 3b^2$ *is divisible by a prime of the form* $p^2 + 3q^2$ *then the quotient can be written in the form* $c^2 + 3d^2$. The main step in the proof of this fact is again the observation that $(pb - aq)(pb + aq) = p^2b^2 + 3q^2b^2 - 3q^2b^2 - a^2q^2 = b^2(p^2 + 3q^2) - q^2(a^2 + 3b^2)$ is divisible by $p^2 + 3q^2$ and therefore, since $p^2 + 3q^2$ is prime, that either $pb - aq$ or $pb + aq$ is divisible by $p^2 + 3q^2$. Therefore $(p^2 + 3q^2)(a^2 + 3b^2) = [p^2 + 3(\pm q)^2](a^2 + 3b^2) = (pa \mp 3qb)^2 + 3(pb \pm aq)^2$ can be divided by $(p^2 + 3q^2)^2$ when the sign is chosen correctly and it follows that $(a^2 + 3b^2)/(p^2 + 3q^2)$ has the desired form.

($4'$) *If a number which can be written in the form* $a^2 + 3b^2$ *has an odd factor which is not of this form then the quotient has an odd factor which is not of this form.* Let $xy = a^2 + 3b^2$ where x is odd. If y is even then by ($2'$) it is divisible by 4 and $x(y/4) = c^2 + 3d^2$. This process can be repeated until $y/4^k$ is odd. Therefore $y = p_1p_2 \ldots p_n$ where each of the p's is either 4 or an odd prime. If all of the odd primes in this factorization of y can be written in the form $c^2 + 3d^2$ then $xy = a^2 + 3b^2$ can be divided successively by each of the p's and ($2'$) and ($3'$) imply that x can be written in the form $c^2 + 3d^2$. Therefore if x does not have this form then y must have an odd factor not of this form.

(5') *If a and b are relatively prime then every odd factor of $a^2 + 3b^2$ is of the form $c^2 + 3d^2$.* Let x be an odd factor of $a^2 + 3b^2$. By division $a = mx \pm c$, $b = nx \pm d$ where $|c| < \frac{1}{2}x$, $|d| < \frac{1}{2}x$ (using the fact that x is odd). Then $c^2 + 3d^2$ is divisible by x, say $c^2 + 3d^2 = xy$ where $y < x$. No common factor of c and d greater than 1 can divide x because then a and b would not be relatively prime. Therefore $c^2 + 3d^2 = xy$ can be divided by the square of the greatest common factor of c and d to give $e^2 + 3f^2 = xz$ where e and f are relatively prime. If x is not of the form $a^2 + 3b^2$ then, by (4'), z has an odd factor, call it w, which is not of this form. Therefore the existence of an odd number x which is a factor of a number of the form $a^2 + 3b^2$ (a and b relatively prime) and which is not itself of the form $c^2 + 3d^2$ would imply the existence of a smaller number w of the same type. By the principle of infinite descent the desired conclusion follows.

Every prime other than 3 is of the form $3n + 1$ or $3n + 2$. A number of the form $3n + 2$ cannot be of the form $a^2 + 3b^2$ because if $a^2 + 3b^2$ is not divisible by 3 then a must not be divisible by 3, $a = 3m \pm 1$, and $a^2 + 3b^2$ is 1 more than a multiple of 3. Thus by (5') an odd prime which divides a number of the form $a^2 + 3b^2$ where a and b are relatively prime cannot be of the form $3n + 2$. If a and b are not relatively prime then $a^2 + 3b^2 = d^2(e^2 + 3f^2)$ where d is their greatest common divisor and where e and f are relatively prime. Thus a number of the form $a^2 + 3b^2$ can be written as a square times a number with no odd prime factors of the form $3n + 2$. The even prime factors of $a^2 + 3b^2$, that is, the number of 2's which divide $a^2 + 3b^2$, comprise, by (2'), a power of 4 and hence a square. Therefore *a necessary condition that a number be of the form $a^2 + 3b^2$ is that its quotient by the largest square it contains contain no prime factors of the form $3n + 2$.* To prove that this condition is also *sufficient* it suffices to prove:

(6') *Every prime of the form $3n + 1$ is of the form $a^2 + 3b^2$.* As before, by Fermat's theorem the prime $p = 3n + 1$ divides the $p - 2$ differences of the numbers $1, 2^{3n}, 3^{3n}, \ldots, (p - 1)^{3n}$. Each of these differences can be factored $a^{3n} - b^{3n} = (a^n - b^n)(a^{2n} + a^n b^n + b^{2n})$ and the second factor, since either a or b is even, can be written in the form $A^2 + A(2B) + (2B)^2 = (A + B)^2 + 3B^2$ with A and B relatively prime. Therefore p must by (5') be of the form $c^2 + 3d^2$ unless it divides the $p - 2$ differences of the numbers $1, 2^n, 3^n, \ldots, (p - 1)^n$. As before, this would imply that p divided $n!$ which is impossible. Therefore p must be of the form $c^2 + 3d^2$ as was to be shown.

If these same arguments are applied to representations in the form $a^2 + 2b^2$, they suffice to prove that *a necessary condition that a number be of the form $a^2 + 2b^2$ is that its quotient by the largest square it contains contain no prime factors of the form $8n + 5$ or $8n + 7$.* To prove that this condition is also sufficient it suffices to prove that all primes of the form

$8n + 1$ or $8n + 3$ can be written in the form $a^2 + 2b^2$. It is this last step, analogous to (6′), that Euler failed to prove; the first proof was given by Lagrange. (A proof is given in Exercises 6 and 7 below.)

EXERCISES

1. Show that if a prime can be represented in the form $a^2 + b^2$ then the representation is unique. [Use the proof of (2).] Is the same true of primes of the form $a^2 + 2b^2$? Of the form $a^2 + 3b^2$?

2. Prove that the nth differences of x^n are all $n!$ and that consequently the higher differences are identically zero. [Prove that the first difference of a polynomial of degree n is a polynomial of degree $n - 1$.]

3. Derive the formula $(a^2 + 3b^2)(c^2 + 3d^2) = (ac - 3bd)^2 + 3(ad + bc)^2$ using the arithmetic of surds. Do the same for the analogous formula for numbers of the form $a^2 + kd^2$.

4. Prove that every factor of a number of the form $a^2 - 2b^2$ with a and b relatively prime is of the same form $c^2 - 2d^2$.

5. It was noted in Section 1.7 that 21 is of the form $a^2 + 5b^2$ but that neither 3 nor 7 is. Where do the techniques of this section fail if one attempts to prove that every factor of a number of the form $a^2 + 5b^2$ with a and b relatively prime must be of the same form $c^2 + 5d^2$?

6. Prove that every prime of the form $8n + 3$ is of the form $a^2 + 2b^2$ as follows. Let $p = 8n + 3$ be prime and let x be the integer $\frac{1}{2}(p + 1)$. Since $x^{8n+2} - 1$ is divisible by p, $x^{8n}(p + 1)^2 - 4$ is divisible by p, and therefore $(x^{4n} - 2)(x^{4n} + 2)$ is divisible by p. Conclude from Exercise 4 that p cannot divide $x^{4n} - 2$. Therefore p divides $x^{4n} + 2$. Conclude that $p = a^2 + 2b^2$.

7. Prove that every prime of the form $8n + 1$ is of the form $a^2 + 2b^2$ as follows. Let $p = 8n + 1$ be prime. The differences of the sequence $1, 2^{8n}, 3^{8n}, 4^{8n}, \ldots$ can be factored $a^{8n} - b^{8n} = (a^{4n} - b^{4n})(a^{4n} + b^{4n})$ and the second factor can be written $(a^{2n} - b^{2n})^2 + 2(a^n b^n)^2$. These observations and the techniques of this section yield a proof.

8. Show that Euler's proof of (2) can be derived from an attempt to perform the division $(a + b\sqrt{-1})/(p \pm q\sqrt{-1})$.

9. Give an alternative proof of (5) by showing that $(x + 1)^{2n} - x^{2n} \equiv 0 \bmod p$ has at most $2n$ distinct roots mod p (in fact at most $2n - 1$). More generally, a congruence $f(x) \equiv 0 \bmod p$, in which $f(x)$ is a polynomial of degree m, has at most m distinct solutions unless all the coefficients of $f(x)$ are zero mod p. [Assume without loss of generality that the leading coefficient of $f(x)$ is not zero mod p. If r is any solution of $f(r) \equiv 0 \bmod p$ then $f(x) = (x - r)q(x) + c$ where $q(x)$ is a polynomial of degree $m - 1$ with the same leading coefficient as $f(x)$ and where c is an integer $c \equiv 0 \bmod p$. Every solution s of $f(s) \equiv 0 \bmod p$ is a solution of $q(s) \equiv 0$ except, possibly, for $s = r$. The proposition to be proved is obviously true when $m = 0$.]

51

10. It was observed in Section 1.8 that the prime divisors of $2^{32} + 1$ must all be 1 mod 64, so that 641 is only the fifth prime that need be tried. Show that in fact the prime factors of $2^{32} + 1$ must be 1 mod 128 and that 641 is only the *second* prime that need be tried. [Let p be a prime divisor of $2^{32} + 1$. Since $p \equiv 1 \bmod 64$, p divides $x^2 - 2$ for some x. Since $x^{p-1} \equiv 1 \bmod p$, the previous argument shows that 64 divides $(p - 1)/2$.]

11. Factor $2^{32} - 1$ into primes and conclude that 257 does not divide $2^{32} + 1$. [No computation is necessary.]

2.5 Remainder of the proof when $n = 3$

Euler's idea of computing with "numbers" of the form $a + b\sqrt{-c}$ is closely related to the use of the formula

$$(x^2 + cy^2)(u^2 + cv^2) = (xu - cyv)^2 + c(xv + yu)^2$$

which has occurred repeatedly above. In words, this formula says that if the integer A is a product of integers B and C, and if B and C can both be written in the form $a^2 + cb^2$, say $B = x^2 + cy^2$, $C = u^2 + cv^2$, then A can also be written in this form by using the formula $a + b\sqrt{-c} = (x + y\sqrt{-c})(u + v\sqrt{-c})$ to define a and b.

The lemma needed to prove Fermat's Last Theorem in the case $n = 3$ states that if a and b are relatively prime and if $a^2 + 3b^2$ is a cube then $a + b\sqrt{-3} = (p + q\sqrt{-3})^3$ for some integers p and q. In order to prove this it is natural to pursue Euler's arguments just one step further to "factor" $a + b\sqrt{-3}$ as follows.

(1) *If a and b are relatively prime and if $a^2 + 3b^2$ is even then $a + b\sqrt{-3}$ can be written in the form*

$$a + b\sqrt{-3} = (1 \pm \sqrt{-3})(u + v\sqrt{-3})$$

where the sign is appropriately chosen and where u and v are integers. Since $a^2 + 3b^2$ is even, a and b must have the same parity, and since they are relatively prime they must both be odd. Therefore each is of the form $4n \pm 1$ and either $a + b$ or $a - b$ must be divisible by 4. If $a + b$ is divisible by 4 then the equation $4(a^2 + 3b^2) = (1^2 + 3 \cdot 1^2)(a^2 + 3b^2) = (a - 3b)^2 + 3(a + b)^2$ can be divided by 4^2 to put $(a^2 + 3b^2)/4$ in the form $u^2 + 3v^2$ where $u = (a - 3b)/4$, $v = (a + b)/4$. These equations can be solved for a and b in terms of u and v by noting that they are equivalent to $u + v\sqrt{-3} = (a + b\sqrt{-3})(1 + \sqrt{-3})/4$, which gives $(1 - \sqrt{-3})(u + v\sqrt{-3}) = a + b\sqrt{-3}$ as desired. Similarly, if $a - b$ is divisible by 4 then $a + b\sqrt{-3} = (1 + \sqrt{-3})(u + v\sqrt{-3})$ for suitable u and v. Note that u and v are relatively prime (otherwise a and b would not be relatively prime) and that $a^2 + 3b^2 = 4(u^2 + 3v^2)$.

(2) *If a and b are relatively prime and if $a^2 + 3b^2$ is divisible by the odd prime P then P can be written in the form $P = p^2 + 3q^2$ with p and q positive integers and $a + b\sqrt{-3}$ can be written in the form $a + b\sqrt{-3}$*

$= (p \pm q\sqrt{-3})(u + v\sqrt{-3})$ *where the sign is appropriately chosen and where u and v are integers.* The first statement, that $P = p^2 + 3q^2$, is just (5′) of the preceding section. As in Euler's proof, either $pb + aq$ or $pb - aq$ is divisible by P. If $pb + aq$ is divisible by P then the equation $P(a^2 + 3b^2) = (p^2 + 3q^2)(a^2 + 3b^2) = (pa - 3qb)^2 + 3(pb + aq)^2$ can be divided by P^2 to write $(a^2 + 3b^2)/P$ in the form $u^2 + 3v^2$ where $u = (pa - 3qb)/P$ and $v = (pb + aq)/P$, that is, where

$$ u + v\sqrt{-3} = (p + q\sqrt{-3})(a + b\sqrt{-3})/P. $$

Then multiplication by $p - q\sqrt{-3}$ gives $(p - q\sqrt{-3})(u + v\sqrt{-3}) = a + b\sqrt{-3}$ as desired. Similarly, if $pb - aq$ is divisible by P then $a + b\sqrt{-3} = (p + q\sqrt{-3})(u + v\sqrt{-3})$. Again u and v are relatively prime and $a^2 + 3b^2 = P(u^2 + 3v^2)$.

(3) *Let a and b be relatively prime. Then $a + b\sqrt{-3}$ can be written in the form*

$$ a + b\sqrt{-3} = \pm\left(p_1 \pm q_1\sqrt{-3}\right)\left(p_2 \pm q_2\sqrt{-3}\right)\cdots\left(p_n \pm q_n\sqrt{-3}\right) $$

where the p's and q's are positive integers and $p_i^2 + 3q_i^2$ is either 4 or an odd prime. If $a^2 + 3b^2$ is even then it is divisible by 4. If $a^2 + 3b^2$ is not 1 then it has a factor P equal to either 4 or an odd prime and either (1) or (2) gives $a + b\sqrt{-3} = (p \pm q\sqrt{-3})(u + v\sqrt{-3})$ where $p^2 + 3q^2 = P$. Then u and v are relatively prime and the problem of taking a factor $p \pm q\sqrt{-3}$ out of $u + v\sqrt{-3}$ is the same as that of taking one out of $a + b\sqrt{-3}$ except that $u^2 + 3v^2 = (a^2 + 3b^2)/P$ is smaller than $a^2 + 3b^2$. Iterating this process must eventually lead to a stage $a + b\sqrt{-3} = (p_1 \pm q_1\sqrt{-3}) \cdots (p_n \pm q_n\sqrt{-3})(u + v\sqrt{-3})$ where $u^2 + 3v^2 = 1$. Then $u = \pm 1$, $v = 0$, $u + v\sqrt{-3} = \pm 1$ and the factorization is complete.

(4) *Let a and b be relatively prime. Then the factors in the above factorization of $a + b\sqrt{-3}$ are completely determined, except for the choice of signs as indicated, by the fact that $(p_1^2 + 3q_1^2)(p_2^2 + 3q_2^2) \cdots (p_n^2 + 3q_n^2) = a^2 + 3b^2$ is a factorization of $a^2 + 3b^2$ into odd primes and 4's. Moreover, if the factor $p + q\sqrt{-3}$ occurs then the factor $p - q\sqrt{-3}$ does not, and conversely.* The thing to be proved in the first statement is that $p^2 + 3q^2 = P$ determines p and q, up to sign, if P is 4 or an odd prime. This is clear in the case $P = 4$. If P is an odd prime and if $a^2 + 3b^2$ were another representation of it then, by (2),

$$ a + b\sqrt{-3} = (p \pm q\sqrt{-3})(u + v\sqrt{-3}), $$

and $P = P(u^2 + 3v^2)$, that is, $u^2 + 3v^2 = 1$, $u = \pm 1$, $v = 0$, $a + b\sqrt{-3} = \pm(p \pm q\sqrt{-3})$ as was to be shown. The second statement is simply the observation that $p + q\sqrt{-3}$ and $p - q\sqrt{-3}$ would combine to give a factor $p^2 + 3q^2$, which is impossible if a and b are relatively prime.

The lemma needed to complete Euler's proof of the case $n = 3$ of Fermat's Last Theorem can now be deduced very easily.

Lemma. *Let a and b be relatively prime numbers such that $a^2 + 3b^2$ is a cube. Then there exist integers p and q such that $a + b\sqrt{-3} = (p + q\sqrt{-3})^3$.*

PROOF. Let $a^2 + 3b^2 = P_1 P_2 \ldots P_n$ be a factorization into 4's and odd primes as in (4). If this factorization contains exactly k factors of 4 then 2^{2k} is the largest power of 2 which divides $a^2 + 3b^2$ and, since $a^2 + 3b^2$ is a cube, it follows that $2k$ and hence k are multiples of 3. Moreover, any odd prime P in the factorization must occur with a multiplicity which is a multiple of 3. Thus n is divisible by 3 and the factors $P_1 P_2 \ldots P_n$ can be arranged in such a way that $P_{3k+1} = P_{3k+2} = P_{3k+3}$. It follows that in the factorization of $a + b\sqrt{-3}$ given by (3) the factors corresponding to each group of three P's are *identical* because the only choice is the choice of sign $p \pm q\sqrt{-3}$ and both signs cannot occur. Taking one factor from each group of three and multiplying them together then gives a number $c + d\sqrt{-3}$ such that $a + b\sqrt{-3} = \pm(c + d\sqrt{-3})^3$. Since $-(c + d\sqrt{-3})^3 = (-c - d\sqrt{-3})^3$ the desired conclusion follows.

EXERCISES

1. Only a slight modification of the proof of this section is needed to prove that if a and b are relatively prime and if $a^2 + 2b^2$ is a cube then there exist integers p and q such that $a + b\sqrt{-2} = (p + q\sqrt{-2})^3$. Use this fact to prove Fermat's statement that the only solution of $x^2 + 2 = y^3$ in integers is $5^2 + 2 = 3^3$.

2. Similarly, if a and b are relatively prime and if $a^2 + b^2$ is a cube then $a + b\sqrt{-1} = (p + q\sqrt{-1})^3$. The proof of this fact in the case $a = 0$, $b = 1$ requires special attention. Prove that the only solution of $x^2 + 4 = y^3$ in which x and y are integers and x *is odd* is $11^2 + 4 = 5^3$. [This proof and the proof of Exercise 1 are essentially Euler's, but Euler appears to have overlooked the fact that the technique for finding solutions of $a^2 + b^2 =$ cube assumes that a and b are relatively prime. His proof that $x = 4, 11$ are the only solutions of $x^2 + 4 = y^3$ therefore appears to be incomplete.]

3. Complete the proof that $11^2 + 4 = 5^3$ and $2^2 + 4 = 3^3$ are the only solutions of $x^2 + 4 = y^3$ by proving that if $x^2 + 4 = y^3$ and $x =$ even then $x = \pm 2$. [The assumed equation leads to $u^2 + 1 = 2v^3$. Use division of $u + \sqrt{-1}$ by $1 + \sqrt{-1}$ to write $v^3 = (u^2 + 1)/2$ as a sum of 2 squares, say $a^2 + b^2$. Since a and b differ by 1 they are relatively prime and one finds $1 = (p + q)(p^2 - 4pq + q^2)$ where p and q are integers. Thus $p + q = p^2 - 4pq + q^2 = \pm 1$ and $-6pq = 0$ or -2. Thus p or q must be 0, which leads back to $x = \pm 2$.]

4. Show that in step (3) at most one of the $p_i^2 + 3q_i^4$ is 4.

5. Find all representations in the form $a^2 + 3b^2$ (a, b not necessarily relatively prime) of the following numbers. (a) 91. (b) 49. (c) 336.

2.6 Addendum on sums of two squares

In a 1654 letter [F3] to Pascal, which was almost certainly unknown to Euler, Fermat stated a number of his theorems, including the theorem that every prime of the form $4n + 1$ is a sum of two squares. The list of theorems that he gives is almost identical to the one that he gives in the letter to Digby mentioned above (Section 2.4) but there is an important difference: he poses in addition the problem of *finding* the decomposition as a sum of two squares, namely, "Given a prime number of this nature, such as 53, find *by a general rule*, the two squares of which it is composed" (italics added). Of course one can always find the decomposition by trial and error (in his example the trial and error solution $53 = 4 + 49$ is almost immediate) but Fermat clearly attaches some importance to being able to solve the problem in a more methodical and efficient way.

Euler's proof is an indirect proof, an argument by contradiction which shows that if a prime of the form $4n + 1$ is not the sum of two squares then an infinitely descending sequence of positive integers could be found; therefore it does not solve Fermat's problem of finding a constructive method. As is frequently the case in such situations, however, a more careful study of the proof by contradiction shows that it can be modified to give a constructive proof and the modification gives the proof a greater clarity.

Step (5) of Euler's proof is quite constructive. It says that if $4n + 1$ is prime then at least one of the numbers $1^{2n} + 2^{2n}, 2^{2n} + 3^{2n}, \ldots, (4n - 1)^{2n} + (4n)^{2n}$ must be divisible by $4n + 1$. In Fermat's example $4n + 1 = 53$ one easily finds that the first number $1 + 2^{26}$ of this list is divisible by 53. Using the notation* of congruences, one finds $2^6 = 64 \equiv 11 \bmod 53$, $2^{12} \equiv 11^2 = 121 \equiv 15 \bmod 53$, $2^{13} \equiv 2 \cdot 15 = 30 \bmod 53$, $2^{26} \equiv 900 \equiv -1 \bmod 53$; hence $2^{26} + 1$ is divisible by 53. The fact that 53 divides $2^{26} + 1$ is used in Euler's proof to show that 53 divides a sum of two squares. The computation above gives an explicit sum of two squares divisible by 53, namely, $30^2 + 1 = 17 \cdot 53$. Euler concludes in (4) that 53 must be a sum of two squares because otherwise it would be possible to construct an infinite descent. The problem is to prove this directly rather than by resorting to a proof by contradiction.

In the general case with $p = 4n + 1$ one can use Euler's method to find $a^2 + b^2 = kp$ where a and b are relatively prime and where $k < p$. The problem is roughly to find some way to divide by k. Euler's proof of (2) shows how to divide by *prime* numbers which are sums of two squares. It is easy to show that any prime factor of $a^2 + b^2$ is either 2 or is of the form $4n + 1$ (see Exercise 2, Section 1.8) so that one can divide by k by

*See Appendix A.1.

factoring k into its prime factors, writing each such factor as a sum of two squares, and dividing by them one-by-one. In the case at hand, $k = 17$ is itself a prime, and it is very easy to write as a sum of two squares, namely, $17 = 1^2 + 4^2$. The division process is to write $17 \cdot 17 \cdot 53 = (1^2 + 4^2)(30^2 + 1^2) = (30 \mp 4)^2 + (1 \pm 120)^2$ and to choose the signs in such a way that division by 17 is possible. This gives $53 = (34/17)^2 + (119/17)^2 = 2^2 + 7^2$ as desired.

This method reduces the problem of writing the prime p as a sum of two squares to the problem of writing smaller primes of the form $4n + 1$, namely, the prime factors of k, as sums of two squares. This may be the method which Fermat himself had in mind when he described his proof by infinite descent in the words, "If a chosen prime number which is one greater than a multiple of 4 were not the sum of two squares there would be a prime number of the same nature less than the given one, and after that yet a third, etc. descending infinitely until you arrived at 5, which is the least of all of this nature, which it would follow was not the sum of two squares, even though it is. From which one must infer, by deduction to the impossible, that all those of this nature are consequently sums of two squares" (letter to Carcavi [F5]).

However, this method is very lengthy because it involves testing k to see whether it is prime—always a difficult process—factoring it if possible, and expressing the factors as sums of two squares. A much better method in general is to divide by k, at the cost of multiplying by a smaller number n in its place, as follows. As in Euler's proof of (4), let the equation $a^2 + b^2 = kp$ be used to find a smaller multiple of p which is a sum of two squares by dividing $a = q_1 k \pm c$, $b = q_2 k \pm d$ and noting that k then must divide $c^2 + d^2$, say $c^2 + d^2 = nk$. Since $|c| \leqslant \frac{1}{2}k$, $|d| \leqslant \frac{1}{2}k$ it follows as before that $n \leqslant \frac{1}{2}k$. Moreover, $nkkp = (c^2 + d^2)(a^2 + b^2) = (ca \mp db)^2 + (cb \pm da)^2$. The objective is to divide this equation by k^2. The previous method of proving that $cb + da$ or $cb - da$ is divisible by k cannot be used in this situation because k is not necessarily prime, but there is a simpler method, namely, to note that $cb \pm da = c(q_2 k \pm d) \pm d(q_1 k \pm c)$ is divisible by k when the signs are chosen in such a way that the two terms $\pm cd \pm dc$ cancel. Therefore the entire equation can be divided by k^2 to give np as a sum of two squares. If $n = 1$ then p has been written as a sum of two squares. Otherwise the process can be repeated to give an integer $m \leqslant \frac{1}{2}n$ and an expression of mp as a sum of two squares. Repeating this process one eventually must arrive at an expression of p as the sum of two squares.

In the example $30^2 + 1^2 = 17 \cdot 53$ one has $c = -4$, $d = 1$, $4^2 + 1^2 = 17$ and this process is the same as before. As a second example consider the case $p = 229$. The first step is to compute 2^{114} mod 229. Now $2^8 = 256 \equiv 27$ mod 229; $2^{16} \equiv 27^2 = 729 \equiv 42$ mod 229; $2^{32} \equiv 42^2 = 1764 \equiv -68$ mod 229; $2^{64} \equiv 68^2 = 4624 \equiv 44$ mod 229; $2^{96} \equiv (-68)(44) = -2992 \equiv -15$ mod 229; $2^{112} \equiv (-15)(42) = -630 \equiv 57$ mod 229; $2^{114} \equiv 4 \cdot 57 \equiv$

$-1 \bmod 229$. Therefore 229 divides $(2^{57})^2 + 1^2$ and there is no need to look further. Next compute $2^{57} \bmod 229$: $2^{48} \equiv (-68)(42) = -2856 \equiv -108$; $2^{56} \equiv (-108)(27) = -2916 \equiv -168 \equiv 61$; and finally $2^{57} \equiv 122 \bmod 229$. Then direct computation gives $122^2 + 1 = 65 \cdot 229$, which begins the process. On the first step $c = 122 - 2 \cdot 65 = -8$ and $d = 1$, from which $8^2 + 1^2 = 65$. Then $65 \cdot 65 \cdot 229 = (8^2 + 1^2)(122^2 + 1^2) = (976 \mp 1)^2 + (8 \pm 122)^2$ from which $229 = (975/65)^2 + (130/65)^2 = 15^2 + 2^2$.*

An exactly analogous procedure can be used to find a representation of a prime of the form $3n + 1$ in the form $a^2 + 3b^2$. The first step is to compute $2^n \bmod p$. If 2^n is one greater than a multiple of p then p divides $2^n - 1$ and $3^n \bmod p$ must be computed. Eventually one must arrive at an integer c such that c^n is not 1 greater than a multiple of p but $(c-1)^n$ is. Then, since p divides $c^{3n} - (c-1)^{3n}$, p must divide $c^{2n} + c^n(c-1)^n + (c-1)^{2n}$, which is of the form $a^2 + 3b^2$. Thus $a^2 + 3b^2 = kp$ and one can arrange for $k \leqslant p$. If a and b are both even then factors of 2 can be cancelled. If both are odd then 4 divides $a^2 + 3b^2$ and the technique for cancelling a 4, namely,

$$\frac{a^2 + 3b^2}{4} = \left(\frac{a \mp 3b}{4} \right)^2 + 3\left(\frac{a \pm b}{4} \right)^2$$

where the signs are chosen so that 4 divides $a \pm b$, can be used. This reduces the problem to the case $a^2 + 3b^2 = kp$ where a and b have opposite parities. If $k = 1$ then the problem is solved. Otherwise a can be reduced to $c \bmod k$ and b reduced to d to find $c^2 + 3d^2 = nk$ where $n < k$ (because k is odd). Then $nkkp = (ac \mp 3bd)^2 + 3(ad \pm bc)^2$ can, when the signs are properly chosen, be divided to give $np = e^2 + 3f^2$. If $n \neq 1$ it can be reduced again until finally $p = g^2 + 3h^2$.

For example, consider the case of $p = 67$. The first step is to compute $2^{22} \bmod 27$. This is a very simple computation $2^6 = 64 \equiv -3$, $2^{12} \equiv 9$, $2^{18} \equiv -27$, $2^{19} \equiv -54 \equiv 13$, $2^{21} \equiv 52 \equiv -15$, $2^{22} \equiv -30$. Thus $2^{22} - 1$ is not divisible by 67 and $2^{44} + 2^{22} + 1$ must be. This is $3(\frac{1}{2} 2^{22})^2 + (\frac{1}{2} 2^{22} + 1)^2 = 3(2^{21})^2 + (2^{21} + 1)^2$. Since $2^{21} \equiv -15$ this implies $3 \cdot (-15)^2 + (-14)^2$ must be divisible by 67. Direct computation gives $3 \cdot 15^2 + 14^2 = 871 = 13 \cdot 67$. Reducing 15 and 14 mod 13 gives $3 \cdot 2^2 + 1^2 \equiv 0 \bmod 13$ and in fact $3 \cdot 2^2 + 1^2 = 13$. Therefore $13 \cdot 13 \cdot 67 = (1^2 + 3 \cdot 2^2)(14^2 + 3 \cdot 15^2) = (14 \mp 3 \cdot 30)^2 + 3(15 \pm 28)^2$, $67 = (104/13)^2 + 3(-13/13)^2 = 8^2 + 3 \cdot 1^2$.

The latter part of the procedure applies in the case of the problem of writing primes in the form $a^2 + 2b^2$. If a number of the form $a^2 + 2b^2$ can be found which is divisible by p then the analogous procedure makes it possible to find a representation of p itself in this form. What is lacking is a method of constructing, given a prime p of one of the forms $p = 8n + 1$ or $8n + 3$, a number of the form $a^2 + 2b^2$ which is divisible by p. A constructive method of achieving this is given in Exercises 6 and 7 of Section 2.4 above.

*For another method of solving $p = a^2 + b^2$ see Exercise 9, Section 7.10.

EXERCISES

1. Write 97 in each of the forms $a^2 + b^2$, $a^2 + 3b^2$, $a^2 + 2b^2$.

2. Write 193 in all three forms as in Exercise 1.

3. Write 7297 in all three forms.

4. Jacobi's paper [J1] contains extensive tables of representations of primes in the forms $a^2 + b^2$, $a^2 + 2b^2$, and $a^2 + 3b^2$. Look up these tables and derive several entries.

From Euler to Kummer 3

3.1 Introduction

When Euler told Goldbach in his letter of August 4, 1753 that he had succeeded in proving Fermat's Last Theorem in the case $n = 3$, he observed that the proof seemed very different from the proof for the case $n = 4$ and that a proof of the general case still seemed remote. In the next ninety years a few—a very few—special cases and partial results toward a proof of Fermat's Last Theorem were obtained, but the general case still seemed quite unapproachable. Then, in the 1840s, Kummer developed his theory of ideal factors and derived from it deep new insights into Fermat's Last Theorem which gave grounds for hope that the general case itself might soon be proved.

This chapter is devoted to the most important results that were obtained during this ninety year period. Section 3.2 is devoted to the statement and proof of Sophie Germain's theorem, Section 3.3 to the proof of the case $n = 5$ of Fermat's Last Theorem by Legendre and Dirichlet, and Section 3.4 to a few remarks about the proofs of Fermat's Last Theorem in the cases $n = 14$ and $n = 7$ by Dirichlet and Lamé respectively. Sophie Germain's theorem is important, and, although it has been generalized and improved, it has not been replaced in the time since it was discovered. The proofs of the cases $n = 5$ and $n = 7$, however, have been replaced by Kummer's proof of Fermat's Last Theorem for "regular primes" (and the case $n = 14$ has been replaced by the more general case $n = 7$) and are of interest only as examples of what can be done by more elementary methods which do not appeal to the theory of ideal factors and as examples of how the theory developed toward Kummer's great discoveries.

Although the progress toward a proof of Fermat's Last Theorem was not great during this period, the progress in number theory overall was

59

enormous. During this period lived three of the greatest number theorists in history—Lagrange, Legendre, and Gauss. Lagrange has already been mentioned above, in connection with both the solution of Pell's equation (Section 1.9) and the proof that every number can be written as a sum of four squares (Section 2.4). Lagrange's great ability was recognized by Euler when Lagrange was still quite young, and the interaction between the two of them was very fruitful. When Euler left the court of Frederick the Great to return to Russia in 1766, Lagrange replaced him in Berlin. On the larger stage, when Euler died in 1783 Lagrange was his obvious successor as the foremost mathematician of Europe. Like Euler, he was an extraordinarily versatile mathematician and made fundamental contributions to celestial mechanics, the calculus of variations, algebra, analysis, and so forth, but, also like Euler, his work shows a special love for the theory of numbers. Legendre—whose name makes it very easy to confuse him with Lagrange —was definitely not in the class of Euler and Lagrange, but he was a fine mathematician who did important work in a wide variety of fields, notably elliptic functions, algebra, and number theory, and, perhaps more importantly, he was a prolific writer whose works covered a great range of topics and reached a wide audience. His *Théorie des Nombres*, first published in 1798, went through several editions and had a profound influence on the mathematical culture of the period. Gauss published his great *Disquisitiones Arithmeticae* in 1801 (he was 24 years old at the time) and won immediate recognition* as a genius of the first order. He too was a great universalist—it has been said that there is not a single development in the mathematics of the nineteenth century which was not anticipated in his work—but he too considered number theory, the higher arithmetic as he preferred to call it, to be the queen of mathematics. In addition to the *Disquisitiones Arithmeticae* he published two classic memoirs on biquadratic reciprocity in 1828 and 1832 which had a great effect on the development of number theory.

The work of these three men does not, for the most part, relate directly to Fermat's Last Theorem or to techniques which were later to be used successfully in the study of Fermat's Last Theorem. Indirectly, however, their work had very great effect on the development of these techniques. Apart from the general effect of causing a whole generation of mathematicians to be educated in the belief that the higher arithmetic is the queen of mathematics, there are at least two very particular effects that will be studied in the chapters which follow—Kummer's development of the theory of ideal factors in order to deal with the higher reciprocity laws and Dirichlet's development of the analytical formula for the number of classes of binary quadratic forms with given determinant—both of which grew out

*Lagrange [L4] wrote to Gauss on 31 May 1804, "Your *Disquisitiones* have placed you immediately in the ranks of the best geometers." (At that time "geometer" meant "pure mathematician.")

of the work of these three number theorists and both of which are essential to the later study of Fermat's Last Theorem.

Thus, the brevity of this chapter should not be taken to mean that the period from Euler to Kummer was one of little progress. On the contrary, this period was in many ways the Golden Age of number theory. The brevity of this chapter should only be taken to mean that during this period the study of Fermat's Last Theorem was for a time relegated to the background while number theory developed in other areas—principally the study of binary quadratic forms and reciprocity laws—which only later would bear fruit in the study of Fermat's Last Theorem.

3.2 Sophie Germain's theorem

One of the very few women to overcome the prejudice and discrimination which have tended to exclude women from the pursuit of higher mathematics up to the present time was Sophie Germain (1776–1831). She had an active correspondence with Gauss in Göttingen and was personally acquainted with Legendre in Paris. In corresponding with Gauss she at first assumed a masculine pseudonym, thinking that otherwise Gauss would not take her seriously. Whether he would have is not clear. What is clear is that she won his respect without appealing to the novelty of the fact that she was a woman mathematician, and that when Gauss learned how he had been deceived he was quite delighted.

> "But how to describe to you my admiration and astonishment at seeing my esteemed correspondent Mr. Leblanc metamorphose himself into this illustrious personage [Sophie Germain, to whom he was writing after having discovered the deception] who gives such a brilliant example of what I would find it difficult to believe. A taste for the abstract sciences in general and above all the mysteries of numbers is excessively rare: it is not a subject which strikes everyone; the enchanting charms of this sublime science reveal themselves only to those who have the courage to go deeply into it. But when a person of the sex which, according to our customs and prejudices, must encounter infinitely more difficulties than men to familiarize herself with these thorny researches, succeeds nevertheless in surmounting these obstacles and penetrating the most obscure parts of them, then without doubt she must have the noblest courage, quite extraordinary talents, and a superior genius. Indeed, nothing could prove to me in so flattering and less equivocal a manner that the attractions of this science, which has enriched my life with so many joys, are not chimerical, as the predilection with which you have honored it."
> (Translated from Gauss's French [G8].)

Relations between Gauss and Legendre were always strained, and it is to the credit of Sophie Germain that she was on the best of terms with both of them. It was Legendre who made her famous* by mentioning her

---

*She also won a prize from the Paris Academy for a paper on the theory of elasticity, but as far as I know this work did not win any lasting reputation.

in his *Theorie des Nombres* [L7] and by attributing to her a very important result on Fermat's Last Theorem which is now known by her name. This theorem is stated below, following a discussion of a few special cases.

Theorem. *If $x^5 + y^5 = z^5$ then one of the numbers x, y, z must be divisible by 5.*

PROOF. Although Fermat's Last Theorem states that $x^5 + y^5 \neq z^5$ for *positive* integers, it is convenient at this point to move z^5 to the other side of the equation $x^5 + y^5 + (-z)^5 \neq 0$ and to state the case $n = 5$ in the more symmetrical form "the equation $x^5 + y^5 + z^5 = 0$ is impossible in nonzero integers x, y, z." The advantage, obviously, of stating the theorem in this form is that the roles of x, y, and z become interchangeable. Since zero is divisible by 5 the theorem to be proved takes the form: *If x, y, z are integers such that $x^5 + y^5 + z^5 = 0$ then one of them must be divisible by 5.*
The first step of the proof is to rewrite the equation in the form

$$- x^5 = (y + z)(y^4 - y^3z + y^2z^2 - yz^3 + z^4).$$

As usual, it can be assumed that x, y, and z are pairwise relatively prime, since otherwise a common factor could be divided out. Then the two factors on the right are relatively prime because if p is any prime which divides $y + z$ then $y \equiv -z \bmod p$, $y^4 - y^3z + y^2z^2 - yz^3 + z^4 \equiv 5y^4 \bmod p$, and if p divides both factors then either p is 5, in which case x is divisible by 5 as was to be shown, or p divides y and $y + z$, in which case y and z are not relatively prime. In other words, if $x^5 + y^5 + z^5 = 0$, if x, y, z are pairwise relatively prime, and if all three are prime to 5, then $y + z$ and $y^4 - y^3z + y^2z^2 - yz^3 + z^4$ are relatively prime. Since their product is a fifth power $y^5 + z^5 = -x^5 = (-x)^5$, this implies that they must each be fifth powers. (The argument of Section 1.4 easily extends to this case.) By symmetry, the same argument applies to $-y^5 = x^5 + z^5$ and $-z^5 = x^5 + y^5$ to show that there are integers $a, \alpha, b, \beta, c, \gamma$ such that

$$
\begin{array}{lll}
y + z = a^5 & y^4 - y^3z + y^2z^2 - yz^3 + z^4 = \alpha^5 & x = -a\alpha \\
z + x = b^5 & z^4 - z^3x + z^2x^2 - zx^3 + x^4 = \beta^5 & y = -b\beta \\
x + y = c^5 & x^4 - x^3y + x^2y^2 - xy^3 + y^4 = \gamma^5 & z = -c\gamma.
\end{array}
$$

It is to be shown that this is impossible.
The key to the proof is the observation that modulo 11 the fifth powers are $-1, 0, 1$. This is clear from Fermat's Theorem, which states that either $x \equiv 0 \bmod 11$ or $(x^5)^2 \equiv 1 \bmod 11$, and hence either $x^5 \equiv 0$ or $x^5 \equiv \pm 1$. (More precisely, one must note that $y^2 \equiv 1$ implies $y^2 - 1 = (y - 1)(y + 1) \equiv 0 \bmod 11$ and therefore, because 11 is prime, implies $y - 1 \equiv 0$ or $y + 1 \equiv 0 \bmod 11$, that is, $y \equiv \pm 1$.) Therefore $x^5 + y^5 + z^5 \equiv 0 \bmod 11$ is possible only if x or y or z is $\equiv 0 \bmod 11$ because $\pm 1 \pm 1 \pm 1 = 0$ is impossible.

Suppose $x^5 + y^5 + z^5 = 0$ where x, y, z are pairwise relatively prime and prime to 5. Then $a, \alpha, b, \beta, c, \gamma$ can be found as above. One of the numbers x, y, z must be divisible by 11. Assume, without loss of generality, that it is x. Then $2x = b^5 + c^5 + (-a)^5$ is divisible by 11 and one of the numbers a, b, c must be divisible by 11. But b can not be divisible by 11 because x is divisible by 11 and this would show that x and z had the common factor 11, contrary to the assumption that they are relatively prime. Similarly c can not be divisible by 11. Thus a must be divisible by 11. But this too is impossible because then $y \equiv -z \bmod 11$, $\alpha^5 \equiv 5y^4 \bmod 11$ while on the other hand $x \equiv 0$, $\gamma^5 \equiv y^4$, which gives $\alpha^5 \equiv 5 \cdot \gamma^5$. Since the fifth powers modulo 11 are $0, \pm 1$ this implies $\alpha \equiv \gamma \equiv 0$ and contradicts the assumption that x and z are relatively prime. This completes the proof of the theorem.

Exactly the same argument as above proves a more general theorem.

Theorem. *If n is an odd prime* and if $2n + 1$ is prime then $x^n + y^n = z^n$ implies that $x, y,$ or z is divisible by n.*

Thus, in order to prove Fermat's Last Theorem for $n = 5$ or $n = 11$ or for numerous other prime values of n it suffices to prove that $x^n + y^n + z^n = 0$ is impossible under the additional assumption that one of the three numbers $x, y,$ or z is divisible by n, because the case in which none is divisible has already been excluded by the theorem. It is traditional, largely because of this theorem, to divide Fermat's Last Theorem into two cases, the first case being that in which none of the three numbers x, y, z is divisible by n and the second case being that in which one and only one of the three numbers is divisible by n. The two cases are traditionally called Case I and Case II in that order, and the theorem above is stated in the form: *if n is an odd prime such that $2n + 1$ is prime, then Case I of Fermat's Last Theorem is true for nth powers.*

Surprisingly enough, Case I of Fermat's Last Theorem is by far the more elementary of the two. Even in cases where the above theorem fails, a slight modification of it frequently succeeds. For example, in the case $n = 7$ it is not true that $2n + 1 = 15$ is prime but it is true that $4n + 1 = 29$ is prime. The 7th powers mod 29 are $0, \pm 1, \pm 12$ (see Exercise 4). Therefore $x^7 + y^7 + z^7 \equiv 0 \bmod 29$ is possible only if one of the numbers is zero mod 29. Thus the argument which was used to prove the theorem above shows that there must then be integers α and γ such that $\alpha^7 \equiv 7\gamma^7 \bmod 29$, but $\alpha \not\equiv 0 \bmod 29$, $\gamma \not\equiv 0 \bmod 29$. Since 29 is prime there is† an integer g such that $\gamma g \equiv 1 \bmod 29$; hence $(\alpha g)^7 \equiv 7 \bmod 29$, contrary to

*If n is the prime 2 then $2n + 1$ is prime and the theorem is still true, as was seen in Section 1.3.

† See Appendix A.1.

the fact that $0, \pm 1, \pm 12$ are the only 7th powers mod 29. This argument is the substance of Sophie Germain's theorem.

Sophie Germain's Theorem. *Let n be an odd prime. If there is an auxiliary prime p with the properties that*
(1) $x^n + y^n + z^n \equiv 0 \bmod p$ implies $x \equiv 0$ or $y \equiv 0$ or $z \equiv 0 \bmod p$, and
(2) $x^n \equiv n \bmod p$ is impossible
then Case I of Fermat's Last Theorem is true for n.

PROOF. Since n is odd, Fermat's Last Theorem for n can be reformulated as the statement that $x^n + y^n + z^n = 0$ is impossible in nonzero integers. Case I of Fermat's Last Theorem for n is the statement that $x^n + y^n + z^n = 0$ is impossible in integers which are not divisible by n. Suppose, therefore, that n and p satisfy the conditions of the theorem and that x, y, z are integers, none divisible by n, such that $x^n + y^n + z^n = 0$. It is to be shown that these assumptions lead to a contradiction.

As usual, it can be assumed that x, y, and z are pairwise relatively prime. The equation $(-x)^n = y^n + z^n = (y + z)(y^{n-1} - y^{n-2}z + y^{n-3}z^2 - \cdots + z^{n-1})$ shows that $y + z$ and $y^{n-1} - y^{n-2}z + \cdots + z^{n-1}$ are both nth powers because the factors are relatively prime.* (If q were a prime which divided them both then $y + z \equiv 0$, $y^{n-1} - y^{n-2}z + \cdots + z^{n-1} \equiv 0$, $y \equiv -z$, $ny^{n-1} \equiv 0 \bmod q$, from which $n \equiv 0$ or $y \equiv 0 \bmod q$. The first is impossible because then $n = q$ would divide x, and the second is impossible because then q would divide both y and $y + z$.) The equations $(-y)^n = x^n + z^n$ and $(-z)^n = x^n + y^n$ can be factored in the same way and it follows that there must be integers $a, \alpha, b, \beta, c, \gamma$ such that

$$
\begin{array}{lll}
y + z = a^n & y^{n-1} - y^{n-2}z + \cdots + z^{n-1} = \alpha^n & x = -a\alpha \\
z + x = b^n & z^{n-1} - z^{n-2}x + \cdots + x^{n-1} = \beta^n & y = -b\beta \\
x + y = c^n & x^{n-1} - x^{n-2}y + \cdots + y^{n-1} = \gamma^n & z = -c\gamma.
\end{array}
$$

Now consider arithmetic modulo p. Since $x^n + y^n + z^n \equiv 0 \bmod p$, the first condition on p implies that x, y, or z must be zero mod p. Assume without loss of generality that $x \equiv 0 \bmod p$. Then $2x = b^n + c^n + (-a)^n \equiv 0 \bmod p$ and, again by the first condition on p, it follows that a, b, or c must be zero mod p. If b or c were zero mod p then $y = -b\beta \equiv 0$ or $z = -c\gamma \equiv 0 \bmod p$, which, together with $x \equiv 0 \bmod p$, would contradict the assumption that x, y, and z are pairwise relatively prime. Therefore $a \equiv 0 \bmod p$. But this implies $y \equiv -z \bmod p$, $\alpha^n \equiv ny^{n-1} \equiv n\gamma^n \bmod p$. Since $\gamma \not\equiv 0 \bmod p$ there is an integer g such that $\gamma g \equiv 1 \bmod p$, from which $(\alpha g)^n \equiv n \bmod p$, contrary to the second assumption on p. This contradiction proves Sophie Germain's theorem.

*Note that again, as in Sections 1.3, 1.5, 1.6, and 2.2, the theorem "if uv is an nth power and if u and v are relatively prime then u and v must both be nth powers" plays a crucial role.

Using this theorem Sophie Germain was able to prove Case I of Fermat's Last Theorem for all primes less than 100. In other words, for each odd prime n less than 100 she was able to find another prime p which satisfied the conditions of the theorem. Legendre extended this result to all odd primes less than 197 and to many other primes as well. With these results, which came before Fermat's Last Theorem was proved even in the case $n = 5$, it became evident that it was time to focus attention on the more recalcitrant Case II.

EXERCISES

1. Prove that the 7th powers mod 29 are 0, ± 1, ± 12. [Since $2^7 = -12$ mod 29, all 5 of these numbers are 7th powers. Of course the fact that these are the only 7th powers can be proved simply by computing $3^7, 4^7, \ldots, 28^7$, mod 29. More efficient than this is to notice that if $x \not\equiv 0$ mod 29 is a 7th power then $x^4 - 1 \equiv 0$ mod 29. Thus it will suffice to prove that if $f(x)$ is a polynomial of degree n with leading coefficient 1 then the congruence $f(x) \equiv 0$ mod p has at most n distinct solutions mod p. This can be done using the Remainder Theorem as in Exercise 9 of Section 2.4. Alternatively one can multiply $f(x)$ by $(x - r_1)(x - r_2) \ldots (x - r_k)$ where r_i ranges over distinct integers which are not solutions of $f(x) \equiv 0$. In this way one could construct a polynomial $g(x)$, with leading coefficient 1 and degree less than p, all of whose values are zero mod p. Euler's differencing argument of Section 2.4 shows that this is impossible.]

2. Prove that condition (1) of Sophie Germain's theorem holds if and only if the list of nth powers mod p does not contain two consecutive nonzero (mod p) integers.

3. Prove that Case I of Fermat's Last Theorem is true for each of the following primes n by finding another prime p for which the conditions of Sophie Germain's theorem are satisfied: $n = 13, 17, 19, 23, 29$.

4. Prove Case I of Fermat's Last Theorem for $n = 5$ by considering congruences modulo 25.

5. Show that the argument of Exercise 4 fails for $n = 7$.

3.3 The case $n = 5$

Credit for the proof of Fermat's Last Theorem for fifth powers is shared by two very eminent mathematicians, the young Dirichlet* who had just turned 20 and was just embarking on his brilliant career at the time, and the aged Legendre who was past 70 and world-famous both as a number theorist and as an analyst.

*Despite the rather French appearance of his name and the fact that he was in Paris at this time, Dirichlet was a German. Except for a few years of what we would call graduate study in Paris, and except for some travel in later years, he was born, raised, and educated in Germany, and lived in Germany all his life.

Fermat's Last Theorem for fifth powers splits into two cases as follows. As was shown in the preceding section, if x, y, z are pairwise relatively prime positive integers such that $x^5 + y^5 = z^5$ then one of the three must be divisible by 5. On the other hand, one of the three obviously must be divisible by 2, since otherwise the equation would have an odd number as the sum of two odd numbers. In this section the *first case* considered will be the case in which the number divisible by 5 is also divisible by 2, and the *second case* considered will be the contrary case in which the even number and the number divisible by 5 are distinct. (This dichotomy should not be confused with the division into Case I and Case II explained in the preceding section. Case I has been eliminated for $n = 5$ and the "first case" and "second case" are subdivisions of Case II.)

In July of 1825, Dirichlet presented a paper before the Paris Academy (of which Legendre w̃as a leading member) in which he proved that the first case is impossible. His proof was essentially the following argument. If x or y is divisible by 5, let the other term on the right hand side of $x^n + y^n = z^n$ be moved to the left, leaving the term divisible by 5 by itself on one side of the equation. Since in the first case this term is also divisible by 2, this puts the equation in the form

$$u^5 \pm v^5 = w^5$$

where u, v, and w are pairwise relatively prime positive integers, and where w is divisible by 10. It is to be shown that this is impossible.

Since u and v are odd (they are relatively prime to the even number w) one can follow Euler's proof of the case $n = 3$ (Section 2.2) by setting $u + v = 2p$ and $u - v = 2q$. Then $u = p + q$, $v = p - q$, and $u^5 \pm v^5 = (p + q)^5 \pm (p - q)^5$ is either $2(p^5 + 10p^3q^2 + 5pq^4)$ or $2(5p^4q + 10p^2q^3 + q^5)$. Interchange p and q if necessary to find an equation of the form

$$2p(p^4 + 10p^2q^2 + 5q^4) = w^5$$

in which p and q are relatively prime positive integers of opposite parity and w is a positive integer divisible by 10. If 5 did not divide p then it would not divide $p^4 + 10p^2q^2 + 5q^4$ and would therefore not divide w^5, contrary to assumption. Therefore 5 divides p, say $p = 5r$, and 5 does not divide q. The equation then becomes

$$5^2 \cdot 2r(q^4 + 2 \cdot 5^2r^2q^2 + 5^3r^4) = w^5.$$

The primes which divide $5^2 \cdot 2r$ are 2, 5, and prime factors of r, none of which divides $q^4 + 2 \cdot 5^2r^2q^2 + 5^3r^4$ (recall that $5 \nmid q$ and q and r have opposite parity). Therefore by the usual theorem

$$5^2 \cdot 2r = \text{fifth power}$$

$$q^4 + 2 \cdot 5^2r^2q^2 + 5^3r^4 = \text{fifth power}.$$

Completion of the square can be used to put the second of these expressions in the form

$$(q^2 + 5^2 r^2)^2 - 5^4 r^4 + 5^3 r^4 = (q^2 + 5^2 r^2)^2 - 5(10 r^2)^2 = P^2 - 5Q^2.$$

The basic idea of Dirichlet's proof is to attempt to follow Euler by writing $P^2 - 5Q^2$ as $(P + Q\sqrt{5})(P - Q\sqrt{5})$ and trying to show that $P + Q\sqrt{5}$ must be a fifth power.

Assume for the moment that this conclusion can be justified, that is, assume it can be shown that there are integers A, B such that $P + Q\sqrt{5} = (A + B\sqrt{5})^5$. Then

$$P = A^5 + 50A^3B^2 + 125AB^4$$
$$Q = 5A^4B + 50A^2B^3 + 25B^5.$$

Since $Q = 10r^2$ and since $5^2 \cdot 2r$ is a fifth power, $(5^2 \cdot 2r)^2 = 5^3 \cdot 2Q$ is a fifth power, and therefore

$$5^4 \cdot 2B[A^4 + 10A^2B^2 + 5B^4] = \text{fifth power}.$$

A and B are relatively prime because P and Q are, they are of opposite parity because P and Q are not both even, and A is prime to 2 and 5 because P is. Therefore

$$5^4 \cdot 2B = \text{fifth power}$$
$$A^4 + 10A^2B^2 + 5B^4 = \text{fifth power}.$$

Completing the square in the last expression puts it in the form $(A^2 + 5B^2)^2 - 25B^4 + 5B^4 = (A^2 + 5B^2)^2 - 5(2B^2)^2 = C^2 - 5D^2$. Assume once again that this can be shown to imply $C + D\sqrt{5} = (a + b\sqrt{5})^5$. Then

$$C = a^5 + 50a^3b^2 + 125ab^4$$
$$D = 5a^4b + 50a^2b^3 + 25b^5.$$

Since $D = 2B^2$ and since $5^4 \cdot 2B$ is a fifth power, $(5^4 \cdot 2B)^2 = 5^8 \cdot 2D$ is a fifth power, that is,

$$5^9 2b[a^4 + 10a^2b^2 + 5b^4] = \text{fifth power},$$

from which, as before,

$$5^9 \cdot 2b = \text{fifth power}$$
$$a^4 + 10a^2b^2 + 5b^4 = \text{fifth power}.$$

The first equation is obviously equivalent to "$5^4 \cdot 2b = \text{fifth power}$" and it is elementary to show that a and b satisfy the same conditions satisfied by A and B, namely, in addition to the condition that $5^4 \cdot 2B$ and $A^4 + 10A^2B^2 + 5B^4$ be fifth powers, the conditions that A and B be relatively prime of opposite parity and that A be prime to both 2 and 5. Therefore the argument can be repeated indefinitely and this leads to an impossible infinite descent, namely, the sequence of *positive* numbers B which decrease by virtue of $2B^2 = D = b(5a^4 + 50a^2b^2 + 25b^4)$, which gives $b > 0$

and $2B^2 > 25b^4$, $B > b$. Therefore, in order to prove that the case under consideration is impossible it will suffice to prove that the implication

$$P^2 - 5Q^2 = \text{fifth power} \quad \Rightarrow \quad P + Q\sqrt{5} = (A + B\sqrt{5}\,)^5$$

is valid in the cases where it is used above.

As Dirichlet observes at the outset of his paper, there are other ways in which $P^2 - 5Q^2$ can be a fifth power. In fact, the solution of Pell's equation $9^2 - 5 \cdot 4^2 = 1$ shows that if P and Q are defined by $P + Q\sqrt{5} = (A + B\sqrt{5}\,)^5(9 + 4\sqrt{5}\,)^k$ for any choice of A, B, k then $P^2 - 5Q^2 = (P + Q\sqrt{5}\,)(P - Q\sqrt{5}\,) = (A^2 - 5B^2)^5$. Therefore the above implication is not justified. Dirichlet's idea was to look for *additional conditions* on P and Q which would suffice to make the implication valid, and he found a very simple one. Note that if $P + Q\sqrt{5} = (A + B\sqrt{5}\,)^5$ then $Q = 5A^4B + 50A^2B^3 + 25B^5$ must be divisible by 5. Thus $5|Q$ is a necessary condition for the desired implication. What Dirichlet proved was that in the case where P and Q are relatively prime and have opposite parity this condition is also sufficient, that is, he proved the following lemma.

Lemma. *Let P and Q be relatively prime integers such that $P^2 - 5Q^2$ is a fifth power, such that $5|Q$, and such that P and Q have opposite parity. Then there exist integers A and B such that $P + Q\sqrt{5} = (A + B\sqrt{5}\,)^5$.*

Note that in the cases above the additional condition $5|Q$ is satisfied. in the first case because the condition "$5^2 \cdot 2r = \text{fifth power}$" implies $5\,|\,r\,|\,Q$ and in the second case because the condition "$5^4 \cdot 2B = \text{fifth power}$" implies $5|B|D$. Moreover, in both of these cases the conditions that P and Q be relatively prime and of opposite parity are satisfied. Therefore, once the lemma is proved it will follow that Fermat's Last Theorem for $n = 5$ is true in the case where the integer in the equation $x^5 + y^5 = z^5$ which is divisible by 5 is also divisible by 2.

The lemma can be proved by almost the same techniques as those explained in Chapter 2 by which Euler found the most general representations of numbers in the forms $a^2 + b^2$, $a^2 + 2b^2$, $a^2 + 3b^2$, but there is one major difference. If one attempts to apply Euler's methods to representations in the form $a^2 - 5b^2$, the attempt fails with the proposition analogous to (4) and (5′) of Section 2.4, namely, the statement that *if a and b are relatively prime then every odd factor of $a^2 - 5b^2$ can be written in the form $p^2 - 5q^2$*. In fact, as was noted at the end of Section 1.7, the analogous statement for representations in the form $a^2 + 5b^2$ is *false*. However, it is a simple consequence of Gauss's theory of binary quadratic forms in the *Disquisitiones Arithmeticae* that this statement is *true* for representations in the form $a^2 - 5b^2$, even though Euler's techniques are not adequate to prove it. This fact is a simple consequence of the theory of Chapter 8 (see Exercise 11 of Section 8.5).

Granted this fact, the techniques of Chapter 2 do suffice to prove that if a and b are relatively prime and of opposite parity and if $a^2 - 5b^2 = P_1^{n_1} P_2^{n_2} \ldots P_k^{n_k}$ is the prime factorization of $a^2 - 5b^2$ then there are integers p_i, q_i $(i = 1, 2, \ldots, k)$ and t, u such that $P_i = p_i^2 - 5q_i^2$, $1 = t^2 - 5u^2$, and

$$a + b\sqrt{5} = \left(p_1 + q_1\sqrt{5}\right)^{n_1} \left(p_2 + q_2\sqrt{5}\right)^{n_2} \cdots \left(p_k + q_k\sqrt{5}\right)^{n_k} (t + u\sqrt{5}).$$

Briefly, the argument is as follows. By the unproved fact granted above there exist p_1, q_1 such that $p_1^2 - 5q_1^2 = P_1$. (P_1 is odd because a and b have opposite parity.) Since P_1 divides both $a^2 - 5b^2$ and $p_1^2 - 5q_1^2$ it also divides $(p_1 b - q_1 a)(p_1 b + q_1 a) = p_1^2 b^2 - q_1^2 a^2 = p_1^2 b^2 - 5q_1^2 b^2 + 5q_1^2 b^2 - q_1^2 a^2 = b^2(p_1^2 - 5q_1^2) - q_1^2(a^2 - 5b^2)$. Therefore P_1 divides either $p_1 b - q_1 a$ or $p_1 b + q_1 a$ and, by changing the sign of q_1 if necessary, it can be assumed that P_1 divides $p_1 b + q_1 a$. Then, because $(a + b\sqrt{5})(p_1 + q_1\sqrt{5}) = (ap_1 + 5bq_1) + (aq_1 + bp_1)\sqrt{5}$ implies $(a^2 - 5b^2)P_1 = (ap_1 + 5bq_1)^2 - 5(aq_1 + bp_1)^2$, the integer $ap_1 + 5bq_1$ is also divisible by P_1, say $(ap_1 + 5bq_1) + (aq_1 + bp_1)\sqrt{5} = P_1(c + d\sqrt{5})$. This gives $(a + b\sqrt{5})(p_1 + q_1\sqrt{5}) = P_1(c + d\sqrt{5})$ and multiplying by $p_1 - q_1\sqrt{5}$ and cancelling P_1 from both sides of the equation gives $a + b\sqrt{5} = (p_1 - q_1\sqrt{5})(c + d\sqrt{5})$. Since $a^2 - 5b^2 = (p_1^2 - 5q_1^2)(c^2 - 5d^2)$ it follows that $c^2 - 5d^2 = (a^2 - 5b^2)/P_1$ and one of the prime factors of $a^2 - 5b^2$ has been removed. The process can then be repeated until all primes have been removed and $a + b\sqrt{5}$ is written as a product of factors of the form $p + q\sqrt{5}$ where there are $n_1 + n_2 + \ldots + n_k + 1$ factors $p + q\sqrt{5}$, one for each prime factor of $a^2 - 5b^2$ and one in which $p^2 - 5q^5 = 1$. Moreover, if the same prime P occurs twice, then the same representation $P = p^2 - 5q^2$ can be used in both cases and the factor $p \pm q\sqrt{5}$ which is removed is the same except possibly for the sign of q. But this too must be the same because if the factors $p + q\sqrt{5}$ and $p - q\sqrt{5}$ both occurred then $a + b\sqrt{5} = (p^2 - 5q^2)(C + D\sqrt{5})$ would result, which would contradict the assumption that a and b are relatively prime.

Therefore if P, Q are relatively prime and of opposite parity, and if $P^2 - 5Q^2$ is a fifth power, it follows that

$$P + Q\sqrt{5} = (A + B\sqrt{5})^5 (t + u\sqrt{5})$$

where $t^2 - 5u^2 = 1$. To prove the lemma it suffices to prove that if Q is divisible by 5 then $t + u\sqrt{5}$ can be reduced to $1 + 0\sqrt{5}$. This naturally leads to an investigation of the possible solutions of $t^2 - 5u^2 = 1$, that is, the solutions of Pell's equation in the case $A = 5$. Since $(9 + 4\sqrt{5})(9 - 4\sqrt{5}) = 1$ it is clear that $\pm(9 + 4\sqrt{5})^k = t + u\sqrt{5}$ gives a solution for every integer k, positive, negative, or zero. Conversely, the arguments outlined in the exercises of Section 1.9 show that *all* solutions are of this form. (This fact will be proved again in Chapter 8. See, for example, Exercise 2 of Section 8.4.) Since every integer k can be written in the form $5q + r$ where $r = 0$, 1, 2, 3, or 4, this shows that $P + Q\sqrt{5} = (A + B\sqrt{5})^5(9 + 4\sqrt{5})^r$ where $0 \leqslant r \leqslant 4$. It suffices then, to show that if $Q \equiv 0$ mod 5 then r cannot be 1, 2, 3, or 4. Let $C + D\sqrt{5} = (A + B\sqrt{5})^5$ and $E + F\sqrt{5} = (9 + 4\sqrt{5})^r$. Then $D \equiv 0$ mod 5 and, since $Q = CF + DE \equiv 0$ mod 5, it follows that $CF \equiv 0$ mod 5. But $C \not\equiv 0$ mod 5 because $C \equiv 0$ mod 5 would imply that P and Q were both divisible by 5. Therefore $F \equiv 0$ mod 5. But if $r \geqslant 1$ then by the binomial theorem $F = r9^{r-1} \cdot 4$ plus terms divisible by 5, and F can not be divisible by 5; therefore $F \equiv 0$ mod 5 implies $r = 0$, as was to be shown. This completes Dirichlet's proof that the *first case* is impossible.

The first complete proof of Fermat's Last Theorem for fifth powers was published shortly thereafter, in September of 1825, by Legendre in his second supplement to the *Théorie des Nombres*. Legendre's proof of the first case is essentially the same as Dirichlet's, a fact which he acknowledges in a footnote* saying, "By an analysis similar to the one we have just used, one can prove the impossibility of the equation $x^5 + y^5 = Az^5$ for a rather large number of values of A; this is what Mr. Lejeune Dieterich [sic!] has done in a Memoire which was presented to the Academy recently and which received its commendation." Legendre then goes on to prove that the *second case* is also impossible, something which Dirichlet had confessed in July that he was unable to do. One must be struck by this counterexample to the common notion that only young men can do important work in mathematics.

Legendre's proof of the second case was rather artificial and involved a great deal of unmotivated manipulation, perhaps a symptom of his great age and long experience. A few months later still, in November of 1825, Dirichlet presented an appendix to his July paper in which he proved the second case in a manner which is a more natural extension of the proof of the first case. This is, in essence, the proof which follows.

In the second case the equation $x^5 + y^5 = z^5$ can be put in the form

$$u^5 \pm v^5 = w^5$$

in which u, v, and w are pairwise relatively prime positive integers with w divisible by 5 and with u and v of opposite parity. Set $p = u + v$ and $q = u - v$. Then

$$u^5 \pm v^5 = \left(\frac{p+q}{2}\right)^5 \pm \left(\frac{p-q}{2}\right)^5$$

and, when p and q are interchanged if necessary as before,

$$2^{-5}2(p^5 + 10p^3q^2 + 5pq^4) = w^5$$

$$p(p^4 + 10p^2q^2 + 5q^4) = 2^4w^5.$$

---

*The second supplement was also published as a Memoire [L7] by the Academy and this publication is, for some reason, dated 1823 (see Dickson, [D2], vol. 2, p. 734), but this footnote shows that it came *after* July 1825.

As before, p must be divisible by 5, say $p = 5r$, and q must prime to 5. Thus

$$5^2 r(q^4 + 50r^2q^2 + 125r^4) = 2^4 w^5.$$

The factors $5^2 r$ and $q^4 + 50r^2q^2 + 125r^4$ are relatively prime and the first factor is odd (p and q are both odd), which implies

$$5^2 r = \text{fifth power}$$
$$(q^2 + 25r^2)^2 - 5(10r^2)^2 = 2^4(\text{fifth power}).$$

The square of any odd number is 1 modulo 8, which shows that $q^2 + 25r^2$ is 2 mod 8, therefore that it is divisible by 2 but not by 2^2. Therefore the second equation can be written

$$\left(\frac{P}{2}\right)^2 - 5\left(\frac{Q}{2}\right)^2 = \text{fifth power}$$

where P and Q are the odd integers $(q^2 + 25r^2)/2$ and $5r^2$ respectively. Note that Q is divisible by 5. It will be shown below that these circumstances imply that there exist odd integers A and B such that

$$\frac{P}{2} + \frac{Q}{2}\sqrt{5} = \left(\frac{A}{2} + \frac{B}{2}\sqrt{5}\right)^5.$$

This gives

$$\frac{Q}{2} = 5\frac{A^4 B}{2^5} + 10\frac{A^2 B^3}{2^5}5 + \frac{B^5}{2^5}5^2$$

and, since $5^2 r$ is a fifth power,

$$(5^2 r)^2 = 5^3 Q = 5^4 B\left[\frac{A^4}{2^4} + 10\frac{A^2 B^2}{2^4} + 5\frac{B^4}{2^4}\right]$$

is a fifth power and

$$5^4 B[A^4 + 10A^2 B^2 + 5B^4] = 2^4 \text{ (fifth power)}.$$

But A and B are relatively prime (both are odd and any odd prime factor they had in common would divide both P and Q) and A is not divisible by 5 (otherwise P would be divisible by 5, from which it would follow that q was divisible by 5), and therefore, as usual,

$$5^4 B = \text{fifth power}$$
$$A^4 + 10A^2 B^2 + 5B^4 = 2^4 \text{ (fifth power)}.$$

The first equation shows that $5|B$ and the second equation can be rewritten

$$\left(\frac{A^2 + 5B^2}{2^2}\right)^2 - 5\left(\frac{B^2}{2}\right)^2 = \text{fifth power}.$$

$A^2 + 5B^2 \equiv 6 \bmod 8$, so this equation is of the form

$$\left(\frac{p}{2}\right)^2 - 5\left(\frac{q}{2}\right)^2 = \text{fifth power}$$

where p and q are relatively prime odd integers such that $5|q$. The lemma to be proved then gives

$$\frac{p}{2} + \frac{q}{2}\sqrt{5} = \left(\frac{a}{2} + \frac{b}{2}\sqrt{5}\right)^5$$

and by the same sequence of steps

$$a^4 + 10a^2b^2 + 5b^4 = 2^4 \text{ (fifth power)}.$$

This gives an infinitely descending (see Exercise 1) sequence of positive fifth powers and hence gives the desired contradiction.

Thus, all that remains to be shown is that if P and Q are relatively prime odd integers such that

$$\left(\frac{P}{2}\right)^2 - 5\left(\frac{Q}{2}\right)^2 = \text{fifth power}$$

(note that $P^2 - 5Q^2 \equiv 4 \bmod 8$ so that $(P/2)^2 - 5(Q/2)^2$ is necessarily an odd integer) and if Q is divisible by 5 then there exist odd integers A and B such that

$$\frac{P}{2} + \frac{Q}{2}\sqrt{5} = \left(\frac{A}{2} + \frac{B}{2}\sqrt{5}\right)^5.$$

This can be proved as follows. Modulo 4 either $P \equiv Q$ or $P \equiv -Q$. In the first case $[(P/2) + (Q\sqrt{5}/2)][(3/2) + (\sqrt{5}/2)] = ((3P + 5Q)/4) + ((P + 3Q)\sqrt{5}/4)$ is of the form $C + D\sqrt{5}$ where C and D are integers and in the second case $[(P/2) + (Q\sqrt{5}/2)][(3/2) - (\sqrt{5}/2)] = C + D\sqrt{5}$ where C and D are integers. Then the equation

$$\left(\frac{P}{2}\right)^2 - 5\left(\frac{Q}{2}\right)^2 = \left(\frac{P}{2} + \frac{Q}{2}\sqrt{5}\right)\left(\frac{P}{2} - \frac{Q}{2}\sqrt{5}\right)$$

$$= (C + D\sqrt{5})\left(\frac{3}{2} \pm \frac{1}{2}\sqrt{5}\right)(C - D\sqrt{5})\left(\frac{3}{2} \mp \frac{1}{2}\sqrt{5}\right)$$

$$= C^2 - 5D^2$$

shows that $C^2 - 5D^2$ is an odd fifth power and the equation $(P/2) + (Q\sqrt{5}/2) = [C + D\sqrt{5}][(3/2) \mp (\sqrt{5}/2)]$ shows that C and D are relatively prime. Therefore, as was shown above, $C + D\sqrt{5} = (c + d\sqrt{5})^5(9 + 4\sqrt{5})^k$ where k is 0, 1, 2, 3, or 4. Simple computation gives $((3/2) + (\sqrt{5}/2))^3 = 9 + 4\sqrt{5}$ so

$$\frac{P}{2} + \frac{Q}{2}\sqrt{5} = (c + d\sqrt{5})^5\left(\frac{3}{2} + \frac{1}{2}\sqrt{5}\right)^{3k}\left(\frac{3}{2} + \frac{1}{2}\sqrt{5}\right)^{\pm 1}$$

and what must be shown is that if Q is divisible by 5, then of the ten values $-1, 1, 2, 4, 5, 7, 8, 10, 11, 13$ possible for $3k \pm 1$ only 5 and 10 can occur. Let $m = 3k \pm 1$, $(\alpha/2) + (\beta\sqrt{5}/2) = ((3/2) + (\sqrt{5}/2))^m$, and $\gamma + \delta\sqrt{5} = (c + d\sqrt{5})^5$. Then δ is divisible by 5 and γ is not, so that $Q = \gamma\beta + \alpha\delta \equiv 0 \bmod 5$ implies $\beta \equiv 0 \bmod 5$. But by the binomial theorem

$$\frac{\beta}{2} = m\left(\frac{3}{2}\right)^{m-1}\left(\frac{1}{2}\right) + \text{terms containing } 5$$

$$2^{m-1}\beta = m3^{m-1} + \text{terms containing } 5$$

which shows that $\beta \equiv 0 \bmod 5$ implies $m \equiv 0 \bmod 5$ and completes the proof (except for the proof of the fact that if A and B are relatively prime then all odd prime factors of $A^2 - 5B^2$ are themselves of the form $p^2 - 5q^2$).

EXERCISES

1. Show that it is impossible for there to be relatively prime odd integers A and B such that 5^4B is a fifth power and such that $A^4 + 10A^2B^2 + 5B^4 = 2^4 \cdot$ (fifth power). [Except for the proof that the sequence of fifth powers actually descends, all of this proof is given in the text.]

2. Show that in order to prove that all odd factors of $A^2 - 5B^2$ (A, B relatively prime) are themselves of the form $p^2 - 5q^2$ it will suffice to prove this in the case of odd *prime* factors.

3.4   The cases $n = 14$ and $n = 7$

Any attempt to prove Fermat's Last Theorem by elementary arguments—say by methods which do not use Kummer's theory of ideal prime divisors—must take into account the fact that the single case $n = 7$ resisted the efforts of the best mathematicians of Europe for many years. Of course it is entirely possible that they were approaching the problem in the wrong way and that there is some simple idea—perhaps discovered by Fermat—which applies to all cases; but, on the other hand, it is more probable that an idea which is valid for *all* n would be found, perhaps in a clumsy form, in an intensive study of *one* n.

In 1832, seven years after his and Legendre's proof of the case $n = 5$, Dirichlet published [D5] a proof of the case $n = 14$. This is of course weaker than the case $n = 7$ (every 14th power is a 7th power but not conversely) and his publication of this proof is in a way a confession of his failure with the case $n = 7$. Another seven years passed before the first proof of the case $n = 7$ was published by Lamé in 1839. These proofs are rather lengthy and technical, and, since they were superseded by Kummer's proof of Fermat's Last Theorem for an entire class of exponents which includes 7, there is no need to study them here except insofar as they shed light on the types of techniques which were tried with only limited success and insofar as they give some insight into the ways in which Kummer might have been led to his discoveries.

Dirichlet's proof of the case $n = 14$ depends on a technique which is essentially the same as the proof for $n = 5$, which, in turn, follows essentially Euler's argument in the case $n = 3$, though of course with considerable computational ingenuity. Not surprisingly, it depends on a lemma which states that if $A^2 + 7B^2$ is a 14th power and if $7|B$ then $A + B\sqrt{-7}$ $= (a + b\sqrt{-7})^{14}$ for some integers a and b. Combining this lemma with some inspired algebraic manipulations and the usual infinite descent argument gives the proof. (See Exercise 1 for more on the details of the argument.)

Lamé's proof is in some ways a vindication of Dirichlet's failure with the case $n = 7$ because Lamé found it necessary to introduce some essentially new techniques in order to solve this problem. The arguments are difficult, unmotivated, and, worst of all, seem to be hopelessly tied to the case $n = 7$. No description or summary of this proof will be given here.* (See Dickson [D2], vol. 2, p. 737 for a very brief summary.) It seems as if this proof led to the edge of an impenetrable thicket; to attack the next case, $n = 11$, by similar methods seemed virtually hopeless and a new assessment of the methods became inevitable. Nothing of significance on the cases $n > 7$ ($n = $ prime) was published for the next eight years, until 1847, but when the great progress of 1847 came it was based on principles which appear to be entirely different from those of Lamé's proof for $n = 7$. These principles are not entirely different from those used in the proofs for $n = 3$, $n = 5$ that are given above, but they are based on another interpretation of them, namely, an interpretation in terms of the nth degree forms $x^n + y^n$ (that is, $x^3 + y^3$ and $x^5 + y^5$) rather than in terms of the quadratic forms $x^2 + 3y^2$ and $x^2 - 5y^2$. As will be seen in the next section, Lamé himself was the instigator, if not the originator, of this new interpretation.

EXERCISE

1. The following is the skeleton of Dirichlet's proof. Fill in the details. Let $x^{14} + y^{14} = z^{14}$. One can assume x, y, z are pairwise relatively prime and positive. z is not divisible by 7 because $a^2 + b^2 \equiv 0$ mod 7 is impossible. Therefore one can assume $z^{14} - x^{14} = y^{14}$ where y is divisible by 7. Rewrite $z^{14} - x^{14}$ as $a(a^6 + 7b^2)$ where $a = z^2 - x^2$, $b = zx(z^4 - z^2x^2 + x^4)$. a, b are relatively prime and of opposite parity. $a = 7c$, $7 \nmid b$ so $7^2c(b^2 + 7(7^2c^3)^2) = y^{14}$ implies 7^2c and $b^2 + 7(7^2c^3)^2$ are 14th powers. Assume it is known — see Section 8.5 — that an odd prime factor of a number of the form $x^2 + 7y^2$ where x and y are relatively prime must itself be of this form, and conclude that $b + 7^2c^3\sqrt{-7}$ $= (d + e\sqrt{-7})^{14}$. Then $2 \cdot 7^2c^3\sqrt{-7} = (d + e\sqrt{-7})^{14} - (d - e\sqrt{-7})^{14}$ can be rewritten as $z^{14} - x^{14}$ was rewritten above to find $7^2c^3 = 2 \cdot de[2^{12}(-7)^3d^6e^6$

*In candor I must say that I have not followed the proof through.

$+ 7f^2]$ where $f = (d^2 + 7e^2)(d^4 - 98d^2e^2 + 49e^4)$. d, e have opposite parity and are relatively prime. f is prime to 2, 7 and d, e, f are relatively prime. $7^2c^3 = 2 \cdot 7 \cdot de[f^2 - (2^6 \cdot 7d^3e^3)^2]$ decomposes as a product of 3 pairwise relatively prime factors $2 \cdot 7 \cdot de, f \pm 2^6 \cdot 7d^3e^3$. Recall that 7^2c is a 14th power and conclude that $2 \cdot 7^5de, f + 2^6 \cdot 7d^3e^3, f - 2^6 \cdot 7d^3e^3$ are all 14th powers. Thus $2^7 \cdot 7d^3e^3$ is on the one hand a difference of 14th powers and on the other hand is 2^4 times a 14th power. In short, a solution of $Z^{14} - X^{14} = 2^4Y^{14}$ has been found. Moreover, X, Y, Z are pairwise relatively prime and Y is divisible by 7. Repeat the whole procedure to find a solution of $\mathcal{Z}^{14} - \mathcal{X}^{14} = 2^{12}\mathcal{Y}^{14}$ where $\mathcal{Y}$ is divisible by 7. More generally, $z^{14} - x^{14} = 2^k y^{14}$ $(k \geqslant 0)$ where x, y, z are relatively prime and y is divisible by 7 leads to $Z^{14} - X^{14} = 2^{4+9k}Y^{14}$ where Y is divisible by 7. Moreover Z is much smaller than z. Therefore $z^{14} - x^{14} = 2^k y^{14}$ $(k \geqslant 0, 7|y)$ is impossible by infinite descent.

4 Kummer's theory of ideal factors

4.1 The events of 1847

The proceedings of the Paris Academy and the Prussian Academy in Berlin for the year 1847 tell a dramatic story in the history of Fermat's Last Theorem. The story begins in the report of the March 1st meeting of the Paris Academy ([A1], p. 310), at which Lamé announced, evidently rather excitedly, that he had found a proof of the impossibility of the equation $x^n + y^n = z^n$ for $n > 2$ and had therefore completely solved this long outstanding problem. The brief sketch of a proof which he gave was, as he no doubt realized later, woefully inadequate, and there is no need to consider it in detail here. However, his basic idea was a simple and compelling one which is central to the later development of the theory. The proofs of the cases $n = 3, 4, 5, 7$ which had been found up to that time all depended on some algebraic factorization such as $x^3 + y^3 = (x + y)(x^2 - xy + y^2)$ in the case $n = 3$. Lamé perceived the increasing difficulty for larger n as resulting from the fact that one of the factors in this decomposition has very large degree, and he noted that this can be overcome by decomposing $x^n + y^n$ *completely* into n linear factors. This can be done by introducing a *complex* number r such that $r^n = 1$ and using the algebraic identity

$$x^n + y^n = (x + y)(x + ry)(x + r^2 y) \cdots (x + r^{n-1} y) \qquad (n \text{ odd}) \quad (1)$$

(For example, if $r = \cos(2\pi/n) + i \sin(2\pi/n) = e^{2\pi i/n}$ then the polynomial $x^n - 1$ has the n distinct roots $1, r, r^2, \ldots, r^{n-1}$ and by elementary algebra $X^n - 1 = (X - 1)(X - r)(X - r^2) \cdots (X - r^{n-1})$; setting $X = -x/y$ and multiplying by $-y^n$ then give the desired identity (1).) Put very briefly, Lamé's idea was to use the techniques which had been used in the past for the cruder factorization of $x^n + y^n$ (in special cases) to this complete factorization. That is, he planned to show that if x and y are such that the factors $x + y, x + ry, \ldots, x + r^{n-1}y$ are relatively prime then $x^n + y^n = z^n$ implies that each of the factors $x + y, x + ry, \ldots$ must itself be an nth power and to derive from this an impossible infinite descent. If $x + y, x + ry, \ldots$ are *not* relatively prime he planned to show that there is a factor m common to *all* of them so that $(x + y)/m, (x + ry)/m, \ldots, (x + r^{n-1}y)/m$ are relatively prime and to apply a similar argument in this case as well.

Not doubting for a moment that the idea of introducing complex numbers in this way was the key that would unlock the door of Fermat's Last Theorem, Lamé enthusiastically told the Academy that he could not claim the entire credit for himself because the idea had been suggested to him in a casual conversation by his colleague Liouville some months before. Liouville for his part, however, did not share Lamé's enthusiasm, and he took the floor after Lamé finished his presentation only to cast some doubts on the proposed proof. He declined any credit for himself in the idea of introducing complex numbers—pointing out that many others, among them Euler, Lagrange,* Gauss, Cauchy, and "above all Jacobi," had used complex numbers in similar ways in the past—and practically said that Lamé's brainchild was among the first ideas that would suggest themselves to a competent mathematician approaching the problem for the first time. What was more, he observed that Lamé's proposed proof had what appeared to him to be a very large gap in it. Would Lamé be justified, he asked, in concluding that each factor was an nth power if all he had shown was that the factors were relatively prime and that their product was an nth power? Of course this conclusion would be valid in the case of ordinary integers, but its proof depends[†] on the factorization of integers into prime factors and it is by no means obvious that the needed techniques can be applied to the complex numbers that Lamé needed them for. Liouville felt that no enthusiasm was justified unless or until these difficult matters had been resolved.

Cauchy, who took the floor after Liouville, seemed to believe there was some likelihood that Lamé would succeed, because he hastened to point out that he himself had presented to the Academy in October of 1846 an idea which he believed might yield a proof of Fermat's Last Theorem but that he had not found the time to develop it further.

*Liouville did not say so, and he may not have known it, but Lagrange actually *explicitly* mentioned the factorization $(x + y)(x + ry) \cdots (x + r^{n-1}y) = x^n + y^n$ in connection with Fermat's Last Theorem ([L3]).

[†]The fact that Liouville spotted this gap instantly and saw instantly that it was related to the problem of proving *unique factorization into primes* for the complex numbers in question seems to indicate that he, and perhaps other mathematicians of the time as well, were well aware of the flaw in Euler's *Algebra* on this point (see Section 2.3). Nonetheless, I do not know of any writer of this period or earlier who criticizes Euler's argument.

The proceedings of the meetings of the weeks following this one show a great deal of activity on the part of Cauchy and Lamé in pursuing these ideas. Lamé admitted the logical validity of Liouville's criticism, but he did not in the least share Liouville's doubts about the truth of the final conclusion. He claimed that his "lemmas" gave him a method of factoring the complex numbers in question and that all of his examples confirmed the existence of unique factorization into primes. He was certain that "there can be no insurmountable obstacle between such a complete verification and an actual proof."

In the meeting of March 15th, Wantzel claimed to have *proved* the validity of unique factorization into primes, but his arguments covered only the cases $n \leqslant 4$ which are easily proved ($n = 2$ is the case of ordinary integers, $n = 3$ is essentially the case proved in Section 2.5, and $n = 4$ was quickly proved by Gauss in his classic paper on biquadratic residues), beyond which he simply said that "one easily sees" that the same arguments can be applied to the cases $n > 4$. One doesn't, and Cauchy said so on March 22nd. Thereafter Cauchy launched into a long series of papers in which he himself attempted to prove a division algorithm for the complex numbers in question—"radical polynomials" as he called them—from which he could conclude that unique factorization was valid.

In the March 22nd proceedings it is recorded that *both* Cauchy and Lamé deposited "secret packets" with the Academy. The depositing of secret packets was an institution of the Academy which allowed members to go on record as having been in possession of certain ideas at a certain time—without revealing them—in case a priority dispute later developed. In view of the circumstances of March 1847, there is little doubt what the subject of these two packets was. As it turned out, however, there was no priority dispute whatever on the subject of unique factorization and Fermat's Last Theorem.

In the following weeks, Lamé and Cauchy each published notices in the proceedings of the Academy, notices that are annoyingly vague and incomplete and inconclusive. Then, on May 24, Liouville read into the proceedings a letter from Kummer in Breslau which ended, or should have ended, the entire discussion. Kummer wrote to Liouville to tell him that his questioning of Lamé's implicit use of unique factorization had been quite correct. Kummer not only asserted that unique factorization *fails*, he also included with his letter a copy of a memoir [K6] he had *published** *three years earlier* in which he had demonstrated the failure of unique factorization in cases where Lamé had been asserting it was valid. However, he went on to say, the theory of factorization can be "saved" by introducing a new kind of complex numbers which he called "ideal complex numbers";

---

*It must be admitted, however, that Kummer chose a very obscure place in which to publish it. Liouville republished it in his *Journal de Mathematiques Pures et Appliquées* in 1847 and this must have been the first time it reached a wide audience.

these results he had *published one year* earlier in the proceedings of the Berlin Academy* in resumé form [K7] and a complete exposition of them was soon to appear in Crelle's Journal [K8]. He had for a long time been occupied with the application of his new theory to Fermat's Last Theorem and said he had succeeded in reducing its proof for a given n to the testing of two conditions on n. For the details of this application and of the two conditions, he refers to the notice he had published that same month in the proceedings of the Berlin Academy (15 April 1847). There he in fact stated the two conditions in full and said that he "had reason to believe" that $n = 37$ did not satisfy them.

The reaction of the learned gentlemen of Paris to this devastating news is not recorded. Lamé simply fell silent. Cauchy, possibly because he had a harder head than Lamé or possibly because he had invested less in the success of unique factorization, continued to publish his vague and inconclusive articles for several more weeks. In his only direct reference to Kummer he said, "What little [Liouville] has said [about Kummer's work] persuades me that the conclusions which Mr. Kummer has reached are, at least in part, those to which I find myself led by the above considerations. If Mr. Kummer has taken the question a few steps further, if in fact he has succeeded in removing all the obstacles, I would be the first to applaud his efforts; for what we should desire the most is that the works of all the friends of science should come together to make known and to propagate the truth." He then proceeded to ignore—rather than to propagate—Kummer's work and to pursue his own ideas with only an occassional promise that he would eventually relate his statements to Kummer's work, a promise he never fulfilled. By the end of the summer, he too fell silent on the subject of Fermat's Last Theorem. (Cauchy was not the silent type, however, and he merely began producing a torrent of papers on mathematical astronomy.) This left the field to Kummer, to whom, after all, it had already belonged for three years.

It is widely believed that Kummer was led to his "ideal complex numbers" by his interest in Fermat's Last Theorem, but this belief is surely mistaken. Kummer's use of the letter λ (lambda) to represent a prime number, his use of the letter α to denote a "λth root of unity"—that is, a solution of $\alpha^\lambda = 1$—and his study[†] of the factorization of prime numbers $p \equiv 1 \bmod \lambda$ into "complex numbers composed of λth roots of unity" all derive directly from a paper of Jacobi [J2] which is concerned with *higher reciprocity laws*. Kummer's 1844 memoir was addressed by the University of Breslau to the University of Königsberg in honor of its jubilee celebra-

---

*This notice was also reprinted in Crelle in 1847. (A flawed translation into English is contained in Smith's *Source Book* [S2].) The Crelle reprint gives the erroneous date 1845 for the original publication; the correct date is 1846.

†Kummer's 1844 memoir dealt only with the factorization of such p. The general problem of factorization was covered by the succeeding papers in 1846 and 1847.

tion, and the memoir was definitely meant as a tribute to Jacobi, who for many years was a professor at Königsberg. It is true that Kummer had studied Fermat's Last Theorem in the 1830s and in all probability he was aware all along that his factorization theory would have implications for Fermat's Last Theorem, but the subject of Jacobi's interest, namely, higher reciprocity laws, was surely more important to him, both at the time he was doing the work and after. At the same time that he was demolishing Lamé's attempted proof and replacing it with his own partial proof, he referred to Fermat's Last Theorem as "a curiosity of number theory rather than a major item," and later, when he published his version of the higher reciprocity law in the form of an unproved conjecture, he referred to the higher reciprocity laws as "the principal subject and the pinnacle of contemporary number theory."

There is even an often told story that Kummer, like Lamé, believed he had proved Fermat's Last Theorem until he was told—by Dirichlet in this story—that his argument depended on the unproved assumption of unique factorization into primes. Although this story does not necessarily conflict with the fact that Kummer's primary interest was in the higher reciprocity laws, there are other reasons to doubt its authenticity. It first appeared in a memorial lecture on Kummer given by Hensel in 1910 and, although Hensel describes his sources as unimpeachable and gives their names, the story is being told at third hand over 65 years later. Moreover, the person who told it to Hensel was not apparently a mathematician and it is very easy to imagine how the story could have grown out of a misunderstanding of known events. Hensel's story would be confirmed if the "draft ready for publication" which Kummer is supposed to have completed and sent to Dirichlet could be found, but unless this happens the story should be regarded with great skepticism. Kummer seems unlikely to have assumed the validity of unique factorization and even more unlikely to have assumed it unwittingly in a paper he intended to publish.*

This chapter is devoted entirely to the theory of factorization of complex numbers $a_0 + a_1\alpha + a_2\alpha^2 + \cdots + a_{\lambda-1}\alpha^{\lambda-1}$ ($a_0, a_1, \ldots, a_{\lambda-1}$ integers) "built up" out of a complex root α of the equation $\alpha^\lambda = 1$, and to the theory of "ideal complex numbers" or divisors which Kummer introduced to "save" unique factorization into primes for such numbers. The next chapter is devoted to a further development of that theory and it is not until the last section of that chapter that the application of the theory to Fermat's Last Theorem will be taken up. At that point Kummer's two conditions can be stated very simply and Fermat's Last Theorem can be proved for all primes which satisfy them. *All* of this work of Kummer was completed by April 11, 1847, a few weeks after Lamé's announcement of March 1.

*For a fuller discussion of this question see [E3] and [E4].

4.2 Cyclotomic integers

As this chapter will attempt to show, Kummer, like all other great mathematicians, was an avid computer, and he was led to his discoveries not by abstract reflection but by the accumulated experience of dealing with many specific computational examples. The practice of computation is in rather low repute today, and the idea that computation can be *fun* is rarely spoken aloud. Yet Gauss once said that he thought it was superfluous to publish a complete table of the classification of binary quadratic forms "because (1) anyone, after a little practice, can easily, without much expenditure of time, compute for himself a table of any particular determinant, if he should happen to want it . . . (2) because the work has a certain charm of its own, so that it is a real pleasure to spend a quarter of an hour in doing it for one's self, and the more so, because (3) it is very seldom that there is any occasion to do it."* One could also point to instances of Newton and Riemann doing long computations just for the fun of it. The material of this chapter, dealing as it does with a more abstract notion of "number" than the positive whole numbers, is necessarily somewhat more difficult than the material of the preceding chapters. Nonetheless, anyone who takes the time to do the computations should find that they and the theory which Kummer drew from them are well within his grasp and he may even, though he need not admit it aloud, find the process enjoyable.

Kummer uses the letter λ to denote a prime number and the letter α to denote an "imaginary" root of the equation $\alpha^\lambda = 1$, that is, a complex root other than 1 of this equation. The problem which he poses is the resolution into prime factors of numbers "built up" (*gebildeten*) from α by repeated operations of addition, subtraction, and multiplication, that is, numbers of the form

$$a_0 + a_1\alpha + a_2\alpha^2 + \cdots + a_{\lambda-1}\alpha^{\lambda-1} \qquad (1)$$

where $a_0, a_1, \ldots, a_{\lambda-1}$ are *integers*. (Here use is made of the equations $\alpha^\lambda = 1$, $\alpha^{\lambda+1} = \alpha$, $\alpha^{\lambda+2} = \alpha^2, \ldots$ to reduce all powers of α greater than $\alpha^{\lambda-1}$. The value of λ is fixed throughout the discussion.) These numbers, which Cauchy called "radical polynomials" and which Kummer and Jacobi considered to be a special type of "complex numbers," are now known as *cyclotomic integers*† because of the geometrical interpretation of α as a point on the circle $|z| = 1$ of the complex z-plane which effects the division of the circle into λ equal parts ("cycl-" = circle, "tom-" = divide)

*Quoted by Smith [S3], p. 261, from a letter to Schumacher.

†A certain linguistic problem arises from the fact that a value of λ must be understood. Thus, it is not meaningful to say that a complex number is a cyclotomic integer, but only that it is a cyclotomic integer for such-and-such a λ. In what follows, a particular value of λ will be assumed to be understood and the shorter term "cyclotomic integer" will be used.

and because of the consequent role which these complex numbers play in Gauss's theory of the division of the circle. Thus, in modern terminology, the problem Kummer—and before him Jacobi—posed was that of *factoring cyclotomic integers*.

Computations with cyclotomic integers are performed in the obvious way, using the commutative, associative, and distributive laws and the equation $\alpha^\lambda = 1$. For example, with $\lambda = 5$, $(\alpha + \alpha^2 + 3\alpha^4)(\alpha^2 - 2\alpha^3) = (\alpha + \alpha^2 + 3\alpha^4)\alpha^2 - (\alpha + \alpha^2 + 3\alpha^4)(2\alpha^3) = \alpha^3 + \alpha^4 + 3\alpha^6 - 2\alpha^4 - 2\alpha^5 - 6\alpha^7 = \alpha^3 + \alpha^4 + 3\alpha - 2\alpha^4 - 2 - 6\alpha^2 = -2 + 3\alpha - 6\alpha^2 + \alpha^3 - \alpha^4$. Moreover, since cyclotomic integers are particular kinds of complex numbers, non-zero terms can be cancelled from both sides of an equation; that is, if $f(\alpha)h(\alpha) = g(\alpha)h(\alpha)$ and $h(\alpha) \neq 0$ then $f(\alpha) = g(\alpha)$.

These computational rules have the slightly surprising consequence that *representations of cyclotomic integers in the form* (1) *are not unique*. For example, $1 + \alpha + \alpha^2 + \cdots + \alpha^{\lambda-1} = \alpha^\lambda + \alpha + \alpha^2 + \cdots + \alpha^{\lambda-1} = \alpha(1 + \alpha + \alpha^2 + \cdots + \alpha^{\lambda-1})$ implies either

$$1 + \alpha + \alpha^2 + \cdots + \alpha^{\lambda-1} = 0 \qquad (2)$$

or $1 = \alpha$. Since it is specifically assumed that $\alpha \neq 1$, the relation (2) is a consequence of the basic assumptions. But the relation (2) implies

$$a_0 + a_1\alpha + \cdots + a_{\lambda-1}\alpha^{\lambda-1}$$
$$= (a_0 + c) + (a_1 + c)\alpha + \cdots + (a_{\lambda-1} + c)\alpha^{\lambda-1} \qquad (3)$$

for any integer c, that is, a cyclotomic integer written in the form (1) is unchanged if the same integer c is added to all its coefficients a_i. This implies (set $c = -a_{\lambda-1}$) that every cyclotomic integer can be written in the form (1) with $a_{\lambda-1} = 0$; in actual computation it proves cumbersome to insist on putting cyclotomic integers in this form, and it is better to work with them in the form (1) bearing in mind the relation (3).

Naturally one must ask whether there are any *other* unforeseen relations among the numbers (1). The answer is that there are not, and it is for this reason that λ was assumed to be *prime*. (If $\lambda = 4$ then $\alpha = \pm i$ and $1 + \alpha^2 = 0$, or $\alpha = -1$ and $1 + \alpha = 0$. More generally, if $\lambda = jk$ then $0 = 1 - \alpha^\lambda = (1 - \alpha^j)(1 + \alpha^j + \alpha^{2j} + \cdots + \alpha^{\lambda-j})$ and one of the factors must be zero.) That is, if $a_0 + a_1\alpha + \cdots + a_{\lambda-1}\alpha^{\lambda-1} = b_0 + b_1\alpha + \cdots + b_{\lambda-1}\alpha^{\lambda-1}$ then necessarily $a_0 - b_0 = a_1 - b_1 = \cdots = a_{\lambda-1} - b_{\lambda-1}$, so that the relation is an instance of (3). This theorem, which is of course fundamental to the study of cyclotomic integers, was proved by Gauss at the outset of his section on cyclotomy in the *Disquisitiones Arithmeticae*. A simple proof is presented in Exercise 15.

Kummer used the notation $f(\alpha)$—or $g(\alpha)$, $\phi(\alpha)$, $F(\alpha)$, etc.—for a cyclotomic integer in the form (1). The great advantage of this notation is that it permits one to write $f(\alpha^2)$ for the cyclotomic integer obtained from $f(\alpha)$ by changing α to α^2, α^2 to α^4, α^3 to α^6, etc., and using $\alpha^\lambda = 1$ to reduce the result to the form (1). In order to show that this is a valid operation on cyclotomic integers it must be shown that if $f(\alpha) = g(\alpha)$ then $f(\alpha^2) = g(\alpha^2)$. This follows from the observation that if $f(\alpha) = g(\alpha)$ then $f(\alpha)$ is identical with $g(\alpha) + c(1 + \alpha + \alpha^2 + \cdots + \alpha^{\lambda-1})$ for some integer c; then $f(\alpha^2)$ is identical with $g(\alpha^2) + c(1 + \alpha^2 + \alpha^4 + \cdots + \alpha^{2\lambda-2})$ where powers α^j beyond α^λ in $1 + \alpha^2 + \alpha^4 + \cdots + \alpha^{2\lambda-2}$ are to be reduced to $\alpha^{j-\lambda}$. Seen in this way, the statement to be proved is that if $\phi(\alpha)$ denotes $1 + \alpha + \alpha^2 + \cdots + \alpha^{\lambda-1}$ then $\phi(\alpha^2)$ is *identical* with $\phi(\alpha)$, a fact which follows easily from the fact that for every integer j there is exactly one integer $j' \bmod \lambda$ such that $2j' \equiv j \bmod \lambda$ (see Appendix A.1.). In exactly the same way, $f(\alpha^3), f(\alpha^4), \ldots, f(\alpha^{\lambda-1})$ are all meaningful. (However $f(\alpha^\lambda)$ is *not* meaningful because, for example, $1 + \alpha = -\alpha^2 - \alpha^3 - \alpha^4$ when $\lambda = 5$, but $1 + \alpha^5 \neq -\alpha^{10} - \alpha^{15} - \alpha^{20}$ because the left side is 2 and the right side is -3.) The cyclotomic integers $f(\alpha)$, $f(\alpha^2)$, $f(\alpha^3), \ldots, f(\alpha^{\lambda-1})$ are called the *conjugates* of $f(\alpha)$. Clearly the relation of conjugacy is an equivalence relation; that is, $f(\alpha)$ is a conjugate of $f(\alpha)$, if $g(\alpha)$ is a conjugate of $f(\alpha)$ then $f(\alpha)$ is a conjugate of $g(\alpha)$, and if $g(\alpha)$ is a conjugate of $f(\alpha)$, and $h(\alpha)$ a conjugate of $g(\alpha)$, then $h(\alpha)$ is a conjugate of $f(\alpha)$, (Exercise 8).

Another view of the conjugations of cyclotomic integers is the following. All that was assumed about α was $\alpha^\lambda = 1$ and $\alpha \neq 1$. If α is any number with these properties then so are $\alpha^2, \alpha^3, \ldots, \alpha^{\lambda-2}$ (provided, of course, that λ is prime). The conjugations can therefore be regarded as operations of changing the choice of the root of the equation $\alpha^\lambda = 1$ ($\alpha \neq 1$) in terms of which the cyclotomic integers are to be expressed.

For any cyclotomic integer $f(\alpha)$, Kummer denotes by $Nf(\alpha)$ the product of the $\lambda - 1$ conjugates of $f(\alpha)$

$$Nf(\alpha) = f(\alpha)f(\alpha^2) \cdots f(\alpha^{\lambda-1}).$$

He calls this the *norm* of $f(\alpha)$, a term he attributes to Dirichlet. The norm $Nf(\alpha)$ of any cyclotomic integer is in fact an *integer*. In order to prove this it suffices to note that any conjugation $\alpha \mapsto \alpha^j$ ($j = 1, 2, \ldots, \lambda - 1$) merely permutes the factors of $Nf(\alpha)$ and therefore leaves $Nf(\alpha)$ unchanged. Thus $Nf(\alpha) = b_0 + b_1\alpha + b_2\alpha^2 + \cdots + b_{\lambda-1}\alpha^{\lambda-1}$ is equal to $b_0 + b_1\alpha^j + b_2\alpha^{2j} + \cdots + b_{\lambda-1}\alpha^{(\lambda-1)j}$ for $j = 2, 3, \ldots, \lambda - 1$. But $b_0 + b_j\alpha^j + \cdots = b_0 + b_1\alpha^j + \cdots$ implies $b_j - b_1 = b_0 - b_0 = 0$. Thus $b_j = b_1$ for $j = 2, 3, \ldots$, $\lambda - 1$ and $Nf(\alpha) = b_0 + b_1(\alpha + \alpha^2 + \cdots + \alpha^{\lambda-1}) = b_0 - b_1$ is an integer. Moreover, $Nf(\alpha)$ is a *positive* integer unless $f(\alpha) = 0$, $Nf(\alpha) = 0$; this follows from the observation that $\alpha^{\lambda-1} = \bar{\alpha}$ (where $\bar{\alpha}$ denotes the complex conjugate of α), from which $\alpha^{\lambda-2} = \bar{\alpha}^2, \ldots, f(\alpha^{\lambda-1}) = \overline{f(\alpha)}, f(\alpha^{\lambda-2}) = \overline{f(\alpha^2)}, \ldots$, and $Nf(\alpha)$ is a product of $\frac{1}{2}(\lambda - 1)$ nonnegative real numbers which are positive unless $f(\alpha^j) = 0$ for some, and hence all, $j = 1, 2, \ldots$, $\lambda - 1$. (See Exercise 10.)

The norm obviously has the property that $f(\alpha)g(\alpha) = h(\alpha)$ implies $Nf(\alpha) \cdot Ng(\alpha) = Nh(\alpha)$. Thus a factorization of the cyclotomic integer $h(\alpha)$

into two cyclotomic integers implies a factorization of the ordinary integer $Nh(\alpha)$ into two ordinary integers. For example, with $\lambda = 7$, the cyclotomic integer $\alpha^5 - \alpha^4 - 3\alpha^2 - 3\alpha - 2$ has norm 1247 (Exercise 5). Since 1247 is a product of just two primes $1247 = 29 \cdot 43$, this shows that the only possible factorizations of $\alpha^5 - \alpha^4 - 3\alpha^2 - 3\alpha - 2$ are into a factor of norm 1 times a factor of norm 1247, or into a factor of norm 29 times a factor of norm 43.

A cyclotomic integer with norm 1 is called a *unit*. If $f(\alpha)$ is a unit, then $f(\alpha^2)f(\alpha^3) \ldots f(\alpha^{\lambda-1})$ times $f(\alpha)$ is 1; therefore $f(\alpha^2)f(\alpha^3) \ldots f(\alpha^{\lambda-1})$ is called the *inverse* of $f(\alpha)$ and is denoted $f(\alpha)^{-1}$. Conversely, if $f(\alpha)$ is a cyclotomic integer for which there is a cyclotomic integer $g(\alpha)$ with the property that $f(\alpha)g(\alpha) = 1$ then it is easily shown (Exercise 14) that $f(\alpha)$ is a unit and $g(\alpha) = f(\alpha^2)f(\alpha^3) \ldots f(\alpha^{\lambda-1})$. A unit $f(\alpha)$ is a factor of *any* cyclotomic integer $h(\alpha)$ because $h(\alpha) = f(\alpha)g(\alpha)$ where $g(\alpha) = f(\alpha)^{-1}h(\alpha)$. Therefore "factors" of norm 1 tell nothing about the number being factored and are not counted as being true factors. Thus, only a factorization of $\alpha^5 - \alpha^4 - 3\alpha^2 - 3\alpha - 2$ into a factor of norm 29 times a factor of norm 43 would be counted as a "factorization" in the example above. A cyclotomic integer $h(\alpha)$ is said to be *irreducible* if it has no true factorizations in this sense, that is, if the only factorizations $h(\alpha) = f(\alpha)g(\alpha)$ are the trivial ones in which $f(\alpha)$ or $g(\alpha)$ is a unit. One might be tempted to call an irreducible cyclotomic integer $h(\alpha)$ "prime," but there is another, stronger property which one should require of a cyclotomic integer before one says that it is "prime". Namely, a cyclotomic integer $h(\alpha)$ is said to be *prime* if it has the property that it divides a product $f(\alpha)g(\alpha)$ only when it divides one of the factors. More precisely, $h(\alpha)$ is said to be prime if there exist cyclotomic integers that it does not divide (that is, it is not a unit) and if the product of any two cyclotomic integers it does not divide is a cyclotomic integer it does not divide. It is easy to see that a prime cyclotomic integer is irreducible (Exercise 18). For ordinary integers, irreducible implies prime (Euclid's *Elements*, Book VII, Proposition 24). The fact that there may be cyclotomic integers which are irreducible but not prime lies at the heart of the failure of unique factorization and of the need for Kummer's theory of ideal factorization.

As was mentioned in the preceding section, Cauchy and others expended considerable effort in trying to find a *division algorithm* for cyclotomic integers, that is, a division-with-remainder process such as the one Euclid used in studying the factorization properties of the positive integers (see Appendix A.1) and the one Gauss used in studying factorization properties of "integers" of the form $a + bi$. Even Kummer in his 1844 paper tried to use such a division-with-remainder for cyclotomic integers. At the same time, Kummer showed how to use the norm for simple division, a process of which Lamé seems to have been unaware in 1847.

Given two cyclotomic integers $f(\alpha)$, $h(\alpha)$, determine whether there is a cyclotomic integer $g(\alpha)$ such that $f(\alpha)g(\alpha) = h(\alpha)$, and if so, find $g(\alpha)$. This is the problem of simple division and Kummer solved it as follows. If $f(\alpha) = 0$ then $g(\alpha)$ exists only if $h(\alpha) = 0$, in which case any $g(\alpha)$ will do. Therefore it suffices to consider the case $f(\alpha) \neq 0$. First consider the case in which $f(\alpha) = a_0 + a_1\alpha + a_2\alpha^2 + \cdots + a_{\lambda-1}\alpha^{\lambda-1}$ is an ordinary integer a_0, that is, can be written in a form in which $a_1 = a_2 = \cdots = a_{\lambda-1} = 0$. Then $f(\alpha)g(\alpha) = h(\alpha)$ implies that all the coefficients of $h(\alpha)$ are multiples of $f(\alpha) = a_0 \neq 0$. However, this condition is not a meaningful condition on $h(\alpha)$ because it depends on the representation of $h(\alpha)$ in the form (1). The condition can be restated in a manner that is independent of the representation of $h(\alpha)$ by saying that *if $h(\alpha) = a_0 g(\alpha)$ then the coefficients of $h(\alpha)$ are all congruent to one another modulo a_0.* But this necessary condition is obviously sufficient because if the coefficients of $h(\alpha) = b_0 + b_1\alpha + \cdots + b_{\lambda-1}\alpha^{\lambda-1}$ are all congruent to one another modulo a_0, $b_i \equiv b_j \bmod a_0$, then all the coefficients of $h(\alpha) = (b_0 - b_{\lambda-1}) + (b_1 - b_{\lambda-1})\alpha + \cdots + (b_{\lambda-2} - b_{\lambda-1})\alpha^{\lambda-2}$ are divisible by a_0 and $h(\alpha)$ can be written in the form $h(\alpha) = a_0 g(\alpha)$. This solves the problem of division in the case $f(\alpha) = a_0$. But, by using the norm, the general problem can be reduced to this special case by the observation that $f(\alpha)g(\alpha) = h(\alpha)$ is equivalent to $Nf(\alpha) \cdot g(\alpha) = h(\alpha)f(\alpha^2)f(\alpha^3) \cdots f(\alpha^{\lambda-1})$; this shows that $f(\alpha)$ divides $h(\alpha)$ if and only if the integer $Nf(\alpha)$ divides $h(\alpha)f(\alpha^2)f(\alpha^3) \cdots f(\alpha^{\lambda-1})$ and when this is the case the quotient $g(\alpha)$ is the same in both cases. This method of division is not entirely satisfactory in that the computation of $Nf(\alpha)$ and $h(\alpha)f(\alpha^2)f(\alpha^3) \cdots f(\alpha^{\lambda-1})$ can be extremely long, but it does show that the divisibility question does have a definite answer which can be achieved in a finite number of steps.

If for given $f(\alpha)$, $h(\alpha)$ there is a $g(\alpha)$ such that $f(\alpha)g(\alpha) = h(\alpha)$ then $f(\alpha)$ is said to *divide* $h(\alpha)$, and $h(\alpha)$ is said to be *divisible* by $f(\alpha)$. The notation $f(\alpha)|h(\alpha)$ means "$f(\alpha)$ divides $h(\alpha)$" and the notation $f(\alpha) \nmid h(\alpha)$ means "$f(\alpha)$ does not divide $h(\alpha)$." The statement that $f(\alpha)$ divides $h(\alpha)$ may also be written $h(\alpha) \equiv 0 \bmod f(\alpha)$. More generally, $h_1(\alpha) \equiv h_2(\alpha) \bmod f(\alpha)$ means that $f(\alpha)$ divides $h_1(\alpha) - h_2(\alpha)$.

This completes the list of the basic aspects of the arithmetic of cyclotomic integers. Before proceeding to the detailed study of the factorization of cyclotomic integers in the next section, it is perhaps worthwhile to pause to consider the philosophical underpinnings of this arithmetic. Kummer invariably refers to cyclotomic integers as "complex numbers" and, to modern ears at least, this suggests a geometrical picture* of them as points in the "complex plane." This view of cyclotomic integers has the advantage that it makes certain properties—notably the property $Nf(\alpha) \geq 0$—easy to

---

*This geometrical picture is explicitly described in Article 38 of Gauss's second paper [G6] on biquadratic reciprocity. It is interesting to note that in this paper, which is the predecessor of Jacobi's as Jacobi's is the predecessor of Kummer's, Gauss explicitly says (note at end of Article 30) that in order to study cubic residues one should consider complex numbers built up from a cube root of unity and that in order to study higher residues one should in the same way "introduce other imaginary quantities."

prove, but at the same time it adds nothing to the understanding of other properties such as the factorization properties of cyclotomic integers which are the primary concern of this chapter. Kronecker, who was a student and close associate of Kummer, suggested many years later (see [K2] and [K3]) that the cyclotomic integers should be approached *abstractly* and *algebraically* as the set of all expressions of the form (1) with addition, subtraction, and multiplication defined in the obvious way subject to the single relation $1 + \alpha + \alpha^2 + \cdots + \alpha^{\lambda-1} = 0$. This approach is more in the spirit of contemporary algebra, and readers who have studied modern algebra will recognize this construction as the quotient of the ring of polynomials in one variable with integer coefficients by the ideal generated by the polynomial $1 + \alpha + \alpha^2 + \cdots + \alpha^{\lambda-1}$. The main virtue of this approach is that it emphasizes the algebraic computational rules of the arithmetic of cyclotomic integers—which are easily learned even if one is not comfortable with complex numbers—and puts all other considerations in the background.

EXERCISES

1. $2 + \alpha + 3\alpha^2 - 2\alpha^3 = -\alpha + \alpha^2 - 4\alpha^3 - 2\alpha^4$ ($\lambda = 5$). Let $f(\alpha)$ be the left side of this equation, $g(\alpha)$ the right side, and prove that $f(\alpha^3) = g(\alpha^3)$. Is $f(\alpha^5) = g(\alpha^5)$?

2. The multiplication of cyclotomic integers can be arranged like the multiplication of polynomials. For example, the product $(\alpha^4 + 7\alpha^2 + 5\alpha + 1)(2\alpha^3 + 3\alpha^2 - \alpha + 2)$ (with $\lambda = 5$) can be arranged

```
                        1       0       7       5       1
                        0       2       3      -1       2
                   ─────────────────────────────────────
                        2       0      14      10       2
                -1       0      -7      -5      -1
         3       0      21      15       3
2        0      14      10       2
───────────────────────────────────────────────────────
2        3      13      33      10      12       9       2
```

to find the product $2\alpha^7 + 3\alpha^6 + 13\alpha^5 + 33\alpha^4 + 10\alpha^3 + 12\alpha^2 + 9\alpha + 2 = (2 + 12)\alpha^2 + (3 + 9)\alpha + (13 + 2) + 33\alpha^4 + 10\alpha^3 = -\alpha^2 - 3\alpha + 0 + 18\alpha^4 - 5\alpha^3$. It is more efficient to build the relation $\alpha^{j+\lambda} = \alpha^j$ into the scheme by writing this multiplication instead in the form

```
    1       0       7       5       1
    0       2       3      -1       2
───────────────────────────────────
    2       0      14      10       2
    0      -7      -5      -1      -1
   21      15       3       3       0
   10       2       2       0      14
───────────────────────────────────
   33      10      14      12      15 = 18   -5   -1   -3   0.
```

Use this multiplication scheme to check the associative law in the case of the triple product $(\alpha^4 + 7\alpha^2 + 5\alpha + 1)(2\alpha^3 + 3\alpha^2 - \alpha + 2)(4\alpha^4 + 2\alpha^3 - \alpha^2 + 5)$, where $\lambda = 5$.

3. Show that when $\lambda = 3$ the norm of any cyclotomic integer is of the form $\frac{1}{4}(A^2 + 3B^2)$ where A and B are integers which are either both odd or both even. Conclude that in this case there are just 6 units and find them all.

4. Show that when $\lambda = 5$ the norm of any cyclotomic integer is of the form $\frac{1}{4}(A^2 - 5B^2)$ where A and B are integers which are either both odd or both even. [In the product $f(\alpha)f(\alpha^2)f(\alpha^3)f(\alpha^4)$ multiply $f(\alpha)f(\alpha^4)$ first. The result is of the form $a + b\theta_0 + c\theta_1$ where $\theta_0 = \alpha + \alpha^4$, $\theta_1 = \alpha^2 + \alpha^3$. Then $f(\alpha^2)f(\alpha^3)$ $= a + b\theta_1 + c\theta_0$ and the norm has the form $(b\theta_0 + c\theta_1)(b\theta_1 + c\theta_0)$. One way of doing it gives $-\frac{1}{4}(A^2 - 5B^2)$, which looks wrong but isn't.] Find an infinite number of units in this case.

5. Let $\lambda = 7$ and $f(\alpha) = \alpha^5 - \alpha^4 - 3\alpha^2 - 3\alpha - 2$. In Exercise 4 of Section 4.4 the formula $Nf(\alpha) = 1247$ will be used. Derive it. [First compute $f(\alpha)f(\alpha^2)f(\alpha^4) = a + b\theta_0 + c\theta_1$. Arrange the computation as in Exercise 2.]

6. Show that if $f(\alpha) = g(\alpha)$ then $f(1) \equiv g(1) \bmod \lambda$. Conclude that $Nf(\alpha) \equiv 0$ or 1 mod λ [Fermat's theorem].

7. Prove that $f(\alpha) + f(\alpha^2) + \cdots + f(\alpha^{\lambda-1}) \equiv -f(1) \bmod \lambda$. This gives an alternative proof of the first part of Exercise 6. $[f(1) + f(\alpha) + f(\alpha^2) + \cdots + f(\alpha^{\lambda-1})$ is easy to write explicitly.]

8. Prove that if $g(\alpha)$ is a conjugate of $f(\alpha)$ and $h(\alpha)$ a conjugate of $g(\alpha)$ then $h(\alpha)$ is a conjugate of $f(\alpha)$. [Note that $\alpha \mapsto \alpha^j$ is not a conjugation if $\lambda | j$.] Prove that if $g(\alpha)$ is a conjugate of $f(\alpha)$ then $f(\alpha)$ is a conjugate of $g(\alpha)$.

9. Show that if $f(\alpha)$ is a *binomial* $A\alpha^j + B\alpha^k$ then the result of Exercise 6 can be strengthened to: With the obvious exceptions of cases where A and B are not relatively prime or $j \equiv k \bmod \lambda$, every *prime factor* of $Nf(\alpha)$ is 0 or 1 mod λ. $[f(\alpha)$ is a unit times a conjugate of $A + B\alpha$ and therefore $(A + B)Nf(\alpha) = A^\lambda + B^\lambda$, $Nf(\alpha) = A^{\lambda-1} - A^{\lambda-2}B + \cdots + B^{\lambda-1}$. If $1 - k + k^2 - \cdots + k^{\lambda-1} \equiv 0 \bmod p$ then $k \equiv -1$ or $k^\lambda \equiv -1 \bmod p$, $p \equiv 1 \bmod \lambda$.]

10. Show that $\alpha^{\lambda-1}$ is the complex conjugate of α. [They have the same product with α.] Fill in the other steps of the proof of $Nf(\alpha) \geqslant 0$ given in the text.

11. Let $\lambda = 5$. Then $38\alpha^3 + 62\alpha^2 + 56\alpha + 29$ is divisible by $3\alpha^3 + 4\alpha^2 + 7\alpha + 1$. Find the quotient. [Use the division method explained in the text and the algorithm of Exercise 2 to do the multiplications.]

12. Prove that $f(\alpha)$ is divisible by $\alpha - 1$ if and only if $f(1) \equiv 0 \bmod \lambda$. [Use the Remainder Theorem.] This gives a third proof of the first part of Exercise 6. Note that in Exercise 11 both divisor and dividend are divisible by $\alpha - 1$. Carry out both divisions and thereby simplify the solution of Exercise 11.

13. Let $\lambda = 5$. Choose two cyclotomic integers and compute their product. Then use the division method of the text to divide the product by one of the factors. Do the same with $\lambda = 7$.

14. Show that if $f(\alpha)$ and $g(\alpha)$ are cyclotomic integers such that $f(\alpha)g(\alpha) = 1$ then $f(\alpha)$ is a unit and $g(\alpha) = f(\alpha^2)f(\alpha^3)\ldots f(\alpha^{\lambda-1})$. [It is simplest to prove $Nf(\alpha) = \pm 1$ and use the fact that norms are nonnegative.]

15. This exercise is devoted to the proof that cyclotomic integers satisfy no relations other than those found in the text. Let $f(\alpha) = 0$ be any relation of the type in question; that is, let $f(X)$ be a polynomial in one variable with integer coefficients which becomes zero when the complex number α is substituted for the variable X. Because $\alpha^\lambda = 1$, $\alpha^{\lambda+1} = \alpha, \ldots$, it can obviously be assumed that $f(X)$ has degree at most $\lambda - 1$. It is to be shown that $f(X)$ must be a multiple of $X^{\lambda-1} + X^{\lambda-2} + \cdots + X + 1$. Let $h(X) = X^{\lambda-1} + X^{\lambda-2} + \cdots + X + 1$. There is an integer a such that $f(X) - ah(X)$ has degree less than $\lambda - 1$ and is zero when $X = \alpha$. Thus it will suffice to show that if $f(X)$ has degree less than $\lambda - 1$ and $f(\alpha) = 0$ then $f(X)$ must be the polynomial 0. Assume not. That is, assume $f(X) \neq 0$, $f(X)$ has degree less than $\lambda - 1$, and $f(\alpha) = 0$. Then $ah(X) = q(X)f(X) + r(X)$ where a is an integer, $q(X)$ and $r(X)$ are polynomials with integer coefficients, and $r(X)$ is of lower degree than $f(X)$. If $r(X) \neq 0$ replace $f(X)$ by $r(X)$ and repeat the process. Thus it can be assumed that $ah(X) = q(X)f(X)$. If $a \neq \pm 1$ then there is a prime p such that $q(X)f(X) \equiv 0 \bmod p$. This implies $q(X) \equiv 0$ or $f(X) \equiv 0 \bmod p$. Therefore $ah(X) = q(X)f(X)$ can be divided by p. Thus it can be assumed that $h(X) = f(X)g(X)$. Consider this relation mod λ. $h(j) \equiv 0$ for $j \equiv 1$, but $h(j) \equiv 1$ for $j \not\equiv 1 \bmod \lambda$. In other words, $h(X)$ and $(X - 1)^{\lambda-1}$ assume the same values mod λ for all integer values of X. Since $h(X) - (X - 1)^{\lambda-1}$ is a polynomial of degree less than λ (in fact less than $\lambda - 1$) whose values for integral X are all $0 \bmod \lambda$, a differencing argument like that of Section 2.4 shows that $h(X) \equiv (X - 1)^{\lambda-1} \bmod \lambda$. A polynomial $F(X)$ is divisible by $X - 1 \bmod \lambda$ if and only if $F(1) \equiv 0 \bmod \lambda$. This is true for either $f(X)$ or $g(X)$ but not for both. Therefore either $f(X)$ or $g(X)$ is divisible by $(X - 1)^{\lambda-1} \bmod \lambda$. Since $f(X)$ has degree less than $\lambda - 1$, $g(X)$ must have degree $\lambda - 1$. Therefore $f(X)$ has degree zero and $f(\alpha) = 0$ implies $f = 0$, which is a contradiction. Fill in the details of this proof.

16. Prove by entirely algebraic means that if $f(\alpha)h(\alpha) = g(\alpha)h(\alpha)$ and $h(\alpha) \neq 0$ then $f(\alpha) = g(\alpha)$. [It suffices to consider the case $g(\alpha) = 0$. Then $f(\alpha)$ satisfies a weaker condition than the $f(\alpha)$ of Exercise 15, but the same argument can be used.] The only property of cyclotomic integers which has not been proved by purely algebraic means in this section is the property $Nf(\alpha) \geqslant 0$ of Exercise 10.

17. Prove that the congruence relation $h_1(\alpha) \equiv h_2(\alpha) \bmod f(\alpha)$ defined in the text has, for fixed $f(\alpha)$, the properties that it is an equivalence relation—that is, it is reflexive [$h(\alpha) \equiv h(\alpha)$], symmetric [$h(\alpha) \equiv k(\alpha)$ implies $k(\alpha) \equiv h(\alpha)$] and transitive [$h(\alpha) \equiv k(\alpha)$ and $k(\alpha) \equiv g(\alpha)$ imply $h(\alpha) \equiv g(\alpha)$]—and it is consistent with addition and multiplication—that is, $h_1(\alpha) \equiv h_2(\alpha)$ and $k_1(\alpha) \equiv k_2(\alpha)$ imply $h_1(\alpha) + k_1(\alpha) \equiv h_2(\alpha) + k_2(\alpha)$ and $h_1(\alpha)k_1(\alpha) \equiv h_2(\alpha)k_2(\alpha)$.

18. Prove that a prime cyclotomic integer is irreducible.

19. Prove that for ordinary integers an irreducible integer is prime.

4.3   Factorization of primes $p \equiv 1 \bmod \lambda$

A natural first step in the study of the factorization of cyclotomic integers is to try to find all the *primes* in this arithmetic. According to the definition of the preceding section, a cyclotomic integer $h(\alpha)$ is said to be *prime* if it is not a unit and if it has the property that it can divide a product of two cyclotomic integers only if it divides one of the factors. In symbols, $h(\alpha) \nmid 1$, and $h(\alpha)|f(\alpha)g(\alpha)$ implies $h(\alpha)|f(\alpha)$ or $h(\alpha)|g(\alpha)$. In the study of Fermat's Last Theorem, first consideration goes to the factors in the equation $(x + y)(x + \alpha y)(x + \alpha^2 y) \cdots (x + \alpha^{\lambda-1}y) = z^\lambda$. Therefore the present section and the one which follows are devoted to the problem of factoring binomials of the form $x + \alpha^j y$ (x and y relatively prime integers, $j = 1, 2, \ldots, \lambda - 1$), and in particular to the problem of finding all possible prime factors of such binomials $x + \alpha^j y$.

(Concentration on the factorization of binomials is justified not only by the fact that they are the first ones to present themselves in Fermat's Last Theorem but also by the fact that, as will be seen in this chapter, the prime factors of such binomials are the simplest primes in the arithmetic of cyclotomic integers. Moreover, these are the prime factors which Kummer studied in his original paper of 1844 on the factorization of cyclotomic integers. His reason for studying these particular factors apparently had nothing to do with Fermat's Last Theorem and stemmed, rather, from Jacobi's work [J2] on higher reciprocity laws.)

The method will be the method of analysis. That is, it will be assumed that a prime factor $h(\alpha)$ of $x + \alpha^j y$ is known and from this assumption enough information will be deduced to make it possible to construct such prime factors $h(\alpha)$ in many cases. The fact that there are some cases in which the method of construction *fails* to produce the expected factor leads first to Kummer's observation that the naive assumption of unique factorization itself fails, and then to Kummer's theory of ideal factorization, which "saves" unique factorization and provides a powerful tool for dealing with the arithmetic of cyclotomic integers.

Assume, therefore, that $h(\alpha)$ is a prime cyclotomic integer which divides $x + \alpha^j y$, where x and y are relatively prime integers and $\alpha^j \neq 1$. Then $h(\alpha)$ also divides $N(x + \alpha^j y)$ which is an ordinary integer. The integer $N(x + \alpha^j y)$ can be written as a product of prime integers, say $N(x + \alpha^j y) = p_1 p_2 \cdots p_n$ (repetitions allowed), and, by the fact that $h(\alpha)$ is prime, one of these prime integers $p_1, p_2, \ldots, p_n$ must be divisible by $h(\alpha)$. Let p be a prime integer which $h(\alpha)$ divides. Then $h(\alpha)$ can divide no integer m relatively prime to p because if m is relatively prime to p then $1 = am + bp$ for some integers a and b, and $h(\alpha)|m$ would imply $h(\alpha)|1$, contrary to the assumption that $h(\alpha)$ is not a unit. Thus *the integers which are divisible by $h(\alpha)$ are precisely the multiples of p.*

The main technique which will be used in finding conditions on $h(\alpha)$ will be to *consider congruences* mod $h(\alpha)$. Specifically, it will be shown that one can determine whether two cyclotomic integers are congruent mod $h(\alpha)$ when just p, x, y, and j are known, even if $h(\alpha)$ itself is not known. This fact is very helpful in the analysis of $h(\alpha)$ and ultimately in the discovery of possible prime factors $h(\alpha)$.

Two cyclotomic integers $f(\alpha)$ and $g(\alpha)$ are said to be congruent modulo a third cyclotomic integer $h(\alpha)$, written $f(\alpha) \equiv g(\alpha)$ mod $h(\alpha)$, if $h(\alpha)$ divides their difference $f(\alpha) - g(\alpha)$. This relation, just like congruence of ordinary integers mod a third integer, is *reflexive, symmetric, transitive, and consistent with addition and multiplication*. Written out, these conditions are $f(\alpha) \equiv f(\alpha)$ mod $h(\alpha)$ (reflexive), $f(\alpha) \equiv g(\alpha)$ mod $h(\alpha)$ implies $g(\alpha) \equiv f(\alpha)$ mod $h(\alpha)$ (symmetric), $f(\alpha) \equiv g(\alpha)$ mod $h(\alpha)$ and $g(\alpha) \equiv \phi(\alpha)$ mod $h(\alpha)$ imply $f(\alpha) \equiv \phi(\alpha)$ mod $h(\alpha)$ (transitive), $f(\alpha) \equiv g(\alpha)$ mod $h(\alpha)$ implies $f(\alpha) + \phi(\alpha) \equiv g(\alpha) + \phi(\alpha)$ mod $h(\alpha)$ for all $\phi(\alpha)$ (consistent with addition) and $f(\alpha) \equiv g(\alpha)$ mod $h(\alpha)$ implies $f(\alpha)\phi(\alpha) \equiv g(\alpha)\phi(\alpha)$ for all $\phi(\alpha)$ (consistent with multiplication). All of them follow immediately from the definition. In short, the usual rules for computing with congruences can be applied in the case of these congruences mod $h(\alpha)$ for any cyclotomic integer $h(\alpha)$.

If $h(\alpha)$, $x + \alpha^j y$, and p are as above, then the statement proved above that the *integers* divisible by $h(\alpha)$ are the multiples of p can also be formulated as the statement that, for *integers* u and v, $u \equiv v$ mod $h(\alpha)$ is equivalent to $u \equiv v$ mod p. This shows that $y \not\equiv 0$ mod p, because $y \equiv 0$ mod p together with $x + \alpha^j y \equiv 0$ mod $h(\alpha)$ would imply $x \equiv 0$ mod $h(\alpha)$, $x \equiv 0$ mod p, contrary to the assumption that x and y are relatively prime. Therefore there is an integer a such that $ay \equiv 1$ mod p. Then $ay \equiv 1$ mod $h(\alpha)$, $0 \equiv a(x + \alpha^j y) \equiv ax + \alpha^j$ mod $h(\alpha)$, $\alpha^j \equiv -ax$ mod $h(\alpha)$. That is, α^j is congruent to an integer mod $h(\alpha)$. Since all powers of α are powers of any given power α^j provided $\alpha^j \neq 1$, this shows that all powers of α are congruent to integers mod $h(\alpha)$. Explicitly, there is an integer i such that $ij \equiv 1$ mod λ (because λ is prime and j is not a multiple of λ), from which it follows that $\alpha = \alpha^{ij} \equiv (-ax)^i$ mod $h(\alpha)$. Let k denote an integer congruent to $(-ax)^i$ mod p. Then k depends only on p, x, y, and j, and it has the property that for any cyclotomic integer $g(\alpha) = a_{\lambda-1}\alpha^{\lambda-1} + \cdots + a_1\alpha + a_0$ the integer $g(k)$ obtained by substituting k for α in $g(\alpha)$ is congruent to $g(\alpha)$ mod $h(\alpha)$. That is, $g(\alpha) \equiv g(k)$ mod $h(\alpha)$. Thus every cyclotomic integer is congruent to an integer mod $h(\alpha)$. Since it is easy to tell when two integers are congruent mod $h(\alpha)$, this makes it easy to tell when two cyclotomic integers are congruent mod $h(\alpha)$ and to prove the following theorem.

Theorem. *Let $h(\alpha)$ be a prime cyclotomic integer which divides both $x + \alpha^j y$ (x, y relatively prime, $\alpha^j \neq 1$) and p (a prime integer). Then there is an integer k, which can be found using only x, y, j, p, such that $\alpha \equiv$*

$k \bmod h(\alpha)$ *and such that, as a result,*

$$f(\alpha) \equiv g(\alpha) \bmod h(\alpha) \quad \Leftrightarrow \quad f(k) \equiv g(k) \bmod p$$

where $f(k)$ and $g(k)$ denote* the integers obtained by setting $\alpha = k$ in the cyclotomic integers $f(\alpha)$ and $g(\alpha)$.

PROOF. $f(\alpha) \equiv f(k) \bmod h(\alpha)$ because $\alpha \equiv k \bmod h(\alpha)$ and because congruences can be added and multiplied. Similarly $g(\alpha) \equiv g(k) \bmod h(\alpha)$. Since $f(k)$ and $g(k)$ are integers, $f(k) \equiv g(k) \bmod h(\alpha)$ is equivalent to $f(k) \equiv g(k) \bmod p$, which is the statement that was to be proved.

The possible values for p and k are greatly limited by the following observations. Since $\alpha^{\lambda-1} + \alpha^{\lambda-2} + \cdots + \alpha + 1 = 0$ is surely divisible by $h(\alpha)$, k and p must satisfy

$$k^{\lambda-1} + k^{\lambda-2} + \cdots + k + 1 \equiv 0 \bmod p \qquad (1)$$

and hence must also satisfy $k^{\lambda} - 1 = (k-1)(k^{\lambda-1} + \cdots + k + 1) \equiv 0$, $k^{\lambda} \equiv 1 \bmod p$. Recall now that, as was seen in the proof of Fermat's theorem in Section 1.8, for any prime p and integer $k \not\equiv 0 \bmod p$ there is a smallest integer d such that $k^d \equiv 1 \bmod p$ and for positive integers j one has $k^j \equiv 1 \bmod p$ if and only if $d | j$. In the present case d can only be 1 or λ because $k^{\lambda} \equiv 1 \bmod p$ and λ is prime. If $d = 1$ then $k \equiv 1 \bmod p$ and (1) shows that $\lambda \equiv 0 \bmod p$; therefore $\lambda = p$. On the other hand, if $\lambda = p$ then $k^{\lambda-1} \equiv 1 \bmod p$ by Fermat's theorem and this combines with $k^{\lambda} \equiv 1 \bmod p$ to give $k \equiv 1 \bmod p$. Thus $k \equiv 1 \bmod p$ if and only if $p = \lambda$. Otherwise $d = \lambda$. Since $k^{p-1} \equiv 1 \bmod p$ by Fermat's theorem, this implies $\lambda | (p-1)$, that is, $p \equiv 1 \bmod \lambda$. In summary, *a prime $h(\alpha)$ which divides a binomial either divides λ or divides a prime $p \equiv 1 \bmod \lambda$. In the first case the integer in the above theorem is $1 \bmod \lambda$; in the second case $k \not\equiv 1 \bmod p$ but $k^{\lambda} \equiv 1 \bmod p$.*

In the first case, $p = \lambda$, $h(\alpha)$ divides $\alpha - 1$ because $1 - 1 \equiv 0 \bmod \lambda$. Therefore $Nh(\alpha)$ divides $N(\alpha - 1)$ (the norm of a product is the product of the norms). Since $N(\alpha - 1) = N(1 - \alpha)$ is the value at $X = 1$ of $(X - \alpha)(X - \alpha^2) \cdots (X - \alpha^{\lambda-1}) = X^{\lambda-1} + X^{\lambda-2} + \cdots + 1$, $N(\alpha - 1) = \lambda$ and, because $Nh(\alpha) \neq 1$, it follows that $Nh(\alpha) = \lambda$ and that $\alpha - 1$ divided by $h(\alpha)$ is a unit. Therefore the only possibility for $h(\alpha)$ is a unit times $\alpha - 1$. This does *not* prove that $\alpha - 1$ satisfies the assumption on $h(\alpha)$, namely, that it is prime. This fact will be proved later in this section.

Consider next the conjugates $h(\alpha^2)$, $h(\alpha^3)$, ..., $h(\alpha^{\lambda-1})$ of $h(\alpha)$. As follows directly from the definition, each of these is also prime. (Because

---

*This notation $f(k)$ should be used cautiously because the cyclotomic integer $f(\alpha)$ does not determine the integer $f(k)$—that is, equality of cyclotomic integers $f(\alpha) = F(\alpha)$ does not imply equality of the integers $f(k)$ and $F(k)$ [see Exercise 11]. However, as the theorem shows, it does imply $f(k) \equiv F(k) \bmod p$, so that, for this p and this k, $f(\alpha)$ does determine $f(k) \bmod p$.

conjugations $\alpha \mapsto \alpha^i$ preserve products, $h(\alpha^j)$ divides $f(\alpha)g(\alpha)$ if and only if $h(\alpha)$ divides $f(\alpha^i)g(\alpha^i)$ where $ij \equiv 1 \bmod \lambda$. Since $h(\alpha)$ is prime, this implies $h(\alpha)$ divides $f(\alpha^i)$ or $g(\alpha^i)$ and therefore that $h(\alpha^j)$ divides $f(\alpha)$ or $g(\alpha)$.) Also each divides a binomial and each divides p; therefore there is a k corresponding to each. When $p = \lambda$ the value of k must be $k \equiv 1 \bmod \lambda$ in all cases. Therefore two cyclotomic integers are congruent mod $h(\alpha^j)$ if and only if they are congruent mod $h(\alpha)$, from which it follows that $h(\alpha^j)$ divides $h(\alpha)$ and $h(\alpha)$ divides $h(\alpha^j)$, that is, $h(\alpha^j)$ is a unit multiple of $h(\alpha)$. Since $h(\alpha)$, if it exists, must be a unit multiple of $\alpha - 1$, this would imply that $\alpha^2 - 1, \alpha^3 - 1, \ldots, \alpha^{\lambda-1} - 1$ are all unit multiples of $\alpha - 1$. This follows directly, without any assumption about $h(\alpha)$, from the formulas $\alpha^j - 1 = (\alpha - 1)(\alpha^{j-1} + \alpha^{j-2} + \cdots + 1)$ and $N(\alpha^j - 1) = N(\alpha - 1)$. When $p \neq \lambda, p \equiv 1 \bmod \lambda$, the situation is very different. The congruence $\alpha \equiv k \bmod h(\alpha)$ implies $\alpha^j \equiv k \bmod h(\alpha^j)$. If some conjugate $h(\alpha^j)$ of $h(\alpha)$ divided some other conjugate $h(\alpha^i)$ then congruence mod $h(\alpha^i)$ would imply congruence mod $h(\alpha^j)$ and it would follow that $\alpha^j \equiv k \equiv \alpha^i \bmod h(\alpha^j)$; this would imply that $h(\alpha^j)$ divided $\alpha^j - \alpha^i$ and therefore that $Nh(\alpha)$ divided $N(\alpha^j - \alpha^i) = N(\alpha^{j-i} - 1)$; since $N(\alpha^{j-i} - 1) = \lambda$ unless $\alpha^j = \alpha^i$, and since $h(\alpha)$ does not divide λ, this shows that *no conjugate of $h(\alpha)$ divides any other*. This implies that when $p \neq \lambda$ the prime factors $h(\alpha), h(\alpha^2), \ldots, h(\alpha^{\lambda-1})$ of p are all distinct. Since $h(\alpha)$ divides p one has $p = h(\alpha)q(\alpha)$. Since $h(\alpha^2)$ divides p but does not divide $h(\alpha)$, and since $h(\alpha^2)$ is prime, it follows that $h(\alpha^2)$ divides $q(\alpha)$, say $q(\alpha) = h(\alpha^2)q_2(\alpha)$. Since $h(\alpha^3)$ divides $p = h(\alpha)h(\alpha^2)q_2(\alpha)$ but divides neither $h(\alpha)$ nor $h(\alpha^2)$ it must divide $q_2(\alpha)$, from which $p = h(\alpha)h(\alpha^2)h(\alpha^3)q_3(\alpha)$. Continuing in this way, one finds $p = Nh(\alpha)q_{\lambda-1}(\alpha)$. Thus the *integer* $Nh(\alpha)$ times the cyclotomic integer $q_{\lambda-1}(\alpha)$ is equal to the *integer* p. This shows first that $q_{\lambda-1}(\alpha)$ must be an *integer* and then, because p is prime, that $q_{\lambda-1}(\alpha) = 1$, $p = Nh(\alpha)$. Thus not only is p the norm of $h(\alpha)$ but the equation $p = Nh(\alpha)$ constitutes a complete factorization of p into distinct prime factors. In summary:

Theorem. *If $h(\alpha)$ is a prime cyclotomic integer which divides a binomial $x + \alpha^j y$ (where x and y are relatively prime integers and $\alpha^j \neq 1$) then $Nh(\alpha)$ is a prime integer which is congruent to 0 or $1 \bmod \lambda$. If $Nh(\alpha) = \lambda$ then $h(\alpha)$ and all of its conjugates are unit multiples of $\alpha - 1$. If $Nh(\alpha) = p \equiv 1 \bmod \lambda$ then $p = Nh(\alpha)$ is a factorization of p as a product of $\lambda - 1$ distinct prime factors; that is, no one of the factors $h(\alpha^j)$ divides any other.*

The analysis of the problem—that is, the deduction of necessary conditions for $h(\alpha)$ to be a prime divisor of a binomial—need be carried no further because the synthesis is now possible:

Theorem. *If $h(\alpha)$ is any cyclotomic integer whose norm is a prime integer, then $h(\alpha)$ is prime and it divides a binomial $x + \alpha^j y$ (x, y relatively prime, $\alpha^j \neq 1$).*

Corollaries. *The cyclotomic integer* $\alpha - 1$ *is prime, as is any unit multiple of it. If* $Nh(\alpha)$ *is prime then it is* 0 *or* 1 *mod* λ. *If* $Nh_1(\alpha) = p = Nh_2(\alpha)$ *is prime then* $h_2(\alpha)$ *is a unit times a conjugate of* $h_1(\alpha)$.

PROOF. The first corollary follows from the theorem because $N(\alpha - 1) = \lambda$ is prime. However, to prove the theorem in the case $Nh(\alpha) = \lambda$ it is simplest to prove directly that $\alpha - 1$ is prime. Since $\alpha - 1$ divides $N(\alpha - 1)$ $= \lambda$ but does not divide 1, it follows as before that an integer is divisible by $\alpha - 1$ only if it is divisible by λ. Moreover, $\alpha \equiv 1$ mod $\alpha - 1$. Therefore if $\alpha - 1$ divides $f(\alpha) g(\alpha)$ it follows that $f(\alpha) g(\alpha) \equiv 0$ mod $\alpha - 1$, $f(1) g(1) \equiv$ 0 mod $\alpha - 1$, $f(1) g(1) \equiv 0$ mod λ (because $f(1) g(1)$ is an integer), $f(1)$ or $g(1) \equiv 0$ mod λ (because λ is prime), $f(1)$ or $g(1) \equiv 0$ mod $\alpha - 1$, $f(\alpha)$ or $g(\alpha) \equiv 0$ mod $\alpha - 1$, and $\alpha - 1$ divides $f(\alpha)$ or $g(\alpha)$. Therefore $\alpha - 1$ is prime. If $h(\alpha)$ is any cyclotomic integer with $Nh(\alpha) = \lambda$ then, because $\alpha - 1$ divides $Nh(\alpha)$ and is prime, $\alpha - 1$ divides one of the conjugates of $h(\alpha)$. Since $\alpha - 1$ and the conjugates of $h(\alpha)$ both have norm λ, the quotient is a unit, and some conjugate of $h(\alpha)$ is a unit times $\alpha - 1$. Then $h(\alpha)$ itself is a unit times a conjugate of $\alpha - 1$ and, since the conjugates of $\alpha - 1$ are all units times $\alpha - 1$, it follows that $h(\alpha)$ itself is a unit times $\alpha - 1$. Since $\alpha - 1$ is prime, it follows that $h(\alpha)$ is prime and the theorem is proved in the case $Nh(\alpha) = \lambda$.

The second corollary follows from the theorem and the preceding one. In this case too, however, it will be simplest to prove this corollary directly, as a part of the proof of the theorem. Because $\alpha - 1$ is the unique prime divisor of λ, in order to evaluate $Nh(\alpha) = p$ mod λ it is natural to consider it mod $\alpha - 1$. Since every conjugate of α is congruent to 1 mod $\alpha - 1$, every conjugate of $h(\alpha)$ is congruent to $h(1)$ mod $\alpha - 1$ and $p = Nh(\alpha) \equiv$ $[h(1)]^{\lambda - 1}$ mod $\alpha - 1$. Since two integers are congruent mod $\alpha - 1$ only if they are congruent mod λ, this gives $p \equiv [h(1)]^{\lambda - 1}$ mod λ, from which $p \equiv 0$ or 1 mod λ by Fermat's theorem.

The main step of the proof is to show that if $Nh(\alpha)$ is a prime then $h(\alpha)$ must divide a binomial. Since, as is to be shown, $h(\alpha)$ must be prime and must divide a binomial, the first theorem of this section shows that $\alpha \equiv k$ mod $h(\alpha)$ for some integer k. That is, $h(\alpha)$ must in fact divide a binomial of the special form $\alpha - k$. It is natural to try to prove this by actually finding an integer k such that $h(\alpha)$ divides $\alpha - k$. This can be done as follows.

Since $\alpha \equiv k$ mod $h(\alpha)$ implies $1 = \alpha^\lambda \equiv k^\lambda$ mod $h(\alpha)$, one must have $k^\lambda - 1 \equiv 0$ mod p, because otherwise there would exist integers a and b such that $1 = a(k^\lambda - 1) + bp$ and this would imply that $1 \equiv 0$ mod $h(\alpha)$, contrary to the assumption that $Nh(\alpha) = p \neq 1$. Therefore, one can restrict consideration to integers k which satisfy $k^\lambda \equiv 1$ mod p. Let γ be a primitive root mod p (see Appendix A.2). Then every nonzero integer mod p can be written in just one way in the form γ^i $(i = 1, 2, \ldots, p - 1)$ and it satisfies $(\gamma^i)^\lambda \equiv 1$ mod p if and only if $p - 1$ divides $i\lambda$. The theorem has already been proved in the case $p = \lambda$ and, since $p \equiv 0$ or $1 \bmod \lambda$, one can assume for the remainder of the proof that $p \equiv 1 \bmod \lambda$, say $p - 1 = \mu\lambda$. Then $(\gamma^i)^\lambda \equiv 1 \bmod p$ if and only if $(p - 1)/\lambda = \mu$ divides i, that is, if and only if i has one of the values $\mu, 2\mu, 3\mu, \ldots, \lambda\mu = p - 1$. Let m be γ^μ. Then $k = m$, $k = m^2$, $k = m^3, \ldots,$ and $k = m^\lambda \equiv 1 \bmod p$ are λ distinct solutions of $k^\lambda \equiv 1 \bmod p$ and every solution is congruent to one of these $\bmod p$. (For an alternative proof that the congruence $k^\lambda \equiv 1 \bmod p$ has precisely λ distinct solutions $\bmod p$ see Exercise 10.) Thus the problem of finding an integer k such that $h(\alpha)$ divides $\alpha - k$ is the same as the problem of finding an integer j such that $h(\alpha)$ divides $\alpha - m^j$ ($j = 1, 2, \ldots, \lambda$). Now if j is such an integer, it follows that $0 \equiv h(\alpha) \equiv h(m^j) \bmod h(\alpha)$, from which, as before, $h(m^j) \equiv 0 \bmod p$. The fact that there is at least one j in the range $j = 1, 2, \ldots, \lambda - 1$ for which this necessary condition is met follows from:

Lemma. $h(m)h(m^2) \cdots h(m^{\lambda-1}) \equiv 0 \bmod p$.

It might appear at first that this lemma is obvious from $Nh(\alpha) = p$ if one merely sets $\alpha = m$ and reduces $\bmod p$. However, as was remarked above, the operation of setting $\alpha = m$ is not well defined for cyclotomic integers. The indicated argument can be justified by treating $h(\alpha)$ as a polynomial in α—say $h(X)$ for emphasis—and dividing the polynomial $h(X)h(X^2) \cdots h(X^{\lambda-1})$ by the polynomial $X^{\lambda-1} + X^{\lambda-2} + \cdots + 1$ to put it in the form $q(X)(X^{\lambda-1} + X^{\lambda-2} + \cdots + 1) + r(X)$ where $r(X)$ is a polynomial of degree $< \lambda - 1$. When one sets $X = \alpha$ in this polynomial identity, one finds $p = q(\alpha) \cdot 0 + r(\alpha)$ from which, because $r(\alpha)$ is a polynomial in α of degree $< \lambda - 1$, it follows from the basic facts about cyclotomic integers in the preceding section that $r(X)$ is the polynomial $r(X) = p$ of degree 0. With $X = m$ one then has that the integer on the left side of the congruence in the lemma is equal to $q(m)(m^{\lambda-1} + m^{\lambda-2} + \cdots + m + 1) + p$. To prove the lemma it will suffice to prove $m^{\lambda-1} + m^{\lambda-2} + \cdots + 1 \equiv 0 \bmod p$, and this follows from $m^{\lambda-1} + m^{\lambda-2} + \cdots + 1 = (m^\lambda - 1)/(m - 1)$ because $m^\lambda \equiv 1 \bmod p$ and $m \not\equiv 1 \bmod p$.

Thus at least one $j = 1, 2, \ldots, \lambda - 1$ satisfies the necessary condition $h(m^j) \equiv 0 \bmod p$ for $h(\alpha)$ to divide $\alpha - m^j$. It will now be shown that this condition is in fact sufficient for $h(\alpha)$ to divide $\alpha - m^j$. By the method of division described in the preceding section, to determine whether $h(\alpha)$ divides $\alpha - m^j$ is the same as to determine whether $p = Nh(\alpha)$ divides $(\alpha - m^j)h(\alpha^2)h(\alpha^3) \cdots h(\alpha^{\lambda-1})$. Let the polynomial $h(X)$ be divided by $X - m^j$ to find $h(X) = q(X)(X - m^j) + r$ where r is an integer. With $X = m^j$ one has $h(m^j) = r$. Thus $r \equiv 0 \bmod p$ by assumption, and $h(\alpha^\nu) \equiv q(\alpha^\nu)(\alpha^\nu - m^j) \bmod p$ for $\nu = 1, 2, \ldots, \lambda - 1$. Therefore $(\alpha - m^j) \cdot h(\alpha^2)h(\alpha^3) \cdots h(\alpha^{\lambda-1}) \equiv (\alpha - m^j) q(\alpha^2)(\alpha^2 - m^j) q(\alpha^3)(\alpha^3 - m^j) \cdots q(\alpha^{\lambda-1})(\alpha^{\lambda-1} - m^j) = N(\alpha - m^j)q(\alpha^2)q(\alpha^3) \cdots q(\alpha^{\lambda-1}) \bmod p$. Since for any integer k the norm of $\alpha - k$ is the value at $X = k$ of $(X - \alpha)(X - $

$\alpha^2) \cdots (X - \alpha^{\lambda-1}) = X^{\lambda-1} + X^{\lambda-2} + \cdots + 1 = (X^\lambda - 1)/(X - 1)$, that is,

$$N(\alpha - k) = \frac{k^\lambda - 1}{k - 1}$$

one has $N(\alpha - m^j) \equiv 0 \bmod p$ and $h(\alpha)$ divides $\alpha - m^j$, as was to be shown.

Therefore, it has been proved that if $Nh(\alpha)$ is a prime then $h(\alpha)$ not only divides a binomial but even divides one of the form $\alpha - k$. To complete the proof of the theorem it remains only to prove that $h(\alpha)$ is prime. For this the same argument as before can be used, namely, that if $h(\alpha)$ divides $f(\alpha)g(\alpha)$ then $f(\alpha)g(\alpha) \equiv 0 \bmod h(\alpha)$, $f(k)g(k) \equiv 0 \bmod h(\alpha)$ (because $\alpha \equiv k \bmod h(\alpha)$), $f(k)g(k) \equiv 0 \bmod p$ (because otherwise 1 could be written as a combination of $f(k)g(k)$ and p, which would imply $1 \equiv 0 \bmod h(\alpha)$), $f(k)$ or $g(k) \equiv 0 \bmod p$ (because p is prime), $f(k)$ or $g(k) \equiv 0 \bmod h(\alpha)$, and finally $f(\alpha)$ or $g(\alpha) \equiv 0 \bmod h(\alpha)$, as was to be shown. This completes the proof of the theorem.

The third corollary follows from the observation that $h_2(\alpha)$ divides $p = Nh_1(\alpha)$ which, because $h_2(\alpha)$ is prime, implies that $h_2(\alpha)$ divides one of the conjugates of $h_1(\alpha)$. Since both have norm p, the quotient has norm 1 and $h_2(\alpha)$ is a unit times a conjugate of $h_1(\alpha)$, as was to be shown.

This theorem will be applied in the following section to find many prime cyclotomic integers for small values of λ. Historically, this is precisely the path which Kummer followed in 1844. As a matter of fact, he made a serious blunder and thought he had even proved that every prime $p \equiv 1 \bmod \lambda$ is the norm of some cyclotomic integer. This statement is false —for example, when $\lambda = 23$ no cyclotomic integer has norm $47 = 2 \cdot 23 + 1$ —as computations in the next section will show. It is somewhat surprising that Kummer did not discover this fact for himself as a result of his computations, but it appears that it was Jacobi who first saw that there was a very obvious reason why $Nh(\alpha) = p$ could not always be solved even if $p \equiv 1 \bmod \lambda$ (see [E4] for more details). Luckily for Kummer, Jacobi warned him of this mistake in time for him to withdraw his paper containing the erroneous theorem before it was published. It is interesting to speculate that it was also lucky for Kummer that he did not discover his error *too soon*. As it was, he had gone far enough into the theory, and perhaps made such an emotional investment in it, that he was able to surmount the newly discovered difficulties and to go on to create a theory which has earned him an important place in the history of mathematics.

EXERCISES

1. Prove that if $h(\alpha)$ is prime then it is irreducible.

2. Prove that an ordinary positive integer n which is irreducible ($n = mk$ implies m or k is 1) is prime ($n|mk$ implies $n|m$ or $n|k$).

3. Prove that if $h(\alpha)$ is irreducible but not prime then there is a cyclotomic integer which can be written in two distinct ways as a product of irreducibles. That is, if "irreducible" does not imply "prime" then unique factorization fails. [Assume that every cyclotomic integer can be written as a product of irreducibles.]

4. Find the norms of the following binomials. (a) When $\lambda = 5$: $x + \alpha y = 1 + \alpha$, $2 + \alpha$, $2 - \alpha$, $3 + \alpha$, $3 - \alpha$, $3 + 2\alpha$, $3 - 2\alpha$, $4 + \alpha$, $5 + 2\alpha$, $5 - 4\alpha$, $7 + \alpha$. (b) When $\lambda = 7$: $x + \alpha y = 2 \pm \alpha$, $3 \pm \alpha$, $3 \pm 2\alpha$, $4 \pm \alpha$, $4 \pm 3\alpha$, $5 + \alpha$, $5 + 2\alpha$, $5 + 3\alpha$, $5 + 4\alpha$. $[N(x + \alpha y) = (x^\lambda + y^\lambda)/(x + y).]$

5. It was shown in the text that if p divides $N(x + \alpha y)$ where x and y are relatively prime and if p has a prime divisor $h(\alpha)$ then $p \equiv 0$ or $1 \bmod \lambda$. Prove that the assumption about a prime divisor $h(\alpha)$ is unnecessary. [Use the formula for $N(x + \alpha y)$.]

6. Using the fact proved in Exercise 5, factor the norms found in Exercise 4.

7. When $\lambda = 5$, find $N(9 - \alpha)$ (a) directly and (b) by noting that it is $N(3 - \alpha)$ $N(3 + \alpha)$.

8. Prove that if $p \equiv 1 \bmod \lambda$ and if $h_1(\alpha)$, $h_2(\alpha)$ are both prime divisors of p then $h_2(\alpha)$ is a unit times a conjugate of $h_1(\alpha)$. In other words, the prime factorization of p, if it exists, is unique.

9. The essence of the last theorem of the text is the statement that *if $h(\alpha)$ has prime norm then $\alpha \equiv k \bmod h(\alpha)$ for some integer k.* Kummer proved this important theorem in his 1844 paper. His proof, however, was altogether different from the one given in the text. Fill in the details: Let $h(\alpha^2)h(\alpha^3) \cdots h(\alpha^{\lambda-1}) = H(\alpha) = A_0 + A_1\alpha + A_2\alpha^2 + \cdots + A_{\lambda-1}\alpha^{\lambda-1}$. Then $\alpha^{-n}H(\alpha) + \alpha^{-2n}H(\alpha^2) + \cdots + \alpha^{-(\lambda-1)n}H(\alpha^{\lambda-1}) = \lambda A_n - (A_0 + A_1 + \cdots + A_{\lambda-1})$ and $\alpha^{-n}(1 - \alpha)H(\alpha) + \alpha^{-2n}(1 - \alpha^2)H(\alpha^2) + \cdots + \alpha^{-(\lambda-1)n}(1 - \alpha^{\lambda-1})H(\alpha^{\lambda-1}) = \lambda(A_n - A_{n-1})$. Since $H(\alpha^j)H(\alpha^k) \equiv 0 \bmod p$ for $j \neq k$ this gives $\lambda^2(A_{n+1} - A_n)^2 - \lambda^2(A_{n+2} - A_{n+1})(A_n - A_{n-1}) \equiv 0 \bmod p$. Therefore the solution ξ of $(A_{n+2} - A_{n+1})\xi \equiv A_{n+1} - A_n$ is the same for all n. Therefore $A_{n+1}\xi - A_n$ is the same $\bmod p$ for all n, $(\xi - \alpha)H(\alpha)$ is zero $\bmod p$, and therefore $h(\alpha)$ divides $\alpha - \xi$.

10. Using the theorems of this section, show that if $p \equiv 1 \bmod \lambda$ is a prime that is divisible by a prime cyclotomic integer $h(\alpha)$ then there are precisely $\lambda - 1$ distinct solutions of $k^\lambda \equiv 1$, $k \equiv 1 \bmod p$ and they are all powers of any one of them. [Consider the integers congruent to $\alpha^j \bmod h(\alpha)$.] Prove that this is true without the assumption about a prime divisor $h(\alpha)$. [Use Fermat's theorem.]

11. With $\lambda = 5$ one has $0 = 1 + \alpha + \alpha^2 + \alpha^3 + \alpha^4$ but, of course, $0 \neq 1 + 7 + 7^2 + 7^3 + 7^4$. Therefore $f(\alpha) = g(\alpha)$ does not imply $f(7) = g(7)$. Modulo which primes p is $f(7) \equiv g(7) \bmod p$?

12. Prove that $f(\alpha) = g(\alpha)$ if and only if $f(X) - g(X) = q(X)(X^{\lambda-1} + X^{\lambda-2} + \cdots + 1)$. Conclude that if $k^{\lambda-1} + k^{\lambda-2} + \cdots + 1 \equiv 0 \bmod p$ then $f(\alpha) = g(\alpha)$ implies $f(k) \equiv g(k) \bmod p$.

4.4 Computations when $p \equiv 1 \bmod \lambda$

There seems to be a deep-seated tendency on the part of mathematicians to assume unconsciously the validity of unique factorization into primes. This tendency no doubt has its origin in experience with the arithmetic of ordinary integers and in the great usefulness of unique factorization for proving such facts as the statement that the product of two relatively prime numbers can be a square only if each of the factors is a square. Evidence of the strength of this tendency is given by Euler's use in his *Algebra* of unique factorization for quadratic integers despite the fact that the counterexample $3 \cdot 7 = (4 + \sqrt{-5})(4 - \sqrt{-5})$ was, in effect, known both to him and, a hundred years before him, to Fermat, in the form of the statement that a prime factor of an integer of the form $a^2 + 5b^2$ need not itself have this form. (See Sections 1.7 and 2.5.) In the case of cyclotomic integers, the tendency is greatly *reinforced* by computational experience, because, in fact, unique factorization is *valid* for cyclotomic integers with $\lambda < 23$. It is hardly surprising, therefore, that Lamé was so confident of the truth of unique factorization for cyclotomic integers: "There can be no insurmountable obstacle between such a complete verification and an actual proof." Kummer too was certainly very hopeful—if not convinced—that unique factorization would hold for cyclotomic integers and it is for this reason that he felt his discovery of its failure for $\lambda = 23$ was lamentable. If Kummer's belief in unique factorization had not been so reinforced by computational experience it seems unlikely that he would have been convinced that some modification of the idea must be valid and unlikely that he would have been able to establish his theory in the form in which we now know it. There is, in fact, some third-hand evidence (but evidence contemporary to Gauss) that Gauss actually had attempted something similar to Kummer's theory in the case of quadratic integers (numbers of the form $a + b\sqrt{D}$ for fixed D) but that he was unable to work out the details and published his results only in the very different and more complicated form of "composition of forms" in the *Disquisitiones Arithmeticae*. (See Section 8.6.)

Lamé's published computations are not at all extensive and in a sober judgement they certainly do not justify his conviction that unique factorization was valid. They do, however, provide a good point of departure for the investigation of some computational examples. By means of the formula $N(x + \alpha y) = (x^\lambda + y^\lambda)/(x + y)$ he found, in [L6], the norms of the following* binomials in the case $\lambda = 5$.

(1) $N(\alpha + 2) = 11$	(6) $N(2\alpha + 5) = 11 \cdot 41$
(2) $N(\alpha - 2) = 31$	(7) $N(\alpha - 4) = 11 \cdot 31$
(3) $N(\alpha + 3) = 61$	(8) $N(\alpha - 9) = 11^2 \cdot 61$
(4) $N(\alpha + 4) = 5 \cdot 41$	(9) $N(4\alpha - 5) = 11 \cdot 191$
(5) $N(\alpha - 3) = 11^2$	(10) $N(\alpha + 7) = 11 \cdot 191$

In the light of the theorems of the preceding section, the first of these equations says that $\alpha + 2$ is *prime*, that a cyclotomic integer $g(\alpha)$ is divisible by $\alpha + 2$ if and only if $g(-2) \equiv 0 \bmod 11$, and that $N(\alpha + 2) = 11$ is a representation of 11 as a

---

*Lamé's notation was quite different but easily recognizable.

product of 4 distinct prime factors each of norm 11. Similarly, equations (2) and (3) give decompositions of 31 and 61 as products of 4 distinct prime factors.

Equation (4) indicates that $\alpha + 4$ is a product of a factor of norm 5 and a factor of norm 41. The only factors of norm 5 $(= \lambda)$ are units times $\alpha - 1$. Therefore divide $\alpha + 4 = \alpha - 1 + 5$ by $\alpha - 1$ to find the quotient $1 + (\alpha^2 - 1)(\alpha^3 - 1)(\alpha^4 - 1)$ [because $5 = N(\alpha - 1)] = 1 + (1 - \alpha^2 - \alpha^3 + 1)(\alpha^4 - 1) = 1 + 2\alpha^4 - \alpha - \alpha^2 - 2 + \alpha^2 + \alpha^3 = -1 - \alpha + \alpha^3 + 2\alpha^4 = \alpha^2 + 2\alpha^3 + 3\alpha^4$ [because $1 + \alpha + \alpha^2 + \alpha^3 + \alpha^4 = 0$]. Thus $\alpha + 4 = (\alpha - 1)\alpha^2(3\alpha^2 + 2\alpha + 1)$. Of course α^2 is a unit. It follows that $3\alpha^2 + 2\alpha + 1$ has norm 41 and is therefore prime. Since $3\alpha^2 + 2\alpha + 1$ divides $\alpha + 4$, $\alpha \equiv -4 \bmod (3\alpha^2 + 2\alpha + 1)$ and $f(\alpha) \equiv g(\alpha) \bmod (3\alpha^2 + 2\alpha + 1)$ is equivalent to $f(-4) \equiv g(-4) \bmod 41$.

Equation (5) shows that each prime divisor of 11 divides some conjugate of $\alpha - 3$. The prime $\alpha + 2$ itself divides $\alpha^3 - 3$ because $(-2)^3 - 3 \equiv 0 \bmod 11$. Therefore $\alpha^2 + 2$ divides $\alpha - 3 = (\alpha^2)^3 - 3$ and the quotient has norm 11. Therefore the quotient is a unit times one of the four prime divisors $\alpha^j + 2$ of 11. Since $\alpha + 2$ does *not* divide $\alpha - 3$, $\alpha^2 - 3$, $\alpha^4 - 3$, the *only* one of its conjugates which divides $\alpha - 3$ is $\alpha^2 + 2$ and it follows that $\alpha - 3 = \text{unit} \cdot (\alpha^2 + 2)^2$.

Equation (6) shows that $2\alpha + 5$ is divisible by a prime divisor of 11 and by a prime divisor of 41. The conjugate of $2\alpha + 5$ that is divisible by $\alpha + 2$ is $2\alpha^3 + 5$ because $2(-2)^3 + 5 \equiv 0 \bmod 11$. Similarly, the conjugate of $2\alpha + 5$ that is divisible by $3\alpha^2 + 2\alpha + 1$ is found by trying $-4, 16, -64 \equiv 18, -4 \cdot 18 = -72 \equiv 10 \bmod 41$ to find $0 \equiv 2 \cdot 18 + 5 \equiv 2(-4)^3 + 5 \bmod 41$; thus $2\alpha^3 + 5$ is the one that is divisible. Therefore $2\alpha^3 + 5 = \text{unit} \cdot (\alpha + 2)(3\alpha^2 + 2\alpha + 1)$, which gives the representation $2\alpha + 5 = \text{unit} \cdot (\alpha^2 + 2)(3\alpha^4 + 2\alpha^2 + 1)$ of $2\alpha + 5$ as a product of primes.

Equations (7) and (8) follow immediately from the decompositions $\alpha^2 - 4 = (\alpha - 2)(\alpha + 2)$, $\alpha^2 - 9 = (\alpha - 3)(\alpha + 3)$ which imply $\alpha - 4 = (\alpha^3 - 2)(\alpha^3 + 2)$, $\alpha - 9 = (\alpha^3 - 3)(\alpha^3 + 3)$. The first of these is a decomposition into prime factors. The second can be factored further by using the factorization $\alpha - 3 = \text{unit} \cdot (\alpha^2 + 2)^2$ to find $\alpha - 9 = \text{unit} \cdot (\alpha + 2)^2(\alpha^3 + 3)$.

Equation (9) shows that $\alpha + 2$ divides some conjugate of $4\alpha - 5$. The one which it divides is $4\alpha^2 - 5$. The quotient is $4\alpha - 8 + 11(\alpha + 2)^{-1} = 4\alpha - 8 + (\alpha^2 + 2)(\alpha^3 + 2)(\alpha^4 + 2) = 4\alpha - 8 + (1 + 2\alpha^2 + 2\alpha^3 + 4)(\alpha^4 + 2) = 4\alpha - 8 + 5\alpha^4 + 2\alpha + 2\alpha^2 + 10 + 4\alpha^2 + 4\alpha^3 = 2 + 6\alpha + 6\alpha^2 + 4\alpha^3 + 5\alpha^4 = -4 - 2\alpha^3 - \alpha^4$. Therefore $\alpha^4 + 2\alpha^3 + 4$ has norm 191 and is consequently prime. Thus $4\alpha - 5 = -(\alpha^3 + 2)(2\alpha^4 + \alpha^2 + 4)$ factors $4\alpha - 5$ into primes. The k for which $\alpha \equiv k \bmod (2\alpha^4 + \alpha^2 + 4)$ satisfies $4k - 5 \equiv 0 \bmod 191$, from which $k \equiv 192k = 48 \cdot 4k \equiv 48 \cdot 5 = 48 \cdot 4 + 48 \equiv 49 \bmod 191$. Thus $g(\alpha)$ is divisible by $2\alpha^4 + \alpha^2 + 4$ if and only if $g(49) \equiv 0 \bmod 191$.

Finally, equation (10) shows that $\alpha + 7$ is divisible by a factor of 11 and one of 191. A brief computation shows that $49^2 \equiv -82 \bmod 191$, $49^3 \equiv -82 \cdot 49 \equiv -7 \bmod 191$. Therefore $\alpha^3 + 7$ is divisible by $2\alpha^4 + \alpha^2 + 4$. On the other hand, $\alpha^2 + 7$ is divisible by $\alpha + 2$. Therefore $\alpha + 7 = \text{unit} \cdot (\alpha^3 + 2)(2\alpha^3 + \alpha^4 + 4)$ is a prime factorization of $\alpha + 7$.

The prime factors found so far are tabulated in Table 4.4.1. The examples above show that it is very useful in testing for divisibility to know not just the value of $k \bmod p$ but its powers as well. These powers, which are the integers congruent to α^2, α^3, and α^4 modulo the prime factor in question, are therefore included in the table. Note that each line of the table in fact gives *four* prime cyclotomic integers when one takes the conjugates of the given prime and permutes the powers of α

`Table 4.4.1

$\lambda = 5$

prime	norm	$\alpha \equiv$	$\alpha^2 \equiv$	$\alpha^3 \equiv$	$\alpha^4 \equiv$
$\alpha + 2$	11	-2	4	3	5
$\alpha - 2$	31	2	4	8	-15
$3\alpha^2 + 2\alpha + 1$	41	-4	16	18	10
$\alpha + 3$	61	-3	9	-27	20
$2\alpha^4 + \alpha^2 + 4$	191	49	-82	-7	39

accordingly—for example, $\alpha^2 + 2$ is prime and for this prime factor $\alpha^2 \equiv -2$, $\alpha^4 \equiv$ 4, $\alpha^6 = \alpha \equiv 3$, $\alpha^8 = \alpha^3 \equiv 5$.

Lamé carried out factorizations like these for each of his equations (1) – (10), though he made many mistakes. He went on to say that by considering binomials $x\alpha + y$ over a more extended range one can factor many more of the primes $p \equiv 1 \bmod 5$. This is on the whole true,* but it is rather misleading because the same process does *not* enable one to factor primes $p \equiv 1 \bmod \lambda$ when $\lambda > 5$. To see how it fails, consider the case $\lambda = 7$.

Here the norms of the first few binomials, found using $N(x\alpha + y) = (x^7 + y^7)/(x + y)$, are

$$N(\alpha + 2) = 43 \qquad\qquad N(\alpha - 2) = 127$$
$$N(\alpha + 3) = 547 \qquad\qquad N(\alpha - 3) = 1093$$
$$N(2\alpha + 3) = 463 \qquad\qquad N(2\alpha - 3) = 2059 = 29 \cdot 71$$
$$N(\alpha + 4) = 3277 = 29 \cdot 113 \quad N(\alpha - 4) = 5461 = 43 \cdot 127$$
$$N(3\alpha + 4) = 2653 = 7 \cdot 379 \quad N(3\alpha - 4) = 14197.$$

(Because the only possible factors are 7, 29, 43, 71, 113, ..., the factorizations shown are easy to find and it is easy to prove that the numbers which are not factored are prime.) When one attempts to imitate Lamé's method one realizes immediately that the success of the method in the case $\lambda = 5$ depended on the fact that in that case the list began with the factorizations of the primes 11, 31, 61 and with a factorization of $5 \cdot 41$ which is tantamount to a factorization of 41. In the case $\lambda = 7$ one finds factorizations of 43 and 127 but not of 29, 71, and 113. The norms are growing rapidly and there is clearly no hope of finding a binomial with norm 29, for example. Therefore one must resort to some other technique in order to factor 29.

The technique which Kummer developed in his 1844 paper was to *look first for the values of k*, after which it is simple to find cyclotomic integers *divisible* by the desired primes and, with luck, to find the desired primes themselves. In the case $\lambda = 7$, $p = 29$, any fourth power $k = a^4$ satisfies $k^7 = a^{29-1} \equiv 1 \bmod 29$, $k \not\equiv 1 \bmod 29$, provided $a \not\equiv 0 \bmod 29$ and $a^4 \not\equiv 1 \bmod 29$. With $a = 2$ one finds $k = 16 \equiv -13 \bmod 29$ and $(-13)^7 \equiv 1 \bmod 29$ follows. The other solutions of $k \not\equiv 1$, $k^7 \equiv 1 \bmod 29$ are $(-13)^2 \equiv -5$, $(-13)^3 \equiv 7$, $(-13)^4 \equiv -4$, $(-13)^5 \equiv -6$, $(-13)^6 \equiv -9$. If 29 *has* a prime factorization then one of the factors $h(\alpha)$ must

---

*It appears that Lamé was trying to give the impression that his computations were more extensive than they actually were. He says that "with few exceptions" every prime $\equiv 1 \bmod 5$ up to 1021 occurs as a factor in $N(x\alpha + y)$ for some pair x, y with $|x| < 12$, $|y| < 12$. The first 23 primes of this form, up to 521, do occur. Of the remaining 18 up to 1021, however, only 6 occur.

have the property that it divides $g(\alpha)$ if and only if $g(-13) \equiv 0$ mod 29. But from the above tabulation of powers of -13 mod 29 it is clear that $g(\alpha) \equiv \alpha^2 - \alpha^4 + 1$ satisfies $g(-13) \equiv 0$ mod 29. Therefore $g(\alpha)$ is divisible by the hypothetical factor of 29. Since the norm of the hypothetical factor is 29, the norm of the quotient is $Ng(\alpha)/29$. A simple computation shows* that $Ng(\alpha) = 29$. Therefore $g(\alpha)$ itself is prime and $Ng(\alpha) = 29$ gives the prime factorization of 29.

The factorizations of 71 and 113 can now be found by dividing $2\alpha - 3$ and $\alpha + 4$ respectively by the appropriate factor of 29. Since $2k - 3 \equiv 0$ mod 29 for $k = -13$, $2\alpha - 3$ is divisible by $-\alpha^4 + \alpha^2 + 1$. The quotient is $(2\alpha - 3) \cdot g(\alpha^2)g(\alpha^3)g(\alpha^4)g(\alpha^5)g(\alpha^6)/29 = (2\alpha - 3)g(\alpha^2)g(\alpha^4)f(\alpha^3)/29$, where $f(\alpha)$ is as in the note below, $= (2\alpha - 3)(-2\alpha^6 - 3\alpha^2 - \alpha)(-1 - 4\theta_1)/29 = (2\alpha - 3)(6\alpha^6 + 20\alpha^5 + 12\alpha^4 + 11\alpha^2 + 13\alpha + 16)/29 = (22\alpha^6 - 36\alpha^5 - 36\alpha^4 + 22\alpha^3 - 7\alpha^2 - 7\alpha - 36)/29 = 2\alpha^6 + 2\alpha^3 + \alpha^2 + \alpha$. Therefore $2\alpha^5 + 2\alpha^2 + \alpha + 1$ is a prime factor of 71. The corresponding k satisfies $2k - 3 \equiv 0$ mod 71, from which $k \equiv 72k = 36 \cdot 2k \equiv 36 \cdot 3 = 36 \cdot 2 + 36 \equiv 37$ mod 71; then $k^2 \equiv 20$, $k^3 \equiv 30$, $k^4 \equiv -26$, $k^5 \equiv 32$, $k^6 \equiv -23$ mod 71 are easily found.

Similarly, to divide $\alpha + 4$ by a factor of 29 one tries $k + 4 \equiv 0$ mod 29 for $k = -13, -5, 7, -4, -6, -9$. The solution is $k = -4$ and it follows that $\alpha^4 + 4$ is divisible by $-\alpha^4 + \alpha^2 + 1$. The majority of the work required for this division was done above. The quotient is $(\alpha^4 + 4)(6\alpha^6 + 20\alpha^5 + 12\alpha^4 + 11\alpha^2 + 13\alpha + 16)/29 = (35\alpha^6 + 93\alpha^5 + 64\alpha^4 + 6\alpha^3 + 64\alpha^2 + 64\alpha + 64)/29 = -\alpha^6 + \alpha^5 - 2\alpha^3 = -\alpha^3(\alpha^3 - \alpha^2 + 2)$. Thus $\alpha^3 - \alpha^2 + 2$ is a prime factor of 113. The corresponding k is easily found using the fact that $\alpha^3 - \alpha^2 + 2$ divides $\alpha^4 + 4$ to conclude that $k^4 \equiv -4$ mod 113. From this $k \equiv k^8 \equiv 16$, $k^2 \equiv 16^2 \equiv 30$, and so forth.

The next prime $\equiv 1$ mod 7 after 127 is 197. If one carries the tabulation of norms $N(x\alpha + y)$ a bit further than was done above one finds $N(\alpha + 6) = 39991 = 7 \cdot 29 \cdot 197$. Division of $\alpha + 6$ by $\alpha - 1$ gives $1 + 7(\alpha - 1)^{-1} = 1 + (\alpha^2 - 1)(\alpha^3 - 1)(\alpha^4 - 1)(\alpha^5 - 1)(\alpha^6 - 1)$. The product $(\alpha^2 - 1)(\alpha^3 - 1) \ldots (\alpha^6 - 1)$ can be found by straightforward multiplication to be $6\alpha^6 + 5\alpha^5 + 4\alpha^4 + 3\alpha^3 + 2\alpha^2 + \alpha$, a for-

---

*It will be useful later to make the following remarks on the computation of $Ng(\alpha) = g(\alpha)g(\alpha^2)g(\alpha^3)g(\alpha^4)g(\alpha^5)g(\alpha^6)$. Since 3 is a primitive root mod 7 (see Appendix A.2) the iterates of the conjugation $\alpha \mapsto \alpha^3$ include all the conjugations. Thus the terms of $Ng(\alpha)$ can be written in the order $g(\alpha)g(\alpha^3)g(\alpha^2)g(\alpha^6)g(\alpha^4)g(\alpha^5)$ in which each term is the conjugate of its predecessor under $\alpha \mapsto \alpha^3$. If one then defines $f(\alpha)$ to be the product of every other term $f(\alpha) = g(\alpha)g(\alpha^2)g(\alpha^4)$ one has $Ng(\alpha) = f(\alpha)f(\alpha^3)$. In this way the computation of $Ng(\alpha)$ is reduced to 3 multiplications. The computation of $f(\alpha)$ can be written

α^6	α^5	α^4	α^3	α^2	α	1
0	0	-1	0	1	0	1
0	0	1	0	0	-1	1
		-1		1		1
	1			-1		-1
	1				-1	
1	1	0	-1	1	-2	1
= 0	0	-1	-2	0	-3	0

α^6	α^5	α^4	α^3	α^2	α	1
0	0	-1	-2	0	-3	0
				-1	1	1
		-1	-2		-3	
	-1	-2		-3		
1	2		3			
1	1	-3	1	-3	-3	0

Thus $f(\alpha) = -3\theta_0 + \theta_1 = -1 - 4\theta_0$ where $\theta_0 = \alpha + \alpha^2 + \alpha^4$ and $\theta_1 = \alpha^3 + \alpha^5 + \alpha^6$. (Because $f(\alpha) = g(\alpha)g(\alpha^2)g(\alpha^4)$ is unchanged when α is changed to α^2, it is clear that for any $g(\alpha)$ the resulting $f(\alpha)$ will be expressible in the form $a\theta_0 + b\theta_1 + c$.) Then $f(\alpha^3) = -1 - 4\theta_1$ and $Ng(\alpha) = (1 + 4\theta_0)(1 + 4\theta_1) = 1 - 4 + 16\theta_0\theta_1$. A simple multiplication gives $\theta_0\theta_1 = \theta_0 + \theta_1 + 3 = 2$ and $Ng(\alpha) = -3 + 32 = 29$.

mula which holds for arbitrary λ—that is, $(\alpha^2 - 1)(\alpha^3 - 1) \ldots (\alpha^{\lambda-1} - 1) = (\lambda - 1)\alpha^{\lambda-1} + \cdots + 2\alpha^2 + \alpha$. (See Exercise 11.) Thus $(\alpha + 6)(\alpha - 1)^{-1} = 6\alpha^6 + 5\alpha^5 + 4\alpha^4 + 3\alpha^3 + 2\alpha^2 + \alpha + 1 = \alpha^2(5\alpha^4 + 4\alpha^3 + 3\alpha^2 + 2\alpha + 1)$. To divide this by a prime factor of 29 one tries, as before, $\alpha \equiv -13, -5, 7, -4, -6, -9$ and reduces mod 29. The result is zero when $k = -6 \equiv (-13)^5 \bmod 29$. Therefore $5\alpha^6 + 4\alpha + 3\alpha^3 + 2\alpha^5 + 1$ is divisible by $-\alpha^4 + \alpha^2 + 1$. The quotient is $(5\alpha^6 + 4\alpha + 3\alpha^3 + 2\alpha^5 + 1)(6\alpha^6 + 20\alpha^5 + 12\alpha^4 + 11\alpha^2 + 13\alpha + 16)/29 = (192\alpha^6 + 163\alpha^5 + 163\alpha^4 + 192\alpha^3 + 105\alpha^2 + 192\alpha + 163)/29 = \alpha(\alpha^5 + \alpha^2 - 2\alpha + 1)$. Thus $5\alpha^4 + 4\alpha^3 + 3\alpha^2 + 2\alpha + 1 = (-\alpha^5 + \alpha^6 + 1) \cdot \alpha^3 \cdot (\alpha + \alpha^6 - 2\alpha^3 + 1)$. In summary, $\alpha + 6 = (\alpha - 1)(\alpha^6 - \alpha^5 + 1)(\alpha^6 - 2\alpha^3 + \alpha + 1) \cdot \alpha^5$ and $\alpha^6 - 2\alpha^3 + \alpha + 1$ is a prime with norm 197 mod which $\alpha \equiv -6$.

This method of factoring binomials does not seem to yield a factor of the next prime $211 \equiv 1 \bmod 7$. The first binomial whose norm is divisible by 211 is $N(3\alpha - 10) = 7 \cdot 211 \cdot 967$ and, of course, one would need to factor 967 before this could be used to factor 211. Therefore one resorts again to Kummer's method of finding possible values of k and from these finding cyclotomic integers *divisible* by the hypothetical factors of 211, in hopes of finding one of norm 211. The first step is to solve the congruence $k \not\equiv 1$, $k^7 \equiv 1 \bmod 211$. For this one can set $k \equiv a^{30}$ provided $a \not\equiv 0$, $a^{30} \not\equiv 1 \bmod 211$. With $a = 2$ one finds $a^{30} \equiv -40 \bmod 211$. Therefore, if 211 has a prime factor it has one mod which $\alpha \equiv -40$, $\alpha^2 \equiv 1600 \equiv -88$, $\alpha^3 \equiv -67$, $\alpha^4 \equiv -63$, $\alpha^5 \equiv -12$, $\alpha^6 \equiv 58$. This factor would therefore divide $\alpha^4 - \alpha^3 - 4$. A computation like those above gives $N(\alpha^4 - \alpha^3 - 4) = 29 \cdot 211$. Therefore $\alpha^4 - \alpha^3 - 4$ can be divided by a prime factor of 29 to find an element with norm 211 and therefore a factorization of 211 into primes. Alternatively, one can try again to find a $g(\alpha)$ which satisfies $g(-40) \equiv 0 \bmod 211$ which has norm 211. A list of the differences of powers of α mod the hypothetical prime (which by the above computation actually exists) of the form $\pm \alpha \pm \alpha^2 \equiv \pm 48, \pm 83$ $(\equiv \pm 128)$; $\pm \alpha \pm \alpha^3 \equiv \pm 27, \pm 104$; $\pm \alpha \pm \alpha^4 \equiv \pm 23, \pm 103; \ldots; \pm \alpha^5 \pm \alpha^6 \equiv \pm 46, \pm 70$ gives 60 integers mod 211 that are congruent to sums or differences of powers of α. The smallest of these is 4, which was used above. However, if one allows another power of α and looks therefore among these 60 integers for one that is near a power of α, one finds cases in which the difference is only 2, for example, $\pm \alpha^3 \pm \alpha^6 \equiv \pm 9, \pm 86$; and $\alpha^2 \equiv -88$ leads to $\alpha^6 - \alpha^3 - \alpha^2 \equiv 213 \equiv 2$. Computation gives $N(\alpha^6 - \alpha^3 - \alpha^2 - 2) = 211$ and this accomplishes the factorization of 211.

This completes the factorization of the first 7 primes $\equiv 1 \bmod 7$. The results are given in Table 4.4.2. The continuation of the process presents no difficulties other than the increasing length of the computations.

Table 4.4.2

$\lambda = 7$

prime	norm	$\alpha \equiv$	$\alpha^2 \equiv$	$\alpha^3 \equiv$	$\alpha^4 \equiv$	$\alpha^5 \equiv$	$\alpha^6 \equiv$
$-\alpha^4 + \alpha^2 + 1$	29	-13	-5	7	-4	-6	-9
$\alpha + 2$	43	-2	4	-8	16	11	-22
$2\alpha^5 + 2\alpha^2 + \alpha + 1$	71	37	20	30	-26	32	-23
$2\alpha^5 + \alpha - 1$	113	16	30	28	-4	49	-7
$\alpha - 2$	127	2	4	8	16	32	-63
$\alpha^6 - 2\alpha^3 + \alpha + 1$	197	-6	36	-19	-83	-93	-33
$\alpha^6 - \alpha^3 - \alpha^2 - 2$	211	-40	-88	-67	-63	-12	58

Kummer carried out such computations and found prime factorizations of all $p \equiv 1 \bmod \lambda$ *in the range* $\lambda \leqslant 19, p < 1000$. His results, which were published in his 1844 paper [K6], are reproduced in Table 4.4.3.

The entries in this table can be found in the same way that the factorizations of 29 and 211 ($\lambda = 7$) were found above. Consider, for example, the case $\lambda = 13, p = 599$. In this case $p = 46\lambda + 1$ and since $2^{46} \equiv 19 \bmod 599$ (an easy computation) one can set $k = 19$, after which $k^2 \equiv -238, k^3 \equiv 270, k^4 \equiv -261, k^5 \equiv -167, k^6 \equiv -178, k^7 \equiv 212, k^8 \equiv -165, k^9 \equiv -140, k^{10} \equiv -264, k^{11} \equiv -224, k^{12} \equiv -63$ can be computed relatively easily. Forming the differences $\pm k \pm k^2 \equiv \pm 257, \pm 219$; $\pm k \pm k^3 \equiv \pm 289, \pm 251$; and so forth, gives 4×66 numbers mod 599; none of these numbers is very small and none is very near a power of k, but many will be found which differ by just one from each other, for example, $k^7 - k \equiv 193, k^2 + k^5 \equiv -405 \equiv 194$. This leads to $1 - \alpha - \alpha^2 - \alpha^5 + \alpha^7$ as possible factor of 599. Now

Table 4.4.3

When $\lambda = 5$ and $\alpha^\lambda = 1$:

$11 = N(2 + \alpha)$	$461 = N(4 - \alpha - \alpha^2)$
$31 = N(2 - \alpha)$	$491 = N(5 + 3\alpha + \alpha^3)$
$41 = N(3 + 2\alpha + \alpha^2)$	$521 = N(5 + \alpha)$
$61 = N(3 + \alpha)$	$541 = N(3 - 3\alpha - \alpha^2)$
$71 = N(3 - \alpha + \alpha^2)$	$571 = N(6 + 5\alpha + 3\alpha^2)$
$101 = N(3 + \alpha - \alpha^2)$	$601 = N(5 + 2\alpha - \alpha^2)$
$131 = N(3 + \alpha - \alpha^4)$	$631 = N(4 - 2\alpha - \alpha^3)$
$151 = N(3 + 2\alpha - \alpha^4)$	$641 = N(5 + 3\alpha + 4\alpha^2)$
$181 = N(4 + 3\alpha)$	$661 = N(5 + \alpha - \alpha^2 + 3\alpha^3)$
$191 = N(4 + \alpha + 2\alpha^2)$	$691 = N(3 - 3\alpha - 2\alpha^2)$
$211 = N(3 - 2\alpha)$	$701 = N(4 - \alpha - 2\alpha^2 + \alpha^3)$
$241 = N(4 - \alpha + \alpha^2)$	$751 = N(6 + 4\alpha + 3\alpha^2)$
$251 = N(5 + 2\alpha + \alpha^4)$	$761 = N(5 - 2\alpha + \alpha^2)$
$271 = N(3 - 3\alpha + \alpha^2)$	$811 = N(3 - 3\alpha - 2\alpha^2 + \alpha^3)$
$281 = N(4 + \alpha - \alpha^2)$	$821 = N(4 - \alpha - 2\alpha^2 + 2\alpha^3)$
$311 = N(3 + 2\alpha + 2\alpha^2 + \alpha^3)$	$881 = N(6 + 2\alpha + \alpha^2)$
$331 = N(4 - 2\alpha + \alpha^2)$	$911 = N(5 + \alpha^2 - 2\alpha^4)$
$401 = N(4 + 3\alpha - \alpha^4)$	$941 = N(4 + 3\alpha - 3\alpha^2 - \alpha^3)$
$421 = N(5 + 2\alpha + 2\alpha^2)$	$971 = N(5 - 2\alpha - \alpha^4)$
$431 = N(4 - 2\alpha - \alpha^4)$	$991 = N(6 + \alpha + \alpha^3)$

When $\lambda = 7$ and $\alpha^\lambda = 1$:

$29 = N(1 + \alpha - \alpha^2)$	$491 = N(3 + \alpha + \alpha^3 - \alpha^5)$
$43 = N(2 + \alpha)$	$547 = N(3 + \alpha)$
$71 = N(2 + \alpha + \alpha^3)$	$617 = N(2 + \alpha + \alpha^2 - \alpha^5)$
$113 = N(2 - \alpha + \alpha^5)$	$631 = N(2 + 2\alpha - \alpha^2 + \alpha^3 + \alpha^6)$
$127 = N(2 - \alpha)$	$659 = N(2 + 2\alpha - \alpha^2 + \alpha^5)$
$197 = N(3 + \alpha + \alpha^5 + \alpha^6)$	$673 = N(4 + 3\alpha + 2\alpha^2 + \alpha^4 + 2\alpha^6)$
$211 = N(3 + \alpha + 2\alpha^2)$	$701 = N(3 + \alpha + \alpha^4 - \alpha^5 + \alpha^6)$
$239 = N(3 + 2\alpha + 2\alpha^2 + \alpha^3)$	$743 = N(3 + 2\alpha - \alpha^3 - \alpha^4)$
$281 = N(2 - \alpha - 2\alpha^3)$	$757 = N(3 + 2\alpha + \alpha^3)$
$337 = N(2 + \alpha - \alpha^2 - \alpha^4)$	$827 = N(2 + 2\alpha - \alpha^4 - \alpha^6)$

Table 4.4.3 (*Cont'd.*)

When $\lambda = 7$ and $\alpha^\lambda = 1$:

$379 = N(3 + 2\alpha + \alpha^2)$ $883 = N(2 - \alpha^2 - 2\alpha^3 - \alpha^5)$

$421 = N(3 + \alpha + \alpha^2)$ $911 = N(3 + 2\alpha - \alpha^3 + \alpha^4)$

$449 = N(2 + \alpha - \alpha^3 - \alpha^6)$ $953 = N(3 + \alpha - \alpha^2 - \alpha^3)$

$463 = N(3 + 2\alpha)$ $967 = N(2 + 2\alpha - \alpha^3 + 2\alpha^5)$

When $\lambda = 11$ and $\alpha^\lambda = 1$:

$23 = N(1 + \alpha + \alpha^9)$ $617 = N(2 + \alpha + \alpha^3 + \alpha^{10})$

$97 = N(1 + \alpha + \alpha^2 + \alpha^4 + \alpha^5)$ $661 = N(1 + \alpha - \alpha^2 + \alpha^4 - \alpha^8)$

$89 = N(1 + \alpha + \alpha^4 + \alpha^6)$ $683 = N(2 + \alpha)$

$199 = N(1 + \alpha - \alpha^2)$ $727 = N(1 + \alpha + \alpha^3 - \alpha^8 - \alpha^9)$

$331 = N(1 - \alpha + \alpha^3 + \alpha^5)$ $859 = N(1 + \alpha + \alpha^2 + \alpha^3 + \alpha^7 - \alpha^8)$

$353 = N(1 + \alpha + \alpha^3 + \alpha^4 - \alpha^7)$ $881 = N(1 + \alpha + \alpha^2 + \alpha^3 - \alpha^4 - \alpha^7 - \alpha^9)$

$397 = N(1 + \alpha + \alpha^6 - \alpha^7)$ $947 = N(2 + \alpha^3 - \alpha^4 - \alpha^6)$

$419 = N(1 + \alpha - \alpha^2 + \alpha^3)$ $991 = N(2 + \alpha + \alpha^3)$

$463 = N(1 - \alpha - \alpha^2 + \alpha^5 + \alpha^6)$

When $\lambda = 13$ and $\alpha^\lambda = 1$:

$53 = N(1 + \alpha + \alpha^3)$ $547 = N(1 - \alpha - \alpha^2 + \alpha^3 + \alpha^6)$

$79 = N(1 - \alpha + \alpha^{10})$ $599 = N(1 + \alpha - \alpha^7 + \alpha^8 + \alpha^{11})$

$131 = N(1 - \alpha + \alpha^{11})$ $677 = N(1 - \alpha - \alpha^4 + \alpha^6 + \alpha^9)$

$157 = N(1 + \alpha + \alpha^2 + \alpha^5)$ $859 = N(1 + \alpha - \alpha^2 - \alpha^5 + \alpha^7)^*$

$313 = N(1 - \alpha + \alpha^3 + \alpha^6)$ $911 = N(1 + \alpha^3 + \alpha^5 - \alpha^7 - \alpha^{11})$

$443 = N(1 + \alpha - \alpha^3 + \alpha^8)$ $937 = N(1 + \alpha^3 - \alpha^7 + \alpha^8 - \alpha^{10})$

$521 = N(1 + \alpha - \alpha^{12})$

When $\lambda = 17$ and $\alpha^\lambda = 1$:

$103 = N(1 + \alpha^2 + \alpha^9)$ $443 = N(1 + \alpha + \alpha^2 + \alpha^3 - \alpha^{15})$

$137 = N(1 + \alpha - \alpha^3)$ $613 = N(1 + \alpha^2 - \alpha^3)$

$239 = N(1 + \alpha + \alpha^3)$ $647 = N(1 + \alpha + \alpha^{13} + \alpha^{15})$

$307 = N(1 - \alpha + \alpha^7)$ $919 = N(1 + \alpha + \alpha^4 + \alpha^5 + \alpha^9)$

$409 = N(1 - \alpha^3 + \alpha^8)$ $953 = N(1 + \alpha + \alpha^9 - \alpha^{13})$

When $\lambda = 19$ and $\alpha^\lambda = 1$:

$191 = N(1 + \alpha + \alpha^{16})$ $571 = N(1 + \alpha + \alpha^2 + \alpha^3 - \alpha^5)$

$229 = N(1 - \alpha - \alpha^5)$ $647 = N(1 - \alpha^2 + \alpha^9)$

$419 = N(1 + \alpha - \alpha^8)$ $761 = N(1 - \alpha^2 + \alpha^{12})$

$457 = N(1 + \alpha + \alpha^3)$

---

*An obvious misprint in Kummer's table has been corrected here.

comes the difficult part of the computation, which is the computation of the norm. The work of this computation can be minimized by organizing it as follows. Use the fact that 2 is a primitive root mod 13 (see Appendix A.2) to write $Nf(\alpha)$ as $f(\alpha)f(\alpha^2)f(\alpha^4)f(\alpha^8)f(\alpha^3)f(\alpha^6)f(\alpha^{12})f(\alpha^{11})f(\alpha^9)f(\alpha^5)f(\alpha^{10})f(\alpha^7)$ in which each factor is obtained from the previous one by substituting α^2 for α. Let $g(\alpha)$ be the product of every fourth one of these 12 factors, that is, $g(\alpha) = f(\alpha)f(\alpha^3)f(\alpha^9)$.

Then clearly $Nf(\alpha) = g(\alpha)g(\alpha^2)g(\alpha^4)g(\alpha^8)$. Let $h(\alpha)$ be the product of every second one of these 4 factors, that is, $h(\alpha) = g(\alpha)g(\alpha^4)$. Then $Nf(\alpha) = h(\alpha)h(\alpha^2)$. The direct computation of $g(\alpha)$ is not too long. The result is $g(\alpha) = -3(\alpha^{12} + \alpha^{10} + \alpha^4) + 2(\alpha^{11} + \alpha^8 + \alpha^7) - 2(\alpha^9 + \alpha^3 + \alpha^1) + 6(\alpha^6 + \alpha^5 + \alpha^2) - 10$. Let $\eta_0 = \alpha + \alpha^3 + \alpha^9$, $\eta_1 = \alpha^2 + \alpha^5 + \alpha^6$, $\eta_2 = \alpha^4 + \alpha^{10} + \alpha^{12}$, $\eta_3 = \alpha^8 + \alpha^7 + \alpha^{11}$. Then substitution of α^2 for α causes a cyclic permutation $\eta_0 \mapsto \eta_1 \mapsto \eta_2 \mapsto \eta_3 \mapsto \eta_0$. Thus

$$h(\alpha) = g(\alpha)g(\alpha^4) = [5\eta_3 + \eta_0 + 9\eta_1 - 7][5\eta_1 + \eta_2 + 9\eta_3 - 7].$$

A multiplication table for the η's can easily be constructed; the only multiplications which need to be carried out are

$$\eta_0^2 = \alpha^2 + \alpha^4 + \alpha^{10} + \alpha^4 + \cdots = \eta_1 + 2\eta_2$$
$$\eta_0\eta_1 = \eta_0 + \eta_1 + \eta_3$$
$$\eta_0\eta_2 = 3 + \eta_1 + \eta_3$$

after which other products such as $\eta_0\eta_3 = \eta_3\eta_0 = \eta_{0+3} + \eta_{1+3} + \eta_{3+3} = \eta_3 + \eta_0 + \eta_2$ can be derived by permuting the η's. Thus

$$h(\alpha) = 25\eta_1\eta_3 + 5\eta_3\eta_2 + 45\eta_3^2 - 35\eta_3 + 5\eta_0\eta_1 + \eta_0\eta_2 + 9\eta_0\eta_3 - 7\eta_0$$
$$+ 45\eta_1^2 + 9\eta_1\eta_2 + 81\eta_1\eta_3 - 63\eta_1 - 35\eta_1 - 7\eta_2 - 63\eta_3 + 49$$
$$= 49 - 7\eta_0 - 98\eta_1 - 7\eta_2 - 98\eta_3 + 25(3 + \eta_2 + \eta_0)$$
$$+ 5(\eta_2 + \eta_3 + \eta_1) + 45(\eta_0 + 2\eta_1) + 5(\eta_0 + \eta_1 + \eta_3)$$
$$+ (3 + \eta_1 + \eta_3) + 9(\eta_3 + \eta_0 + \eta_2) + 45(\eta_2 + 2\eta_3)$$
$$+ 9(\eta_1 + \eta_2 + \eta_0) + 81(3 + \eta_2 + \eta_0)$$
$$= 370 + 167\eta_0 + 12\eta_1 + 167\eta_2 + 12\eta_3$$
$$= 370 + 167\theta_0 + 12\theta_1 = 358 + 155\theta_0$$

where $\theta_0 = \eta_0 + \eta_2 = \alpha + \alpha^4 + \alpha^3 + \alpha^{12} + \alpha^9 + \alpha^{10}$ and $\theta_1 = \eta_1 + \eta_3 = \alpha^2 + \alpha^8 + \alpha^6 + \alpha^{11} + \alpha^5 + \alpha^7 = -1 - \theta_0$. Then $\theta_0\theta_1 = 3\theta_0 + 3\theta_1 = -3$, $h(\alpha^2) = 358 + 155\theta_1$, and finally $Nf(\alpha) = (358 + 155\theta_0)(358 + 155\theta_1) = 358^2 + 358 \cdot 155(\theta_0 + \theta_1) - 3 \cdot 155^2 = 128164 - 55490 - 72075 = 599$ as desired. Thus $1 - \alpha - \alpha^2 - \alpha^5 + \alpha^7$ is a prime factor of 599. (Kummer's factor in this case is erroneous. See Exercise 8.)

Consider finally the case $\lambda = 23$. The first prime $p \equiv 1 \bmod 23$ is $p = 47$. In this case k can be taken to be any square mod 47, say $k = 4$. Then the powers $k, k^2, k^3, \ldots$ mod 47 are easily found to be $4, 16, 17, 21, -10, 7, -19, 18, -22, 6, 24, 2, 8, -15, -13, -5, -20, 14, 9, -11, 3, 12$, in that order. Of course $k^{23} \equiv 1$. An obvious attempt at a factor of 47 which divides $\alpha - 4$ would be $1 - \alpha + \alpha^{21}$ because $1 - 4 + 3 = 0$. In order to follow the procedure used above, it is necessary to compute $N(1 - \alpha + \alpha^{21})$. Since $22 = 11 \cdot 2$ this computation can be reduced to a multiplication of 11 factors followed by a multiplication of two factors. A primitive root mod 23 is -2 so that the reduction is, explicitly, $Nf(\alpha) = G(\alpha)G(\alpha^{-2})$ where

$$G(\alpha) = f(\alpha)f(\alpha^4)f(\alpha^{-7})f(\alpha^{-5})f(\alpha^3)f(\alpha^{-11})f(\alpha^2)f(\alpha^8)f(\alpha^9)f(\alpha^{-10})f(\alpha^6).$$

(Here α^{-2} means α^{21}, α^{-7} means α^{16}, etc.) The computation of $G(\alpha)$ is a bit long. It can be organized as follows. Set $g(\alpha) = f(\alpha)f(\alpha^4)$ and $h(\alpha) = g(\alpha)f(\alpha^{-7})$ so that

104

$G(\alpha) = h(\alpha)h(\alpha^{-5})h(\alpha^2)g(\alpha^{-10})$. Then

$$g(\alpha) = \alpha^{21} - \alpha^{16} + \alpha^{15} + \alpha^{13} + \alpha^5 - \alpha^4 - \alpha^2 - \alpha + 1$$

$$h(\alpha) = \alpha^{20} + \alpha^{19} + \alpha^{17} - 3\alpha^{16} + \alpha^{13} + \alpha^{12} + \alpha^9 - \alpha^8 - \alpha^7 + \alpha^5 - \alpha^2 - \alpha + 1$$

$$h(\alpha^{-5}) = \alpha^{15} + \alpha^{20} + \alpha^7 - 3\alpha^{12} + \alpha^4 + \alpha^9 + \alpha - \alpha^6 - \alpha^{11} + \alpha^{21} - \alpha^{13} - \alpha^{18} + 1$$

$$h(\alpha^2) = \alpha^{17} + \alpha^{15} + \alpha^{11} - 3\alpha^9 + \alpha^3 + \alpha + \alpha^{18} - \alpha^{16} - \alpha^{14} + \alpha^{10} - \alpha^4 - \alpha^2 + 1$$

$$g(\alpha^{-10}) = \alpha^{20} - \alpha + \alpha^{11} + \alpha^8 + \alpha^{19} - \alpha^6 - \alpha^3 - \alpha^{13} + 1$$

and the computation of $G(\alpha)$ requires two medium-sized multiplications, followed by the rather long multiplication of the two results. The final result is $G(\alpha) = -44 + 15\theta_0 - 13\theta_1$ where $\theta_0 = \alpha + \alpha^4 + \alpha^{-7} + \cdots + \alpha^6$ and $\theta_1 = -1 - \theta_0$. Because $\theta_0\theta_1$ is easily seen by direct multiplication to be $11 + 5\theta_0 + 5\theta_1 = 6$ this gives $G(\alpha)G(\alpha^{-2}) = (-31 + 28\theta_0)(-31 + 28\theta_1) = 31^2 - 31 \cdot 28(\theta_0 + \theta_1) + 6 \cdot 28^2 = 961 + 868 + 4704 = 6533 = 47 \cdot 139$. Thus the procedure does not on this attempt yield a factor of 47. In fact, it is easy to see that it can *never* yield a factor of 47 because, whatever the original $f(\alpha)$ might be, the analog of $G(\alpha)$ can be put in the form $a + b\theta_0$ so that the desired result is $47 = (a + b\theta_0)(a + b\theta_1) = a^2 - ab + 6b^2$. The impossibility of this equation can be proved by applying Gauss's theory of binary quadratic forms or, more simply, by writing it in the form $4 \cdot 47 = 4a^2 - 4ab + 24b^2$, $188 = (2a - b)^2 + 23b^2$. Only two trials are necessary to verify that 188 is not a square plus 23 times a square ($188 - 23$ and $188 - 23 \cdot 4$ are not squares).

This proves that in the case $\lambda = 23$ no cyclotomic integer has norm 47. Since $N(\alpha - 4) \equiv 0 \bmod 47$, any prime factor of 47 would have to divide one of the conjugates of $\alpha - 4$ and therefore, by the theorem of the preceding section, would have to have norm 47. Thus 47 *has no prime factors at all*, much less a unique factorization into primes. The anomaly can be viewed in a more elementary way as well, one which does not appeal to the theorem of the preceding section. Since $1 - \alpha + \alpha^{-2}$ divides $47 \cdot 139$ but divides neither 47 nor 139 (because its norm divides neither $N(47) = 47^{22}$ nor $N(139) = 139^{22}$) it is not prime and, if the usual factorization properties are to hold, it should decompose into factors. But, since its norm is $47 \cdot 139$, the only possible decomposition into factors would be into a factor of norm 47 and one of norm 139, whereas it was just seen that a norm of 47 is impossible. [This is exactly like the example of the failure of unique factorization $3 \cdot 7 = N(4 + \sqrt{-5})$ in quadratic integers $x + y\sqrt{-5}$ cited above. Since $4 + \sqrt{-5}$ divides $3 \cdot 7$ but not 3 or 7—because its norm does not divide $N(3) = 3^2$ or $N(7) = 7^2$—it is not prime and must be the product of a factor of norm 3 and a factor of norm 7. But $N(x + y\sqrt{-5}) = x^2 + 5y^2 = 3$ is impossible.]

Of the 8 primes $p \equiv 1 \bmod 23$ less than 1000, Kummer gave for 3 of them solutions $h(\alpha)$ of $Nh(\alpha) \equiv p$ (see Table 4.4.4) and for the remaining 5, including $p = 47$ and $p = 139$, he showed by the above argument that such an $h(\alpha)$ is impossible. For each of these five p's he gave a cyclotomic integer $h(\alpha)$ with the two properties that $h(\alpha^{-1}) = h(\alpha)$ and the product of the 11 conjugates of $h(\alpha)$ under $\alpha \mapsto \alpha^4$ is p. (In other words,

$$p = h(\alpha)h(\alpha^4)h(\alpha^{-7})h(\alpha^{-5})h(\alpha^3)h(\alpha^{-11})h(\alpha^2)h(\alpha^8)h(\alpha^9)h(\alpha^{-10})h(\alpha^6).$$

Then because $h(\alpha) = h(\alpha^{-1})$, $Nh(\alpha) = p^2$.) These 5 cyclotomic integers are given in Table 4.4.5. This gives very explicitly two entirely different ways of factoring

105

Table 4.4.4

$\lambda = 23$ and $\alpha^\lambda = 1$
$599 = N(1 + \alpha^{15} - \alpha^{16})$
$691 = N(1 + \alpha + \alpha^5)$
$829 = N(1 + \alpha^{11} + \alpha^{20})$

Table 4.4.5

$\lambda = 23$ and $\alpha^\lambda = 1$
$47^2 = N(\alpha^{10} + \alpha^{-10} + \alpha^8 + \alpha^{-8} + \alpha^7 + \alpha^{-7})$
$139^2 = N(\alpha^{10} + \alpha^{-10} + \alpha^8 + \alpha^{-8} + \alpha^4 + \alpha^{-4})$
$277^2 = N(2 + \alpha + \alpha^{-1} + \alpha^7 + \alpha^{-7})$
$461^2 = N(\alpha + \alpha^{-1} + \alpha^{10} + \alpha^{-10} + \alpha^8 + \alpha^{-8} + \alpha^9 + \alpha^{-9})$
$967^2 = N(2 + \alpha^{11} + \alpha^{-11} + \alpha^4 + \alpha^{-4})$

$47 \cdot 139$, the one as $N(1 - \alpha + \alpha^{-2})$ and the other as Kummer's factorization of 47 times his factorization of 139. In the first one there are 22 factors of norm $47 \cdot 139$, and in the second there are 11 factors of norm 47^2 and 11 factors of norm 139^2. All these factors are irreducible because a norm of 47 or 139 is impossible.

The kernel of Kummer's theory of ideal complex numbers is the simple observation that the test for divisibility by the hypothetical factor of 47, namely, the test whether $g(4) \equiv 0$ mod 47, is perfectly meaningful even though there is no actual factor of 47 for which it tests. One can choose to regard it as a test for divisibility by an *ideal* prime factor of 47 and this, in a nutshell, is the idea of Kummer's theory. Before Kummer could take this decisive step, however, it was necessary for him first to master the factorization of primes $p \not\equiv 0, 1$ mod λ and it was to such factorizations that he turned his attention in the period from 1844 to 1846. Here too he found methods of testing for divisibility by prime factors of p, tests which continued to be defined even when there was no actual factor for which they tested. He took these tests to be—by definition—tests for divisibility by "ideal prime factors" of p and built his theory of ideal complex numbers on the basis of these ideal prime factors.

The next four sections are devoted to the study of prime factors of primes $p \not\equiv 0, 1$ mod λ, generalizing the above study of the case $p \equiv 1$ mod λ. In the fifth section the experience gained in these four sections is used to motivate the very natural definition of *ideal prime factors*, which in this book will be called *prime divisors*.

EXERCISES

1. Verify the following computations ($\lambda = 5$).
   (a) $(3 - \alpha)(2 + \alpha)(2 + \alpha^3)(2 + \alpha^4) = 11(\alpha^4 + \alpha^3 + 2)$.
   (b) $(2 + \alpha^2)^2$ divides $3 - \alpha$ and the quotient is a unit. Find the unit and its inverse.
   (c) $(2 + \alpha^2)(3\alpha^4 + 2\alpha^2 + 1)$ divides $5 + 2\alpha$ and the quotient is a unit.

(d) $2 + \alpha$ divides $5 - 4\alpha^2$. What is the norm of the quotient?

(e) $2\alpha^3 + \alpha^4 + 4, 2\alpha^2 + \alpha + 4$, and $2\alpha + \alpha^3 + 4$ are not divisible by $2\alpha^4 + \alpha^2 + 4$.

(f) $(\alpha^3 + 2)(2\alpha^3 + \alpha^4 + 4)$ divides $\alpha + 7$ and the quotient is a unit.

(g) $\alpha + 2$ divides $\alpha^4 - 5$.

2. Verify the following statements in the case $\lambda = 7$.

(a) $2\alpha^5 + 2\alpha^2 + \alpha + 1$ has norm 71 and divides $3 - 2\alpha$.

(b) $4 + \alpha$ is divisible by one of the conjugates of $1 + \alpha^2 - \alpha^4$. Find the quotient.

(c) $N(2 + 2\alpha^2 + \alpha^6) = 197$.

(d) $N(3 - \alpha^4 - \alpha^6) = 8 \cdot 197$.

(e) Find an element with norm 8. Note that all prime factors of the norm need not be 1 mod 7.

3. Factor the following cyclotomic integers ($\lambda = 5$).

(a) $5\alpha^4 - 3\alpha^3 - 5\alpha^2 + 5\alpha - 2$.

(b) $-4\alpha^3 - 11\alpha^2 + 8\alpha + 15$.

4. With $\lambda = 7$ factor (a) $\alpha^2 - 3\alpha - 4$, and (b) $\alpha^5 - \alpha^4 - 3\alpha^2 - 3\alpha - 2$.

5. In the case $\lambda = 7, p = 71$, the factor in Kummer's table differs markedly from the one found in the text. Write one as a unit times a conjugate of the other.

6. Find an element with norm p in the case $\lambda = 19, p = 191$. Compare your answer to Kummer's in Table 4.4.3.

7. As in Exercise 6, construct more entries in Kummer's tables.

8. Show that Kummer's factor in the case $\lambda = 13, p = 599$ is erroneous. [It does not satisfy $f(k) \equiv 0 \bmod p$ for any k.]

9. Let $\lambda = 23$. The condition $f(\alpha^{-1}) = f(\alpha)$ coupled with $f(4) \equiv 0 \bmod 47$ gives a condition of the form $a_0 + 16a_1 + \cdots - 21a_{11} \equiv 0 \bmod 47$ on the 12 variables $a_0, a_1, \ldots, a_{11}$. Develop a few simple candidates for solutions of $Nf(\alpha) = 47^2, f(\alpha) = f(\alpha^{-1})$. Show that a conjugate of Kummer's factor $\alpha^{10} + \alpha^{13} + \alpha^8 + \alpha^{15} + \alpha^7 + \alpha^{16}$ satisfies the condition. Compute $Nf(\alpha)$ for this candidate.

10. Prove that if λ divides $Ng(\alpha)$ then $\alpha - 1$ divides $g(\alpha)$.

11. Prove that $(\alpha^2 - 1)(\alpha^3 - 1) \ldots (\alpha^{\lambda-1} - 1) = 1 + 2\alpha + 3\alpha^2 + \cdots + \lambda\alpha^{\lambda-1} = \alpha + 2\alpha^2 + \cdots + (\lambda - 1)\alpha^{\lambda-1}$.

4.5 Periods

In the computations of the preceding section it was useful to choose a primitive root γ mod λ and to write the factors of the norm $Ng(\alpha)$ of a cyclotomic integer $g(\alpha)$ in the order $Ng(\alpha) = g(\alpha)g(\alpha^\gamma)g(\alpha^{\gamma\gamma}) \ldots$ rather than in the order $Ng(\alpha) = g(\alpha)g(\alpha^2)g(\alpha^3) \ldots$. In order to avoid writing superscripts on superscripts it is useful to introduce a notation for the

107

conjugation $\alpha \mapsto \alpha^\gamma$, say σ. Then $\sigma g(\alpha) = g(\alpha^\gamma)$, $\sigma^2 g(\alpha) = g(\alpha^{\gamma\gamma})$, etc., and $Ng(\alpha) = g(\alpha) \cdot \sigma g(\alpha) \cdot \sigma^2 g(\alpha) \cdots \sigma^{\lambda-2} g(\alpha)$. Note that $\sigma^{\lambda-1}$ is the identity conjugation $\alpha \mapsto \alpha$ and that the conjugations $\sigma, \sigma^2, \ldots, \sigma^{\lambda-1}$ are all distinct because each carries α to a different power of α.

The method used to compute norms in the preceding section was to use a factor e of $\lambda - 1$ to write $Ng(\alpha) = G(\alpha) \cdot \sigma G(\alpha) \cdots \cdots \sigma^{e-1} G(\alpha)$ where*
$G(\alpha) = g(\alpha) \cdot \sigma^e g(\alpha) \cdot \sigma^{2e} g(\alpha) \cdots \cdots \sigma^{-e} g(\alpha)$. In this way the number of multiplications required to find $Ng(\alpha)$ could be reduced. For example, in the case $\lambda = 13$ the primitive root $\gamma = 2$ was used with $e = 4$ to find $Ng(\alpha) = G(\alpha) G(\alpha^2) G(\alpha^4) G(\alpha^8)$ where $G(\alpha) = g(\alpha) g(\alpha^3) g(\alpha^9)$. Moreover, the fact that $\sigma^e G(\alpha) = G(\alpha)$ (the conjugation σ^e merely permutes the factors of $G(\alpha)$) means that $G(\alpha)$ must have a very special form. In the example $\lambda = 13$, $\gamma = 2$, $e = 4$, it has the form $a + b\eta_0 + c\eta_1 + d\eta_2 + e\eta_3$ where $\eta_0 = \alpha + \alpha^3 + \alpha^9$, $\eta_1 = \alpha^2 + \alpha^6 + \alpha^5$, $\eta_2 = \alpha^4 + \alpha^{12} + \alpha^{10}$, and $\eta_3 = \alpha^8 + \alpha^{11} + \alpha^7$. In the same way, in the general case $G(\alpha)$ is of the form $a + b\eta_0 + c\eta_1 + \cdots + d\eta_{e-1}$ where $a, b, c, \ldots, d$ are integers, where $\eta_0 = \alpha + \sigma^e \alpha + \sigma^{2e} \alpha + \cdots + \sigma^{-e} \alpha$, and where $\eta_{i+1} = \sigma \eta_i$. This is true simply because the equation $\sigma^e G(\alpha) = G(\alpha)$ implies that the coefficient of α in $G(\alpha)$ is equal to the coefficient of $\sigma^e \alpha$ and, more generally, the coefficient of α^j is equal to the coefficient of $\sigma^e \alpha^j$. The cyclotomic integers $\eta_0, \eta_1, \ldots, \eta_{e-1}$ defined in this way for given λ, γ, e, are called *periods*, a name given them by Gauss in the seventh section of his *Disquisitiones Arithmeticae* (Art. 343). Periods are of importance not only in the rather minor application of the preceding section but also in a vast range of applications in number theory and linear algebra, of which Gauss's use of them in the study of the construction of regular polygons was only the first example. Kummer was thoroughly familiar with Gauss's use of periods, and he made extensive use of them himself in the development of the theory of factorization of cyclotomic integers.

Let λ, γ, e be as above (that is, let λ be an odd prime, γ a primitive root mod λ, and e a divisor of $\lambda - 1$). Let $f = (\lambda - 1)/e$. Then $\eta_0 = \alpha + \sigma^e \alpha + \sigma^{2e} \alpha + \cdots + \sigma^{-e} \alpha$ has f terms because $\sigma^{ef} \alpha = \sigma^{\lambda-1} \alpha = \alpha$, $\sigma^{-e} \alpha = (\sigma^e)^{f-1} \alpha$. Therefore, $\eta_0, \eta_1, \ldots, \eta_{e-1}$ all contain f terms. For this reason they are called *periods of length f*. Note that it is natural to define $\eta_0 = \eta_e$ (because η_e should be $\sigma \eta_{e-1} = \sigma^2 \eta_{e-2} = \cdots = \sigma^e \eta_0 = \eta_0$) and $\eta_1 = \eta_{e+1}$, $\eta_2 = \eta_{e+2}, \ldots$, as well as $\eta_{-1} = \eta_{e-1}$, $\eta_{-2} = \eta_{e-2}, \ldots$, so that η_i is defined for all integers i and $\sigma \eta_i = \eta_{i+1}$. For a given λ, and for a given f which divides $\lambda - 1$, the periods of length f are defined relative to a choice of the primitive root γ. (Set $\sigma(\alpha) = \alpha^\gamma$ and $e = (\lambda - 1)/f$ in the formulas defining the periods.) It is valid to speak of *the periods of length f* because the periods are independent of the choice of the primitive root γ, although their order is not necessarily independent of this choice. This can be seen as follows.

*Here σ^{-e} denotes $\sigma^{\lambda-1-e}$, that is, the inverse of the conjugation σ^e.

Let γ' be another primitive root mod λ, let σ' denote the corresponding conjugation $\alpha \to \alpha^{\gamma'}$, and let η_i' denote the periods of length f defined using σ' instead of σ. To show that the periods $\eta_1, \eta_2, \ldots, \eta_e$ coincide, though possibly in a different order, with $\eta_1', \eta_2', \ldots, \eta_e'$, it is necessary and sufficient to show that if α^j is any power of α, and if η_i is the period which contains it, then η_i also contains $\sigma'^e \alpha^j$. (Then it also contains $\sigma'^e \sigma'^e \alpha^j = \sigma'^{2e} \alpha^j, \sigma'^{3e} \alpha^j, \ldots, \sigma'^{-e} \alpha^j$; that is, it contains all the terms of the period η_k' which contains α^j.) For this it will suffice to prove that σ'^e is a power of σ^e. But since every conjugation is a power of σ (every nonzero integer mod λ is congruent to a power of γ) σ' is a power of σ and it follows immediately that σ'^e is a power of σ^e. This proves that the periods η coincide with the periods η'. To prove that the order may be different it suffices to find an example where this is the case. Such an example is the case $\lambda = 13, f = 3$. The primitive roots mod 13 are $\pm 2, \pm 6$. With $\gamma = 2$ the periods of length 3 are

$$\eta_1 = \alpha^2 + \alpha^6 + \alpha^5$$
$$\eta_2 = \alpha^4 + \alpha^{-1} + \alpha^{-3}$$
$$\eta_3 = \alpha^{-5} + \alpha^{-2} + \alpha^{-6}$$
$$\eta_0 = \eta_4 = \alpha^3 + \alpha^{-4} + \alpha.$$

(Note that the columns on the right, read from top to bottom and left to right, are just the powers $\alpha^2, \alpha^4, \alpha^8 = \alpha^{-5}, \alpha^{16} = \alpha^3, \alpha^{32} = \alpha^6, \ldots$ of α in the order dictated by the primitive root 2 that was chosen.) With $\gamma' = -2$ the period η_1' must contain α^{-2} and therefore $\eta_1' = \eta_3 \neq \eta_1$.

A cyclotomic integer is said to be *made up of periods of length f* if it is of the form $a_0 + a_1 \eta_1 + a_2 \eta_2 + \cdots + a_e \eta_e$ where $a_0, a_1, \ldots, a_e$ are integers and $\eta_1, \eta_2, \ldots, \eta_e$ are periods of length f. As was noted above, $G(\alpha)$ is made up of periods of length f if and only if $\sigma^e G(\alpha) = G(\alpha)$. This implies, because σ^e is a conjugation and therefore carries products to products, that the product of two cyclotomic integers made up of periods is itself made up of periods. For example, for the periods $\eta_0, \eta_1, \eta_2, \eta_3$ of length 3 when $\lambda = 13$, the multiplication table

$$\eta_0^2 = \eta_1 + 2\eta_2$$
$$\eta_0 \eta_1 = \eta_0 + \eta_1 + \eta_3$$
$$\eta_0 \eta_2 = 3 + \eta_1 + \eta_3$$

was found in the preceding section. Application of σ to these identities gives $\eta_1^2 = \eta_2 + 2\eta_3$, $\eta_2^2 = \eta_3 + 2\eta_0$, $\eta_3^2 = \eta_0 + 2\eta_1$, $\eta_1 \eta_2 = \eta_1 + \eta_2 + \eta_0$, and so forth.

It will be noted that in some contexts η_0 is used above, and in others η_e is used. Kummer consistently uses η_0 and writes a typical cyclotomic integer made up of periods as $a_0\eta_0 + a_1\eta_1 + \cdots + a_{e-1}\eta_{e-1}$, making use of the identity $1 + \eta_0 + \eta_1 + \cdots + \eta_{e-1} = 0$ to eliminate any integers from the expression. In some contexts in this book Kummer's notation will be followed, but in others it will be more convenient to use the form $a_0 + a_1\eta_1 + a_2\eta_2 + \cdots + a_e\eta_e$ for a cyclotomic integer made up of periods of length $f = (\lambda - 1)/e$.

EXERCISES

1. In the case $\lambda = 5$, find the periods of length 2 and a multiplication table for them. Find the formula for the norm of a cyclotomic integer made up of periods of length 2 ($\lambda = 5$). Find a cyclotomic integer whose norm is divisible by a prime that is neither 1 nor 0 mod 5. Conclude that if this cyclotomic integer has a prime factor it is not of the type found in the preceding section.

2. Show that when $\lambda = 7$ the norm of any cyclotomic integer is of the form $\frac{1}{4}[A^2 + 7B^2]$ where A and B are integers which are either both odd or both even. [Let θ_0, θ_1 be the periods of length 3. Find $\theta_0\theta_1$.]

3. Find a general formula for $(\theta_0 - \theta_1)^2$ where θ_0, θ_1 are the periods of length $\frac{1}{2}(\lambda - 1)$ by induction, that is, by evaluating this integer for several values of λ.

4. Prove the formula found in Exercise 3.

5. For all cases there is a formula for $Ng(\alpha)$ analogous to $Ng(\alpha) = \frac{1}{4}[A^2 + 7B^2]$ in the case $\lambda = 7$. Find this formula.

6. Find the periods and the multiplication table in the case $\lambda = 17, f = 4$.

7. Under what conditions on p and f are the periods of length f invariant under the conjugation $\alpha \mapsto \alpha^p$ where p is a prime?

8. Show that for given f and λ (λ = prime, $f|(\lambda - 1)$) the period η_0 is independent of the choice of primitive root γ.

4.6 Factorization of primes $p \not\equiv 1 \bmod \lambda$

Let $h(\alpha)$ be any prime cyclotomic integer (for a given fixed λ). The arguments of Section 4.3 suffice to show that there is a prime integer p with the property that $h(\alpha)$ divides an integer u if and only if $u \equiv 0 \bmod p$; in other words, $u \equiv v \bmod h(\alpha)$ for *integers* u and v is equivalent to $u \equiv v \bmod p$. For this it suffices to note first that $h(\alpha)$ divides the integer $Nh(\alpha)$ and therefore divides at least one of the prime integer factors of $Nh(\alpha)$, say $h(\alpha)|p$. For any given integer u, then, $h(\alpha)|u$ implies that $h(\alpha)$ divides the greatest common divisor d of u and p because $d = ap + bu$; on the other hand p is prime so that $d = 1$ or p, and, since $h(\alpha)$ is not a unit, it follows that $d \neq 1$, $d = p$, $u \equiv 0 \bmod p$. Conversely, of course, $u \equiv 0 \bmod p$ implies $u \equiv 0 \bmod h(\alpha)$ and the two statements are equivalent.

In the case of the prime factors $h(\alpha)$ studied in Section 4.3 this fact could be made the basis of a test for divisibility by $h(\alpha)$ because for them every cyclotomic integer was congruent mod $h(\alpha)$ to an ordinary integer. This applies *only* to primes $h(\alpha)$ which divide prime integers $p \equiv 0$ or $1 \bmod \lambda$, however, and other methods are needed in order to find prime factors $h(\alpha)$ of prime integers $p \not\equiv 0$ or $1 \bmod \lambda$.

It is useful to ask, for a given prime $h(\alpha)$ dividing p, which cyclotomic integers $g(\alpha)$ are congruent to integers mod $h(\alpha)$? A necessary condition for $g(\alpha) \equiv u \bmod h(\alpha)$ follows immediately from Fermat's theorem $u^p \equiv u \bmod p$ because this implies $g(\alpha)^p \equiv u^p \equiv u \equiv g(\alpha) \bmod h(\alpha)$. In testing whether this necessary condition is satisfied it is very helpful to take note of the fact that

$$g(\alpha)^p \equiv g(\alpha^p) \bmod p.$$

This remarkable identity is actually nothing more than a generalization of Fermat's theorem. It can be proved as follows.

Consider $(X + Y)^p - X^p - Y^p$ as a polynomial in two variables with integer coefficients. Each term is of degree p and there are no terms in X^p or Y^p, so the polynomial is of the form $c_1 X^{p-1} Y + c_2 X^{p-2} Y^2 + \cdots + c_{p-1} X Y^{p-1}$ where the c_i are positive integers. When $Y = 1$ and X is an integer, say $X = u$, the value of the polynomial is on the one hand equal to $c_1 u^{p-1} + c_2 u^{p-2} + \cdots + c_{p-1} u$ and on the other hand equal to $(u + 1)^p - u^p - 1^p \equiv (u + 1) - u - 1 = 0 \bmod p$. By Euler's differencing argument of Section 2.4, the only way that a polynomial of degree less than p can have all its values for integer values of the variable congruent to zero mod p is for all its coefficients to be zero mod p; thus* $c_1 \equiv c_2 \equiv \cdots \equiv c_{p-1} \equiv 0 \bmod p$. Thus for any two cyclotomic integers $g_0(\alpha), g_1(\alpha)$ it follows that $(g_0(\alpha) + g_1(\alpha))^p - g_0(\alpha)^p - g_1(\alpha)^p = c_1 g_0(\alpha)^{p-1} g_1(\alpha) + \cdots + c_{p-1} g_0(\alpha) g_1(\alpha)^{p-1} \equiv 0 \bmod p$ and consequently

$$(g_0(\alpha) + g_1(\alpha))^p \equiv g_0(\alpha)^p + g_1(\alpha)^p \bmod p.$$

Therefore, if $g(\alpha) = a_0 + a_1 \alpha + \cdots + a_{\lambda-1} \alpha^{\lambda-1}$, one has $g(\alpha)^p \equiv a_0^p + (a_1 \alpha + a_2 \alpha^2 + \cdots + a_{\lambda-1} \alpha^{\lambda-1})^p \equiv a_0^p + a_1^p \alpha^p + (a_2 \alpha^2 + \cdots + a_{\lambda-1} \alpha^{\lambda-1})^p \equiv \cdots \equiv a_0^p + a_1^p \alpha^p + a_2^p \alpha^{2p} + \cdots + a_{\lambda-1}^p (\alpha^p)^{\lambda-1}$. Since $a_i^p \equiv a_i \bmod p$, this says $g(\alpha)^p \equiv g(\alpha^p) \bmod p$, as was to be shown.

---

*In other words, the binomial coefficients $\binom{p}{j}$ are divisible by p for $j = 1, 2, \ldots, p - 1$. An alternative method for proving Fermat's theorem is to deduce this from the formula $\binom{p}{j} = p!/j!(p - j)!$. (Because p is prime, the factor of p in the numerator cannot be cancelled by the denominator.) It follows that $(x + y)^p \equiv x^p + y^p \bmod p$ for all integers x and y. Therefore $(x + y + z)^p \equiv (x + y)^p + z^p \equiv x^p + y^p + z^p \bmod p$ and, more generally, $(\sum x_i)^p \equiv \sum x_i^p \bmod p$ for any finite sum of integers $\sum x_i$. In particular, if $x = 1 + 1 + \cdots + 1$ then $x^p = (1 + 1 + \cdots + 1)^p \equiv 1^p + 1^p + \cdots + 1^p = x \bmod p$, which is Fermat's theorem.

Thus $g(\alpha)^p \equiv g(\alpha^p) \bmod h(\alpha)$, and a necessary condition for $g(\alpha) \equiv u \bmod h(\alpha)$ is $g(\alpha^p) \equiv g(\alpha) \bmod h(\alpha)$. A very simple way to find cyclotomic integers which satisfy this condition is to consider cyclotomic integers $g(\alpha)$ for which $g(\alpha^p)$ is *equal* to $g(\alpha)$, that is, cyclotomic integers $g(\alpha)$ that are invariant under the conjugation $\alpha \mapsto \alpha^p$. It is easy to derive necessary and sufficient conditions for $g(\alpha) = g(\alpha^p)$ as follows. Let τ denote the conjugation $\alpha \mapsto \alpha^p$, and let f be the least positive integer for which τ^f is the identity conjugation $\alpha \mapsto \alpha$; in other words, let f be the least positive integer such that $p^f \equiv 1 \bmod \lambda$. This integer f is called *the exponent of p mod λ*. Then $f|(\lambda - 1)$. It will be shown that $g(\alpha) = g(\alpha^p)$ *if and only if* $g(\alpha)$ *is a cyclotomic integer made up of periods of length f* where f is the exponent of p mod λ. Suppose first that $g(\alpha)$ is made up of periods of length f. Then, as was seen in the preceding section, $\sigma^e g(\alpha) = g(\alpha)$ where $e = (\lambda - 1)/f$ and where σ is the conjugation $\alpha \mapsto \alpha^\gamma$ for some primitive root γ mod λ. Since every conjugation is a power of σ, $\tau = \sigma^k$ for some k. Since $\tau^f = \sigma^{kf}$ is the identity, kf is divisible by $\lambda - 1 = ef$. Thus e divides k, τ is a power of σ^e, and $\tau g(\alpha) = g(\alpha)$, as was to be shown. Conversely, suppose $\tau g(\alpha) = g(\alpha)$. As was just shown, $e|k$. By definition $e|(\lambda - 1)$. Thus e is a common divisor of k and $\lambda - 1$. It is the greatest common divisor, because if d divides both, say $k = qd$ and $\lambda - 1 = df'$, then $\tau^{f'} = \sigma^{kf'} = \sigma^{qdf'} = (\sigma^{\lambda-1})^q$ is the identity, from which it follows that $f' \geq f$ and $d \leq e$. Thus $e = ak + b(\lambda - 1)$ for some integers a and b, $\sigma^e = \sigma^{ak}\sigma^{b(\lambda-1)} = \tau^a$, and $\sigma^e g(\alpha) = \tau^a g(\alpha) = g(\alpha)$, as was to be shown..

This shows, in particular, that cyclotomic integers made up of periods of length f satisfy the above necessary condition $g(\alpha) \equiv g(\alpha^p) \bmod h(\alpha)$ for $g(\alpha) \equiv$ integer mod $h(\alpha)$. Kummer proved that they are, in fact, congruent to integers mod $h(\alpha)$.

Theorem. *Let $h(\alpha)$ be a prime cyclotomic integer, let p be the prime integer which $h(\alpha)$ divides, let f be the exponent of p mod λ, and let η_i be a period of length f. Then there is an integer u_i such that $\eta_i \equiv u_i \bmod h(\alpha)$. Consequently, every cyclotomic integer made up of periods of length f is congruent to an integer mod $h(\alpha)$.*

PROOF. This theorem too depends on a generalization of Fermat's theorem, namely, the congruence
$$X^p - X \equiv [X - 1][X - 2] \cdots [X - p] \bmod p,$$
which means that the polynomial $X^p - X - (X - 1)(X - 2) \cdots (X - p)$ has all its coefficients divisible by p. Since this polynomial is of degree less than p, in order to prove the congruence it will suffice, as before, to prove that all its values for integer values of X are divisible by p. But all values of $X^p - X$ for integer values of X are zero mod p by Fermat's theorem, whereas all values of $[X - 1][X - 2] \cdots [X - p]$ are products of p consecutive integers—one of which must be a multiple of p—and are therefore zero mod p.

It follows that for any cyclotomic integer $g(\alpha)$, $g(\alpha)^p - g(\alpha) \equiv [g(\alpha)-1] \cdot$ $[g(\alpha) - 2] \cdots [g(\alpha) - p]$ mod p. Therefore $g(\alpha^p) - g(\alpha) \equiv [g(\alpha) - 1] \cdot$ $[g(\alpha) - 2] \cdots [g(\alpha) - p]$ mod p and *a fortiori* mod $h(\alpha)$. If $g(\alpha)$ is made up of periods of length f then $g(\alpha^p) - g(\alpha) = 0$ and it follows that $[g(\alpha) - 1][g(\alpha) - 2] \cdots [g(\alpha) - p] \equiv 0$ mod $h(\alpha)$. Since $h(\alpha)$ is prime, one of the factors $g(\alpha) - 1, g(\alpha) - 2, \ldots, g(\alpha) - p$ must be divisible by $h(\alpha)$; that is, $g(\alpha)$ is congruent to an integer mod $h(\alpha)$, as was to be shown. In particular, each period η_i is itself congruent to an integer; following Kummer, this integer will be denoted u_i.

The multiplication table for the periods η_i imply relations among the integers u_i mod p just as the relation $1 + \alpha + \alpha^2 + \cdots + \alpha^{\lambda-1} = 0$ implied in Section 4.3 that the integer k had to satisfy $1 + k + k^2 + \cdots + k^{\lambda-1} \equiv 0$ mod p. In the next section several specific examples are considered. In each of them it is possible to find all possible solutions $u_1, u_2, \ldots, u_e$ of these congruences mod p ($e = (\lambda - 1)/f$ is the number of distinct periods of length f) and therefore to determine which cyclotomic integers made up of periods of length f are divisible by a hypothetical prime factor $h(\alpha)$ of p without ever knowing $h(\alpha)$. In most cases it will even be possible to find a cyclotomic integer $g(\alpha)$ made up of periods of length f which is divisible by the hypothetical $h(\alpha)$ and which satisfies $Ng(\alpha) = p^f$. It will then follow that *if p has a prime factor $h(\alpha)$ then $h(\alpha)$ divides one of the conjugates of $g(\alpha)$*; since $Nh(\alpha)$ divides $Np = p^{\lambda-1}$ and since $Nh(\alpha) \equiv h(1)^{\lambda-1} \equiv 0$ or 1 mod $(\alpha - 1)$, $Nh(\alpha) \equiv 0$ or 1 mod λ, the norm of $h(\alpha)$ must be a power of p which is congruent to 1 mod λ, that is, $Nh(\alpha)$ must be a power of p^f. Therefore it follows from $h(\alpha)|g(\alpha)$ that $Nh(\alpha) = Ng(\alpha) = p^f$. In short, *if p has a prime divisor $h(\alpha)$ then $h(\alpha)$ is a unit times a conjugate of $g(\alpha)$*.

In the terminology of Section 4.3, this completes the *analysis*—it gives very strong necessary conditions satisfied by prime cyclotomic integers $h(\alpha)$, so strong that in the examples of the next section it will be possible to find all cyclotomic integers $h(\alpha)$ which could possibly be prime divisors of a given p. The analog of the *synthesis* of Section 4.4 would be to prove that these possible prime divisors are in fact prime; that is, if $g(\alpha)$ is a cyclotomic integer made up of periods of length f for which $Ng(\alpha) = p^f$ where p is a prime whose exponent mod λ is f, then $g(\alpha)$ is prime. Although this theorem is perfectly true, its proof is considerably more difficult than the proof of the special case $f = 1$ of Section 4.3 and this does not seem to be the appropriate point of the development to enter into this proof. Instead it will be postponed to Section 4.12. This means that in the next two sections the cyclotomic integers $h(\alpha)$ satisfying $Nh(\alpha) = p^f$ will not have been proved to be prime, although they are in fact prime. It will be convenient, nonetheless, to refer to them as "primes" with this understanding that the proof that they are prime has been postponed.

EXERCISES

1. By using Pascal's triangle, find all binomial coefficients $\binom{n}{j}$ for $n \leqslant 13$. Verify directly that when n is prime and $0 < j < n$, the coefficient is divisible by n in all cases $n \leqslant 13$.

2. Calculate the polynomial $X(X - 1)(X - 2) \cdots (X - p + 1)$ for $p = 3, 5, 7, 11$. (Each calculation is part of the following one.) Verify in each case that it is $\equiv X^p - X \bmod p$.

3. Verify the formula $\eta^p \equiv \eta \bmod p$ in the case $\lambda = 7, p = 2$ by finding explicitly f, all periods of length f, and the pth power of each. Do the same for $\lambda = 13, p = 3$.

4. Show that if λ is any prime and k any integer then the prime factors of $k^{\lambda-1} + k^{\lambda-2} + \cdots + 1$ are all either 0 or 1 mod λ.

5. Show that in the case $\lambda = 5, p = 19$ if p has a prime factor then the integers u_i provided by the theorem of the text must satisfy $u^2 + u - 1 \equiv 0 \bmod 19$. Find the solutions u of this congruence. Show then that $\eta_0 - u$ divides 19 and find a factorization of 19. Show that the factors of this factorization are irreducible. (They are in fact prime, as will be proved later in this chapter.)

6. Using the procedure of the preceding exercise, find a factorization of 3 into irreducible factors in the case $\lambda = 13$. (Again, these factors are actually prime.)

7. Prove that if f is the exponent of p mod λ and if $g(\alpha)$ is a cyclotomic integer for which $Ng(\alpha) = p^f$ then $g(\alpha)$ is irreducible.

4.7 Computations when $p \not\equiv 1 \bmod \lambda$

Since the exponent f of a prime p mod λ must divide $\lambda - 1$, in the case $\lambda = 5$ the only possible values of f are 1, 2, 4. The case $f = 1$ was treated in Section 4.4. In the case $f = 4$ the assumption $h(\alpha)|p$, $Nh(\alpha)|p^4$, together with the fact that $Nh(\alpha) \equiv [h(1)]^4 \bmod \alpha - 1$, $Nh(\alpha) \equiv 0$ or 1 mod λ, implies $p^f|Nh(\alpha)|p^4$, $Nh(\alpha) = Np$, from which it follows that $h(\alpha)$ is a unit times p. In short, *the only possible prime divisors of primes $p = 2, 3, 7, 13, \ldots$ of exponent 4 mod 5 are units times p itself*. Only the case $f = 2$ remains in which to find possible prime factors of p. These are the primes $19, 29, 59, 79, 89, \ldots$, that are 4 mod 5. (1 has exponent 1, 4 has exponent 2, and 2 and 3 have exponent 4 mod 5.)

In the case $\lambda = 5, f = 2$, the periods of interest are $\eta_0 = \alpha + \alpha^4, \eta_1 = \alpha^2 + \alpha^3$. (The numbering of the periods is the same, regardless of whether one chooses $\gamma = 2$ or $\gamma = 3$.) In cases such as this one where $f = \frac{1}{2}(\lambda - 1), e = 2$, the periods will be denoted θ_0, θ_1 rather than η_0, η_1. Because $1 + \theta_0 + \theta_1 = 0$, any cyclotomic integer $a + b\theta_0 + c\theta_1$ made up of periods of length 2 can be written in the form $A + B\theta_0$ ($A = a - c, B = b - c$) and its norm can be written in the form $(A + B\theta_0)(A + B\theta_1)(A + B\theta_0)(A + B\theta_1) = [A^2 + AB(\theta_0 + \theta_1) + B^2\theta_0\theta_1]^2 = [A^2 - AB - B^2]^2$ since, as a simple computation shows, $\theta_0\theta_1 = -1$. As was explained at the end of the preceding section, the objective is to find, for a given p with exponent 2 mod 5, first the integers u_0, u_1 for which $\theta_0 \equiv u_0, \theta_1 \equiv u_1 \bmod h(\alpha)$ are possible, and then to find a cyclotomic integer $g(\alpha) = a + b\theta_0 + c\theta_1$ which satisfies $a + bu_0 + cu_1 \equiv 0 \bmod p$ and which satisfies $Ng(\alpha) = p^2$. In particular, then, the objective is to find

Table 4.7.1

A	B	$A^2 - AB - B^2$
2	1	1
2	-1	5
3	1	5
3	-1	11
3	2	-1
3	-2	11
4	1	11
4	-1	19
4	3	-5
4	-3	19
5	1	19
5	-1	29

for each $p = 19, 29, 59, \ldots$ a cyclotomic integer $A + B\theta_0$ for which $A^2 - AB - B^2$ $= \pm p$.

As with the case $f = 1$, one can make a certain amount of progress by simple trial and error. Table 4.7.1 shows the values of $A^2 - AB - B^2$ for some small values of A and B. (One may as well assume that A and B are relatively prime, that A is the larger of the two in absolute value, and that $A > 0$.) Since $(4 - \theta_0)(4 - \theta_1)$ $= 19$, the arguments of the preceding section show that the only possible prime factors of 19 are units times conjugates of $4 - \theta_0$. Modulo $4 - \theta_0$ one has, obviously, $\theta_0 \equiv 4$ and $\theta_1 = -1 - \theta_0 \equiv -1 - 4 = -5$. Therefore $a + b\theta_0 + c\theta_1 \equiv a$ $+ 4b - 5c$ and $a + b\theta_0 + c\theta_1$ is divisible by $4 - \theta_0$ if and only if $a + 4b - 5c \equiv$ 0 mod 19. Similarly, the only possible prime factors of 29 are units times conjugates of $5 - \theta_0$, and $5 - \theta_0$ divides $a + b\theta_0 + c\theta_1$ if and only if $a + 5b - 6c \equiv$ 0 mod 29. This gives prime* factorizations $19 = (4 - \theta_0)(4 - \theta_1)$ and $29 = (5 - \theta_0)(5 - \theta_1)$ of 19 and 29, together with tests for divisibility by the prime factors.

Table 4.7.1 would need to be extended quite far in order to reach the next prime 59 with exponent 2 mod 5. Therefore the method suggested in the preceding section should be tried. This method is to observe that if the prime factor were known then there would be integers u_0 and u_1 to which the periods θ_0 and θ_1 were congruent modulo the factor. Since $1 + \theta_0 + \theta_1 = 0$ and $\theta_0\theta_1 = -1$ these integers would have to satisfy $1 + u_0 + u_1 = 0$, $u_0 u_1 \equiv -1$ mod the factor of 59 and therefore mod 59. These two congruences in two unknowns can be reduced to one in one unknown, $u_0(-1 - u_0) \equiv -1$, $u_0^2 + u_0 - 1 \equiv 0$ mod 59. A solution of this congruence can be found as follows. $u_0(u_0 + 1) = 1 + 59n$ for some n. Since $u_0(u_0 + 1)$ is even, n must be odd. Since $u_0(u_0 + 1)$ is positive or zero ($u_0 =$ integer) n must be positive. Trying $n = 1, 3, 5, \ldots$, one finds $60 = 2^2 \cdot 3 \cdot 5$, $178 = 2 \cdot 89$, $296 = 2^3 \cdot 37$, $414 = 2 \cdot 3^2 \cdot 23$, $532 = 2^2 \cdot 7 \cdot 19$, $650 = 2 \cdot 5^2 \cdot 13, \ldots$. The first of these numbers which can be written in the form $u_0(u_0 + 1)$ is $650 = 25 \cdot 26$. This gives the solution $u_0 = 25$, $u_1 = -26$. The congruence $u^2 + u - 1 \equiv 0$ mod 59 has *at most* 2 solutions; therefore it has only the solutions $25, -26$. Since $u_0 \equiv 25$ implies

---

*As was explained at the end of the preceding section, the proof that these factors actually are prime will be given in Section 4.12.

115

$u_1 \equiv -26$ and $u_0 \equiv -26$ implies $u_1 \equiv 25$, these are the *only* possible values of u_0, u_1 mod 59. A cyclotomic integer of the form $a + b\theta_0 + c\theta_1$ is divisible by the hypothetical factor of 59 mod which $u_0 \equiv 25$ if and only if $a + 25b - 26c \equiv 0$ mod 59. For example, $25 - \theta_0$ is divisible by it. Now $N(25 - \theta_0) = [625 + 25 - 1]^2 = 11^2 \cdot 59^2$. Therefore the desired factor of 59 can be found by dividing by two suitable factors of 11. Modulo the prime factor $\alpha + 2$ of 11 (see Table 4.4.1) one has $\theta_0 = \alpha + \alpha^4 \equiv -2 + (-2)^4 \equiv 3$ and $\theta_1 = \alpha^2 + \alpha^3 \equiv -4$. Therefore $25 - \theta_0$ is divisible by $\alpha + 2$ because $25 - 3 \equiv 0$ mod 11. It is also divisible by $\alpha^4 + 2$ because the conjugation $\alpha \mapsto \alpha^4$ leaves θ_0 and θ_1 invariant. Therefore the desired factor must be

$$\frac{25 - \theta_0}{(\alpha + 2)(\alpha^4 + 2)} = \frac{25 - \theta_0}{1 + 2\theta_0 + 4} = \frac{(25 - \theta_0)(5 + 2\theta_1)}{(5 + 2\theta_0)(5 + 2\theta_1)}$$

$$= \frac{125 - 5\theta_0 + 50\theta_1 + 2}{25 - 10 - 4} = \frac{132 + 55\theta_1}{11} = 12 + 5\theta_1.$$

Thus $59 = (12 + 5\theta_1)(12 + 5\theta_0)$ is the desired prime factorization of 59.

The next prime to be factored is 79. The technique used in the case of 59 can be used to find $11 \cdot 79 + 1 = 870 = 29 \cdot 30$. The divisibility test is therefore $a + 29b - 30c \equiv 0$ mod 79. A solution of this congruence in relatively small integers is $b = 3$, $a = -8$, $c = 0$. Since $(-8 + 3\theta_0)(-8 + 3\theta_1) = 64 + 24 - 9 = 79$, this gives a factorization of 79. The next prime 89 is very easy to factor: $89 + 1 = 9 \cdot 10$, the divisibility test is $a + 9b - 10c \equiv 0$ mod 89, $\theta_0 - 9$ is divisible, and $(\theta_0 - 9)(\theta_1 - 9) = 89$. This completes the factorization of the primes less than 100. The results are shown in Table 4.7.2 (and in Table 4.4.1).

When $\lambda = 7$ the exponent f of p mod λ must be 1, 2, 3, or 6 because $f | (\lambda - 1) = 6$. The case $f = 1$ was treated in Section 4.4 and the case $f = 6$ is of no interest because in these cases p can have no factorization. The case $f = 3$ occurs when $p \equiv 2$ or 4 mod 7. Thus $p = 2, 11, 23, 37, 53, \ldots$. The case $f = 2$ occurs when $p \equiv -1$ mod 7, $p = 13, 41, 83, \ldots$. The case $f = 3$ will be considered first.

The periods of length 3 are $\theta_0 = \alpha + \alpha^2 + \alpha^4$, $\theta_1 = \alpha^3 + \alpha^5 + \alpha^6$ (no matter whether $\gamma = 3$ or $\gamma = 5$ is used). $1 + \theta_0 + \theta_1 = 0$ and $\theta_0\theta_1 = \alpha^4 + \alpha^6 + 1 + \alpha^5 + 1 + \alpha + 1 + \alpha^2 + \alpha^3 = 2$. Therefore every cyclotomic integer made up of periods of length 3 is of the form $A + B\theta_0$ and its norm is $(A + B\theta_0)^3(A + B\theta_1)^3 = [A^2 - AB + 2B^2]^3$. In this case it turns out that all the small primes can be factored very simply by trial and error. What is being sought is a pair of integers A, B such that $A^2 - AB + 2B^2 = p$ where $p = 2, 11, 23, 37, \ldots$. The solution for $p = 2$ is immediate and the corresponding factorization of 2 is simply $2 = \theta_0\theta_1$. For odd

Table 4.7.2

$\lambda = 5, f = 2$

prime	norm	$\theta_0 \equiv$	$\theta_1 \equiv$
$\theta_0 - 4$	19^2	4	-5
$\theta_0 - 5$	29^2	5	-6
$5\theta_0 + 12$	59^2	-26	25
$3\theta_0 - 8$	79^2	29	-30
$\theta_0 - 9$	89^2	9	-10

Table 4.7.3

A	B	$A^2 - AB + 2B^2$
1	2	7
1	4	29
1	6	67
3	2	11
3	4	29
5	2	23
5	4	37
5	6	67
7	2	43
7	4	53
7	6	79
9	2	71
9	4	77

primes it is clear that A must be odd and B must be even. Since they must also be relatively prime, this limits the trials to those shown in Table 4.7.3 (for $p < 100$). The desired factorizations can then be read off from the table; $67 = (1 + 6\theta_0)(1 + 6\theta_1)$, $11 = (3 + 2\theta_0)(3 + 2\theta_1)$, and so forth. To find the integers to which θ_0 and θ_1 are congruent modulo these primes one need only solve a simple linear congruence. For example, if $\theta_0 \equiv k \bmod 1 + 6\theta_0$ then $1 + 6k \equiv 0 \bmod 1 + 6\theta_0$, $6k \equiv -1 \bmod 67$, $66k \equiv -11 \bmod 67$, $k \equiv 11 \bmod 67$. These factorizations are tabulated in Table 4.7.4.

Now consider the case $\lambda = 7$, $f = 2$. The periods of length 2 are $\eta_0 = \alpha + \alpha^6$, $\eta_1 = \alpha^3 + \alpha^4$, $\eta_2 = \alpha^2 + \alpha^5$ when the primitive root $\gamma = 3$ is used. (If $\gamma = 5$ is used, the periods are the same but their names are changed.) They satisfy the equations $1 + \eta_0 + \eta_1 + \eta_2 = 0$, $\eta_0^2 = 2 + \eta_2$, and $\eta_0 \eta_1 = \eta_1 + \eta_2$, as is easily found by direct computation. Other relations they satisfy can be found by taking the conjugates of these, which amounts to performing a cyclic permutation of the η's—$\eta_1^2 = 2 + \eta_0$, $\eta_2^2 = 2 + \eta_1$, $\eta_1 \eta_2 = \eta_2 + \eta_0$, $\eta_2 \eta_0 = \eta_0 + \eta_1$. The first prime to be factored is 13. If u_0, u_1, u_2 are the integers to which η_0, η_1, η_2 are congruent modulo a prime factor

Table 4.7.4

$\lambda = 7$, $f = 3$

prime	norm	$\theta_0 \equiv$	$\theta_1 \equiv$
θ_0	2^3	0	1
$2\theta_0 + 3$	11^3	4	-5
$2\theta_0 + 5$	23^3	9	-10
$4\theta_0 + 5$	37^3	8	-9
$4\theta_0 + 7$	53^3	-15	14
$6\theta_0 + 1$	67^3	11	-12
$6\theta_0 + 7$	79^3	12	-13

of 13 (assuming there is one) then

$$1 + u_0 + u_1 + u_2 \equiv 0 \bmod 13$$

$$u_0^2 \equiv 2 + u_2 \bmod 13$$

$$u_0 u_1 \equiv u_1 + u_2 \bmod 13$$

and, similarly, congruences obtained from these by cyclic permutations of the subscripts hold. Thus

$$u_0^3 \equiv 2u_0 + u_0 u_2 \equiv 2u_0 + u_0 + u_1$$

$$\equiv -1 + 2u_0 - u_2$$

$$\equiv -1 + 2u_0 + 2 - u_0^2$$

$$u_0^3 + u_0^2 - 2u_0 - 1 \equiv 0 \bmod 13.$$

This congruence is not satisfied by $u_0 = 0$, ± 1, ± 2, or 3, but it is satisfied by $u_0 = -3$. Then the other two values $u_2 \equiv u_0^2 - 2 \equiv 7 \equiv -6 \bmod 13$ and $u_1 \equiv -1 - u_0 - u_2 \equiv -1 + 3 + 6 = 8 \equiv -5$ are forced. Note that, as expected, $u_0 = -5$ and $u_0 = -6$ are also solutions of the congruence for u_0 and if either of them is used the other two are forced. Therefore, because a congruence of degree 3 has at most 3 solutions modulo a prime, the *only possible* solutions of the congruences satisfied by u_0, u_1, u_2 are $-3, -5, -6$; $-5, -6, -3$; and $-6, -3, -5$. Therefore $a + b\eta_0 + c\eta_1 + d\eta_2$ is divisible by a prime divisor of 13 only if it or one of its conjugates satisfies $a - 3b - 5c - 6d \equiv 0 \bmod 13$. For example, $1 - \eta_1 + \eta_2$ or one of its conjugates would be divisible by a prime factor of 13. The product of the 3 distinct conjugates is

$$(1 - \eta_1 + \eta_2)(1 - \eta_2 + \eta_0)(1 - \eta_0 + \eta_1)$$

$$= (1 - \eta_1 + \eta_2)(1 - \eta_0 + \eta_1 - \eta_2 + \eta_0 + \eta_1 - \eta_2 - \eta_0 + \eta_0 - 2 - \eta_2 + \eta_1 + \eta_2)$$

$$= (1 - \eta_1 + \eta_2)(-1 + 3\eta_1 - 2\eta_2)$$

$$= -1 + 3\eta_1 - 2\eta_2 + \eta_1 - 3(2 + \eta_0) + 2(\eta_2 + \eta_0) - \eta_2 + 3(\eta_2 + \eta_0) - 2(2 + \eta_1)$$

$$= -11 + 2\eta_0 + 2\eta_1 + 2\eta_2 = -13.$$

This gives the factorization $13 = (\eta_1 - \eta_2 - 1)(\eta_2 - \eta_0 - 1)(\eta_0 - \eta_1 - 1)$.

The next prime with exponent 2 mod 7 is 41. As for 13, u must satisfy $u_0^3 + u_0^2 - 2u_0 - 1 \equiv 0 \bmod 41$. The values $u_0 \equiv 0, \pm 1, \pm 2, \pm 3, 4$ are quickly rejected, but $u_0 \equiv -4$ does satisfy the congruence. Then $u_2 \equiv u_0^2 - 2 \equiv 14$ and $u_1 \equiv -1 - u_0 - u_2 \equiv -11$ are forced. Again $-4, -11, 14$ are the only solutions of the congruence satisfied by u_0, after which the values of u_1 and u_2 are forced. Therefore, if $a + b\eta_0 + c\eta_1 + d\eta_2$ is divisible by a prime divisor of 41, it or one of its conjugates must satisfy $a - 4b - 11c + 14d \equiv 0 \bmod 41$. For example, one might try $4 + \eta_0$. This choice succeeds because

$$(4 + \eta_0)(4 + \eta_1)(4 + \eta_2) = (4 + \eta_0)(16 + 4\eta_1 + 4\eta_2 + \eta_2 + \eta_0)$$

$$= (4 + \eta_0)(12 - 3\eta_0 + \eta_2) = 48 - 12\eta_0 + 4\eta_2 + 12\eta_0 - 3(2 + \eta_2) + (\eta_0 + \eta_1)$$

$$= 42 + \eta_0 + \eta_1 + \eta_2 = 41$$

factors 41.

The only other primes less than 100 with exponent 2 mod 7 are 83 and 97. The solution of the congruence $u_0^3 + u_0^2 - 2u_0 - 1 \equiv 0 \bmod 83$ by simple trial and error requires some patience. Eventually one finds the solution $u_0 \equiv 10$, after which $u_2 \equiv 100 - 2 \equiv 15$, $u_1 \equiv -1 - 10 - 15 = -26$ are the other 2 solutions. This leads

Table 4.7.5

$\lambda = 7, f = 2$

prime	norm	$\eta_0 \equiv$	$\eta_1 \equiv$	$\eta_2 \equiv$
$\eta_1 - \eta_2 - 1$	13^2	-3	-5	-6
$\eta_0 + 4$	41^2	-4	-11	14
$\eta_0 - \eta_2 + 5$	83^2	10	-26	15
$4\eta_0 - 3$	97^2	25	30	41

to the consideration of those $a + b\eta_0 + c\eta_1 + d\eta_2$ for which $a + 10b - 26c + 15d \equiv 0 \bmod 83$, for example, $5 + \eta_0 - \eta_2$. A brief computation shows that $(5 + \eta_0 - \eta_2)(5 + \eta_1 - \eta_0)(5 + \eta_2 - \eta_1) = 83$ and $(4\eta_0 - 3)(4\eta_1 - 3)(4\eta_2 - 3) = 97$. This completes the factorization of primes $p < 100$ when $\lambda = 7$. The results are given in Tables 4.7.5, 4.7.4, and 4.4.2.

As another example, consider the case $\lambda = 13, p = 29$. In this case $p \equiv 3, p^2 \equiv 9, p^3 \equiv 1 \bmod \lambda$. Therefore $f = 3$. As was found in Section 4.4, the periods of length 3 are $\eta_0 = \alpha + \alpha^3 + \alpha^9$, $\eta_1 = \alpha^2 + \alpha^5 + \alpha^6$, $\eta_2 = \alpha^4 + \alpha^{10} + \alpha^{12}$, $\eta_3 = \alpha^8 + \alpha^7 + \alpha^{11}$ when either $\gamma = 2$ or 6 is used as the primitive root mod 13. (If $\gamma = -2$ or -6 then the periods are the same but their names are changed.) Moreover, the relations they satisfy are

$$1 + \eta_0 + \eta_1 + \eta_2 + \eta_3 = 0$$
$$\eta_0^2 = \eta_1 + 2\eta_2$$
$$\eta_0\eta_1 = \eta_0 + \eta_1 + \eta_3$$
$$\eta_0\eta_2 = 3 + \eta_1 + \eta_3$$
$$\eta_0\eta_3 = (\text{from } \eta_0\eta_1)$$

and relations obtained from these by cyclic permutations $\eta_0 \mapsto \eta_1 \mapsto \eta_2 \mapsto \eta_3 \mapsto \eta_0$. If u_0, u_1, u_2, u_3 are the integers to which $\eta_0, \eta_1, \eta_2, \eta_3$ are congruent modulo a prime factor of 29 (assuming 29 has a prime factor) then

$$1 + u_0 + u_1 + u_2 + u_3 \equiv 0 \bmod 29$$
$$u_0^2 \equiv u_1 + 2u_2 \bmod 29$$
$$u_0u_1 \equiv u_0 + u_1 + u_3 \bmod 29$$
$$u_0u_2 \equiv 3 + u_1 + u_3$$

all hold, as do congruences obtained from these by cyclic permutation of the u's. Then

$$u_0^3 \equiv u_0u_1 + 2u_0u_2 \equiv u_0 + u_1 + u_3 + 2(3 + u_1 + u_3)$$
$$\equiv 6 + u_0 + 3u_1 + 3u_3 \equiv 3 - 2u_0 - 3u_2$$
$$u_0^4 \equiv 3u_0 - 2(u_1 + 2u_2) - 3(3 + u_1 + u_3)$$
$$\equiv -9 + 3u_0 - 5u_1 - 4u_2 - 3u_3$$
$$\equiv -6 + 6u_0 - 2u_1 - u_2$$
$$u_0^4 + 2u_0^2 \equiv -6 + 6u_0 + 3u_2 \equiv -6 + 6u_0 + (3 - 2u_0 - u_0^3)$$
$$u_0^4 + u_0^3 + 2u_0^2 - 4u_0 + 3 \equiv 0 \bmod 29.$$

One quickly finds that $u_0 \equiv 0$, ± 1, ± 2, 3 do not satisfy this congruence but that $u_0 \equiv -3$ does. Then $3u_2 \equiv 3 - 2u_0 - u_0^3 \equiv 3 + 6 - 2 = 7$, $30u_2 \equiv 70 \equiv 12$, $u_2 \equiv 12$, $u_1 \equiv u_0^2 - 2u_2 \equiv 9 - 24 \equiv 14$, and $u_3 \equiv -1 - u_0 - u_1 - u_2 \equiv -1 + 3 - 12 - 14 \equiv 5$ are all forced. The only possible solutions of the congruences satisfied by the u's are therefore $u_0 \equiv -3$, $u_1 \equiv 14$, $u_2 \equiv 12$, $u_3 \equiv 5 \bmod 29$, and cyclic permutations of these, because these 4 integers are the only solutions of the congruence $u_0^4 + u_0^3 + 2u_0^2 - 4u_0 + 3 \equiv 0 \bmod 29$, after which the values of the other 3 are forced. Therefore if $a + b\eta_0 + c\eta_1 + d\eta_2 + e\eta_3$ is divisible by a prime factor of 29, either it or one of its conjugates satisfies $a - 3b + 14c + 12d + 5e \equiv 0 \bmod 29$. For example, one might try $3 + \eta_0$. Now $(3 + \eta_0)(3 + \eta_2) = 9 + 3\eta_2 + 3\eta_0 + 3 + \eta_1 + \eta_3 = 12 + 3\theta_0 + \theta_1 = 11 + 2\theta_0$ where $\theta_0 = \eta_0 + \eta_2$, $\theta_1 = \eta_1 + \eta_3$. The product of the 4 conjugates of $3 + \eta_0$ is therefore $(11 + 2\theta_0)(11 + 2\theta_1) = 121 - 22 + 4\theta_0\theta_1 = 121 - 34 = 87 = 3 \cdot 29$, because $\theta_0\theta_1 = -3$. In order to find a factor of 29 it is necessary to divide $3 + \eta_0$ by an appropriate factor of 3.

Most of the work required to factor 3 has already been done. The exponent of 3 mod 13 is 3 and the factorization depends on solving the congruence $u_0^4 + u_0^3 + 2u_0^2 - 4u_0 + 3 \equiv 0 \bmod 3$. $u_0 \equiv 0$ is a solution. Then $u_1 + u_3 \equiv 0$, $u_2 \equiv -1 - u_0 - u_1 - u_3 \equiv -1$, $u_1 \equiv u_0^2 - 2u_2 \equiv -1$, $u_3 \equiv -u_1 \equiv 1$ are all forced. Therefore one looks among cyclotomic integers $a + b\eta_0 + c\eta_1 + d\eta_2 + e\eta_3$ for which $a - c - d + e \equiv 0 \bmod 3$ for a factor of 3. For example, let $b = 1$, $a = c = d = e = 0$. Then $\eta_0\eta_2 = 3 + \theta_1$, $\eta_0\eta_1\eta_2\eta_3 = (3 + \theta_0)(3 + \theta_1) = 9 - 3 - 3 = 3$. Therefore η_0 is a prime divisor of 3 mod which $\eta_0 \equiv 0$, $\eta_1 \equiv -1$, $\eta_2 \equiv -1$, $\eta_3 \equiv 1$. Therefore $3 + \eta_0 \equiv 3 \equiv 0 \bmod \eta_0$. Division of $3 + \eta_0 = \eta_0\eta_1\eta_2\eta_3 + \eta_0$ by η_0 gives $\eta_1\eta_2\eta_3 + 1 = \eta_2(3 + \eta_0 + \eta_2) + 1 = 3\eta_2 + 3 + \eta_1 + \eta_3 + \eta_3 + 2\eta_0 + 1 = 4 + 2\eta_0 + \eta_1 + 3\eta_2 + 2\eta_3 = 2 - \eta_1 + \eta_2$ as a factor of 29 mod which $\eta_0 \equiv -3$, $\eta_1 \equiv 14$, $\eta_2 \equiv 12$, $\eta_3 \equiv 5$.

In cases with $f > 1$, just as in cases with $f = 1$, there are cases in which this method fails to produce a factorization of p. An example of this is provided by the case $\lambda = 31$, $p = 2$. Here $f = 5$ and when the periods of length 5 are constructed using the primitive root 3 mod 31 they satisfy

$$\eta_0 = \alpha + \alpha^{-15} + \alpha^8 + \alpha^4 + \alpha^2 \qquad\qquad \eta_0^2 = \eta_0 + 2\eta_1 + 2\eta_2$$
$$\eta_1 = \alpha^3 + \alpha^{-14} + \alpha^{-7} + \alpha^{12} + \alpha^6 \qquad \eta_0\eta_1 = \eta_0 + \eta_2 + 2\eta_4 + \eta_5$$
$$\eta_2 = \alpha^9 + \alpha^{-11} + \alpha^{10} + \alpha^5 + \alpha^{-13} \qquad \eta_0\eta_2 = \eta_1 + \eta_2 + \eta_4 + 2\eta_5$$
$$\eta_3 = \alpha^{-4} + \alpha^{-2} + \alpha^{-1} + \alpha^{15} + \alpha^{-8} \qquad \eta_0\eta_3 = 5 + \eta_0 + \eta_1 + \eta_3 + \eta_4$$
$$\eta_4 = \alpha^{-12} + \alpha^{-6} + \alpha^{-3} + \alpha^{14} + \alpha^7 \qquad \eta_0\eta_4 = (\text{from } \eta_0\eta_2)$$
$$\eta_5 = \alpha^{-5} + \alpha^{13} + \alpha^{-9} + \alpha^{11} + \alpha^{-10} \qquad \eta_0\eta_5 = (\text{from } \eta_0\eta_1)$$

Suppose that u_0, u_1, u_2, u_3, u_4, u_5 are integers which satisfy the equations relating the η's as congruences mod 2. One can assume that $u_i = 0$ or 1. Regarding the u's as cyclically ordered, one sees that the solution must contain either two consecutive zeros or two consecutive ones because 0, 1, 0, 1, 0, 1 and 1, 0, 1, 0, 1, 0 do not satisfy $u_0u_3 \equiv 5 + u_0 + u_1 + u_3 + u_4 \bmod 2$. If one assumes $u_0 = u_1 = 0$ then $0 \equiv 0 + u_2 + 2u_4 + u_5$, $0 \equiv 0 + u_2 + u_4 + 2u_5$, and $0 \equiv 5 + 0 + 0 + u_3 + u_4$ imply $u_2 \equiv u_5 \equiv u_4 \equiv 1 + u_3$. From $\eta_2\eta_4 = \eta_3 + \eta_4 + \eta_0 + 2\eta_1$ it follows that $u_2u_4 \equiv 1 \bmod 2$ and the solution 0, 0, 1, 0, 1, 1 is forced. Similarly, if $u_0 \equiv 1$, $u_1 \equiv 1$ then 1, 1, 0, 0, 1, 0 is forced. Thus, up to cyclic permutations, there is only one possible set of u's mod 2. (This does *not*, however, prove that this set of u's satisfies *all* relations among the η's as congruences mod 2. That they in fact do follows from the proposition of Section 4.9 below.) Therefore if $a + b\eta_0 + c\eta_1 + d\eta_2 + e\eta_3 + f\eta_4 +$

$g\eta_5$ is divisible by a factor of 2 then it or one of its conjugates satisfies $a + d + f + g \equiv 0 \bmod 2$. A simple candidate is η_0. However, $\eta_0\eta_2\eta_4 = \eta_0(\eta_3 + \eta_4 + \eta_0 + 2\eta_1) = 5 + \eta_0 + \eta_1 + \eta_3 + \eta_4 + \eta_5 + \eta_0 + \eta_2 + 2\eta_3 + \eta_0 + 2\eta_1 + 2\eta_2 + 2(\eta_0 + \eta_2 + 2\eta_4 + \eta_5) = 5 + 5\eta_0 + 3\eta_1 + 5\eta_2 + 3\eta_3 + 5\eta_4 + 3\eta_5 = 2 + 2\theta_0$ (where $\theta_0 = \eta_0 + \eta_2 + \eta_4$) and $\eta_0\eta_1\eta_2\eta_3\eta_4\eta_5 = (2 + 2\theta_0)(2 + 2\theta_1) = 4 - 4 + 4\theta_0\theta_1 = 32$ because $\theta_0\theta_1 = 8$. Therefore not only does this proposed factor fail, but any proposed factor of this type fails because the product of 3 of its conjugates will have the form $X + Y\theta_0$ and the product of all 6 will have the form $(X + Y\theta_0)(X + Y\theta_1) = X^2 - XY + 8Y^2 = (4X^2 - 4XY + 32Y^2)/4 = [(2X - Y)^2 + 31Y^2]/4$ for some integers X and Y. Since 8 cannot be written in the form $a^2 + 31b^2$ for integers a and b, this shows that the product of the 6 conjugates cannot be 2. Unlike the case $f = 1$, however, in this case it is not possible to conclude in a simple way that 2 has *no* prime factorization but only that it has none in which the factors are made up of periods of length 5.

EXERCISES

1. Table 4.7.5 shows that, when $\lambda = 7$, $\eta_2 - \eta_1 + 1$ divides $\eta_0 + 3$. Show that the quotient is a unit and find it and its inverse explicitly.

2. Derive the last line of Table 4.7.5, that is, the factorization of 97 when $\lambda = 7$.

3. In the case $\lambda = 13$ find the one prime p remaining, after 3 and 29, which has exponent 3 and is less than 100. Write it as a product of 4 factors each of norm p^3. [This computation is rather long, but comparable to the case $p = 29$ treated in the text. The congruence for u is solved by $u = 10$ and for no u smaller than 10 in absolute value.]

4. Extend Table 4.7.2 to cover the next prime with exponent 2 mod 5.

5. For $\lambda = 13$, what are the possible exponents mod λ? Find the exponent of each prime less than 100. For each exponent greater than 2, factor the first prime which has that exponent.

4.8   Extension of the divisibility test

As before, let p and λ be distinct primes, $\lambda > 2$, and let f be the exponent of p mod λ. It was shown in Section 4.6 that if p has a prime factor $h(\alpha)$ in cyclotomic integers built up from a λth root of unity α then there is a simple way to test whether a given cyclotomic integer $a_0 + a_1\eta_1 + \cdots + a_e\eta_e$ made up of periods of length f is divisible by $h(\alpha)$; namely, there exist integers $u_1, u_2, \ldots, u_e$ such that $a_0 + a_1\eta_1 + \cdots + a_e\eta_e$ is divisible by $h(\alpha)$ if and only if $a_0 + a_1u_1 + \cdots + a_eu_e \equiv 0 \bmod p$. Otherwise stated, $\eta_i \equiv u_i \bmod h(\alpha)$ and for integers x, $x \equiv 0 \bmod h(\alpha)$ is equivalent to $x \equiv 0 \bmod p$. The objective of this section is to show how this divisibility test can be extended to a test which applies to all cyclotomic integers $g(\alpha)$, not just to those made up of periods of length f.

The device by which Kummer accomplished this extension is based on the following idea which comes from Gauss's study of cyclotomy (*Disquisitiones Arithmeticae*, Art. 348). Let $\eta_0 = \eta_e$ denote the period $\eta_0 = \alpha + \sigma^e\alpha + $

$\sigma^{2e}\alpha + \cdots + \sigma^{-e}\alpha$ of length f which contains α and let $P(X)$ be the polynomial in X with cyclotomic integer coefficients obtained by expanding the product $\prod(X - \alpha^j)$ where α^j ranges over all powers of α contained in η_0; in other words

$$P(X) = \prod_{i=1}^{f} (X - \sigma^{ei}\alpha)$$

where, as usual, $e = (\lambda - 1)/f$ and σ is a conjugation $\alpha \mapsto \alpha^\gamma$ in which γ is a primitive root mod λ. For example, in the case $\lambda = 13$, $f = 3$ one has $\eta_0 = \alpha + \alpha^3 + \alpha^{-4}$ and $P(X) = (X - \alpha)(X - \alpha^3)(X - \alpha^{-4}) = [X^2 - (\alpha + \alpha^3)X + \alpha^4][X - \alpha^{-4}] = X^3 - (\alpha + \alpha^3 + \alpha^{-4})X^2 + (\alpha^4 + \alpha^{-3} + \alpha^{-1})X - 1 = X^3 - \eta_0 X^2 + \eta_2 X - 1$. (Although σ depends on the choice of the primitive root γ, the polynomial $P(X)$ depends only on which powers of α lie in the same period η_0 as α itself, and, as was shown in Section 4.5, this is independent of the choice of γ. Therefore, in the example, the coefficients 1, $-\eta_0$, η_2, -1 must all be independent of the choice of γ. Indeed, for either of the possible choices $\gamma = \pm 2$ one has $\eta_2 = \alpha^4 + \alpha^{-3} + \alpha^{-1}$, even though η_1 and η_3 do depend on the choice of γ.) It is easy to see that in all cases, as in this one, the coefficients of $P(X)$ are cyclotomic integers made up of periods of length f, that is, $P(X)$ has the form* $P(X) = X^f + \phi_1(\eta)X^{f-1} + \cdots + \phi_{f-1}(\eta)X + \phi_f(\eta)$ where $\phi_1(\eta), \phi_2(\eta), \ldots, \phi_f(\eta)$ are cyclotomic integers made up of periods of length f; for this it suffices to note that σ^e applied to $P(X)$ merely effects a cyclic permutation of the factors and therefore leaves $P(X)$, and consequently all its coefficients, unchanged.

Now set $X = \alpha$. Since $P(\alpha) = \prod(\alpha - \alpha^j)$ contains the factor $\alpha - \alpha$, this cyclotomic integer is 0, that is,

$$\alpha^f + \phi_1(\eta)\alpha^{f-1} + \cdots + \phi_f(\eta) = 0.$$

This relation will be called *the equation satisfied by α over the periods of length f*. In the example above, this equation is $\alpha^3 - \eta_0\alpha^2 + \eta_2\alpha - 1 = 0$, an equation which can be verified by writing out the periods $\alpha^3 - (\alpha + \alpha^3 + \alpha^9)\alpha^2 + (\alpha^4 + \alpha^{10} + \alpha^{12})\alpha - 1 = \alpha^3 - \alpha^3 - \alpha^5 - \alpha^{11} + \alpha^5 + \alpha^{11} + 1 - 1 = 0$.

By making use of the equation $P(\alpha) = 0$ satisfied by α over the periods of length f, *one can express an arbitrary cyclotomic integer $g(\alpha)$ in the form*

$$g(\alpha) = g_1(\eta)\alpha^{f-1} + g_2(\eta)\alpha^{f-2} + \cdots + g_f(\eta) \qquad (1)$$

in which $g_1(\eta), g_2(\eta), \ldots, g_f(\eta)$ are cyclotomic integers made up of periods of length f. (By counting dimensions one can prove that this representation is in fact *unique*, a fact which will not be needed in what follows.) To accomplish this one can successively replace the highest power of α in $g(\alpha)$ by lower powers using $\alpha^f = -\phi_1(\eta)\alpha^{f-1} - \phi_2(\eta)\alpha^{f-2} - \cdots$

---

*The notation $\phi(\eta)$, $f(\eta)$, $g(\eta)$, etc., for a cyclotomic integer made up of periods is taken from Kummer. Note that it is quite different from the notation $f(\alpha)$, in that $f(\eta)$ involves several η's, in fact e of them.

$-\phi_f(\eta)$ until the highest power of α in $g(\alpha)$ is less than f. This process is the same as performing division of polynomials to write $g(\alpha) = q(\alpha)P(\alpha) + r(\alpha)$ where $g(\alpha)$, $q(\alpha)$, $P(\alpha)$, $r(\alpha)$ are treated as polynomials in α with coefficients that are cyclotomic integers made up of periods of length f. (Division is possible because the leading coefficient of $P(\alpha)$ is 1.) As cyclotomic integers, $P(\alpha) = 0$ and $g(\alpha) = r(\alpha)$, where $r(\alpha)$ is of the required form; that is, $r(\alpha)$ is a polynomial of degree less than f with coefficients that are cyclotomic integers made up of periods of length f.

Assume now that a prime factor $h(\alpha)$ of p is given. Then each period η_j of length f is congruent mod $h(\alpha)$ to an integer u_j. Therefore each $g_i(\eta)$ in the above representation of $g(\alpha)$ is congruent mod $h(\alpha)$ to an integer $g_i(u)$, where $g_i(u)$ denotes the integer obtained substituting u_1 for η_1, u_2 for $\eta_2, \ldots, u_e$ for η_e in $g_i(\eta)$. Moreover, since congruence mod p implies congruence mod $h(\alpha)$, each of the integers $g_i(u)$ is congruent mod $h(\alpha)$ to a nonnegative integer less than p. Therefore every cyclotomic integer $g(\alpha)$ is congruent mod $h(\alpha)$ to one of the form $a_1\alpha^{f-1} + a_2\alpha^{f-2} + \cdots + a_f$ where the a_i are integers in the range $0 \leqslant a_i < p$. Thus the given problem, which is to determine whether $g(\alpha) \equiv 0$ mod $h(\alpha)$, will be solved if it can be determined whether $a_1\alpha^{f-1} + a_2\alpha^{f-2} + \cdots + a_f \equiv 0$ mod $h(\alpha)$ for just these p^f specific cyclotomic integers. The problem is therefore solved by the following theorem stating that of these p^f cyclotomic integers only 0 is $\equiv 0$ mod $h(\alpha)$.

Theorem. *Let $h(\alpha)$ be a prime factor of p, where $p \neq \lambda$ is a prime with exponent f mod λ. Let S be the set of p^f cyclotomic integers of the form $a_1\alpha^{f-1} + a_2\alpha^{f-2} + \cdots + a_f$ in which the a_i are integers in the range $0 \leqslant a_i < p$. Then, by the method described above, for any given cyclotomic integer $g(\alpha)$ one can find a cyclotomic integer $\bar{g}(\alpha)$ in the set S for which $g(\alpha) \equiv \bar{g}(\alpha)$ mod $h(\alpha)$. Two elements of S are congruent mod $h(\alpha)$ if and only if they are identical. This gives a test for divisibility by $h(\alpha)$ because it shows that $\bar{g}(\alpha)$ in S can be found for which $g(\alpha) \equiv \bar{g}(\alpha)$ mod $h(\alpha)$ and that $g(\alpha) \equiv 0$ mod $h(\alpha)$ if and only if $\bar{g}(\alpha) = 0$.*

Corollary. *To test $g(\alpha)$ for divisibility by $h(\alpha)$ it is not necessary to know $h(\alpha)$ but only the integers $u_1, u_2, \ldots, u_e$ for which $\eta_i \equiv u_i$ mod $h(\alpha)$ holds $(i = 1, 2, \ldots, e)$.*

PROOF. The corollary follows of course from the fact that the reduction of $g(\alpha)$ to $\bar{g}(\alpha)$—which is to write $g(\alpha)$ in the form (1) and then to reduce $g_i(u)$ mod p—uses only the u's and not $h(\alpha)$. The method of proof of the theorem is simple *counting* and, specifically, showing that *there must be at least p^f incongruent cyclotomic integers* mod $h(\alpha)$. Then, since every cyclotomic integer is congruent to one of the p^f cyclotomic integers in S, all p^f elements of S are incongruent mod $h(\alpha)$, which is the assertion of the theorem.

Note first that the number of incongruent elements mod $h(\alpha)$ is a power of p, say p^n. In terms of group theory this follows immediately from the fact that the additive group of cyclotomic integers mod $h(\alpha)$ is a quotient group of the additive group of cyclotomic integers mod p and that this latter group has $p^{\lambda-1}$ elements; the number of elements in the quotient group divides $p^{\lambda-1}$ and therefore, because p is prime, must be a power of p. The appeal to group theory can very easily be avoided by constructing directly a set of cyclotomic integers $g_1(\alpha), g_2(\alpha), \ldots, g_n(\alpha)$ such that every cyclotomic integer is congruent mod $h(\alpha)$ to exactly one of the p^n cyclotomic integers $b_1 g_1(\alpha) + b_2 g_2(\alpha) + \cdots + b_n g_n(\alpha)$ $(0 \leqslant b_i < p)$. (See Exercise 4.)

Note next that the number of incongruent cyclotomic integers mod $h(\alpha)$ is at least $\lambda + 1$ because $0, \alpha, \alpha^2, \ldots, \alpha^\lambda = 1$ are all incongruent mod $h(\alpha)$. This is true because any cyclotomic integer divisible by $h(\alpha)$ must have norm divisible by p, whereas the monomials $\alpha^j - 0$ have norm 1 and the binomials $\alpha^i - \alpha^j$ $(i \not\equiv j \bmod \lambda)$ have norm equal to $N(\alpha - 1) = \lambda$, neither of which is divisible by p.

Note finally that the number of *nonzero* incongruent cyclotomic integers mod $h(\alpha)$ is divisible by λ. This can be proved as follows. If $\alpha, \alpha^2, \ldots, \alpha^\lambda = 1$ exhaust the nonzero cyclotomic integers mod $h(\alpha)$ then there are exactly λ of them. Otherwise there is a cyclotomic integer $\psi(\alpha)$ such that $\psi(\alpha) \not\equiv 0 \bmod h(\alpha)$ and $\psi(\alpha) \not\equiv \alpha^j \bmod h(\alpha)$ $(j = 1, 2, \ldots, \lambda)$. Then $\psi(\alpha)\alpha, \psi(\alpha)\alpha^2, \ldots, \psi(\alpha)\alpha^\lambda = \psi(\alpha)$ are all nonzero mod $h(\alpha)$ (because $h(\alpha)$ is prime), distinct mod $h(\alpha)$ (because $\psi(\alpha)\alpha^j \equiv \psi(\alpha)\alpha^k$ would imply $\alpha^j \equiv \alpha^k$) and distinct from $\alpha, \alpha^2, \ldots, \alpha^\lambda$ (because $\psi(\alpha)\alpha^j \equiv \alpha^i$ would imply $\psi(\alpha) \equiv \alpha^k$). If those found so far exhaust the nonzero cyclotomic integers mod $h(\alpha)$ then there are just 2λ of them. Otherwise there is a cyclotomic integer $\phi(\alpha)$ such that the 3λ cyclotomic integers $\alpha^i, \psi(\alpha)\alpha^j$, $\phi(\alpha)\alpha^k$ are nonzero and distinct mod $h(\alpha)$. If these exhaust the possibilities there are just 3λ of them and otherwise the process can be continued. Ultimately the possibilities must be exhausted and the process terminates with a list of $m\lambda$ distinct nonzero cyclotomic integers mod $h(\alpha)$ and $m\lambda = p^n - 1$.

Thus $p^n \equiv 1 \bmod \lambda$. By the definition of the exponent f of p mod λ this implies $n \geqslant f$. (Note that $p^n \geqslant \lambda + 1$ implies $n \neq 0$.) Thus the number p^n of incongruent elements mod $h(\alpha)$ is at least p^f, as was to be shown.

EXERCISES

1. In each of the following cases find the equation satisfied by α over the periods of length f and verify the equation directly. (a) $\lambda = 5, f = 2$. (b) $\lambda = 7, f = 2$. (c) $\lambda = 11, f = 5$. (d) $\lambda = 13, f = 3$. (e) $\lambda = 31, f = 5$.

2. If $h(\alpha)$ is a prime divisor of p for which the u_i are known, then one can find a polynomial of degree f with leading coefficient 1 which is divisible by $h(\alpha)$ simply by substituting u_i for η_i in the equation satisfied by α over the periods of

length f. For example, in the case $\lambda = 13$, $p = 29$ the values $u_0 = -3$, $u_1 = 14$, $u_2 = 12$, $u_3 = 5$ found in Section 4.7 in the equation $\alpha^3 - \eta_0 \alpha^2 + \eta_2 \alpha - 1 = 0$ of this section give $\alpha^3 + 3\alpha^2 + 12\alpha - 1$ as a cyclotomic integer divisible by the $h(\alpha)$ in question. The 4 distinct conjugates of this $h(\alpha)$ have u's given by cyclic permutations of those above. Therefore the product $(\alpha^3 + 3\alpha^2 + 12\alpha - 1)(\alpha^3 - 14\alpha^2 + 5\alpha - 1)(\alpha^3 - 12\alpha^2 - 3\alpha - 1)(\alpha^3 - 5\alpha^2 + 14\alpha - 1)$ is divisible by all 4 prime divisors of 29 (assuming 29 has prime divisors) and therefore should be expected to be divisible by 29. Prove this directly by proving that this product is congruent to $\alpha^{12} + \alpha^{11} + \alpha^{10} + \cdots + 1 = 0$ mod 29. [Multiply the polynomials.]

3. The technique of Exercise 2 can be regarded as a method of *factoring the polynomial* $X^{\lambda-1} + X^{\lambda-2} + \cdots + X + 1$ mod p. Use this technique to factor $X^4 + X^3 + X^2 + X + 1$ mod 19, mod 59, and mod 41. Also factor $X^6 + X^5 + \cdots + 1$ mod 3, mod 13, mod 29. Finally, factor $X^{30} + X^{29} + \cdots + 1$ mod 2. In each case one finds e factors of degree f, where f is the exponent of p mod λ. As the theorem of the next section will imply, these factors are *irreducible* mod p, that is, they cannot be factored further mod p.

4. Prove that the number of incongruent cyclotomic integers mod $h(\alpha)$ is a power of p. [The integers $0, 1, 2, \ldots, p-1$ are p incongruent elements mod $h(\alpha)$, because for integers congruence mod $h(\alpha)$ is equivalent to congruence mod p. If there is some $g_1(\alpha)$ not congruent to any of these then $a + bg_1(\alpha)$ are incongruent to each other for $0 \leqslant a < p$, $0 \leqslant b < p$. If these exhaust the possibilities then there are p^2. Otherwise one can go on to construct p^3 incongruent ones, and so forth.]

5. Determine which of the primes in Table 4.7.2 divide $7\alpha^3 + 45\alpha^2 + 83\alpha + 90$. [Note that each line of the table gives *two* primes, the listed one and its conjugate.]

6. Show that for a prime factor $h(\alpha)$ of a prime p of exponent f mod λ there is a polynomial $Q_h(\alpha)$ of degree f in α, with integer coefficients and leading coefficient 1, and with the property that $h(\alpha)$ divides a cyclotomic integer $g(\alpha)$ if and only if division of polynomials $g(X) = q(X)Q_h(X) + r(X)$ leaves a remainder $r(X)$ whose coefficients are all zero mod p.

4.9 Prime divisors

Let λ be a prime greater than 2 and let $p \neq \lambda$ be a prime whose exponent mod λ is f. In Section 4.6 it was shown that if p has a prime factor $h(\alpha)$—that is, if there is a prime cyclotomic integer $h(\alpha)$ which divides p—then each of the $e(=(\lambda-1)/f)$ periods $\eta_1, \eta_2, \ldots, \eta_e$ of length f is congruent mod $h(\alpha)$ to an integer. That is, given a prime divisor $h(\alpha)$ of p there exist integers $u_1, u_2, \ldots, u_e$ for which $\eta_i \equiv u_i$ mod $h(\alpha)$. Therefore, any cyclotomic integer made up of periods of length f is congruent to an integer mod $h(\alpha)$ and, because two integers are congruent mod $h(\alpha)$ if and only if they are congruent mod p, a knowledge of the u's makes it possible to determine whether two cyclotomic integers made up of periods of length

f are congruent mod $h(\alpha)$. In short, $g(\eta) \equiv \phi(\eta)$ mod $h(\alpha)$ if and only if*
$g(u) \equiv \phi(u)$ mod p. In particular, the identities $\eta_1\eta_j = c_0 + c_1\eta_1 + \cdots +$
$c_e\eta_e$ imply congruences $u_1u_j \equiv c_0 + c_1u_1 + \cdots + c_eu_e$ mod p satisfied by
the u's. It was seen in the examples of Section 4.7 that these congruences
could be solved to find *all possible* sets of u's in the cases considered.
Finally, in Section 4.8, it was shown that a knowledge of the u's makes it
possible to determine whether two cyclotomic integers are congruent mod
$h(\alpha)$; given $g(\alpha)$, $\phi(\alpha)$, they can be written in the form $g(\alpha) = g_1(\eta)\alpha^{f-1} +$
$g_2(\eta)\alpha^{f-2} + \cdots + g_f(\eta)$, $\phi(\alpha) = \phi_1(\eta)\alpha^{f-1} + \phi_2(\eta)\alpha^{f-2} + \cdots + \phi_f(\eta)$
and $g(\alpha) \equiv \phi(\alpha)$ mod $h(\alpha)$ if and only if $g_1(u) \equiv \phi_1(u)$, $g_2(u) \equiv$
$\phi_2(u), \ldots, g_f(u) \equiv \phi_f(u)$ mod p. In summary, then, *a prime factor $h(\alpha)$ of p
determines integers $u_1, u_2, \ldots, u_e$ mod p and a knowledge of these integers
suffices to determine the relation of congruence* mod $h(\alpha)$.

The fundamental idea of Kummer's theory of ideal factorization can
now be stated as follows. *Prove that the congruences mod p which the u's
satisfy always have a solution and use this solution to define "congruence
modulo a prime factor of p" even in cases where there is no actual prime
factor of p.* This is the program of the present section. Thus the first step is:

Proposition. *Given λ, p, f, e as above, there exist integers $u_1, u_2, \ldots, u_e$ with
the property that every equation satisfied by the periods η of length f is
satisfied as a congruence modulo p by the integers u. That is,
$u_1, u_2, \ldots, u_e$ are such that if $F(\eta_1, \eta_2, \ldots, \eta_e)$ is any polynomial expression in the η's with integer coefficients and if the cyclotomic integer
$F(\eta_1, \eta_2, \ldots, \eta_e)$ is zero then the integer $F(u_1, u_2, \ldots, u_e)$ obtained by
substituting u_i in place of η_i ($i = 1, 2, \ldots, e$) satisfies $F(u_1, u_2, \ldots, u_e) \equiv
0$ mod p.*

If it were known that the u's came from an actual prime factor $h(\alpha)$ of p
then the technique of the preceding section would enable one to use them
to determine whether given cyclotomic integers are congruent mod $h(\alpha)$.
The main idea is that the *congruence relation* can be defined even when
there is no actual $h(\alpha)$.

Theorem 1. *Let λ, p, f, e, and $u_1, u_2, \ldots, u_e$ be as in the proposition. Then it
is possible to define one and only one congruence relation on cyclotomic
integers with $\alpha^\lambda = 1$ with all the usual properties (it is reflexive, symmetric,
transitive, and consistent with addition and multiplication) in which η_i is*

*Here $g(u)$ is the integer obtained by substituting u_i for η_i in $g(\eta)$ and similarly for $\phi(u)$. As in
Section 4.3, this notation must be used cautiously because this is not a well-defined operation
on cyclotomic integers made up of periods, that is, $g(\eta) = \phi(\eta)$ does not imply $g(u) = \phi(u)$.
However, the particular u's in question do have the property that $g(\eta) \equiv \phi(\eta)$ mod $h(\alpha)$
implies $g(u) \equiv \phi(u)$ mod p—by the ordinary rules for computing with congruences mod $h(\alpha)$
—and *a fortiori* $g(\eta) = \phi(\eta)$ implies $g(u) \equiv \phi(u)$ mod p. Therefore, even though the integers
$g(u)$ and $\phi(u)$ are not well defined, they are well defined mod p in the case of these particular
u's.

congruent to u_i, p is congruent to 0, *and* 1 *is not congruent to* 0. *Moreover, this congruence relation is* prime *in the sense that a product can be congruent to* 0 *only if one of the factors is. Finally, the number of incongruent elements is exactly* p^f.

Definition. The congruence relation whose existence is asserted by Theorem 1 will be called *congruence modulo the prime divisor of p corresponding to* $u_1, u_2, \ldots, u_e$. If $g(\alpha)$ is congruent to 0 modulo this relation then $g(\alpha)$ will be said to be *divisible by the prime divisor of p corresponding to* $u_1, u_2, \ldots, u_e$.

Note that "the prime divisor of p corresponding to $u_1, u_2, \ldots, u_e$" has not been defined, but only *congruence* modulo such a prime divisor and *divisibility* by such a prime divisor. If p has an actual prime divisor $h(\alpha)$ then, as has been shown, congruence modulo $h(\alpha)$ coincides with "congruence modulo a prime divisor of p" as just defined, namely, congruence modulo the prime divisor of p corresponding to $u_1, u_2, \ldots, u_e$ where $u_i \equiv \eta_i$ modulo $h(\alpha)$. Therefore the new definition coincides with the old one whenever the old one is valid. However, the new one is valid in cases where the old one is not. For example, in the case $\lambda = 23$, *congruence modulo* the prime divisor of 47 corresponding to $\alpha \equiv 4, \alpha^2 \equiv 4^2, \ldots, \alpha^{22} \equiv 4^{22}$ mod 47 is now defined, even though no prime cyclotomic integer $h(\alpha)$ divides 47. For special emphasis, congruence modulo a prime divisor may occasionally be called congruence modulo an *ideal* prime divisor—or divisibility by a prime divisor may be called divisibility by an *ideal* prime divisor—in cases where there is no actual prime divisor of p. However, statements involving congruence modulo prime divisors or divisibility by prime divisors are perfectly meaningful, by virtue of the above definition, even in cases where there is no actual prime divisor.

If the integers $u_1, u_2, \ldots, u_e$ have the property described in the proposition then so do the integers $u_2, u_3, \ldots, u_e, u_1$ and, consequently, so do all e of the cyclic permutations of the u's. This follows immediately from the fact that there is a conjugation $\alpha \mapsto \alpha^\gamma$ which effects a cyclic permutation $\eta_1 \mapsto \eta_2 \mapsto \cdots \mapsto \eta_e \mapsto \eta_1$ of the periods and which therefore carries a relation of the form $F(\eta_1, \eta_2, \ldots, \eta_e) = 0$ to the relation $F(\eta_2, \eta_3, \ldots, \eta_e, \eta_1) = 0$. Thus the proposition guarantees the existence of not just 1 but e sets of u's. In the examples of Section 4.7 it was found that there were always *precisely* e solutions $u_1, u_2, \ldots, u_e$ and that all of them could be derived from any one by cyclic permutations. Moreover, in all but one of the examples it was possible to show that if p had a prime factorization then it necessarily had one of the form $p = \pm \phi(\eta) \cdot \sigma\phi(\eta) \cdots \sigma^{e-1}\phi(\eta)$ in which $\phi(\eta)$ was prime, that is, p must be a unit times the product of its e distinct prime factors. The following theorem states that when these properties are put in terms of divisibility by prime divisors—in the wider sense—they are always true.

Theorem 2. *For given* λ, p, f, e, *let* $u_1, u_2, \ldots, u_e$ *and* $\bar{u}_1, \bar{u}_2, \ldots, \bar{u}_e$ *be any two sets of integers with the property specified in the proposition. Then there is one and only one integer* k, $0 \leqslant k < e$ *such that congruence modulo the prime divisor of* p *corresponding to* $\bar{u}_1, \bar{u}_2, \ldots, \bar{u}_e$ *coincides with congruence modulo the prime divisor of* p *corresponding to* u_{1+k}, $u_{2+k}, \ldots, u_k$ *where, as usual,* $u_j = u_{j+e}$ *by definition. Thus there are precisely* e *distinct congruence relations of the type described in Theorem 1. Divisibility by* p *is equivalent to divisibility by all* e *of these prime divisors of* p; *that is, a cyclotomic integer* $g(\alpha)$ *is divisible by* p *in the ordinary sense if and only if it is divisible by all* e *prime divisors of* p *in the sense just defined.*

The remainder of this section is devoted to the proofs of the proposition and of the two theorems. These proofs are essentially the ones given by Kummer, but the central idea—that of the construction of the cyclotomic integer $\Psi(\eta)$—comes not from his original 1847 treatise [K8] on the theory of ideal factorization but from a brief note [K16] written 10 years later. It is a curious fact that his original proofs appear to be wholly inadequate and that for the first 10 years of its existence Kummer's theory stood on a faulty foundation. The gap at the crucial point—which is the proof of the proposition above—was beginning to be noticed by others (see [E3]) around 1856, but the 1857 paper appeared before the flaw became at all well known. For some reason Kummer was not very forthright in acknowledging the gap in his work. There are in fact two gaps in his original proof, the first being the proof that the congruence of degree e in u that is satisfied by all of the u's has e solutions mod p (counted with multiplicities) and the second being the proof that these e solutions can be ordered in such a way that they have the property described in the proposition. With respect to the first gap, Kummer still insisted in 1857 that he had "rigorously proved" it in his earlier work even though he did not give any explanation of this "proof"; the fact is that although it is not very difficult to fill this first gap (see Exercise 6) it does not appear to be possible to do so in the way suggested in Kummer's highly abbreviated "proof" (see Exercise 7). With respect to the second gap, which seems more difficult to fill, Kummer perhaps was being more guarded when he said that he believed it required "clarification and a more complete foundation" and that he was presenting the new method of proof for this reason. The new method was certainly a great simplification and was worthy of presentation even if Kummer truly believed, as he implied, that his earlier method was not inadequate. However, it seems more plausible that he was by that time aware of the inadequacies of the earlier method but that he felt that to call attention to them would be foolish.

PROOF OF THE PROPOSITION. Let a cyclotomic integer $\Psi(\eta)$ made up of periods of length f be constructed as follows. Consider the ep cyclotomic integers $j - \eta_i$ $(j = 1, 2, \ldots, p$ and $i = 1, 2, \ldots, e)$. At the first step, choose one of these ep cyclotomic integers which is not divisible by p. (It is obvious—see Exercise 9—that there is always at least one which is not divisible.) At the nth step, choose one of the remaining $ep - (n-1)$ cyclotomic integers $j - \eta_i$ in such a way that the product of all n of those that have been chosen is not divisible by p. If it is impossible to choose an nth factor with this property then the construction terminates and $\Psi(\eta)$ is defined to be the product of the $n - 1$ factors that have been chosen.* Since, as was shown in Section 4.6, $(\eta_i - 1)(\eta_i - 2) \cdots (\eta_i - p) \equiv \eta_i^p - \eta_i \equiv 0 \bmod p$, the construction of $\Psi(\eta)$ terminates before the ep factors are exhausted and, in fact, for each i at least one factor $j - \eta_i$ is not in $\Psi(\eta)$. Let $u_i - \eta_i$, for $i = 1, 2, \ldots, e$, denote a factor of the form $j - \eta_i$ that is not in $\Psi(\eta)$. Then $(u_i - \eta_i)\Psi(\eta) \equiv 0 \bmod p$ since otherwise the construction of $\Psi(\eta)$ would not have terminated. On the other hand, if $j - \eta_i$ is *any* factor of this form not included in $\Psi(\eta)$ then $(j - \eta_i)\Psi(\eta) \equiv 0 \equiv (u_i - \eta_i)\Psi(\eta) \bmod p$, which implies $(j - u_i)\Psi(\eta) \equiv 0 \bmod p$. If $j \neq u_i$ then, because both j and u_i lie in the range $1 \leqslant u_i \leqslant p$, $1 \leqslant j \leqslant p$, $j - u_i$ would be invertible $\bmod p$, say $b(j - u_i) \equiv 1 \bmod p$, from which $\Psi(\eta) \equiv b(j - u_i)\Psi(\eta) \equiv 0 \bmod p$, contrary to assumption. Therefore $j = u_i$ and u_i is the *only* integer between 1 and p for which $u_i - \eta_i$ is not in $\Psi(\eta)$. Thus $\Psi(\eta)$ contains exactly $ep - e$ factors, namely, all of the $j - \eta_i$ except for the factors $u_i - \eta_i$.

Now $\eta_i \Psi(\eta) = u_i \Psi(\eta) \bmod p$ for $i = 1, 2, \ldots, e$. Therefore $F(\eta_1, \eta_2, \ldots, \eta_e)\Psi(\eta) \equiv F(u_1, u_2, \ldots, u_e)\Psi(\eta) \bmod p$ for any polynomial $F(\eta_1, \eta_2, \ldots, \eta_e)$ in the periods. This follows immediately from the fact that congruence $\bmod p$ is consistent with addition and multiplication. In particular, if $F(\eta_1, \eta_2, \ldots, \eta_e) = 0$ then $F(u_1, u_2, \ldots, u_e)\Psi(\eta) \equiv 0 \bmod p$. Since $\Psi(\eta) \not\equiv 0 \bmod p$ by construction, it follows that the integer $F(u_1, u_2, \ldots, u_e)$ is not invertible $\bmod p$, that is, $F(u_1, u_2, \ldots, u_e) \equiv 0 \bmod p$. Therefore the integers $u_1, u_2, \ldots, u_e$ have the property required by the proposition, and the proposition is proved.

PROOF OF THEOREM 1. For this proof it will be convenient to assume not only that the integers $u_1, u_2, \ldots, u_e$ have the property of the proposition but also that they are given by the construction in the proof of the proposition—explicitly that the cyclotomic integer $\Psi(\eta)$ which is the product of the $ep - e$ factors $j - \eta_i$ with $j \neq u_i$ is not divisible by p. It will be shown in the proof of Theorem 2 that any other set of integers $\bar{u}_1, \bar{u}_2, \ldots, \bar{u}_e$ for which the proposition is true must be a cyclic permutation of these and therefore must have this property. The method of the proof of Theorem 1 will be to show that the congruence relation $g(\alpha)\Psi(\eta) \equiv \phi(\alpha)\Psi(\eta) \bmod p$ satisfies the conditions of the theorem. Therefore, for

---

*This differs slightly from Kummer's actual construction. Rather than use all ep factors as was done above, he uses only the factors $j - \eta_i$ in which j satisfies $(j - \eta_1)(j - \eta_2) \cdots (j - \eta_e) \equiv 0 \bmod p$. If it were a question of actually computing $\Psi(\eta)$, this would be more efficient. However, not $\Psi(\eta)$ but the existence of the integers u is at issue here. Once they are known to exist there are much easier ways to find them than this construction.

cyclotomic integers $g(\alpha)$, $\phi(\alpha)$ let $g(\alpha) \equiv \phi(\alpha)$ mean $g(\alpha)\Psi(\eta) \equiv \phi(\alpha)\Psi(\eta) \bmod p$. Then the relation $\equiv$ is clearly reflexive, symmetric, transitive, and consistent with addition and multiplication. Moreover, $p \equiv 0$ and, as was seen in the proof of the proposition, $\eta_i \equiv u_i$. Since $\Psi(\eta) \not\equiv 0 \bmod p$, $1 \not\equiv 0$.

Now let $\sim$ denote any other congruence relation that is reflexive, symmetric, transitive, and consistent with addition and multiplication, for which $\eta_i \sim u_i$, $p \sim 0$, $1 \not\sim 0$. Since every cyclotomic integer $g(\alpha)$ can be written in the form $g_1(\eta)\alpha^{f-1} + g_2(\eta)\alpha^{f-2} + \cdots + g_f(\eta)$ it follows that $g(\alpha) \sim a_1\alpha^{f-1} + a_2\alpha^{f-2} + \cdots + a_f$ where the a_i are integers $0 \leqslant a_i < p$. Therefore there are *at most p^f* incongruent cyclotomic integers modulo the congruence relation $\sim$. *It will be shown that $\sim$ must be prime and must have at least p^f incongruent cyclotomic integers.* From this it will follow that $\sim$ is completely determined, because for any given $g(\alpha)$ and $\phi(\alpha)$ one can find $g(\alpha) \sim a_1\alpha^{f-1} + a_2\alpha^{f-2} + \cdots + a_f$ and $\phi(\alpha) \sim b_1\alpha^{f-1} + b_2\alpha^{f-2} + \cdots + b_f$ where the a_i and b_i are integers $0 \leqslant a_i < p$, $0 \leqslant b_i < p$ and, by the transitivity of $\sim$ and the fact that $\sim$ has p^f incongruent cyclotomic integers, $g(\alpha) \sim \phi(\alpha)$ if and only if the cyclotomic integers $a_1\alpha^{f-1} + a_2\alpha^{f-2} + \cdots + a_f$ and $b_1\alpha^{f-1} + b_2\alpha^{f-2} + \cdots + b_f$ are *equal*. Since $\equiv$ has all the properties assumed for $\sim$ it will then follow that $\equiv$ coincides with $\sim$ and the first statement of the theorem will follow.

If it could be shown that $\sim$ was prime,* then the argument of the preceding section would show that there were at least p^f incongruent cyclotomic integers modulo $\sim$ because, as before, the number of incongruent cyclotomic integers would be a power of p (because it is a quotient group of a group with $p^{\lambda-1}$ elements) greater than 1 (because $1 \not\sim 0$) which is 1 mod λ (because the subsets of the form $g(\alpha)$, $g(\alpha)\alpha$, $g(\alpha)\alpha^2, \ldots, g(\alpha)\alpha^\lambda$ are nonoverlapping sets containing exactly λ incongruent nonzero cyclotomic integers); therefore it is a positive power of p^f and must be *at least p^f*. Therefore not just the first statement but the entire theorem will be proved if it is proved that $\sim$ is prime.

The method of proof will be to show that if $\sim$ is not prime then there is another congruence relation, say $\approx$, which has all the properties assumed for $\sim$ but which is *coarser* than $\sim$, that is, $g(\alpha) \sim \phi(\alpha)$ implies $g(\alpha) \approx \phi(\alpha)$ but there are some cyclotomic integers $g(\alpha)$, $\phi(\alpha)$ for which $g(\alpha) \approx \phi(\alpha)$ and $g(\alpha) \not\sim \phi(\alpha)$. The theorem will then follow by infinite descent: If $\approx$ is not prime then, because $\approx$ has all the properties of $\sim$, the process can be repeated to give a third congruence relation, say $\approx_3$, coarser than $\approx$ but having all the properties assumed for $\sim$. If $\approx_3$ is not prime then there is a fourth congruence relation $\approx_4$ coarser than $\approx_3$, and so forth. Since the number of incongruent cyclotomic integers was at most p^f to start with and decreases at each step, this process must eventually result in a

*The congruence relation of the preceding section was prime by assumption, since it was congruence mod $h(\alpha)$ where $h(\alpha)$ was assumed to be a prime divisor of p. The whole point of the theorems of this section is to eliminate the assumption of the existence of such an $h(\alpha)$.

congruence relation $\approx_n$ which has all the properties of $\sim$ and which in addition is *prime*. Then, by the argument of the preceding section that was summarized above, there are at least p^f incongruent cyclotomic integers modulo $\approx_n$. Therefore no coarsening of the original relation $\sim$ was in fact possible and this relation was prime and had p^f incongruent cyclotomic integers.

Thus the proof of the theorem is reduced to showing that if $\sim$ is not prime then there is a coarser congruence relation $\approx$ which has the same properties as $\sim$. This can be seen very simply as follows. If $\sim$ is not prime then there are cyclotomic integers $\psi_1(\alpha)$, $\psi_2(\alpha)$ such that $\psi_1(\alpha)\psi_2(\alpha) \sim 0$ but $\psi_1(\alpha) \not\sim 0$, $\psi_2(\alpha) \not\sim 0$. Define $g(\alpha) \approx \phi(\alpha)$ to mean $g(\alpha)\psi_1(\alpha) \sim \phi(\alpha)\psi_1(\alpha)$. Then $\approx$ is reflexive, symmetric, transitive, consistent with addition and multiplication, and $\eta_i \approx u_i$ (because $\eta_i \sim u_i$ implies $\eta_i\psi_1(\alpha) \sim u_i\psi_1(\alpha)$) and $p \approx 0$. Moreover, $1 \not\approx 0$ because $\psi_1(\alpha) \not\sim 0$ and $\approx$ is coarser than $\sim$ because $\psi_2(\alpha) \approx 0$ but $\psi_2(\alpha) \not\sim 0$. This completes the proof of the theorem.

PROOF OF THEOREM 2. Let $u_1, u_2, \ldots, u_e$ be a set of integers yielded by the construction in the proof of the proposition and let $\bar{u}_1, \bar{u}_2, \ldots, \bar{u}_e$ be any set of integers satisfying the conditions of the proposition. To prove the first statement of Theorem 2 it will suffice to prove that for this particular $u_1, u_2, \ldots, u_e$ there is a unique integer k, $0 \leqslant k < e$ for which $\bar{u}_i \equiv u_{i+k} \bmod p$ because this will prove that the only sets of integers $\bar{u}_1, \bar{u}_2, \ldots, \bar{u}_e$ which satisfy the conditions of the proposition are cyclic permutations of the u's; then, since two cyclic permutations of the same set are cyclic permutations of each other, the desired conclusion will follow.

The uniqueness statement of the theorem is the statement that for every $k = 1, 2, \ldots, e - 1$ there is an $i = 1, 2, \ldots, e$ for which $u_i \not\equiv u_{i+k} \bmod p$ where, as usual, $u_i = u_{i+e}$ by definition; in short, no cyclic permutation of the u's leaves them unchanged mod p other than the identity permutation. This fact is the core of the theorem. It can be proved as follows.

Consider the actual equations satisfied by the periods. They are of the form $\eta_1\eta_i = c_0 + c_1\eta_1 + c_2\eta_2 + \cdots + c_e\eta_e$. In computing the coefficients c in these equations one merely writes out all f^2 terms of the product on the left and notes that, since the product is invariant under σ^e, two powers of α which lie in the same period η_k must occur the same number of times; this number is c_k. This shows that $f^2 = c_0 + c_1 f + c_2 f + \cdots + c_e f$ because these are just two ways of counting the number of terms in the product. Now $c_0 \neq 0$ only if at least one term α^μ of η_i is the inverse of a term $\alpha^{-\mu}$ in η_1. But when this is the case η_i contains the inverse $\sigma^{\nu e}\alpha^\mu$ of every term $\sigma^{\nu e}\alpha^{-\mu}$ in η_1. Therefore $c_0 = f$ for this one value of j and $c_0 = 0$ for all other values of j. For one value of j, then, $c_0 = f$, $c_1 + c_2 + \cdots + c_e = (f^2 - f)/f = f - 1$, whereas for all others $c_0 = 0$, $c_1 + c_2 + \cdots + c_e = f$. (One can in fact say precisely which value of j is the exceptional one, namely, $j = 0$ if f is even and $j = \frac{1}{2}e$ if f is odd. See Exercise 11. However, this fact will not be needed.) Therefore $\eta_1\eta_j + \eta_2\eta_{j+1} + \cdots + \eta_e\eta_{j-1} = \eta_1\eta_j +$

$\sigma(\eta_1\eta_j) + \cdots + \sigma^{e-1}(\eta_1\eta_j) = ec_0 + c_1(\eta_1 + \eta_2 + \cdots + \eta_e) + \cdots + c_e(\eta_e + \eta_1 + \cdots + \eta_{e-1}) = ec_0 - (c_1 + c_2 + \cdots + c_e) = -f$ in all but one case; in the exceptional case it is $ef - (f-1) = ef + 1 - f = \lambda - f$. This proves the useful identity

$$f + \eta_1\eta_j + \eta_2\eta_{j+1} + \cdots + \eta_e\eta_{j-1} = \begin{cases} \lambda & \text{in one case,} \\ 0 & \text{in all others.} \end{cases}$$

Therefore, by the defining property of the u's,

$$f + u_1u_j + u_2u_{j+1} + \cdots + u_eu_{j-1} \begin{cases} \not\equiv 0 \bmod p & \text{in one case,} \\ \equiv 0 \bmod p & \text{in all others.} \end{cases}$$

If there were a k such that $u_i \equiv u_{i+k} \bmod p$ for all i then $f + u_1u_j + u_2u_{j+1} + \cdots + u_eu_{j-1} \equiv f + u_1u_{j+k} + u_2u_{j+k+1} + \cdots + u_eu_{j+k-1} \bmod p$; if j is chosen to be the unique value such that the integer on the left is not zero mod p then this congruence implies $j \equiv j + k \bmod e$, that is, $k \equiv 0 \bmod e$, as was to be shown.

This proves that the e cyclic permutations of the u's are all distinct. Next consider $M = \Psi(\eta) + \sigma\Psi(\eta) + \cdots + \sigma^{e-1}\Psi(\eta)$ where σ is the conjugation $\alpha \mapsto \alpha^\gamma$ which carries η_i to η_{i+1}. Since $\sigma M = M$, it is clear that M is an ordinary integer. Now

$$M\Psi(\eta) = \Psi(\eta)\Psi(\eta) + [\sigma\Psi(\eta)]\cdot\Psi(\eta) + \cdots + [\sigma^{e-1}\Psi(\eta)]\cdot\Psi(\eta)$$

$$\equiv \Psi(u)\Psi(\eta) + [\sigma\Psi(u)]\Psi(\eta) + \cdots + [\sigma^{e-1}\Psi(u)]\cdot\Psi(\eta) \bmod p$$

where $\sigma^k\Psi(u)$ denotes the integer obtained by applying σ^k to $\Psi(\eta)$ and setting $\eta_1 = u_1, \eta_2 = u_2, \ldots, \eta_e = u_e$ in the result; in short, $\sigma^k\Psi(u) = \Pi(j - u_{i+k})$ where $i = 1, 2, \ldots, e$ and j ranges over all integers from 1 to p except u_i. By what was just shown, $\sigma^k\Psi(u)$ contains a factor of 0 (namely, a factor $u_{i+k} - u_{i+k}$) unless $k = 0$, in which case $\sigma^k\Psi(u) = \Psi(u)$. Therefore $M\Psi(\eta) \equiv \Psi(u)\Psi(\eta) \not\equiv 0 \bmod p$ because $\Psi(u) \not\equiv 0 \bmod p$. [$\Psi(u)$ is a product of $ep - e$ integers, none of them divisible by p.] Therefore $M \not\equiv 0 \bmod p$.

If $\bar{u}_1, \bar{u}_2, \ldots, \bar{u}_e$ satisfy the condition of the proposition then $M \equiv \Psi(\bar{u}) + \sigma\Psi(\bar{u}) + \cdots + \sigma^{e-1}\Psi(\bar{u}) \bmod p$. Since $M \not\equiv 0 \bmod p$ this implies $\sigma^k\Psi(\bar{u}) \not\equiv 0 \bmod p$ for at least one k. That is, $\Pi(j - \bar{u}_{i+k}) \not\equiv 0 \bmod p$ for at least one k, where $i = 1, 2, \ldots, e$ and j assumes all values $1, 2, \ldots, p$ except u_i. This implies that $u_i \equiv \bar{u}_{i+k} \bmod p$ and the $\bar{u}$'s are a cyclic permutation of the u's as was to be shown.

Finally, it remains to show that to say $g(\alpha) \equiv 0 \bmod p$ is the same as saying that $g(\alpha)$ is divisible by all e of the prime divisors that have now been defined. Since p is divisible by all e of them, one half of this statement is a consequence of the fact that congruence modulo a prime divisor is consistent with multiplication. Conversely, it was shown in the proof of Theorem 1 that divisibility of $g(\alpha)$ by the prime divisor of p corresponding to $u_1, u_2, \ldots, u_e$ is equivalent to $g(\alpha)\Psi(\eta) \equiv 0 \bmod p$ where $\Psi(\eta)$ is the product of the $ep - e$ factors $j - \eta_i$ in which $j \neq u_i$. Therefore divisibility of $g(\alpha)$ by the prime divisor of p corresponding to u_{k+1}, $u_{k+2}, \ldots, u_k$ is equivalent to $g(\alpha)\sigma^{-k}\Psi(\eta) \equiv 0 \bmod p$ because

$$\sigma^{-k}\Psi(\eta) = \sigma^{-k}\left[\prod_{i=1}^{e}\prod_{j=1}^{p}(j-\eta_i)/\prod_{i=1}^{e}(u_i-\eta_i)\right]$$

$$= \prod_{i=1}^{e}\prod_{j=1}^{p}(j-\eta_{i-k})/\prod_{i=1}^{e}(u_i-\eta_{i-k})$$

$$= \prod_{i=1}^{e}\prod_{j=1}^{p}(j-\eta_i)/\prod_{i=1}^{e}(u_{i+k}-\eta_i),$$

that is, $\sigma^{-k}\Psi(\eta)$ is the $\Psi(\eta)$ that results from replacing $u_1, u_2, \ldots, u_e$ with $u_{1+k}, u_{2+k}, \ldots, u_k$. Therefore if $g(\alpha)$ is divisible by all e prime divisors of p then $g(\alpha)\sigma^{-k}\Psi(\eta) \equiv 0 \bmod p$ for all $k = 0, 1, \ldots, e-1$. The sum of these congruences gives $g(\alpha) \cdot M \equiv 0 \bmod p$ which, because $M \not\equiv 0 \bmod p$, implies $g(\alpha) \equiv 0 \bmod p$, as was to be shown.

EXERCISES

1. Construct $\Psi(\eta) \bmod 2$ for the prime divisor of 2 in the case $\lambda = 31$ that was found in Section 4.7. [Of course $\Psi(\eta)$ need only be known $\bmod p$ for purposes of the proofs given in the text.]

2. Construct $\Psi(\eta) \bmod 3$ for the prime divisor of 3 in the case $\lambda = 13$ that was found in Section 4.7.

3. Prove that if p has exponent $\lambda - 1 \bmod \lambda$ then p is a prime cyclotomic integer. [This is principally a matter of checking that the proofs of this section are valid when $e = 1$.]

4. Prove that the factors found in the examples of Section 4.7—that is, cyclotomic integers $\phi(\eta)$ made up of periods of length f which have norm p^f where p is a prime whose exponent $\bmod \lambda$ is f—are prime. [A proof of this is given in Section 4.11.]

5. If $\eta_1, \eta_2, \ldots, \eta_e$ are the e periods of length f $(ef = \lambda - 1)$ show that $\phi(X) = (X - \eta_1)(X - \eta_2) \cdots (X - \eta_e)$ is a polynomial in X with *integer* coefficients. Find this polynomial in all cases with $\lambda \leqslant 13$.

6. The first step in Kummer's faulty "proof" of the proposition is to attempt to show that the polynomial $\phi(X)$ of Exercise 5 is such that the congruence $\phi(u) \equiv 0 \bmod p$ has e distinct solutions $\bmod p$. A correct proof of this fact can be given as follows.

$$\phi(X-1)\phi(X-2) \cdots \phi(X-p)$$
$$\equiv [(X-\eta_1)^p - (X-\eta_1)][(X-\eta_2)^p - (X-\eta_2)] \cdots [(X-\eta_e)^p - (X-\eta_e)]$$
$$\equiv (X-1)^e(X-2)^e \cdots (X-p)^e \bmod p.$$

Therefore $\phi(X)$ is a product of factors of the form $X - k$ (k = integer) $\bmod p$

and this is the *meaning* of the statement that $\phi(X)$ has e distinct roots mod p. Fill in the details of this proof paying special attention to defining the notion of two coincident roots of a congruence $g(u) \equiv 0$ mod p.

7. Kummer's "proof" of the statement in Exercise 6 is to say that for every integer y one has $(y - \eta_i - 1)(y - \eta_i - 2) \cdots (y - \eta_i - p) \equiv (y - \eta_i)^p - (y - \eta_i) \equiv (y^p - y) - (\eta_i^p - \eta_i) \equiv y^p - y \equiv 0$ mod p, from which $\phi(y - 1)\phi(y - 2) \cdots \phi(y - p) \equiv 0$ mod p^e; he then says "one easily concludes" that $\phi(y) \equiv 0$ mod p has e distinct solutions. Refute this by finding a p, an e, and a polynomial $\phi(X)$ such that $\phi(y - 1)\phi(y - 2) \cdots \phi(y - p) \equiv 0$ mod p^e for all integers y but for which $\phi(y)$ has degree less than e. [One can take $p = 2$, $e = 3$.]

8. In the examples of Section 4.7 (and in Section 4.4) the main step in constructing the prime divisor was to find one solution u of $\phi(u) \equiv 0$ mod p, after which the remaining $e - 1$ solutions and their correct order were relatively easy to find. Explain what computational circumstances made this possible and would be needed in order to go from one solution of $\phi(u) \equiv 0$ to a complete set of u's with property of the proposition. [If u_0 is known then $u_1, u_2, \ldots, u_{e-1}$ satisfy $e - 1$ linear congruences in $e - 1$ unknowns. Properties of solutions of these are like those of linear equations. These give *necessary* conditions on the u's, but even if they are satisfied it is not clear that the u's have *all* the required properties.] This is the second gap in Kummer's original "proof."

9. Show that at least one of the ep cyclotomic integers $j - \eta_i$ ($j = 1, 2, \ldots, p$ and $i = 1, 2, \ldots, e$) is not divisible by p.

10. In the language of modern algebra, the construction of a prime divisor of p amounts to the construction of a field with p^f elements and a homomorphism of the cyclotomic integers into that field. This in turn amounts to the construction of a polynomial of degree f which is irreducible mod p and which, mod p, divides the polynomial $X^{\lambda-1} + X^{\lambda-2} + \cdots + X + 1$ which defines the cyclotomic integers. Perform this construction *ab initio*. Find the polynomial in the cases of prime divisors that have been constructed in Section 4.4 and Section 4.7. [See Exercises 2 and 3 of Section 4.8.]

11. Let j be the integer $0 \leqslant j < e$ for which α^{-1} lies in η_j. Show that $j = 0$ if f is even and $j = \frac{1}{2}e$ if f is odd.

4.10 Multiplicities and the exceptional prime

The objective of Kummer's theory of ideal factorization was to "save" the property of unique factorization into primes for cyclotomic integers. Briefly stated, the objective was to show that, up to unit multiples, a cyclotomic integer is determined by its prime divisors. The bulk of the work needed to realize this objective is contained in the preceding section, in which the notion of "prime divisor" is extended to include cases in which divisibility by a prime divisor may not coincide with divisibility by any actual cyclotomic integer. However, these definitions must be supplemented in two ways—one major and one minor—before the full theory can be developed. The major way in which they must be supplemented is that a definition must be given of the *multiplicity* with which a given prime divisor divides a given cyclotomic integer. After all, even for ordinary positive integers it is not true that an integer is determined by its prime divisors but only that it is determined by its prime divisors *and* the multiplicities with which it is divisible by each. (For example, $12 = 2^2 \cdot 3$ and $18 = 2 \cdot 3^2$ have the same prime divisors.) The minor way in which they must be supplemented is that the prime divisors of the integer λ—the one prime integer not covered by the theorems of the preceding section—must be investigated.

If divisibility by a prime divisor of p coincides with divisibility by an actual cyclotomic integer $h(\alpha)$ then it is easy to assign multiplicities to divisibility by the prime divisor. If $g(\alpha)$ is divisible by the prime divisor, one can actually carry out division by $h(\alpha)$ to find a quotient $g(\alpha)/h(\alpha)$ which is a cyclotomic integer; if this quotient is again divisible then division can again be carried out to find $g(\alpha)/h(\alpha)^2$, and so forth. One says, in this case, that $g(\alpha)$ is divisible μ times by the prime divisor in question if $g(\alpha)$ is divisible by $h(\alpha)^\mu$. If there is no actual $h(\alpha)$, however, then some other method of defining multiplicities must be devised. The idea which underlies the following definition is that the cyclotomic integer $\Psi(\eta)$ used in the construction of the preceding section is divisible by all prime divisors of p *except* the one in question, so that $g(\alpha)$ times a high power of $\Psi(\eta)$ will be divisible by p as many times as $g(\alpha)$ is divisible by the prime divisor of p that is missing from $\Psi(\eta)$.

Definition. A cyclotomic integer $g(\alpha)$ is said to be divisible μ times by the prime divisor of p corresponding to $u_1, u_2, \ldots, u_e$ if $g(\alpha)\Psi(\eta)^\mu$ is divisible by p^μ, where $\Psi(\eta)$ is the product of the $ep - e$ factors $j - \eta_i$ $(i = 1, 2, \ldots, e; j = 0, 1, \ldots, p - 1; j \neq u_i)$ that are not divisible by this prime divisor. It is said to be divisible *exactly* μ times by the prime divisor if it is divisible μ times but not divisible $\mu + 1$ times.

Proposition. *The notion of "divisible one time" coincides with the notion of "divisible" as defined in the preceding section. The notion of "divisible exactly zero times" coincides with "not divisible." The integer p is divisible exactly once; the integer 1 is not divisible at all. If $g_1(\alpha)$ and $g_2(\alpha)$ are both divisible μ times then so is $g_1(\alpha) + g_2(\alpha)$. If $g_1(\alpha)$ is divisible exactly μ times and $g_2(\alpha)$ is divisible exactly ν times then $g_1(\alpha)g_2(\alpha)$ is divisible exactly $\mu + \nu$ times. Finally, if $g(\alpha) \neq 0$ then there is a unique integer $\mu \geqslant 0$ such that $g(\alpha)$ is divisible exactly μ times.*

PROOF. It was shown in the preceding section that $g(\alpha)$ is divisible by the prime divisor if and only if $g(\alpha)\Psi(\eta) \equiv 0 \bmod p$, that is, if and only if $g(\alpha)$ is divisible one time. This proves the first two statements of the proposition. The third statement says that p^2 does not divide $p[\Psi(\eta)]^2$ or, what is the same, that $\Psi(\eta)\Psi(\eta) \not\equiv 0 \bmod p$. This was proved in the preceding section. The statement that 1 is not divisible is just the statement that $\Psi(\eta) \not\equiv 0 \bmod p$. The next statement—that $g_1(\alpha) + g_2(\alpha)$ is divisible μ times whenever $g_1(\alpha)$ and $g_2(\alpha)$ are—follows immediately from the definition. Now if $g_1(\alpha)$ is divisible exactly μ times then $g_1(\alpha)\Psi(\eta)^\mu$ is divisible by p^μ but $g_1(\alpha)\Psi(\eta)^{\mu+1}$ is not divisible by $p^{\mu+1}$. Therefore $\phi_1(\alpha) = p^{-\mu}g_1(\alpha)\Psi(\eta)^\mu$ is a cyclotomic integer. If $\phi_1(\alpha)$ were divisible by the prime divisor in question then, since $\Psi(\eta)$ is divisible by the remaining $e - 1$ prime divisors of p $[\Psi(\eta)\sigma^k\Psi(\eta) \equiv 0 \bmod p]$, it would follow that $\phi_1(\alpha)\Psi(\eta)$ was divisible by all e prime divisors of p; hence, by the theorem of the preceding section, it would follow that p divided $\phi_1(\alpha)\Psi(\eta)$, and hence that $g_1(\alpha)$ was divisible $\mu + 1$ times by the prime divisor in question, contrary to assumption. Therefore $\phi_1(\alpha)$ is not divisible by the prime divisor in question. Similarly, if $g_2(\alpha)$ is divisible exactly ν times then $\phi_2(\alpha) = p^{-\nu}g_2(\alpha)\Psi(\eta)^\nu$ is a cyclotomic integer that is not divisible by the prime divisor of p in question. Therefore $g_1(\alpha)g_2(\alpha)\Psi(\eta)^{\mu+\nu}$ is divisible by $p^{\mu+\nu}$ and the quotient is a product $\phi_1(\alpha)\phi_2(\alpha)$ of two cyclotomic integers neither of which is divisible by the prime divisor in question. Because this prime divisor is *prime*, $\phi_1(\alpha)\phi_2(\alpha)$ is not divisible by the prime divisor in question. Since $\Psi(\eta)$ is not divisible by this prime divisor $[\Psi(\eta)\Psi(\eta) \not\equiv 0 \bmod p]$ it follows that $\phi_1(\alpha)\phi_2(\alpha)\Psi(\eta) = g_1(\alpha)g_2(\alpha)\Psi(\eta)^{\mu+\nu+1}p^{-\mu-\nu}$ is not divisible by the prime divisor of p, and *a fortiori* that is not divisible by p. Therefore $g_1(\alpha)g_2(\alpha)\Psi(\eta)^{\mu+\nu+1}$ is not divisible by $p^{\mu+\nu+1}$ and $g_1(\alpha)g_2(\alpha)$ is divisible with multiplicity exactly $\mu + \nu$, as was to be shown.

Let $g(\alpha)$ be a given cyclotomic integer, $g(\alpha) \neq 0$. If there is an integer $k \geqslant 0$ such that $g(\alpha)$ is not divisible k times by the prime divisor of p in question, then, directly from the definition, there must be at least one integer μ, $0 \leqslant \mu < k$, such that $g(\alpha)$ is divisible exactly μ times. As was seen above, this implies that $g(\alpha)\Psi(\eta)^\mu/p^\mu$ is not divisible at all. Therefore, for any $j \geqslant 0$, $g(\alpha)\Psi(\eta)^{\mu+j}/p^\mu$ is not divisible at all and, consequently, $g(\alpha)$ is not divisible with multiplicity $\mu + j$ for $j > 0$. In particular, $g(\alpha)$ is not divisible with multiplicity m whenever $m > k$. To prove the last statement of the proposition it will suffice to prove that there is an integer k such that $g(\alpha)$ is not divisible k times, because then the exact multiplicity μ with which $g(\alpha)$ is divisible is uniquely determined as the largest integer μ such that $g(\alpha)$ is divisible μ times. To this end, note that if $g(\alpha)$ is divisible k times, so is $Ng(\alpha)$. Let $Ng(\alpha) = p^n q$ where $n \geqslant 0$ and q is not divisible by p. Then $Ng(\alpha)$ is divisible exactly n times and can not, therefore, be divisible k times whenever $k > n$, and the proof is complete.

The other definition required for the fundamental theorem is the definition of the single prime divisor of λ which is simply $\alpha - 1$.

Definitions. A *prime divisor* in the arithmetic of cyclotomic integers (for a fixed prime $\lambda > 2$) is either (a) one of the e prime divisors of some prime $p \neq \lambda$ that have been defined above or (b) the prime divisor $\alpha - 1$

of λ. The prime divisors of the first type are described by giving a prime $p \neq \lambda$ and the integers $u_1, u_2, \ldots, u_e$ as in the preceding section. Given a cyclotomic integer $g(\alpha) \neq 0$ and a prime divisor there is a *multiplicity* $\mu \geq 0$ with which $g(\alpha)$ is divisible by the prime divisor. For prime divisors of the first type this multiplicity is defined to be the *exact* multiplicity with which $g(\alpha)$ is divisible by the prime divisor in the sense just defined above. For the single prime divisor of the second type it is defined to be the number of times that $\alpha - 1$ divides $g(\alpha)$.

Proposition. *Congruence* mod $\alpha - 1$ *is reflexive, symmetric, transitive, and consistent with addition and multiplication. Moreover,* $\alpha - 1$ *is prime; that is,* $g_1(\alpha) g_2(\alpha) \equiv 0$ mod $\alpha - 1$ *implies that either* $g_1(\alpha) \equiv 0$ *or* $g_2(\alpha) \equiv 0$ mod $\alpha - 1$. *An integer is divisible by* $\alpha - 1$ *if and only if it is divisible by* λ. *The multiplicity with which* $g(\alpha) \neq 0$ *is divisible by* $\alpha - 1$ *is well defined; that is, there is a* $\mu \geq 0$ *such that* $(\alpha - 1)^\mu$ *divides* $g(\alpha)$ *but* $(\alpha - 1)^{\mu+1}$ *does not. If* $\alpha - 1$ *divides both* $g_1(\alpha)$ *and* $g_2(\alpha)$ *with multiplicity at least* μ *then it divides* $g_1(\alpha) + g_2(\alpha)$ *with multiplicity at least* μ. *If* $\alpha - 1$ *divides* $g_1(\alpha)$ *with multiplicity exactly* μ *and* $g_2(\alpha)$ *with multiplicity exactly* ν *then it divides* $g_1(\alpha) g_2(\alpha)$ *with multiplicity exactly* $\mu + \nu$. *Finally,* λ *divides* $g(\alpha)$ *if and only if* $\alpha - 1$ *divides* $g(\alpha)$ *with multiplicity at least* $\lambda - 1$.

PROOF. It was seen in Section 4.3 that $\alpha - 1$ divides $g(\alpha)$ if and only if $g(1) \equiv 0$ mod λ. This shows that $\alpha - 1$ is prime. All statements but the last one are elementary. The last statement follows from the fact that $\lambda = (\alpha - 1)(\alpha^2 - 1) \cdots (\alpha^{\lambda-1} - 1) = (\alpha - 1)^{\lambda-1} \cdot$ unit.

4.11 The fundamental theorem

Theorem. *Let* λ *be a given prime greater than 2 and let* $g(\alpha)$ *and* $h(\alpha)$ *be nonzero cyclotomic integers built up from a* λth *root of unity* $\alpha \neq 1$. *Then* $g(\alpha)$ *divides* $h(\alpha)$ *if and only if every prime divisor which divides* $g(\alpha)$ *also divides* $h(\alpha)$ *with multiplicity at least as great.*

PROOF. If $g(\alpha)$ divides $h(\alpha)$, say $h(\alpha) = q(\alpha) g(\alpha)$, then certainly every prime divisor which divides $g(\alpha)$ also divides $h(\alpha)$ with multiplicity at least as great. It is the converse of this statement which requires proof. Now $g(\alpha)$ divides $h(\alpha)$ if and only if the integer $Ng(\alpha) = g(\alpha) g(\alpha^2) \cdots g(\alpha^{\lambda-1})$ divides $h(\alpha) g(\alpha^2) \cdots g(\alpha^{\lambda-1})$. If every prime divisor which divides $g(\alpha)$ also divides $h(\alpha)$ with multiplicity at least as great then, by the way in which multiplicities combine under multiplication, every prime divisor which divides the integer $Ng(\alpha)$ also divides $h(\alpha) g(\alpha^2) g(\alpha^3) \cdots g(\alpha^{\lambda-1})$ with multiplicity at least as great. Therefore in order to prove the theorem it will suffice to prove it in the special case in which $g(\alpha)$ is an integer.

137

Now if $g(\alpha)$ is a prime integer $p \neq \lambda$ then the statement to be proved is that if $h(\alpha)$ is divisible by each of the e prime divisors of p then $h(\alpha)$ is divisible by p. This was proved in Section 4.9. If $g(\alpha) = \lambda$ then the statement is that if $(\alpha - 1)^{\lambda - 1}$ divides $h(\alpha)$ then λ divides $h(\alpha)$. This is immediate from $\lambda = (\alpha - 1)^{\lambda - 1} \cdot$ unit. Therefore in order to prove the theorem it will suffice to prove that if the theorem is true for division by $g_1(\alpha)$ and true for division by $g_2(\alpha)$ then it is true for division by $g_1(\alpha) g_2(\alpha)$. This is very simple to prove. If every prime divisor which divides $g_1(\alpha) g_2(\alpha)$ also divides $h(\alpha)$ with multiplicity at least as great and if the theorem is true for division by $g_1(\alpha)$ then $g_1(\alpha)$ divides $h(\alpha)$, say $h(\alpha) = g_1(\alpha) h_1(\alpha)$. By the way in which multiplicities combine it follows that every prime divisor which divides $g_2(\alpha)$ also divides $h_1(\alpha)$ with multiplicity at least as great. If the theorem is true for division by $g_2(\alpha)$ it follows that $h_1(\alpha) = g_2(\alpha) h_2(\alpha)$ for some $h_2(\alpha)$. Therefore $h(\alpha) = g_1(\alpha) g_2(\alpha) h_2(\alpha)$ and $h(\alpha)$ is divisible by $g_1(\alpha) g_2(\alpha)$, as was to be shown. This completes the proof of the fundamental theorem.

This theorem "saves" the property of unique factorization into primes in the following sense.

Corollary. *If two cyclotomic integers $g(\alpha)$ and $h(\alpha)$ are divisible by exactly the same prime divisors with exactly the same multiplicities, then they differ only by a unit multiple, $g(\alpha) = unit \cdot h(\alpha)$.*

PROOF. By the fundamental theorem $g(\alpha)$ divides $h(\alpha)$ and $h(\alpha)$ divides $g(\alpha)$. Therefore $h(\alpha)/g(\alpha)$ and $g(\alpha)/h(\alpha)$ are cyclotomic integers. Since their product is 1 they must both be units.

Of course something is lost. Although the "factorization" of a cyclotomic integer determines the cyclotomic integer to within unit multiples, it is no longer true, as it is in the case of the factorization of ordinary integers, that the "factorization" can be prescribed arbitrarily. For example, when $\lambda = 23$, it is impossible to find a cyclotomic integer which is divisible once by one prime divisor of 47 and not divisible by any other prime divisor. A very natural and very important question then arises, namely, the question of *which* "factorizations" are actually possible. This question leads very naturally, as will be shown in Chapter 5, to the notion of equivalence and to the *divisor class group*, which is a finite group whose structure gives a means of describing very subtle facts about the arithmetic of cyclotomic integers. The remainder of this chapter is devoted to developing the notation and terminology for expressing Kummer's theory in a convenient form.

EXERCISES

1. Prove that if $p \neq \lambda$ is a prime with exponent f mod λ and if $g(\alpha)$ is a cyclotomic integer with norm p^f then $g(\alpha)$ is divisible by one prime divisor of p and is not

divisible by the others. Conclude that $g(\alpha)$ is *prime* and that it divides $h(\alpha)$ μ times if and only if $h(\alpha)$ is divisible with multiplicity μ by the prime divisor of p that divides $g(\alpha)$.

2. Prove that if $Ng(\alpha) = p_1^{\mu_1} p_2^{\mu_2} \cdots p_k^{\mu_k}$ is the ordinary prime factorization of the integer $Ng(\alpha)$ then, for each i, $g(\alpha)$ is divisible by exactly μ_i/f_i prime divisors of p_i—counted with multiplicities—where f_i is the exponent of $p_i \bmod \lambda$ when $p_i \neq \lambda$ and where $f_i = 1$ when $p_i = \lambda$.

4.12 Divisors

As usual, let λ be a fixed prime integer greater than 2 and let "cyclotomic integer" mean a cyclotomic integer built up from a λth root of unity $\alpha \neq 1$. The *divisor of a nonzero cyclotomic integer* $g(\alpha)$ is a list, with multiplicities, of all the prime divisors which divide $g(\alpha)$; that is, it is a list of prime divisors which gives all prime divisors which divide $g(\alpha)$, listing each one a number of times equal to the multiplicity with which it divides $g(\alpha)$. A *divisor* is any finite list of prime divisors. A given prime divisor may occur more than once in a divisor—that is, it may occur with a multiplicity. Also the *empty* list is considered to be a divisor; it is the divisor of the cyclotomic integer 1. The *product* of two divisors is simply the juxtaposition of the two lists, that is, the list containing all prime divisors, counted with multiplicities, that are in the two given lists. In this way an arbitrary divisor can be regarded as a product of prime divisors. A divisor is said to be *divisible* by a given divisor if it can be written as a product of that divisor with another divisor. A cyclotomic integer is said to be divisible by a divisor if its divisor is divisible by that divisor, and, similarly, a divisor is said to be divisible by a cyclotomic integer if it is divisible by the divisor of the cyclotomic integer. A divisor (as opposed to *the* divisor) of a cyclotomic integer is any divisor which divides it, or, when the meaning is clear from the context, any cyclotomic integer which divides it.

With these definitions the fundamental theorem becomes the statement that $g(\alpha)$ divides $h(\alpha)$ if and only if the divisor of $g(\alpha)$ divides the divisor of $h(\alpha)$ (provided $g(\alpha)$ and $h(\alpha)$ are both nonzero). Note also that the way in which multiplicities combine under multiplication can be summarized by saying that the divisor of a product is the product of the divisors. Since 0 is divisible by every nonzero cyclotomic integer $g(\alpha)$, it is sometimes convenient to consider 0 to have the divisor which contains every prime divisor an infinite number of times. The theorem that factorization into prime divisors is unique is the statement that two cyclotomic integers $g(\alpha)$, $h(\alpha)$ which have the same divisor differ by a unit multiple, $g(\alpha) = \text{unit} \cdot h(\alpha)$. The failure of unique factorization into *actual* prime factors, when it occurs, is a reflection of the fact that a given divisor may not be the divisor of a cyclotomic integer. For example, when $\lambda = 23$ the divisor of $47 \cdot 139$ is the product of the 22 prime divisors of 47 and the 22 prime divisors of 139.

None of these 44 prime divisors is the divisor of an actual cyclotomic integer. The two factorizations of $47 \cdot 139$ that were described at the end of Section 4.4 show that there are two quite different ways of arranging these 44 divisors into 22 pairs so that each pair is the divisor of an actual cyclotomic integer.

Since the basis of the entire theory is the notion of a prime divisor, it is useful to have a more concise way of designating prime divisors than as "the prime divisor of p corresponding to the set of integers $u_1, u_2, \ldots, u_e$ with the property described in the proposition of Section 4.9." (The one exceptional prime divisor can of course be designated simply by $\alpha - 1$.) Kummer never introduced a briefer designation, and this failure may have impeded the acceptance of his theory. However, he did make use of the description of prime divisors in terms of $\psi(\eta)$ that is described below and it enabled him to deal with prime divisors without noticeable inconvenience or awkwardness.

Theorem. *Given a prime divisor of $p \neq \lambda$ there is a cyclotomic integer $\psi(\eta)$ made up of periods of length f (the exponent of p mod λ) which is divisible exactly once by that prime divisor of p and which is not divisible by the remaining $e - 1$ prime divisors of p. Consequently, a cyclotomic integer $g(\alpha)$ is divisible with multiplicity μ by the given prime divisor of p if and only if $g(\alpha)[\sigma\psi(\eta)]^{\mu}[\sigma^2\psi(\eta)]^{\mu} \cdots [\sigma^{e-1}\psi(\eta)]^{\mu}$ is divisible by p^{μ}.*

PROOF. Let $\Psi(\eta)$ correspond as before to the given prime divisor of p. That is, let $u_1, u_2, \ldots, u_e$ be the integers such that $u_i - \eta_i$ is divisible by the given prime divisor and let $\Psi(\eta)$ be the product of the $ep - e$ factors $j - \eta_i$ in which $j \neq u_i$. Then $\Psi(\eta)$ is divisible by all e prime divisors of p except the given one. Then $\phi(\eta) = \sigma\Psi(\eta) + \sigma^2\Psi(\eta) + \cdots + \sigma^{e-1}\Psi(\eta)$ is divisible by the given prime divisor of p (each term of the sum is divisible by it) but not divisible by the remaining $e - 1$ prime divisors of p (for each of these, all but one of the terms are divisible by it but that one is not divisible). If $\phi(\eta)$ is divisible with multiplicity exactly one by the given prime divisor of p then $\psi(\eta) = \phi(\eta)$ has the required properties. If $\phi(\eta)$ is divisible with multiplicity greater than one, set $\psi(\eta) = \phi(\eta) + p$. Then $\psi(\eta)$ is not divisible by the $e - 1$ prime divisors other than the given one (because if it were then $\phi(\eta) = \psi(\eta) - p$ would be), it is divisible by the given prime divisor of p (because $\phi(\eta)$ is), and it is not divisible with multiplicity greater than one (because if it were then $p = \psi(\eta) - \phi(\eta)$ would be). This completes the construction of $\psi(\eta)$. The second statement of the theorem is an immediate consequence of the fundamental theorem.

Theorem. *The cyclotomic integers called "prime" in the examples of Section 4.7 are indeed prime. More generally, if $p \neq \lambda$ is a prime whose exponent mod λ is f and if $g(\alpha)$ is a cyclotomic integer whose norm is p^f then $g(\alpha)$ is prime. [If $g(\alpha) = g(\eta)$ is made up of periods of length f then the condition*

$Ng(\alpha) = p^f$ *can also be stated in the form* $g(\eta) \cdot \sigma g(\eta) \cdot \ldots \cdot \sigma^{e-1} g(\eta) = \pm p.$]

PROOF. Let $\nu_1, \nu_2, \ldots, \nu_e$ be the exact multiplicities with which the prime divisors of p divide $g(\alpha)$. Then $\sigma g(\alpha)$ is divisible with the multiplicities $\nu_2, \nu_3, \ldots, \nu_e, \nu_1$ exactly, $\sigma^2 g(\alpha)$ with the multiplicities $\nu_3, \nu_4, \ldots, \nu_2$ exactly, and so forth. Therefore each prime divisor of p divides $Ng(\alpha)$ with multiplicity exactly $(\nu_1 + \nu_2 + \cdots + \nu_e)f$. If $Ng(\alpha) = p^f$ then one ν_i must be 1 and the others 0. Then, by the fundamental theorem, divisibility by $g(\alpha)$ coincides with divisibility by a prime divisor and it follows that $g(\alpha)$ is prime.

Notations. Let a prime divisor of p be given and let $\psi(\eta)$ be a cyclotomic integer as in the theorem, that is, a cyclotomic integer made up of periods of length f (the exponent of p mod λ) which is divisible just once by the prime divisor of p in question and not divisible by any of the remaining $e - 1$ prime divisors of p. Then the given prime divisor will be designated $(p, \psi(\eta))$. Since this divisor divides p and $\psi(\eta)$, both with multiplicity exactly one, and no other prime divisor divides both, $(p, \psi(\eta))$ is the *greatest common divisor* of p and $\psi(\eta)$. Thus the notation $(p, \psi(\eta))$ is in accord with a common notation for the greatest common divisor of two numbers. If it is possible to find such a $\psi(\eta)$ with the additional property that its divisor *coincides* with the prime divisor of p in question then this prime divisor will also be denoted $(\psi(\eta))$. The unique prime divisor of λ will be denoted $(\alpha - 1)$ (as opposed to $\alpha - 1$, without parentheses, which denotes the cyclotomic integer of which $(\alpha - 1)$ is the divisor). Divisors will normally be designated by capital Latin letters $A, B, C \ldots$ and the product of divisors will be designated as usual by simple juxtaposition. Thus AB denotes the product of the divisor A and the divisor B. Moreover, A^n, where n is a positive integer, will designate the product of A with itself n times. An arbitrary divisor A can then be written in the form

$$(p_1, \psi_1)^{\mu_1} (p_2, \psi_2)^{\mu_2} \cdots (p_m, \psi_m)^{\mu_m}$$

where, for each $i = 1, 2, \ldots, m$, p_i is a prime integer, μ_i is a positive integer, and ψ_i is a cyclotomic integer made up of periods of length f_i (where f_i is the exponent of p_i mod λ or, if $p_i = \lambda$, where $f_i = 1$) whose divisor contains one prime divisor of p_i with multiplicity exactly one and contains no other prime divisors of p_i at all. If the divisor of $\psi(\eta)$ is *equal* to a prime divisor of p_i then p_i may be omitted and the divisor $(p_i, \psi_i)^{\mu_i}$ may instead be written as $(\psi_i)^{\mu_i}$. The empty divisor will be denoted by I and an exponent of 0 on a divisor A will be defined to mean $A^0 = I$.

EXERCISES

1. Kummer states ([K8], p. 333) that if the list of integers $u_1, u_2, \ldots, u_e$ contains one integer u_j with multiplicity one then $u_j - \eta_0$ is divisible by one of the e prime

divisors of p with multiplicity exactly 1 and is not divisible by the remaining $e - 1$ prime divisors at all. This statement is slightly in error. Prove that $u_j + kp - \eta_0$ has the required property for exactly $p - 1$ of the p possible values of $k \bmod p$. Prove, more generally, that if $\phi(\eta)$ is divisible by exactly one prime divisor of p then so is $\phi(\eta) + kp$ for all integers k and the multiplicity with which this one divides it is exactly 1 for $p - 1$ of the possible values of $k \bmod p$, greater than 1 for the remaining value of $k \bmod p$.

2. In the case $\lambda = 31$ write the prime divisors of 2 in the form $(2, \psi)$.

3. In the notation introduced in this section what are the prime divisors of 47 in the case $\lambda = 23$?

4. Using Exercise 1, give a necessary and sufficient condition for the "factorization" of $p \neq \lambda$ to take the form $(p) = (p, \eta_0 - u)(p, \eta_1 - u) \ldots (p, \eta_{e-1} - u)$ for some integer u.

5. Prove that for any given divisor A the number of incongruent cyclotomic integers mod A is finite. [Find an integer n such that congruence mod A implies congruence mod n.]

6. Let $\psi(\eta)$ be a cyclotomic integer made up of periods of length f, and let p be a prime whose exponent mod λ is f. Show that $\psi(\eta)$ is divisible by a single prime divisor of p, and by that one with multiplicity 1, if and only if $N\psi(\eta)$ is divisible by p^f but not by p^{f+1}.

4.13 Terminology

Kummer called cyclotomic integers "complex numbers" or, when he was being more specific, "complex numbers built up from a complex λth root of unity." He called divisors "ideal complex numbers." This latter was a poor choice of terminology for several reasons. In the first place, although a "complex number" (cyclotomic integer) does determine an "ideal complex number" (divisor), many different complex numbers determine in this way the same ideal complex number because cyclotomic integers which differ by a unit multiple have the same divisor. Thus an "ideal complex number" is *not* a generalized sort of "complex number" as Kummer's terminology suggests. Moreover, there is no meaningful way in which divisors can be added, and it is surely misleading to call things "numbers" which cannot be added. Another drawback of Kummer's terminology is the semantic problem posed by the fact that an ideal complex number *may* (in his usage) be an actual complex number, so that it may be necessary to distinguish between actual ideal complex numbers and ideal ideal complex numbers. Furthermore, although Kummer had in mind some beautiful and clear analogies when he introduced the term (see in particular the first page of his announcement [K7] and the concluding remarks of Section 10 of his main paper [K8] on the theory of ideal factorization), these analogies do not appear to have been appealing or helpful to his successors. Finally, the terminology "ideal complex number" has the great psychological disadvantage that it leads almost inevitably to the question "What *is* an ideal complex number?" This question opens the same Pandora's box of metaphysical issues as the questions "What is a natural number?" or "What is a real number?" and it is no more susceptible of a definitive answer than are these ancient conundrums. (Of course, logically speaking, there was no need for Kummer to deal with this question at all, and he did not. What he did was to describe in detail how ideal complex numbers are to be represented and how computations with them are to be performed, which, after all, is the answer that a practising mathematician gives to the question "What is a number?"—see [E1], especially "The Parable of the Logician and the Carpenter.") For all of these reasons, the modern terminology of divisors is preferable to Kummer's terminology of ideal complex numbers.

The term "divisor" is well chosen in that it describes the main context in which divisors are used. The idea of a "prime divisor" was defined in Section 4.9 in the context of defining what it means to say that a given cyclotomic integer is *divisible* by such a prime divisor. More generally, for such a prime divisor A, the statement "$g_1(\alpha) \equiv g_2(\alpha) \bmod A$" was defined. For any divisor A, prime or not, one can define $g_1(\alpha) \equiv g_2(\alpha) \bmod A$ to mean that $g_1(\alpha) - g_2(\alpha)$ is divisible by A in the sense defined in the preceding section. It is primarily in the context of the statement "$g_1(\alpha) \equiv g_2(\alpha) \bmod A$" that divisors A have their meaning. In view of this, the following theorem has the important consequence that a knowledge of the relation $g_1(\alpha) \equiv g_2(\alpha) \bmod A$ is sufficient to determine the divisor A.

Theorem. *Let A and B be divisors. If every cyclotomic integer divisible by A is also divisible by B then A is divisible by B.*

Corollary. *If A and B are divisors and if the relation "$g_1(\alpha) \equiv g_2(\alpha) \bmod A$" is equivalent to "$g_1(\alpha) \equiv g_2(\alpha) \bmod B$" (that is, one relation holds if and only if the other does) then $A = B$.*

PROOF. The corollary follows from the observation that if A divides B and B divides A then $A = B$. Let $p_1, p_2, \ldots, p_n$ be the prime integers that are divisible by the prime divisors which divide A (in other words, the p's are the prime factors of the norm of A—see the next section) and let $A = A_1 A_2 \cdots A_n$ be a decomposition of A as a product of divisors A_i such that A_i contains all prime divisors in A which divide p_i. Since a sufficiently high power of $p_1 p_2 \cdots p_n$ is divisible by A it must, by assumption, be divisible by B. This shows that every prime divisor in B divides one of the p_i and therefore that B can be decomposed as a product $B = B_1 B_2 \cdots B_n$ in which B_i contains all prime divisors in B which divide p_i. (Some of the B_i may be I.) Since the order of the p's is arbitrary, it will suffice to prove that B_1 divides A_1. If $p_1 = \lambda$ then $A_1 = (\alpha - 1)^\mu$ for some μ. Then the cyclotomic integer $(\alpha - 1)^\mu (p_2 p_3 \cdots p_n)^\nu$ is divisible by A for all sufficiently large ν. Therefore it is also divisible by B for large ν, and this implies, because B_1 does not divide $(p_2 p_3 \cdots p_n)^\nu$ at all, that B_1 divides the cyclotomic integer $(\alpha - 1)^\mu$. Therefore B_1 divides the divisor of $(\alpha - 1)^\mu$, which is A_1. If $p_1 \neq \lambda$ let f be the exponent of p_1 mod λ and let ψ be a cyclotomic integer made up of periods of length f which is divisible by one of the $e = (\lambda - 1)/f$ prime divisors of p_1 with multiplicity 1 and is not divisible at all by the remaining $e - 1$. Then one can form a product of conjugates of ψ, call it x, with the property that the divisor of x is A_1 times prime divisors which do not divide p_1. Now $x(p_2 p_3 \cdots p_n)^\nu$ is divisible by A for large ν; therefore it is divisible by B_1. Since B_1 does not divide $(p_2 p_3 \cdots p_n)^\nu$ at all, it follows that B_1 divides the divisor of x. Since the divisor of x is A_1 times a factor that contains none of the prime divisors of p_1 (and *a fortiori* none of the prime divisors of B_1) B_1 divides A_1 as was to be shown.

Richard Dedekind (1831–1916), who did important work in generalizing Kummer's theory, did not preserve Kummer's terminology. He was very much absorbed with the philosophical problem "What is a number?" and in the case of Kummer's "ideal complex numbers" he gave an answer which has had a lasting impact on the mathematical vocabulary. He took advantage of the fact proved above, that a divisor is determined by the set of all cyclotomic integers it divides, to *identify* the ideal complex number with the set of all things that it divides. Such a set he called, in a rather bizarre adaptation of Kummer's terminology, an *ideal*. He then showed that "ideals" are characterized, among all subsets of the cyclotomic integers, by two simple properties:

(i) The sum of any two cyclotomic integers in a given ideal is again in that ideal.
(ii) The product of a cyclotomic integer in a given ideal with any cyclotomic integer is again in the given ideal.

That ideals have these properties is obvious—$g_1(\alpha) \equiv 0$, $g_2(\alpha) \equiv 0$ mod A imply $g_1(\alpha) + g_2(\alpha) \equiv 0$ mod A and $g_1(\alpha)\phi(\alpha) \equiv 0$ mod A for all $\phi(\alpha)$. What Dedekind proved was that, conversely, *any* subset of the cyclotomic integers which has these two properties is an ideal, that is, is the set of all cyclotomic integers divisible by A for some divisor A. [For this to be true in complete generality one must define the divisor of 0 to have the property that *only* 0 is divisible by it. See Exercise 3.] This observation makes it possible to characterize the prime divisors without ever finding them explicitly and consequently makes it possible to describe the theory abstractly without going through the actual construction of the prime divisors. This approach is especially helpful in studying the extension of Kummer's theory to other types of algebraic integers than cyclotomic integers; the achievement of this extension was one of Dedekind's major contributions.

Dedekind's terminology and in particular the notion of an "ideal" are very important in modern abstract algebra. However, in the remainder of this book the more constructive and computational notion of a "divisor" will be used to the exclusion of ideals. In other words, Kummer's more concrete and explicit formulation will be adopted even though his terminology will not be.

EXERCISES

1. What third property added to (i) and (ii) of the text characterizes *prime* ideals, that is, ideals which arise from prime divisors?

2. Let $\mathcal{G}$ be any subset of the cyclotomic integers which has properties (i) and (ii). Show that there is a finite set of cyclotomic integers $g_1(\alpha), g_2(\alpha), \ldots, g_n(\alpha)$ such that $\mathcal{G}$ consists of all cyclotomic integers which can be written in the form $b_1(\alpha) g_1(\alpha) + b_2(\alpha) g_2(\alpha) + \cdots + b_n(\alpha) g_n(\alpha)$ for some cyclotomic integers $b_1(\alpha), b_2(\alpha), \ldots, b_n(\alpha)$. [If $\mathcal{G}$ consists of 0 alone set $n = 1$ and $g_1(\alpha) = 0$. Otherwise choose $g_1(\alpha) \neq 0$ in $\mathcal{G}$. Show there are only finitely many incongruent elements mod $g_1(\alpha)$. From each class mod $g_1(\alpha)$ which contains an element of $\mathcal{G}$ choose one and include it among $g_2(\alpha), g_3(\alpha), \ldots, g_n(\alpha)$. The g's constructed in this way have the desired property.]

3. Use Exercise 2 to find what the divisor corresponding to $\mathcal{G}$ must be. [If $\mathcal{G} = \{0\}$ then there is no divisor unless the divisor of zero is defined to be the "divisor" which contains *all* prime divisors with *infinite* multiplicities, the justification for this being that 0 is divisible by everything and divides nothing but itself.] The actual proof that every subset $\mathcal{G}$ which has properties (i) and (ii) is the set of things divisible by some divisor is studied in Exercise 3 of Section 4.14.

4.14 Conjugations and the norm of a divisor

Just as conjugations $\alpha \mapsto \alpha^j$ act on cyclotomic integers, there is a natural way in which they act on *divisors*. Specifically, let σ be a conjugation $\alpha \mapsto \alpha^\gamma$ in which γ is a primitive root mod λ, so that conjugations can be written as powers of σ without using double superscripts. To describe the action of a typical conjugation σ^i on divisors, it suffices to describe the action of σ itself on divisors because σ^i is σ composed with itself i times. One way to do this is as follows.

Since by definition $\sigma(AB)$ will be $\sigma(A) \cdot \sigma(B)$, it will suffice to describe the action of σ on *prime* divisors. A simple way to do this is to write prime divisors in the form $(p, \psi(\eta))$ as in Section 4.12 and to define $\sigma(p, \psi(\eta))$ to be $(p, \sigma\psi(\eta))$. (The special divisor $(\alpha - 1)$ is equal to all of its conjugates because $\alpha^\gamma - 1$ has the same divisor as $\alpha - 1$.) This defines the action of σ on divisors. It is easily checked that this action has the following natural property.

145

Proposition 1. *Let σ^i be a conjugation of cyclotomic integers, let $g_1(\alpha)$, $g_2(\alpha)$ be cyclotomic integers, and let A be a divisor. Then to say $g_1(\alpha) \equiv g_2(\alpha)$ mod $\sigma^i A$ is equivalent to saying $\sigma^{-i} g_1(\alpha) \equiv \sigma^{-i} g_2(\alpha)$ mod A.*

PROOF. Exercise 1.

Proposition 1 describes congruence mod $\sigma^i A$ in terms of concepts that were already defined in the preceding sections and therefore, because a divisor is determined by the corresponding congruence relation, the proposition gives an alternative way of *defining $\sigma^i A$*.

Just as the norm of a cyclotomic integer $g(\alpha)$ is defined to be $g(\alpha) \cdot \sigma g(\alpha) \cdot \sigma^2 g(\alpha) \cdot \ldots \cdot \sigma^{\lambda-2} g(\alpha)$, the norm of a divisor A can be defined to be $A \cdot \sigma A \cdot \sigma^2 A \cdot \ldots \cdot \sigma^{\lambda-2} A$. The norm of a cyclotomic integer is a cyclotomic integer which, because it is invariant under σ, is in fact an ordinary integer. As was seen in Section 4.2, the norm of a cyclotomic integer is not only an integer but a nonnegative one. The norm of a divisor is a *divisor*. However, the following proposition shows that it can be considered to be a positive integer in a natural way.

Proposition 2. *If A is any divisor, there is a positive integer whose divisor is equal to the norm of A.*

PROOF. Exercise 2.

The norm of a divisor A will be denoted $N(A)$ or NA. It is, in the first instance, a divisor, but in certain situations—particularly in Chapter 6—it is useful to consider $N(A)$ to be the positive integer of which it is the divisor. This positive integer can also be described in the following simple way.

Theorem. *When $N(A)$ is regarded as a positive integer it is equal to the number of incongruent cyclotomic integers mod A. In other words, one can find a set of $N(A)$ cyclotomic integers with the property that every cyclotomic integer is congruent mod A to one and only one element of the set.*

PROOF. If A is a prime divisor, then its norm is p^f, where p is the prime integer it divides, and f is the exponent of p mod λ; the fact that this is also equal to the number of incongruent elements mod A was proved in Theorem 1 of Section 4.9. (If A is the exceptional prime divisor $(1 - \alpha)$ then every cyclotomic integer is congruent to one and only one of the λ integers $0, 1, 2, \ldots, \lambda - 1$.)

Consider next the case in which A is a power of a prime divisor $A = P^n$. Let ψ be a cyclotomic integer which is divisible by P with multiplicity exactly 1 and not divisible at all by any of the conjugates of P. It will be shown that each cyclotomic integer is congruent to one of the form

$a_0 + a_1\psi + a_2\psi^2 + \cdots + a_{n-1}\psi^{n-1}$ mod P^n and that two cyclotomic integers of this form are congruent mod P^n if and only if the coefficients $a_0, a_1, \ldots, a_{n-1}$ are the same mod P; this will prove, then, that the number of classes mod P^n is equal to the number of ways of choosing $a_0, a_1, \ldots, a_{n-1}$ mod P, which is $N(P)^n = N(P^n)$, as is to be shown. When $n = 1$ this obvious. Suppose it has been proved for $n - 1$. Then for a given cyclotomic integer x there are cyclotomic integers $a_0, a_1, \ldots, a_{n-2}$ for which $x \equiv a_0 + a_1\psi + \cdots + a_{n-2}\psi^{n-2}$ mod P^{n-1} and the coefficients $a_0, a_1, \ldots, a_{n-2}$ are uniquely determined mod P. Let $y = x - a_0 - a_1\psi - \cdots - a_{n-2}\psi^{n-2}$. Then $y \equiv 0$ mod P^{n-1}. What is to be shown is that $y \equiv a\psi^{n-1}$ mod P^n for some a and that a is uniquely determined mod P. Let Ψ denote the product of the $e - 1$ distinct conjugates of ψ so that $\psi\Psi = pk$ where k is an integer relatively prime to p. Then the desired congruence $y \equiv a\psi^{n-1}$ mod P^n is equivalent to $y\Psi^{n-1} \equiv ap^{n-1}k^{n-1}$ mod P^n which, in turn, is equivalent to $y\Psi^{n-1}m^{n-1} \equiv ap^{n-1}$ mod P^n where m is an integer such that $mk \equiv 1$ mod p. Since $y \equiv 0$ mod P^{n-1} by assumption, $y\Psi^{n-1}m^{n-1}$ is divisible by p^{n-1} and the desired congruence is equivalent to the statement that the quotient is congruent to a mod P, which shows, as required, that there is one and only one such a mod P. This completes the proof in the case $A = P^n$. The general case now follows easily from the generalized Chinese remainder theorem:

Chinese Remainder Theorem. *Let A and B be relatively prime divisors and let a and b be cyclotomic integers. Then there is a solution x of the congruences $x \equiv a$ mod A and $x \equiv b$ mod B.*

Using this theorem, the proof of the previous theorem can be completed as follows. The number of incongruent elements mod AB is never more than the number of incongruent elements mod A times the number of incongruent elements mod B because $x \equiv x'$ mod A and $x \equiv x'$ mod B implies $x \equiv x'$ mod AB (A and B both divide $x - x'$, and they are relatively prime, which implies that AB divides $x - x'$), that is, the class of x mod AB is determined by its class mod A and its class mod B. The Chinese remainder theorem shows that all possible combinations of classes mod A and mod B occur, so that the number of classes mod AB is *equal* to the number of classes mod A times the number of classes mod B. By induction, if $A, B, C, \ldots, D$ are relatively prime divisors then the number of classes mod $ABC \cdots D$ is equal to the number of classes mod A times the number of classes mod B times the number of classes mod C times $\cdots$ times the number of classes mod D. If $A, B, C, \ldots, D$ are powers of prime divisors, then it was shown above that this number is equal to $N(A)N(B)N(C) \cdots N(D) = N(ABC \cdots D)$. Since any divisor can be written in the form $ABC \cdots D$ where $A, B, C, \ldots, D$ are relatively prime powers of prime divisors, it follows that the number of classes mod any divisor is the norm of that divisor, as was to be shown.

PROOF OF THE CHINESE REMAINDER THEOREM. The theorem implies that *if A and B are relatively prime divisors then there is a cyclotomic integer g satisfying $g \equiv 1 \bmod A$ and $g \equiv 0 \bmod B$.* Conversely, if this statement is known to be true then, by symmetry, there is also a cyclotomic integer h such that $h \equiv 1 \bmod B$ and $h \equiv 0 \bmod A$; then $ga + hb$ satisfies the congruences of the Chinese remainder theorem. Therefore this statement is equivalent to the Chinese remainder theorem.

Consider first the case where A and B are prime divisors. If they divide different prime integers, say p_A and p_B where $p_A \neq p_B$, then by the ordinary Chinese remainder theorem for integers there is an integer k satisfying $k \equiv 1 \bmod p_A$ and $k \equiv 0 \bmod p_B$. Then $k \equiv 1 \bmod A$ and $k \equiv 0 \bmod B$, as desired. If $p_A = p_B$ then, in particular, $p_A \neq \lambda$ because λ has only one prime divisor and by assumption $A \neq B$. The theorem of Section 4.12 then shows that there is a cyclotomic integer ψ which satisfies $\psi \equiv 0 \bmod B$ and $\psi \not\equiv 0 \bmod A$. Now, since A is prime, *nonzero elements are invertible* mod A, that is, $\psi \not\equiv 0 \bmod A$ implies there is a ϕ such that $\psi\phi \equiv 1 \bmod A$. This can be proved as follows. Let $a_1, a_2, \ldots, a_\nu$, where $\nu = p_A^f$, be a complete set of representatives mod A, that is, a set of cyclotomic integers such that each cyclotomic integer is congruent mod A to one and only one of the a_i. Then $\psi a_1, \psi a_2, \ldots, \psi a_\nu$ are all distinct mod A because $\psi a_i \equiv \psi a_j \bmod A$ implies $\psi(a_i - a_j) \equiv 0 \bmod A$, which, because $\psi \not\equiv 0 \bmod A$ and A is prime, implies $a_i - a_j \equiv 0 \bmod A$ and therefore $a_i = a_j$. Because each ψa_i is congruent to one and only one a_j, and because no two are congruent to the same a_j, each a_j is congruent to one and only one ψa_i and, in particular, 1 is congruent to ψa_i for some i. Then $\psi a_i \equiv 1 \bmod A$ and $\psi a_i \equiv 0 \bmod B$, as desired.

Consider next the case in which A and B are prime powers, say $A = P^n$, $B = Q^m$ where P and Q are distinct prime divisors and m and n are positive integers. By what was shown above, there is a cyclotomic integer g such that $g \equiv 1 \bmod P$ and $g \equiv 0 \bmod Q$. Let $h = g^m$. Then $h \equiv 0 \bmod B$. Mod A, one can write $h \equiv a_0 + a_1\psi + \cdots + a_{n-1}\psi^{n-1}$ where ψ is divisible by P with multiplicity exactly 1 and not divisible at all by any of the conjugates of P. Moreover, because $h \equiv 1 \bmod P$, one has $a_0 \equiv 1 \bmod P$ and, in the construction of $a_0 + a_1\psi + \cdots + a_{n-1}\psi^{n-1}$ given above, one can set $a_0 = 1$ and solve for $a_1, a_2, \ldots, a_{n-1}$. One can then find a cyclotomic integer f for which $hf \equiv 1 \bmod A$ simply by writing $f = b_0 + b_1\psi + \cdots + b_{n-1}\psi^{n-1}$, $hf \equiv b_0 + (b_1 + a_1 b_0)\psi + (b_2 + a_1 b_1 + a_2 b_0)\psi^2 + \cdots + (b_{n-1} + a_1 b_{n-2} + \cdots + a_{n-1} b_0)\psi^{n-1} \bmod A$ and using the equations $b_0 = 1$, $b_1 + a_1 b_0 = 0$, $b_2 + a_1 b_1 + a_2 b_0 = 0, \ldots, b_{n-1} + a_1 b_{n-2} + \cdots + a_{n-1} b_0$ to define $b_0, b_1, b_2, \ldots, b_{n-1}$ successively. Then $hf \equiv 1 \bmod A$, $hf \equiv 0 \bmod B$, as desired.

Consider finally the case in which $A = A_1 A_2 \ldots A_\mu$, $B = B_1 B_2 \ldots B_\nu$ where the A_i and the B_j are all powers of prime divisors and are all relatively prime to one another. Then, as was just shown, for $C = A_2, A_3, \ldots, A_\mu, B_1, B_2, \ldots, B_\nu$ there is a cyclotomic integer which is 1 mod A_1 and 0 mod C. The product of these $\mu + \nu - 1$ cyclotomic integers is then 1 mod A_1 and 0 mod $A_2, A_3, \ldots, A_\mu, B_1, B_2, \ldots, B_\nu$. Let g_1 denote this cyclotomic integer. In the same way, one can find cyclotomic integers $g_2, g_3, \ldots, g_\mu$ with the property that $g_i \equiv 1$ mod A_i but $g_i \equiv 0$ mod all of the divisors $A_1, \ldots, A_\mu, B_1, \ldots B_\nu$ other than A_i. Then $g_1 + g_2 + \cdots + g_\mu$ is 1 mod $A_1, A_2, \ldots, A_\mu$, and therefore 1 mod A, at the same time that it is 0 mod $B_1, B_2, \ldots, B_\nu$ and therefore 0 mod B. This completes the proof of the Chinese remainder theorem.

EXERCISES

1. Prove Proposition 1. [Show that it will suffice to prove that $g(\alpha) \equiv 0$ mod σA implies $\sigma^{-1} g(\alpha) \equiv 0$ mod A. Prove this directly in the case where A is a power of a prime divisor. Then deduce the general case.]

2. Prove Proposition 2.

3. Prove the following generalization of the fact that the greatest common divisor d of two integers m and n can be written in the form $d = am + bn$ for some integers a and b: If $g_1(\alpha), g_2(\alpha), \ldots, g_n(\alpha)$ are given cyclotomic integers then a cyclotomic integer $\phi(\alpha)$ can be written in the form $\phi(\alpha) = b_1(\alpha) g_1(\alpha) + b_2(\alpha) g_2(\alpha) + \cdots + b_n(\alpha) g_n(\alpha)$ if and only if $\phi(\alpha)$ is divisible by the greatest common divisor of $g_1(\alpha), g_2(\alpha), \ldots, g_n(\alpha)$. Together with Exercise 2 of the preceding section this proves Dedekind's theorem that "ideals" can be described either in terms of properties (i) and (ii) or in terms of divisors. [Clearly if ϕ is a combination of the g's it is divisible by their g.c.d. It is the converse which requires proof. If $n = 1$ this is the fundamental theorem. Show that if it is true for $n - 1$ then it is true for n. Let A be the g.c.d. and let AB be the g.c.d. of the first $n-1$ of the g's. What must be shown is that if $\phi(\alpha) \equiv 0$ mod A then the congruence $\phi(\alpha) \equiv b_n(\alpha) g_n(\alpha)$ mod AB can be solved for $b_n(\alpha)$. Prove this by considering the prime power divisors of AB separately and applying the Chinese remainder theorem.]

4.15 Summary

The theory of divisors can be described as a mapping

$$\left\{ \begin{array}{c} \text{nonzero} \\ \text{cyclotomic} \\ \text{integers} \end{array} \right\} \to \{ \text{divisors} \}.$$

The main step in the construction of the theory is the definition of the *prime* divisors. Each prime divisor divides a prime integer p; the prime λ has one prime divisor and a prime $p \neq \lambda$ has e prime divisors where

$ef = \lambda - 1$ and f is the exponent of p mod λ. A divisor is any product of the prime divisors. Thus *by definition* unique factorization into primes holds for divisors. The mapping indicated by the arrow above assigns to each nonzero cyclotomic integer its divisor, that is, the prime divisors, counted with multiplicities, which divide it. The basic properties of this mapping are:

(i) The divisor of λ is the $(\lambda - 1)$st power of the prime divisor which divides it. For other prime integers p the divisor of p is the product of the e prime divisors which divide it (each to the first power).

(ii) The divisor of a product is the product of the divisors.

(iii) If $g_1(\alpha) \equiv g_2(\alpha)$ mod A is defined to mean that the divisor of $g_1(\alpha) - g_2(\alpha)$ is divisible by A (where $g_1(\alpha)$, $g_2(\alpha)$ are cyclotomic integers and A is a divisor) then this congruence relation has the usual properties; that is, it is reflexive, symmetric, transitive, and consistent with addition and multiplication.

(iv) Two divisors which define the same congruence as in (iii) are necessarily identical.

(v) The fundamental theorem: If the divisor of $g(\alpha)$ divides the divisor of $h(\alpha)$ then $g(\alpha)$ divides $h(\alpha)$.

It follows from the fundamental theorem that *two cyclotomic integers which have the same divisor are the same except for a unit multiple*. Thus, if unit multiples are disregarded, a cyclotomic integer is determined by its divisor. The divisor can be regarded as a "factorization" of the cyclotomic integer, but into prime divisors, not into prime cyclotomic integers. If every prime divisor is the divisor of a cyclotomic integer—that is, if the above mapping is *onto* the set of all divisors—then unique factorization into prime cyclotomic integers holds (Exercise 1). However, when $\lambda = 23$ the prime divisors of $p = 47$ are not divisors of any cyclotomic integers. In this case and in any case where there is a prime divisor not the divisor of a cyclotomic integer, unique factorization into prime cyclotomic integers *fails* (Exercise 2).

The further study of the factorization of cyclotomic integers depends on the study of the question "Which divisors are the divisors of cyclotomic integers?" or, in other words, "What is the *image* of the above mapping?" This is the subject of the next chapter.

EXERCISES

1. Show that the mapping is onto the set of all divisors if and only if every prime divisor is the divisor of some cyclotomic integer. Show that when this is the case every cyclotomic integer can be written as a product of prime cyclotomic integers and two such decompositions of a given cyclotomic integer are the same except for unit multiples.

2. Show that if there is a prime divisor which is not the divisor of a cyclotomic integer then there is a cyclotomic integer which is not divisible by any prime cyclotomic integer. Show, moreover, that in this case there is a cyclotomic integer which can be decomposed in two distinct ways as a product of irreducible cyclotomic integers — ways that are distinct even if unit multiples are disregarded. [Use the cyclotomic integer $\psi(\eta)$ that is divisible by exactly one prime divisor of p with multiplicity exactly one.]

5

Fermat's Last Theorem for regular primes

5.1 Kummer's remarks on quadratic integers

It frequently happens that great innovations are made not by revolutionaries bent on change but by men who have a great respect for what has gone before and who are motivated by the wish to conserve and fulfill the traditions of their predecessors. This was certainly the case with Kummer. As K.-R. Biermann points out in his biography of Kummer in the *Dictionary of Scientific Biography*, Kummer was by nature very conservative, not in any narrow political sense, but in the sense that he was dedicated to building on the basis of existing traditions. In understanding the motivation of Kummer's work, it is important to realize that he had no wish to introduce new abstract structures for their own sake but rather, as he said at the beginning of his announcement [K7] of the new theory, his goal was "to complete and simplify" existing structures.

This chapter is devoted to Kummer's proof of Fermat's Last Theorem for a large class of prime number exponents λ which are now known as the *regular* primes. This proof requires another important innovation of Kummer's, namely, the notion of *equivalence* between two divisors (two ideal complex numbers). In keeping with Kummer's personality, it is to be expected that this new notion of equivalence was motivated by some very compelling consideration; Kummer would not have introduced it simply because it was an "interesting" possibility. Although it is tempting to suppose that it was the very attempt to prove Fermat's Last Theorem that motivated the definition of equivalence of divisors, the fact is that this definition was included as a very prominent part of Kummer's initial announcement of the theory of ideal divisors in 1846, well before Lamé's premature announcement prodded him into working out in detail the consequences of his theory for Fermat's Last Theorem. Thus it seems unlikely that this work had a major role in motivating the original definition of equivalence.

There appear to be at least two considerations which did motivate the definition. The first is that in applying divisor theory to actual problems—for example to the problems in cyclotomy which Kummer considers in his 1846 announcement—one is very soon confronted with the problem "when is a given divisor the divisor of an actual cyclotomic integer?" The solution of this problem leads, as will be shown in the next section, very naturally to the notion of equivalence. But the second motive appears from Kummer's statements to be almost as important to him as the first. It is the fact that this notion of equivalence is very closely related to Gauss's notion of the equivalence of binary quadratic forms.

Here again a "conservative" interpretation can be placed on Kummer's innovation. The study of Pell's equation and of other equations of degree two in two variables leads very naturally to the concept of *equivalence* of binary quadratic forms (see Section 8.2). Gauss in the *Disquisitiones Arithmeticae* was led to introduce the stronger notion of *proper equivalence* (see Section 8.3). Kummer observes that this notion always seems forced and artificial in the theory of binary quadratic forms—it requires, for example, that one regard the two forms $ax^2 + 2bxy + cy^2$ and $cx^2 + 2bxy + ay^2$ as not being properly equivalent—despite the fact that "it must be recognized that the Gaussian classification corresponds more closely to the essential nature of the matter." Thus, Kummer implicitly concludes, the Gaussian notion of proper equivalence is something which needs to be *saved* from this appearance of artificiality. He says that the theory of ideal factorization accomplishes this because "the entire theory of binary quadratic forms can be interpreted as the theory of complex numbers of the form $x + y\sqrt{D}$ [D is Gauss's notation for the *determinant* $b^2 - ac$ of the quadratic form $ax^2 + 2bxy + cy^2$] and as a result of this interpretation leads necessarily to ideal complex numbers [divisors] of the same type." He then goes on to say, in effect, that the notion of equivalence which he has defined for ideal complex numbers of the form $a_0 + a_1\alpha + a_2\alpha^2 + \cdots + a_{\lambda-1}\alpha^{\lambda-1}$ also applies to ideal complex numbers of the form $x + y\sqrt{D}$ and that, when these latter are interpreted as binary quadratic forms, the notion of equivalence coincides with Gauss's notion of proper equivalence. This, he then concludes, is the "true basis" of the Gaussian notion of equivalence.

Quite mysteriously, Kummer *never published* any details on this connection between binary quadratic forms and ideal complex numbers (divisors) of the form $x + y\sqrt{D}$. The few informal remarks in his 1846 announcement and a few more sketchy indications in later treatises ([K8, p. 366] and [K11, p. 114]) constitute all that he said on the subject or, at any rate, all that has survived. Thus, although it seems quite certain that the analogy with Gauss's theory played some role in the genesis of the notion of

153

equivalence of divisors, its exact role cannot be known. In Sections 5.2 and 5.3 the notion of equivalence of divisors is developed without any reference at all to Gauss's theory. As will be seen in Chapters 7 and 8, the motivating question of Section 5.2 is, however, very closely related to Gauss's theory. In my personal opinion the development that is given here is probably very close to the one that Kummer actually followed, but all I want to claim for it is that it makes the notion of equivalence seem a very natural and useful one.

The remainder of the chapter then gives two applications of Kummer's theory. The first and more important one is his proof of Fermat's Last Theorem for all prime exponents λ which satisfy certain conditions (A) and (B). The second is a proof of the famous law of quadratic reciprocity. This proof should be regarded rather as an aside. It is a natural extension of the line of thought of Section 5.2 which leads not only to a proof of quadratic reciprocity but even to a discovery of the *statement* of quadratic reciprocity, but it is here completely anachronistic because Kummer's work came 50 years after the first proof of quadratic reciprocity and this application of the theory to quadratic reciprocity was never, as far as we know, made by Kummer.

5.2 Equivalence of divisors in a special case

Kummer says in his 1846 announcement of the theory of ideal factorization that his definition of "equivalence" for ideal complex numbers (divisors) can be reinterpreted and specialized to give a definition of "equivalence" for certain binary quadratic forms with determinant λ and that when this is done the resulting definition coincides with Gauss's definition of *proper* equivalence. One can be certain that the special case which Kummer had in mind was the case of cyclotomic integers which, in the notation of Chapter 4, are of the form $a + b\theta_0 + c\theta_1$. (Here θ_0, θ_1 are the two periods of length $\frac{1}{2}(\lambda - 1)$ and a, b, c are integers.) For quadratic integers of this form the *norm* $N(a + b\theta_0 + c\theta_1)$ is essentially a binary quadratic form and this establishes the connection between cyclotomic integers and Gauss's theory.

As was stated in Chapter 4, the basic problem which leads to the introduction of the notion of equivalence is the problem of determining *which divisors are divisors of cyclotomic integers*, that is, the problem of determining the image of the mapping of Section 4.15. In this section a very special case of this problem will be considered, specifically: *When $\lambda = 23$, which divisors are divisors of cyclotomic integers of the form $a + b\theta_0 + c\theta_1$?* (Here θ_0 and θ_1 are the two periods of length 11, θ_0 being the one which contains α and θ_1 being the other. Thus $1 + \theta_0 + \theta_1 = 0$, $\theta_0 = \alpha + \alpha^4 + \alpha^{-7} + \alpha^{-5} + \alpha^3 + \alpha^{-11} + \alpha^2 + \alpha^8 + \alpha^9 + \alpha^{-10} + \alpha^6$, and $\theta_1 = \alpha^{-2} + \alpha^{-8} + \cdots + \alpha^{11}$.) The study of this one special case will lead in a natural way to the idea of equivalence of divisors.

A useful first step in attacking the problem is to compile a list of many cyclotomic integers $a + b\theta_0 + c\theta_1$ and their divisors. By virtue of $1 + \theta_0 + \theta_1 = 0$, every such cyclotomic integer can be written in the form $a + b\theta_0$. Its norm is a product of 22 factors, 11 of which are $a + b\theta_0$ and 11 of which are $a + b\theta_1$; thus its norm is the 11th power of $(a + b\theta_0)(a + b\theta_1) = a^2 + ab(\theta_0 + \theta_1) + b^2\theta_0\theta_1 = a^2 - ab + 6b^2$. (The equation $\theta_0\theta_1 = 6$ when $\lambda = 23$ was noted in Section 4.4.) It is the fact that the norm of $a + b\theta_0$ is so closely related to the binary quadratic form $a^2 - ab + 6b^2$ that provides the connection between this special case of Kummer's theory and Gauss's theory that was alluded to above. For more about this connection, which will not play a role in the remainder of this section, see Chapters 7 and 8.

Table 5.2.1 contains a list of values of $a^2 - ab + 6b^2$ for various values of a and b. (For each positive a, all positive values of b relatively prime to a are given for which $a^2 - ab + 6b^2 < 150$.) Using this list it is easy to compile a list of the divisors of the corresponding cyclotomic integers $a + b\theta_0$. For example, the first line $(1 + \theta_0)(1 + \theta_1) = 2 \cdot 3$ of the table shows that each prime divisor of 2 divides $(1 + \theta_0)(1 + \theta_1)$ with multiplicity exactly 1 and therefore divides $1 + \theta_0$ or $1 + \theta_1$ but not both. If it divides $1 + \theta_0$ then its conjugate under $\alpha \mapsto \alpha^{-2}$ divides $1 + \theta_1$ and conversely. Therefore, there must be an even number of prime divisors of 2, half of which divide $1 + \theta_0$ and half of which divide $1 + \theta_1$. (Actually it is simple to show that the exponent of 2 mod 23 is 11 so that 2 has just two prime divisors. However, this more detailed knowledge will not be needed in what follows.) In Table 5.2.2 the symbol $(2, 1)$ denotes the divisor which is the product of all prime divisors of 2 which divide $1 + \theta_0$. Similarly, 3 must have an even number of prime divisors, half of which divide $1 + \theta_0$ and half of which divide $1 + \theta_1$. The symbol $(3, -1)$ denotes, for a reason that will be explained shortly, the product of those prime divisors of 3 which divide $1 + \theta_0$. Since $(1 + \theta_0)(1 + \theta_1) = 2 \cdot 3$ shows that no prime divisor of any prime p other than 2 or 3 divides $1 + \theta_0$, and shows that those which do divide it do so with multiplicity exactly 1, it is clear that $(2, 1)(3, -1)$ is the divisor of $1 + \theta_0$, as is shown in Table 5.2.2.

The second line of the tables is special because it involves the special prime 23 which, instead of having a divisor which is a product of distinct prime divisors, has as its divisor the 22nd power of the divisor $(\alpha - 1)$. Since the norm of $1 + 2\theta_0$ is 23^{11}, its divisor is some power of $(\alpha - 1)$ and, in fact, the 11th power, as is shown in Table 5.2.2.

It is in connection with the third line that the notation can be most conveniently explained. From $(1 + 3\theta_0)(1 + 3\theta_1) = 2^2 \cdot 13$ it follows that every prime divisor of 2 divides either $1 + 3\theta_0$ or $1 + 3\theta_1$. No prime divisor of 2 can divide both because then it would divide their sum $2 + 3\theta_0 + 3\theta_1 = -1$, which is impossible. Thus 2 must have an even number of prime divisors, half of which divide $1 + 3\theta_0$ and half of which divide $1 + 3\theta_1$. This splitting up of the prime divisors of 2 is the same as that which occurred in line 1, because if P is a prime divisor of 2 then $2 \equiv 0 \bmod P$, $2\theta_0 \equiv$

Table 5.2.1

a	b	$(a + b\theta_0)(a + b\theta_1)$	a	b	$(a + b\theta_0)(a + b\theta_1)$
1	1	$6 = 2 \cdot 3$	5	4	$101 = \text{prime}$
1	2	$23 = \text{prime}$	6	1	$36 = 2^2 \cdot 3^2$
1	3	$52 = 2^2 \cdot 13$	6	5	$156 = 2^2 \cdot 3 \cdot 13$
1	4	$93 = 3 \cdot 31$	7	1	$48 = 2^4 \cdot 3$
1	5	$146 = 2 \cdot 73$	7	2	$59 = \text{prime}$
2	1	$8 = 2^3$	7	3	$82 = 2 \cdot 41$
2	3	$52 = 2^2 \cdot 13$	7	4	$117 = 3^2 \cdot 13$
2	5	$144 = 2^4 \cdot 3^2$	8	1	$62 = 2 \cdot 31$
3	1	$12 = 2^2 \cdot 3$	8	3	$94 = 2 \cdot 47$
3	2	$27 = 3^3$	9	1	$78 = 2 \cdot 3 \cdot 13$
3	4	$93 = 3 \cdot 31$	9	2	$87 = 3 \cdot 29$
3	5	$144 = 2^4 \cdot 3^2$	9	4	$141 = 3 \cdot 47$
4	1	$18 = 2 \cdot 3^2$	10	1	$96 = 2^5 \cdot 3$
4	3	$58 = 2 \cdot 29$	10	3	$124 = 2^2 \cdot 31$
4	5	$146 = 2 \cdot 73$	11	1	$116 = 2^2 \cdot 29$
5	1	$26 = 2 \cdot 13$	11	2	$123 = 3 \cdot 41$
5	2	$39 = 3 \cdot 13$	11	3	$142 = 2 \cdot 71$
5	3	$64 = 2^6$	12	1	$138 = 2 \cdot 3 \cdot 23$

$0 \bmod P$, and P divides $1 + \theta_0$ if and only if it divides $1 + \theta_0 + 2\theta_0 = 1 + 3\theta_0$. In other lines of the tables such as $(1 + 5\theta_0)(1 + 5\theta_1)$, $(2 + \theta_0)(2 + \theta_1)$, $(4 + 3\theta_0)(4 + 3\theta_1)$, and so forth, the same splitting up of the prime divisors of 2 into two subsets occurs. To devise a notation for the two factors into which the divisor of 2 is decomposed in all these cases it is natural to note that, in all these cases, the equation shows that modulo any prime divisor of 2 both θ_0 and θ_1 are congruent to integers and the splitting up is according to these integers. For example, if P is a prime divisor of 2 which divides $1 + \theta_0$ then $1 + \theta_0 \equiv 0 \bmod P$, $\theta_0 \equiv -1 \equiv 1 \bmod P$, $\theta_1 = -1 - \theta_0 \equiv -1 - 1 \equiv 0 \bmod P$. If P divides $1 + \theta_1$, on the other hand, then $\theta_0 \equiv 0$, $\theta_1 \equiv 1 \bmod P$. Thus in the first case $3 + \theta_0 \equiv 3 + 1 \equiv 0 \bmod P$ and in the second case $3 + \theta_0 \equiv 3 + 0 \not\equiv 0 \bmod P$, so that P divides $3 + \theta_0$ in the first case but not in the second. When $(2, 1)$ is used to denote the product of those prime divisors of 2 mod which $\theta_0 \equiv 1$, then each factor of $(2, 1)$ divides $1 + 3\theta_0$ with multiplicity exactly 2 because it does not divide $1 + 3\theta_1$ at all $(1 + 3 \cdot 0 = 1 \not\equiv 0)$ and divides $(1 + 3\theta_0)(1 + 3\theta_1)$ with multiplicity exactly 2. The prime divisors of 13 divide $1 + 3\theta_0$ or $1 + 3\theta_1$ but not both. Those which divide $1 + 3\theta_0$ are those mod which $1 + 3\theta_0 \equiv 0$, $4 + 12\theta_0 \equiv 0$, $4 - \theta_0 \equiv 0$, $\theta_0 \equiv 4$. This way of identifying them makes it easy to determine that, for example, these prime divisors of 13 are the same as those which divide $5 + 2\theta_0$ (because mod them $5 + 2\theta_0 \equiv 5 + 2 \cdot 4 \equiv 13 \equiv 0$) and different from those which divide $2 + 3\theta_0$ (because $2 + 3\theta_0 \equiv 2 + 3 \cdot$

Table 5.2.2

Cyclotomic integer	Divisor	Cyclotomic integer	Divisor
$1 + \theta_0$	$(2, 1)(3, -1)$	$5 + 4\theta_0$	$(101, 24)$
$1 + 2\theta_0$	$(\alpha - 1)^{11}$	$6 + \theta_0$	$(2, 0)^2(3, 0)^2$
$1 + 3\theta_0$	$(2, 1)^2(13, 4)$	$6 + 5\theta_0$	$(2, 0)^2(3, 0)(13, 4)$
$1 + 4\theta_0$	$(3, -1)(31, -8)$	$7 + \theta_0$	$(2, 1)^4(3, -1)$
$1 + 5\theta_0$	$(2, 1)(73, 29)$	$7 + 2\theta_0$	$(59, 26)$
$2 + \theta_0$	$(2, 0)^3$	$7 + 3\theta_0$	$(2, 1)(41, -16)$
$2 + 3\theta_0$	$(2, 0)^2(13, -5)$	$7 + 4\theta_0$	$(3, -1)^2(13, -5)$
$2 + 5\theta_0$	$(2, 0)^4(3, -1)^2$	$8 + \theta_0$	$(2, 0)(31, -8)$
$3 + \theta_0$	$(2, 1)^2(3, 0)$	$8 + 3\theta_0$	$(2, 0)(47, 13)$
$3 + 2\theta_0$	$(3, 0)^3$	$9 + \theta_0$	$(2, 1)(3, 0)(13, 4)$
$3 + 4\theta_0$	$(3, 0)(31, 7)$	$9 + 2\theta_0$	$(3, 0)(29, 10)$
$3 + 5\theta_0$	$(2, 1)^4(3, 0)^2$	$9 + 4\theta_0$	$(3, 0)(47, -14)$
$4 + \theta_0$	$(2, 0)(3, -1)^2$	$10 + \theta_0$	$(2, 0)^5(3, -1)$
$4 + 3\theta_0$	$(2, 0)(29, -11)$	$10 + 3\theta_0$	$(2, 0)^2(31, 7)$
$4 + 5\theta_0$	$(2, 0)(73, -30)$	$11 + \theta_0$	$(2, 1)^2(29, -11)$
$5 + \theta_0$	$(2, 1)(13, -5)$	$11 + 2\theta_0$	$(3, -1)(41, 15)$
$5 + 2\theta_0$	$(3, -1)(13, 4)$	$11 + 3\theta_0$	$(2, 1)(71, 20)$
$5 + 3\theta_0$	$(2, 1)^6$	$12 + \theta_0$	$(2, 0)(3, 0)(\alpha - 1)^{11}$

$4 \equiv 1 \not\equiv 0$). When $(13, 4)$ is used to denote the prime divisors of 13 mod which $\theta_0 \equiv 4$, the divisor of $1 + 3\theta_0$ can be written $(2, 1)^2(13, 4)$.

The remaining lines of Table 5.2.2 can now be derived quite easily. In each case the notation (p, u) is used to denote all prime divisors of p which divide $\theta_0 - u$. This consists of just half the prime divisors of p because if P divided both $\theta_0 - u$ and $\theta_1 - u$, then its conjugate under $\alpha \mapsto \alpha^{-2}$ would also divide both, from which it would follow that all the conjugates of P divided both; the conjugates of P under $\alpha \mapsto \alpha^{-2}$ include *all* prime divisors of p and it would follow that p divided both $\theta_0 - u$ and $\theta_1 - u$, contrary to the fact that it obviously divides neither. (In actual fact it is easy to prove that the divisors (p, u) in the table are all *prime* divisors except for $(47, 13)$ and $(47, -14)$. This fact is not relevant to the problem at hand.)

Thus Table 5.2.2 gives a list of many divisors which are divisors of cyclotomic integers of the form $a + b\theta_0 + c\theta_1$. The number of entries in the table can be doubled by taking conjugates under $\alpha \mapsto \alpha^{-2}$. This carries θ_0 to $\theta_1 = -1 - \theta_0$ and carries the divisor (p, u) to the divisor $(p, -1 - u)$. Thus the divisors $(2, 0)(3, 0)$, $(2, 0)^2(13, -5)$, $(3, 0)(31, 7)$, ... are all divisors of cyclotomic integers of the form $a + b\theta_0 + c\theta_1$. (Actually the number of entries is not quite doubled because $(\alpha - 1)^{11}$ is its own conjugate.) A divisor which is the divisor of something is called a *principal*

157

divisor for historical reasons that are explained in Section 8.5. In the present context a divisor will be called *principal* if it is the divisor of a cyclotomic integer of the form $a + b\theta_0 + c\theta_1$. The list of principal divisors given above is very long and is easy to extend further. The problem is to determine, for a given divisor, whether it is principal.

There are first some very simple necessary conditions for a divisor A to be principal. Suppose first that $(\alpha - 1)$ divides A and that A is the divisor of $a + b\theta_0$. Then $(\alpha - 1)$ divides the integer $(a + b\theta_0)(a + b\theta_1)$, from which it follows that λ divides $(a + b\theta_0)(a + b\theta_1)$, say $(a + b\theta_0)(a + b\theta_1) = \lambda^j k$ where k is an integer not divisible by λ, and $j \geqslant 1$. Since $(\alpha - 1)$ divides $a + b\theta_0$ with multiplicity μ if and only if it divides $a + b\theta_1$ with multiplicity μ, it follows immediately that $(\alpha - 1)$ divides $a + b\theta_0$ with multiplicity exactly $11j$. *Thus a necessary condition for A to be principal is that it have the form $A = (\alpha - 1)^{11j}B$ where $j \geqslant 0$ and B is not divisible by $(\alpha - 1)$.*

Now let P be any prime divisor other than the exceptional one $(\alpha - 1)$, and let $P|A$ where A is principal. Let p be the prime integer which P divides and let $P_1, P_2, \ldots, P_e$ be the prime divisors of p where $P = P_1$, P_{j+1} is the conjugate of P_j under $\alpha \mapsto \alpha^{-2}$, and $f = (\lambda - 1)/e$ is the exponent of p mod 23. Since A is the divisor of $a + b\theta_0$ and since $a + b\theta_0$ is invariant under $\alpha \mapsto \alpha^4$, the assumption $P_1|A$ implies that $P_3, P_5, P_7, \ldots$ all divide A. If e is odd then this list includes *all* the prime divisors of p, and A is divisible by the divisor (p) of p. *Thus a further necessary condition for A to be principal is that it be of the form $A = (\alpha - 1)^{11j}(p_1)(p_2) \cdots (p_k)B$ where B is a divisor all of whose prime divisors have an even number e of conjugates.* Finally, if e is even then $P_1 P_3 \cdots P_{e-1}$ divides A. The claim is that this divisor is of the form (p, u). To prove this it suffices to note that the periods of length $f = 22/e = 22/2k$ are all congruent to integers mod P_1 or, for that matter, mod any prime divisor of p. Here $k = e/2$ by definition. Since θ_0 and θ_1 are periods of length $11 = k(22/2k) = kf$ they are sums of k periods of length f and are therefore congruent to integers mod P_1, say $\theta_0 \equiv u \bmod P_1$. Then $P_1, P_3, \ldots, P_{e-1}$ all divide $\theta_0 - u$. On the other hand none of $P_2, P_4, \ldots, P_e$ divide $\theta_0 - u$, because if one did they all would and this would imply $p|(\theta_0 - u)$, which is absurd.

This shows, then, that a necessary condition for A to be principal is that it have the form $A = (\alpha - 1)^{11j}(p_1')(p_2') \cdots (p_r')(p_1, u_1)(p_2, u_2) \cdots (p_s, u_s)$ where j is a nonnegative integer, where $p_1', p_2', \ldots, p_r'$ are primes* and where (p_i, u_i) for $i = 1, 2, \ldots, s$ denotes the product of those prime divisors of p_i which divide $\theta_0 - u_i$, the integers p_i and u_i being such that p_i is a prime with an even number of prime divisors, exactly half of which divide $\theta_0 - u_i$. Now the divisor $(\alpha - 1)^{11j}(p_1')(p_2') \cdots (p_r')$ is obviously principal, being the divisor of $(1 + 2\theta_0)^j p_1' p_2' \cdots p_r'$. Thus if $(p_1, u_1) \cdot$

---

*One could make the further assumption that $p_1', p_2', \ldots, p_r'$ all have an odd number of prime divisors, but if a divisor (p') in which p' has an even number of prime divisors—that is, $(p') = (p', u)(p', -1 - u)$—can be factored out of A, this simplifies the problem.

$(p_2, u_2) \cdots (p_s, u_s)$ is principal, multiplication of the cyclotomic integer $a + b\theta_0$ of which it is the divisor by $(1 + 2\theta_0)^j p_1' p_2' \cdots p_r'$ shows that A is principal. Conversely, if A is principal—say it is the divisor of $a + b\theta_0$—then, by the fundamental theorem, $(1 + 2\theta_0)^j p_1' p_2' \cdots p_r'$ divides $a + b\theta_0$; moreover, the quotient has divisor $(p_1, u_1)(p_2, u_2) \cdots (p_s, u_s)$ and is of the form $c + d\theta_0$ because it can be obtained by multiplication by $(1 + 2\theta_1)^j$ followed by division by $\lambda^j p_1' p_2' \cdots p_r'$. In short, A is principal if and only if $(p_1, u_1)(p_2, u_2) \cdots (p_s, u_s)$ is principal. The essence of the problem, therefore, is to *determine which divisors of the form* $(p_1, u_1) \cdot (p_2, u_2) \cdots (p_s, u_s)$ *are principal.*

An inspection of the table of principal divisors that has been compiled so far does not yield any obvious answers. A good approach at this point is to compile lists of divisors that are surely *not* principal divisors to see if they complement those which are. It was already noted in Section 4.4 that $(a + b\theta_0)(a + b\theta_1) = 47$ is impossible and therefore that $(47, 13)$ and $(47, -14)$ are not principal. A very simple way to prove this is to note that $(a + b\theta_0)(a + b\theta_1) = 2$ is impossible—this would imply $a^2 - ab + 6b^2 = 2$, $4a^2 - 4ab + 24b^2 = 8$, $(2a - b)^2 + 23b^2 = 8$, which is obviously impossible—and therefore that $(2, 0)$ is not principal because if $a + b\theta_0$ had divisor $(2, 0)$ then $(a + b\theta_0)(a + b\theta_1)$ would have the same divisor as 2 and would be a positive integer (because its 11th power is the norm of a cyclotomic integer and is therefore positive). Then, because $(2, 0)$ is not principal but $(2, 0)(47, 13)$ is principal (from Table 5.2.2), it follows that $(47, 13)$ is not principal: if $(47, 13)$ were the divisor of $a + b\theta_0$ then $a + b\theta_0$ would divide $8 + 3\theta_0$ and the quotient $(8 + 3\theta_0)/(a + b\theta_0) = (8 + 3\theta_0) \cdot (a + b\theta_1)/47$ would have divisor $(2, 0)$, which is impossible. In the same way, if A is any divisor with the property that $(2, 0)A$ is principal then A cannot be principal. Using Table 5.2.2 one can then write down immediately the nonprincipal divisors given in Table 5.2.3.

Table 5.2.3

$(3, 0)$	$(2, 0)^5$
$(2, 0)(13, -5)$	$(2, 0)(3, 0)^2$
$(73, -30)$	$(2, 0)(3, 0)(13, 4)$
$(2, 0)^2$	$(2, 0)^3(3, 0)$
$(2, 0)(13, -5)$	$(41, 15)$
$(2, 0)^3(3, -1)^2$	$(31, -8)$
$(2, 0)(3, -1)$	$(47, 13)$
$(2, 0)^3(3, -1)^2$	$(3, -1)(13, -5)$
$(3, -1)^2$	$(2, 0)^4(3, -1)$
$(29, -11)$	$(2, 0)(31, 7)$
$(73, -30)$	$(2, 0)(29, 10)$
$(13, 4)$	$(71, -21)$

Similarly, if $(2, 1)A$ is principal then A is not principal. This gives another list of nonprincipal divisors, which are simply the conjugates of the divisors in Table 5.2.3. Also, since $(3, 0)$ is not principal, the divisors A for which $(3, 0)A$ are principal are all nonprincipal. Applying this idea again to the principal divisors in Table 5.2.2 one finds the nonprincipal divisors in this list have already been listed because they are in fact all conjugates of divisors in Table 5.2.3. More generally, further tries will always show that for a given nonprincipal divisor B, the divisors A for which BA is principal coincide with either the divisors A in Table 5.2.3 or their conjugates.

Another way of saying the same thing is to say that if B is any divisor and if there is a divisor A in Table 5.2.3 for which BA is principal then BA is principal for *all* divisors A in Table 5.2.3. Once this phenomenon has been observed it is easily proved to hold in general: If A and A' are in Table 5.2.3 and if BA is principal then $(2, 0)A$, $(2, 0)A'$, and BA are all principal, say they are the divisors of $a + b\theta_0$, $c + d\theta_0$, and $e + f\theta_0$ respectively. Then by the fundamental theorem $a + b\theta_0$ divides $(c + d\theta_0)(e + f\theta_0)$ and the quotient has divisor BA'. The quotient is invariant under $\alpha \mapsto \alpha^4$ and is therefore of the form $g + h\theta_0$. Thus BA' is principal, as was to be shown. This shows that the divisors A in the table are all *equivalent* in the following sense.

Definition. Two divisors* A and A' are said to be *equivalent*, written $A \sim A'$, if for all divisors B, AB is principal when and only when $A'B$ is principal. In other words, in any divisor divisible by A, one can replace A by A' and the new divisor will be principal if and only if the old one was.

From the definition itself it is obvious that this relation of equivalence is reflexive, symmetric, transitive (if A can be replaced by A' and if A' can be replaced by A'' then A can be replaced by A''), and consistent with multiplication (if A can be replaced by A' then AC can be replaced by $A'C$).

It was shown above that if there is a single divisor C such that CA and CA' are both principal then $A \sim A'$. (In the proof above, $C = (2, 0)$.) Therefore not only are all the divisors in Table 5.2.3 equivalent to each other ($C = (2, 0)$), but all of their conjugates are equivalent to each other ($C = (2, 1)$), and all principal divisors equivalent to each other ($C = I$). A little experimentation shows that every divisor of the form $(p_1, u_1)(p_2, u_2) \cdots (p_s, u_s)$ seems to lie in one of these three equivalence classes. It will be shown now not only that this is true but also that it is a simple computation to determine, for a given divisor of this form, which of the three classes it belongs to. Since the original problem was to determine

---

*The divisors under discussion are those which are of the form $(\alpha - 1)^{11j} \cdot (p_1')(p_2') \cdots (p_r')(p_1, u_1)(p_2, u_2) \cdots (p_s, u_s)$ and which therefore might be divisors of cyclotomic integers of the form $a + b\theta_0$.

whether a given divisor of this form is principal—that is, whether it is equivalent to I—this is even more than a solution of the original problem.

Because equivalence of divisors is consistent with multiplication, the computation of the class of the product of two divisors AB can be simplified by replacing A or B or both by equivalent divisors. For example, if A and B are both in Table 5.2.3 then they are both equivalent to $(3, 0)$, their product is equivalent to $(3, 0)^2$, and this is in the equivalence class conjugate to Table 5.2.3. More generally, $(3, 0)$ lies in the class of Table 5.2.3, $(3, 0)^2$ lies in its conjugate, and $(3, 0)^3$ is principal; therefore if A and B both lie in the union of these three classes then both A and B are equivalent to powers of $(3, 0)$, and, since $(3, 0)^3$ is principal, AB must therefore be in one of the 3 classes. Moreover, if the classes of A and B are known, the class of AB is found merely by addition modulo 3 of the exponents.

Therefore, in order to prove that every product of (p, u)'s is in one of the three classes it suffices to show that every one of the (p, u)'s is in one of them. For as far as the tables go this is easy to verify: $(2, 0)$ and $(2, -1)$ do not appear in any of the three, but this is only because the fact that $(2, 0)(2, 1)$ is principal was overlooked when Table 5.2.3 was constructed. In fact, $(2, 1) \sim (3, 0)$, $(2, 0) \sim (3, 0)^2$. Then, in order of increasing p, $(3, 0) \sim (3, 0)$, $(13, 4) \sim (3, 0)$, $(31, -8) \sim (3, 0)$, $(41, 15) \sim (3, 0)$, $(47, 13) \sim (3, 0)$, $(59, 26) \sim I$, $(71, -21) \sim (3, 0)$, $(73, -30) \sim (3, 0)$, $(101, 24) \sim I$. Of course the conjugate of any divisor equivalent to $(3, 0)$ is equivalent to $(3, -1) \sim (3, 0)^2$.

Consider now an arbitrary divisor of the form (p, u). It divides $\theta_0 - u$. Since it also divides p, one can reduce the integer $u \bmod p$ to the smallest possible absolute value, in which case $|u| \leqslant \frac{1}{2} p$. Then $(\theta_0 - u)(\theta_1 - u) = u^2 + u + 6$ is a positive integer (its 11th power is a norm) which is at most $\frac{1}{4} p^2 + \frac{1}{2} p + 6$. If p is at all large this is less than p^2. In fact, it is less than p^2 for all primes p except 2 and 3. Since (p, u) and its conjugate both divide $(\theta_0 - u)(\theta_1 - u)$, this shows that $(\theta_0 - u)(\theta_1 - u) = pk$ where $k < p$ (when $p = 2, 3$ are excluded). The divisor of $\theta_0 - u$ is of the form $(\alpha - 1)^{11j}(p_1')(p_2')$ $(p_r')(p, u)(p_1, u_1)(p_2, u_2) \cdots (p_s, u_s)$. Moreover, $r = 0$ because $r > 0$ would imply that p_1' divided $\theta_0 - u$, contrary to the fact that no integer greater than 1 divides $\theta_0 - u$. Then $pk = 23^j pp_1 p_2 \cdots p_s$. On the other hand, $(p, -1 - u) \sim (\alpha - 1)^{11j}(p_1, u_1) \cdot (p_2, u_2) \cdots (p_s, u_s)$ because both become principal when they are multiplied by (p, u). Since $(\alpha - 1)^{11j} \sim I$ (it is the divisor of $(1 + 2\theta_1)^j$) this gives $(p, u) \sim (p_1, -1 - u_1) \cdot (p_2, -1 - u_2) \cdots (p_s, -1 - u_s)$. In words, *any divisor of the form (p, u) in which $p > 3$ is equivalent to a product of divisors of the same form in which the p's are all strictly smaller than the original p.* If any of the new p's are greater than 3 the same process can be used to reduce them until finally one arrives at an equivalence of the form $(p, u) \sim (p_1'', u_1'') \cdot (p_2'', u_2'') \cdots (p_t'', u_t'')$ in which all the p_j'' are 2 or 3. This proves that (p, u) is equivalent to $(3, 0)$, or $(3, 0)^2$, or $(3, 0)^3 \sim I$.

In summary, when equivalence of divisors of the form $(\alpha - 1)^{11j} \cdot$ $(p_1')(p_2') \cdots (p_r')(p_1, u_1)(p_2, u_2) \cdots (p_s, u_s)$ is defined as above, every such divisor is equivalent to one and only one of the divisors I, $(3, 0)$, or $(3, 0)^2$. Moreover, an explicit construction has been given for determining which of the three possible equivalences holds. The solution of the given problem then is that *a divisor is principal if and only if it has the above form and is equivalent to I.*

EXERCISE

1. In each of the cases $\lambda = 31, 39, 43$ find the formula for $(a + b\theta_0)(a + b\theta_1)$, construct tables analogous to Tables 5.2.1 and 5.2.2, and find a set of divisors like I, $(3, 0)$, $(3, 0)^2$ in the case $\lambda = 23$ such that (1) every divisor is equivalent to one of the divisors of the set and (2) no two divisors of the set are equivalent to each other.

5.3 The class number

The preceding section gives the definitions of "principal" and "equivalent" for divisors that divide cyclotomic integers of the form $a + b\theta_0 + c\theta_1$ in the case $\lambda = 23$. The general definitions are obvious from this special case. A divisor (for some fixed λ) is said to be *principal** if there is some cyclotomic integer of which it is the divisor. Two divisors A and B are said to be *equivalent,* denoted $A \sim B$, if it is true that a divisor of the form AC is principal when and only when BC is principal. That is, $A \sim B$ means that A can be replaced by B in any divisor divisible by A and the new divisor will be principal if and only if the original one was. It is easy to prove the following properties of these definitions (Exercise 1):

(1) If A and B are both principal then so is AB.
(2) If A and B are divisors such that A and AB are both principal then B is principal.
(3) A is principal if and only if $A \sim I$ where I is the empty divisor, that is, the divisor of 1.
(4) $A \sim B$ if and only if there is a third divisor C such that AC and BC are both principal. (This is the way in which Kummer defines equivalence of divisors.)
(5) The relation of equivalence is reflexive, symmetric, and transitive. That is, $A \sim A$, if $A \sim B$ then $B \sim A$, and if both $A \sim B$ and $B \sim C$ then $A \sim C$.
(6) Multiplication of divisors is consistent with the equivalence relation. That is, $A \sim B$ implies $AC \sim BC$ for all divisors C.

*The etymology of this strange adjective is explained in Section 8.5. It would be more correct, but longer, to say that such a divisor is "in the principal class" where a "class" of divisors is defined by the equivalence relations defined here and where the principal class is the class containing I.

(7) Given any divisor A, there is another divisor B such that $AB \sim I$.

(8) $A \sim B$ if and only if there exist principal divisors M and N such that $AM = BN$. (This is an exact analogy of the definition of $a \equiv b \bmod k$ given in Appendix 1, to wit, there exist positive integers m and n divisible by k such that $a + m = b + n$.)

Kummer proved that in all cases (that is, for all primes $\lambda > 2$) it is possible to find a *finite number* of divisors $A_1, A_2, \ldots, A_k$ such that every divisor is equivalent to one of the A_i (just as, in the preceding section, every divisor of the form $a + b\theta_0 + c\theta_1$ in the case $\lambda = 23$ is equivalent to one of three divisors I, $(3, 0)$, or $(3, 0)^2$). The proof of this important fact is given at the end of this section.

A *representative set* of divisors is a set of divisors $A_1, A_2, \ldots, A_k$ with the above property that every divisor is equivalent to one of the A_i and with the additional property that no two divisors of the set are equivalent to each other. Such a representative set makes it possible to *classify* divisors. Every divisor is equivalent to one and only one of the A_i and two divisors are in the same *equivalence class*—are equivalent to each other—if and only if they are equivalent to the same A_i. Thus the number of divisors in a representative set is the number of distinct equivalence classes. This number is called the *class number*.

To find a representative set, and hence to determine the class number, would seem to be a simple matter of taking a set $A_1, A_2, \ldots, A_k$ such as is provided by Kummer's theorem and of eliminating duplicates; that is, if $A_1, A_2, \ldots, A_k$ is not a representative set then $A_i \sim A_j$ for some i and j, so that one of these two can be eliminated without losing the property that every divisor is equivalent to one of the divisors of the set; the continuation of this process must eventually lead to a representative set. This "construction" is illusory, however, because it involves solving the problem of deciding whether two given divisors are equivalent, a problem which requires additional techniques and which in fact comes down to the basic problem which motivated the entire theory of equivalence in the first place.

The fact is that Kummer's theorem is not sufficient to prove constructively the existence of a representative set and that, as Kummer says in the last section of his original 1847 exposition of the theory of ideal factorization, the computation of the class number requires "essentially different principles from the ones contained in the present work." These new principles and the determination of the class number, which Kummer published later in 1847, are the subject of Chapter 6. In the present chapter all that will be used about the class number is its *meaning*: to say that the class number is h for a given λ means that it is possible to construct a set of h divisors $A_1, A_2, \ldots, A_h$ (for this λ) with the property that every divisor is equivalent to one and only one of the A_i ($i = 1, 2, \ldots, h$).

The remainder of this section is devoted to the proof of Kummer's theorem that there exists a finite set $A_1, A_2, \ldots, A_k$ of divisors with the property that every divisor is equivalent to one of the A_i. The proof is similar to the proof of the preceding section that every prime divisor is equivalent to a product of prime divisors of smaller norm. Specifically, what will be shown is that given any divisor A (in the preceding section only divisors (p, u) were considered) there is a cyclotomic integer ($\theta_0 - u$ in the preceding section) which is divisible by A and whose norm divided by the norm of A is less than some constant K independent of A. That is, there is a divisor B with norm less than K such that AB is principal. The number of divisors with norm less than K is finite. Therefore it will follow that there is a finite set of divisors $B_1, B_2, \ldots, B_k$ such that, for every given A, AB_i is principal for some i. For each B_i choose an A_i such that $A_i B_i$ is principal. Then every given A will be equivalent to one of the A_i, and a set $A_1, A_2, \ldots, A_k$ of divisors with the required property will have been found.

What is to be shown, then, is that there is a constant K with the property that if A is any given divisor and if n is the norm of A then there is a cyclotomic integer that is divisible by A, the norm of which is less than Kn. This suggests seeking a cyclotomic integer that is divisible by A and that has norm as small as possible. This is made very simple by the theorem of Section 4.14 stating that the number of incongruent cyclotomic integers mod A is $N(A)$, because this shows that *any set of more than n distinct cyclotomic integers must contain two whose difference is divisible by A*. Since the size of the norm is related to the size of the coefficients of a cyclotomic integer, a natural way to construct a cyclotomic integer that is divisible by A and has small norm is to take a set of more than n cyclotomic integers with small coefficients. This suggests deriving a specific relationship between the size of the coefficients and the size of the norm, which can be done as follows.

Let $g(\alpha) = a_0 + a_1\alpha + \cdots + a_{\lambda-1}\alpha^{\lambda-1}$. Then

$$Ng(\alpha) = g(\alpha)g(\alpha^2)\cdots g(\alpha^{\lambda-1}) = \left[g(\alpha)g(\alpha^{-1})\right]\left[g(\alpha^2)g(\alpha^{-2})\right]\cdots$$

and each of the $\frac{1}{2}(\lambda - 1)$ terms of this product can be regarded as the modulus squared of a complex number $g(\alpha^j)$. As such, each of these $\frac{1}{2}(\lambda - 1)$ terms is clearly no more than the square of $|a_0| + |a_1| + \cdots + |a_{\lambda-1}|$. Therefore if all the a_i satisfy $|a_i| \leqslant c$ then

$$Ng(\alpha) \leqslant (\lambda^2 c^2)^{(\lambda-1)/2} = \lambda^{\lambda-1}c^{\lambda-1}.$$

The set of all $g(\alpha)$ in which $a_0 = 0$ and $0 \leqslant a_i \leqslant c$ for $i > 0$ contains more than n elements provided $(c + 1)^{\lambda-1} > n$. Therefore the difference of some two of them is divisible by A and has norm at most $\lambda^{\lambda-1}c^{\lambda-1}$. If c is as small as possible subject to $(c + 1)^{\lambda-1} > n$ then of course $c^{\lambda-1} \leqslant n$ and a cyclotomic integer has been found that is divisible by A and has norm at most $\lambda^{\lambda-1}n$. This proves the theorem and gives the value $\lambda^{\lambda-1}$ for K. (Kummer's proof is based on a more adroit estimate of $Ng(\alpha)$; it gives the better value $(\lambda - 1)^{(\lambda-1)/2}$ for K. See Exercise 3.)

EXERCISES

1. Deduce properties (1)–(8) from the definitions of "principal" and "equivalent."

2. Prove that, for a given λ, unique factorization is valid if and only if the corresponding class number is 1.

3. Kummer estimates $Ng(\alpha)$ by deriving the identity $g(\alpha)g(\alpha^{-1}) + g(\alpha^2)g(\alpha^{-2}) + \cdots + g(\alpha^{\lambda-1})g(\alpha^{-\lambda+1}) = \lambda(a_1^2 + a_2^2 + \cdots + a_{\lambda-1}^2) - (a_1 + a_2 + \cdots + a_{\lambda-1})^2$ (when $a_0 = 0$) and by applying the theorem which states that the geometric mean $\sqrt[n]{c_1 c_2 \ldots c_n}$ of a set of positive numbers is always less than the arithmetic mean $n^{-1}(c_1 + c_2 + \cdots + c_n)$. Carry this through to find the value $K = (\lambda^{\lambda-1})^{1/2}$. Note that both of these proofs depend on the interpretation of $g(\alpha)$ as a complex number.

5.4 Kummer's two conditions

It was remarked several times above that the central idea which occurs in the proofs of Fermat's Last Theorem is the idea of proving that if u and v are such that their product is a λth power $uv = w^\lambda$ (where λ is the prime exponent for which Fermat's Last Theorem $x^\lambda + y^\lambda \neq z^\lambda$ is to be proved) and if u and v are relatively prime, then u and v must each be λth powers. This conclusion would be justified if unique factorization were valid in the strictest sense, but even if λ is such that the corresponding cyclotomic integers have unique factorization—which amounts to saying that the corresponding class number is 1—this conclusion is not valid because "unique" factorization normally means "unique except for unit multiples." Thus, even in the simplest cases, this conclusion is not justified without further conditions on u and v. What are the natural conditions to place on them?

Note first that unique factorization in the strictest sense *is* valid for *divisors* and that $uv = w^\lambda$ does imply that if u and v have no ideal prime factors in common then their *divisors* are each λth powers. That is, if A and B are the divisors of u and v respectively, if $uv = w^\lambda$, and if A and B have no common factors, then there exist divisors C and D such that $A = C^\lambda$ and $B = D^\lambda$. Since the natural definition of "u and v are relatively prime" is "no ideal prime factor divides both u and v" this reduces the above question to the following: Let u be a cyclotomic integer whose divisor A is the λth power of another divisor $A = C^\lambda$. Under what additional conditions on λ and u can one conclude that u itself is a λth power?

165

This leads naturally to the question of whether C is principal, that is, of whether $C^\lambda = A \sim I$ (because A is the divisor of u) implies $C \sim I$. Kummer found a very natural condition on λ such that this implication is valid without any assumption on u. This condition is based on the simple observation that *if C is any divisor and if h is the class number corresponding to λ then $C^h \sim I$*. This is obviously a generalization* of Fermat's theorem and its proof is easy. To say that the class number is h means that there is a representative set $A_1, A_2, \ldots, A_h$ with h elements. Each of the powers $C, C^2, C^3, \ldots$ of C is equivalent to one and only one of the A_i. Obviously different powers of C must be equivalent to the same A_i and therefore must be equivalent to each other, say for example $C^j \sim C^{j+k}$. Then $C^k \sim I$, that is, some power of C is equivalent to I. Thus one can test (using the representative set) all powers of C prior to this one and find the *first* power of C that is equivalent to I, say $C^d \sim I$. Then $I, C, C^2, \ldots, C^{d-1}$ are inequivalent to one another ($C^j \sim C^{j+k}$ implies $C^k \sim I$) and are therefore equivalent to d of the A_i. If this exhausts the A_i then $d = h$. Otherwise there is an A_i which is not in this set of d, call it B. Then $B, BC, BC^2, \ldots, BC^{d-1}$ give d more divisors inequivalent to each other and inequivalent to any of the first set of d. This gives another set of d divisors of the representative set. Continuing in this way effects a partition of the h elements of the representative set into disjoint sets of d elements each. Thus $d|h$ and $C^d \sim I$ implies $C^h \sim I$ as was to be shown.

Now if $C^\lambda \sim I$ as well, then $C \sim I$ unless $\lambda|h$. This follows from the observation that, because λ is prime, if h is not divisible by λ then $mh = n\lambda + 1$ for suitable integers m and n, from which $I \sim (C^h)^m = C^{mh} = C^{n\lambda+1} = (C^\lambda)^n C \sim I \cdot C \sim C$. Therefore the desired implication $C^\lambda \sim I \Rightarrow C \sim I$ is valid whenever λ satisfies Kummer's condition:

(A) The exponent λ should have the property that it does not divide the corresponding class number h.

If u is a cyclotomic integer whose divisor A is a λth power $A = C^\lambda$ and if condition (A) is satisfied then $C \sim I$ (because C^λ is the divisor of u which implies $C^\lambda \sim I$); say C is the divisor of the cyclotomic integer x. Then u and x^λ have the same divisor and it follows, by the fundamental theorem, that $u = ex^\lambda$ where e is a cyclotomic integer which is a unit. In short, condition (A) implies that every cyclotomic integer u whose divisor is a λth power is of the form $u = ex^\lambda$ where e and x are cyclotomic integers and e is a unit. The problem is to find conditions under which the stronger conclusion $u = x^\lambda$ is valid.

The central idea of Dirichlet's proof of the case $\lambda = 5$ (see Section 3.3) was, as he himself said, to find an *additional condition* under which

---

*A generalization of both is the theorem which states that if H is a subgroup of the finite group G then the order of H divides the order of G.

"$P^2 + 5Q^2$ = fifth power" and "P, Q relatively prime" imply

$$P + Q\sqrt{5} = (A + B\sqrt{5})^5.$$

This conclusion is *not* valid without a further assumption on P and Q. The additional condition which Dirichlet introduced is derived from the observation that $(A + B\sqrt{5})^5$ has 5's in all terms of its binomial expansion except the first, so that $P + Q\sqrt{5} = (A + B\sqrt{5})^5$ implies $Q \equiv 0$ mod 5. He shows that this *necessary* condition, when added to the other conditions $(P^2 + 5Q^2)$ = fifth power and P, Q relatively prime, is also *sufficient*.

The corresponding necessary condition in the case of the cyclotomic integers is easy to find. It was noted in Section 4.5 that modulo λ the operation of raising to the λth power carries sums to sums, that is, $(a + b)^\lambda \equiv a^\lambda + b^\lambda$ mod λ. Thus $u = x^\lambda$ implies

$$u = (a_0 + a_1\alpha + a_2\alpha^2 + \cdots + a_{\lambda-1}\alpha^{\lambda-1})^\lambda$$
$$\equiv a_0^\lambda + a_1^\lambda + \cdots + a_{\lambda-1}^\lambda = \text{integer mod } \lambda.$$

That is, a necessary condition for $u = x^\lambda$ is that u be an integer mod λ. Following Dirichlet, then, it is natural to try to show that $u \equiv$ integer mod λ, added to the condition that the divisor of u be a λth power, is also sufficient.

The condition as it stands is obviously not sufficient because it can be satisfied in a trivial way if $u \equiv 0$ mod λ. This simple observation is, as will be seen in the next section, the crux of the distinction between Case I and Case II of Fermat's Last Theorem. In the simpler case $u \not\equiv 0$ mod λ the conditions "$u \equiv$ integer mod λ" and "the divisor of u is a λth power" imply, when condition (A) holds, that $u = ex^\lambda$ for some cyclotomic integers e and x with e a unit. The question of whether u is a λth power is the question of whether the unit e is a λth power. The assumptions on u guarantee that $e \equiv$ integer mod λ because $u \equiv ex^\lambda \equiv e \cdot b$ mod λ for some integer b and $b \not\equiv 0$ mod λ (otherwise $u \equiv 0$ mod λ), so that division of $u \equiv$ integer by b gives $e \equiv$ integer mod λ. Therefore the desired conclusion $u = x^\lambda$ will be valid if it is valid in the special case $u = e =$ unit. This is Kummer's second condition.

(B) The exponent λ should have the property that the units of the corresponding cyclotomic integers are such that the necessary condition $e \equiv$ integer mod λ for a unit e to be a λth power is also sufficient. In short, if e is a unit and if $e \equiv$ integer mod λ then $e = (e')^\lambda$ for some cyclotomic integer e'. (Of course e' is a unit; its inverse is $(e')^{\lambda-1}e^{-1}$.)

Kummer found that this condition was satisfied for all the primes λ he tested, at least those for which he could show that (A) was satisfied. These two conditions (A) and (B), when they are satisfied, imply the following theorem about the equation $u = x^\lambda$ in cyclotomic integers for the corresponding λ.

Theorem. *Suppose λ satisfies conditions (A) and (B), and let u be given. The problem is to determine whether u is a λth power. If $u \not\equiv 0$ mod λ (Case I) then u is a λth power if and only if its divisor is a λth power and $u \equiv integer$ mod λ. If $u \equiv 0$ mod λ (Case II) then u is a λth power if and only if its divisor is a λth power and its quotient by the largest power of $\alpha - 1$ it contains is congruent mod λ to an integer.*

The proof of this theorem follows immediately from the remarks which motivate the conditions (A) and (B). A prime λ is called *regular* if it satisfies conditions (A) and (B). It is for such primes that Kummer's general proof of Fermat's Last Theorem $x^\lambda + y^\lambda \neq z^\lambda$, given in the next section, is valid.

Kummer made three noteworthy conjectures about conditions (A) and (B) at the time that he first stated them. The first was that not all primes satisfy them—that is, there exist *irregular* primes—the second was that (A) implies (B) so that (B) is superfluous, and the third was that the regular primes are infinite in number. The first two of these conjectures are true, as Kummer himself proved in a very few months. (The proofs are in Chapter 6.) The third, however, is *still* an unsolved problem and Kummer later retracted the conjecture, saying he did not know whether there were infinitely many regular primes. (Ironically, it *has* been proved that there are infinitely many irregular primes.)

EXERCISES

1. Prove the theorem stated in this section.

2. Prove that a cyclotomic integer $u = b_0 + b_1\alpha + \cdots + b_{\lambda-1}\alpha^{\lambda-1}$ satisfies $u \equiv$ integer mod λ if and only if $b_1 \equiv b_2 \equiv \cdots \equiv b_{\lambda-1}$ mod λ.

5.5 The proof for regular primes

Much of the literature on Fermat's Last Theorem tends to give the impression that its proof would be easy if unique factorization into primes were valid for cyclotomic integers. The extent to which this impression is false may be judged from the proof which follows. The proof would not be any easier if Kummer's assumption (A) were replaced by the stronger assumption of unique factorization. And even when the further property (B) is used (it will be shown in Chapter 6 that (B) is in fact a subtle consequence of (A)) a good deal of ingenuity is still needed in order to provide a proof.

Kummer's proof makes great use of a particular property of the units, a property which is not at all obvious but one which is basically quite elementary and would have been common knowledge in 1847 to Kummer, Kronecker, Dirichlet, and others interested in the structure of the units of the cyclotomic integers. The property in question is the following one: If e is any cyclotomic integer which is a unit and if $\bar{e}$ is its *complex conjugate* (obtained from e by replacing α by α^{-1} in e) then $e/\bar{e} = \alpha^r$ for some integer r. This statement is plausible enough. The operation of complex conjugation obviously carries $e/\bar{e}$ to its inverse, which shows that $e/\bar{e}$, considered as a complex number, is on the unit circle. This leads to the conjecture that $e/\bar{e}$ might be of the form α^r, or at least of the form $\pm \alpha^r$, but the proof is not immediate. A proof is given at the end of this section.

The first step of the proof of Fermat's Last Theorem for regular primes is of course to write the equation $x^\lambda + y^\lambda = z^\lambda$ in the form $(x + y)(x + \alpha y)(x + \alpha^2 y) \cdots (x + \alpha^{\lambda-1} y) = z^\lambda$. One must first ask whether the factors on the left are relatively prime. Now if $x + \alpha^j y$ and $x + \alpha^{j+k} y$ have a factor in common then this same factor divides both

$$(x + \alpha^{j+k} y) - (x + \alpha^j y) = \alpha^j (\alpha^k - 1)y = \text{unit} \cdot (\alpha - 1) \cdot y$$

and

$$(x + \alpha^{j+k} y) - \alpha^k (x + \alpha^j y) = \text{unit} \cdot (\alpha - 1) \cdot x.$$

Since x and y are relatively prime integers, they have no common factors—not even ideal factors—and the only possible common factor is $\alpha - 1$. Now the same equations show that if $\alpha - 1$ divides any one of the factors $x + \alpha^j y$ then it divides *all* the others. Thus the question naturally divides itself into the two usual cases—the one in which λ divides z and consequently $\alpha - 1$ divides all the factors of $x^\lambda + y^\lambda$, and the one in which z is prime to λ and consequently the factors of $x^\lambda + y^\lambda$ are relatively prime. Because λ is odd, if x is divisible by λ then the equation can be written in the form $x^\lambda = z^\lambda + (-y)^\lambda$, and if y is divisible by λ it can be written in the form $y^\lambda = z^\lambda + (-x)^\lambda$. Therefore if any one of the three is divisible by λ one can assume that it is z, and the cases are:

Case I. $x^\lambda + y^\lambda = z^\lambda$ where x, y, z are pairwise relatively prime and all prime to λ.

Case II. $x^\lambda + y^\lambda = z^\lambda$ where x, y, z are pairwise relatively prime and where $\lambda | z$.

PROOF OF CASE I. In this case the factors $x + y, x + \alpha y, \ldots, x + \alpha^{\lambda-1} y$ are relatively prime and their product is a λth power. Therefore the divisor of each factor is a λth power, and, by property (A) of the prime λ, each factor $x + \alpha^j y$ is a unit times a λth power. Kummer uses only the case $j = 1$ in his proof; that is, he uses only the fact that there exist a unit e and a cyclotomic integer t such that $x + \alpha y = et^\lambda$. Replacement of α by α^{-1} puts this equation in the form $x + \alpha^{-1} y = \bar{e} \bar{t}^\lambda$. Now $e/\bar{e} = \alpha^r$ by the above property of units and $t^\lambda \equiv C \equiv \bar{C} \equiv \bar{t}^\lambda$ mod λ because mod λ every λth power is an integer and integers are invariant under $\alpha \mapsto \alpha^{-1}$. Thus $x + \alpha^{-1}y = \alpha^{-r}e\bar{t}^\lambda \equiv \alpha^{-r}et^\lambda = \alpha^{-r}(x + \alpha y)$ mod λ. The remaining idea* in the proof is to write all terms in the congruence

$$\alpha^r(x + \alpha^{-1}y) \equiv x + \alpha y \text{ mod } \lambda$$

in terms of integers times powers of $\alpha - 1$ and to see what conditions the congruence imposes on the integers x, y, and r. Note first that r is only determined modulo λ and that $r \equiv 0$ mod λ is impossible because that would imply

$$x + \alpha^{-1}y \equiv x + \alpha y \text{ mod } \lambda, \qquad 0 \equiv (\alpha^2 - 1)y \text{ mod } (\alpha - 1)^{\lambda-1}$$

from which it would follow that $\alpha - 1$ divided y, contrary to assumption. Therefore one can assume $0 < r < \lambda$. The congruence can then be written

$$\alpha^{r-1}(\alpha x + y) \equiv x + \alpha y \text{ mod } \lambda$$

$$[1 + (\alpha - 1)]^{r-1}[x + y + x(\alpha - 1)] \equiv x + y + y(\alpha - 1) \text{ mod } (\alpha - 1)^{\lambda-1}.$$

Now a congruence of the form $a_0 + a_1(\alpha - 1) + a_2(\alpha - 1)^2 + \cdots + a_{\lambda-2}(\alpha - 1)^{\lambda-2} \equiv 0$ mod $(\alpha - 1)^{\lambda-1}$ easily implies (Exercise 1) that $a_0 \equiv 0$, $a_1 \equiv 0, \ldots, a_{\lambda-1} \equiv 0$ mod λ. Therefore, in the above congruence, coefficients of corresponding powers of $(\alpha - 1)$ must be congruent mod λ provided all powers are less than the $(\lambda - 1)$st. The above congruence is impossible when $1 < r < \lambda - 1$ because then the highest order term on the left is $x(\alpha - 1)^r$ where $2 \leqslant r \leqslant \lambda - 2$, and this would give $x \equiv 0$ mod λ contrary to assumption. If $r = \lambda - 1$ then the next-to-last-term on the left is $(\alpha - 1)^{r-1}(x + y) + (r - 1)(\alpha - 1)^{r-2}x(\alpha - 1) = [x + y + (\lambda - 2)x](\alpha - 1)^{\lambda-2}$, which gives $-x + y \equiv 0$ mod λ, $x \equiv y$ mod λ. (Examination of other terms easily shows that this case $r = \lambda - 1$ is also impossible, but the conclusion $x \equiv y$ mod λ is sufficient to reach the desired contradiction.) If $r = 1$ then the same conclusion $x \equiv y$ mod λ follows immediately.

Now Case I is symmetrical in the three variables. It is natural, therefore, to write the equation in the form $x^\lambda + y^\lambda + z^\lambda = 0$ in this case. What has just been shown is that such an equation implies, when $x \not\equiv 0$, $y \not\equiv 0$, and $z \not\equiv 0$ mod λ, that $x \equiv y$ mod λ. By symmetry $x \equiv z$, $y \equiv z$ mod λ as well. But $x^\lambda \equiv x$, $y^\lambda \equiv y$, $z^\lambda \equiv z$ mod λ by Fermat's theorem. Therefore $0 = x^\lambda + y^\lambda + z^\lambda \equiv 3x$ mod λ. Since $x \not\equiv 0$ mod λ this implies $3 \equiv 0$ mod λ, which implies $\lambda = 3$. Thus Case I is impossible for $\lambda \neq 3$. Since Case I has already been proved impossible for $\lambda = 3$—say by the simple theorem of Section 3.2 together with the observation that $2 \cdot 3 + 1 = 7$ is prime—this completes the proof that Case I is impossible for regular primes. (In fact, only property (A) of regular primes is needed.)

*This part of the proof is an interpolation. Kummer merely writes the congruence in the form $(1 - \alpha^r)x + (\alpha - \alpha^{r-1})y \equiv 0$ mod λ and states the conclusion $r \equiv 1$, $x \equiv y$ mod λ. He may have had a simpler argument than the one given here.

PROOF OF CASE II. In this case *all* the factors of $x^\lambda + y^\lambda$ are divisible by $\alpha - 1$ and the quotients are relatively prime. Since the product of the quotients is $z^\lambda(\alpha - 1)^{-\lambda}$, the product is a λth power and it follows from property (A) that $(\alpha - 1)^{-1}(x + \alpha^j y) = e_j t_j^\lambda$ for some cyclotomic integers e_j, t_j with e_j a unit. Moreover, the t_j are pairwise relatively prime. The special prime $\alpha - 1$ divides at most one of the t_j and in fact it divides t_0 because the facts that $\alpha - 1$ divides $x + y$ and that x and y are integers imply that in fact $(\alpha - 1)^{\lambda - 1}$ divides $x + y$. Let $t_0 = (\alpha - 1)^k w$ where w is not divisible by $\alpha - 1$. Then $k \geqslant 1$. Among the λ equations derived above, one has in particular

$$x + \alpha^{-1}y = (\alpha - 1)e_{-1}t_{-1}^\lambda$$

$$x + y = (\alpha - 1)e_0(\alpha - 1)^{k\lambda}w^\lambda$$

$$x + \alpha y = (\alpha - 1)e_1 t_1^\lambda.$$

These 3 equations can be used to eliminate the two unknowns x and y, which gives first

$$(\alpha - 1)y = (\alpha - 1)\left[e_1 t_1^\lambda - e_0(\alpha - 1)^{k\lambda}w^\lambda \right]$$

$$\alpha^{-1}(\alpha - 1)y = (\alpha - 1)\left[e_0(\alpha - 1)^{k\lambda}w^\lambda - e_{-1}t_{-1}^\lambda \right]$$

and then

$$0 = e_1 t_1^\lambda - e_0(\alpha - 1)^{k\lambda}w^\lambda - \alpha e_0(\alpha - 1)^{k\lambda}w^\lambda + \alpha e_{-1}t_{-1}^\lambda.$$

Since $1 + \alpha$ is a unit (it is $\alpha^2 - 1$ divided by $\alpha - 1$) this equation can be put in the form

$$E_0(\alpha - 1)^{k\lambda}w^\lambda = t_1^\lambda + E_{-1}t_{-1}^\lambda$$

where E_0, E_{-1} are units. Now property (B) of the prime λ can be used to get rid of the factor E_{-1} because mod λ the above equation reads $0 \equiv C_1 + E_{-1}C_{-1}$ where C_1, C_{-1} are integers which are not zero mod λ because t_1 and t_{-1} are relatively prime to $t_0 = (\alpha - 1)^k w$. Thus $E_{-1} \equiv$ integer mod λ, $E_{-1} = e^\lambda$ for some unit e, and $E_0(\alpha - 1)^{k\lambda}w^\lambda = t_1^\lambda + (et_{-1})^\lambda$. This equation is of very nearly the same form as the original equation $z^\lambda = x^\lambda + y^\lambda$ and the same argument, with only minor modifications, can be applied to obtain a third equation of the same form.

Specifically, if one starts with an equation of the form

$$x^\lambda + y^\lambda = e(\alpha - 1)^{k\lambda}w^\lambda$$

where e is a unit, k is a positive integer, and x, y, w; and $\alpha - 1$ are pairwise relatively prime *cyclotomic* integers, one can proceed as follows. (Case II of Fermat's Last Theorem is the special case where x, y, w are *integers*, $z = \lambda^m w$, $k = m(\lambda - 1)$, $e = [\lambda(\alpha - 1)^{-\lambda+1}]^{m\lambda}$.) At least one of the factors $x + \alpha^j y$ of the left side is divisible by $\alpha - 1$, from which it follows as above that all are divisible by $\alpha - 1$ and that the quotients are relatively prime.

The previous argument that $x + y$ is divisible by $(\alpha - 1)^2$ no longer applies, but it is still true that at least one of the factors (and hence exactly one) is divisible by $(\alpha - 1)^2$ because, mod $(\alpha - 1)^2$

$$x \equiv a_0 + a_1(\alpha - 1) \bmod (\alpha - 1)^2$$
$$y \equiv b_0 + b_1(\alpha - 1) \bmod (\alpha - 1)^2$$

for some integers a_0, a_1, b_0, b_1 (Exercise 1), from which

$$x + \alpha^j y \equiv \left[a_0 + a_1(\alpha - 1)\right] + \left[1 + (\alpha - 1)\right]^j \left[b_0 + b_1(\alpha - 1)\right]$$
$$\equiv a_0 + b_0 + \left[a_1 + b_1 + jb_0\right](\alpha - 1) \bmod (\alpha - 1)^2.$$

Since $\alpha - 1$ divides $x + \alpha^j y$ for all j, this shows that $a_0 + b_0 \equiv 0 \bmod \lambda$. Moreover, it shows that $(\alpha - 1)^2$ divides $x + \alpha^j y$ if and only if $a_1 + b_1 + jb_0 \equiv 0 \bmod \lambda$, a condition which holds for one and only one value of j mod λ provided $b_0 \not\equiv 0 \bmod \lambda$; but $b_0 \not\equiv 0 \bmod \lambda$ because otherwise $\alpha - 1$ would divide y, contrary to assumption. *Therefore $k > 1$*, say $k = K + 1$, where K is a positive integer. Moreover, since y can be replaced by $\alpha^j y$ without changing the form of the original equation, it can be assumed that $x + y$ is the factor of $x^\lambda + y^\lambda$ that is divisible by $(\alpha - 1)^2$. Then the $k\lambda$ factors of $(\alpha - 1)$ in $x^\lambda + y^\lambda$ consist of one factor in each term $x + \alpha^j y$ ($j = 1, 2, \ldots, \lambda - 1$) and $1 + (k - 1)\lambda = 1 + K\lambda$ factors in the term $x + y$. Thus

$$x + \alpha^{-1} y = (\alpha - 1)e_{-1}t_{-1}^\lambda$$
$$x + y = (\alpha - 1)e_0(\alpha - 1)^{K\lambda}w^\lambda$$
$$x + \alpha y = (\alpha - 1)e_1 t_1^\lambda$$

and exactly the same sequence of steps as before leads to an equation of the form

$$X^\lambda + Y^\lambda = E(\alpha - 1)^{K\lambda}W^\lambda$$

in which X, Y, W, and $\alpha - 1$ are pairwise relatively prime cyclotomic integers, E is a unit, and $K = k - 1$. Now this is obviously impossible because repetition of this process will eventually lead to an equation in which $k = 1$, whereas it was just shown that $k = 1$ is impossible. This completes the proof of Case II.

Kummer also claims to have proved the impossibility of $x^\lambda + y^\lambda = z^\lambda$ for cyclotomic integers, not just for integers, but there appears to be a gap in his proof of this theorem. The proof just given shows that if x, y, and z are pairwise relatively prime cyclotomic integers and if one of them is divisible by $\alpha - 1$ then $x^\lambda + y^\lambda = z^\lambda$ is indeed impossible. Moreover, Kummer was able to modify his proof of Case I so that it applied to relatively prime cyclotomic integers x, y, z (see Exercise 3). Where he appears to have left a gap is in his assumption that it suffices to prove the theorem in the case where x, y, z are relatively prime. In other words, he did prove that $x^\lambda + y^\lambda = z^\lambda$ is impossible in relatively prime cyclotomic integers but the general case does not reduce to the case where x, y, z are relatively prime. The problem, of course, is that x, y, z might have a common *ideal* divisor—a divisor which divides them all but is not principal —which could not be divided out of $x^\lambda + y^\lambda = z^\lambda$. (For a proof of the impossibility of $x^\lambda + y^\lambda = z^\lambda$ in which this gap is filled, see Hilbert [H3].) That the inventor of the theory of ideal divisors should commit this particular *faux pas* is indeed remarkable.

It remains to show that if e is any unit then $e/\bar{e} = \alpha^r$ for some r. Choose a particular polynomial $E(X) = a_0 + a_1 X + \cdots + a_{\lambda-1} X^{\lambda-1}$ in one unknown X with integer coefficients so that $E(\alpha)$ is the unit $e/\bar{e}$. Let the polynomial $E(X^{\lambda-1})E(X)$ be divided by $X^\lambda - 1$ to give $E(X^{\lambda-1}) \cdot E(X) = Q(X)(X^\lambda - 1) + R(X)$ where $R(X)$ is a polynomial of degree less than λ, say $R(X) = A_0 + A_1 X + \cdots + A_{\lambda-1} X^{\lambda-1}$. With $X = 1$ in this equation one finds $(a_0 + a_1 + \cdots + a_{\lambda-1})^2 = A_0 + A_1 + \cdots + A_{\lambda-1}$. With $X = \alpha$ one finds $A_0 + A_1\alpha + \cdots + A_{\lambda-1}\alpha^{\lambda-1} = E(\alpha^{-1})E(\alpha) = [e(\alpha^{-1})/\bar{e}(\alpha^{-1})][e(\alpha)/\bar{e}(\alpha)] = [\bar{e}/e][e/\bar{e}] = 1$, $(A_0 - 1) + A_1\alpha + A_2\alpha^2 + \cdots + A_{\lambda-1}\alpha^{\lambda-1} = 0$, from which $A_0 - 1 = A_1 = A_2 = \cdots = A_{\lambda-1}$. Let k be their common value. Then $(a_0 + a_1 + \cdots + a_{\lambda-1})^2 = 1 + \lambda k \equiv 1 \bmod \lambda$ and $a_0 + a_1 + \cdots + a_{\lambda-1} \equiv \pm 1 \bmod \lambda$. By adding or subtracting the same integer from all the a_i one can arrange to have $a_0 + a_1 + \cdots + a_{\lambda-1} = \pm 1$. This changes the original $E(X)$ and gives a new set of A's for which $1 + \lambda k = 1$, $k = 0$, $A_0 = 1$, $A_1 = A_2 = \cdots = A_{\lambda-1} = 0$. On the other hand, a study of the way that the A's are computed shows that $A_0 = a_0^2 + a_1^2 + \cdots + a_{\lambda-1}^2$. One can prove this formula by noting that a typical term $a_i X^{\lambda i} \cdot a_j X^j$ of $E(X^{\lambda-1})E(X)$ can be written in the form $a_i a_j X^{q\lambda+r} = a_i a_j X^r \cdot (X^{q\lambda} - 1) + a_i a_j X^r = Q_{i,j}(X)(X^\lambda - 1) + a_i a_j X^r$ where $r \equiv j - i \bmod \lambda$, $0 \leqslant r < \lambda$, and $Q_{i,j}(X) = a_i a_j X^r (X^{q\lambda - \lambda} + \cdots + X^\lambda + 1)$ if $q > 1$, $Q_{i,j}(X) = a_i a_j X^r$ if $q = 1$, and $Q_{i,j}(X) = 0$ if $q = 0$. When these equations are summed over all i and j from 0 to $\lambda - 1$ they show that $R(X) = \sum_{r=0}^{\lambda-1} (\sum_{j-i\equiv r} a_i a_j) X^r$. In particular $A_0 = a_0^2 + a_1^2 + \cdots + a_{\lambda-1}^2$. Thus $A_0 = 1$ implies that exactly one a_i is nonzero and that this one is ± 1. Therefore $E(\alpha) = \pm \alpha^r$ for some r. It will suffice, then, to prove that $E(\alpha) = -\alpha^r$ is impossible. If it were possible then, because either r or $r + \lambda$ is even, $E(\alpha) = -\alpha^{2s}$ would be possible, from which $e\alpha^{-s} = -\bar{e}\alpha^s$. Let $F(\alpha) = b_0 + b_1\alpha + \cdots + b_{\lambda-1}\alpha^{\lambda-1}$ denote this unit $e\alpha^{-s}$. Then $F(\alpha)$ would be a unit such that $F(\alpha) = -F(\alpha^{-1})$. Assume without loss of generality that $b_0 = 0$. Then $F(\alpha) = b_1(\alpha - \alpha^{-1}) + b_2(\alpha^2 - \alpha^{-2}) + \cdots$. This shows that $F(\alpha)$ would be divisible by $\alpha - \alpha^{-1}$ which is impossible because $\alpha - \alpha^{-1}$ is not a unit. This completes the proof.

EXERCISES

1. Prove that every cyclotomic integer is congruent mod $\alpha - 1$ to an ordinary integer. (Recall the test for divisibility by $\alpha - 1$.) Conclude that for any given cyclotomic integer x and any given integer $k > 0$ there exist integers

$a_0, a_1, \ldots, a_{k-1}$ such that $x \equiv a_0 + a_1(\alpha - 1) + \cdots + a_{k-1}(\alpha - 1)^{k-1} \bmod (\alpha - 1)^k$. Finally, prove that if $k \leqslant \lambda - 1$ then the integers $a_0, a_1, \ldots, a_{k-1}$ are uniquely determined mod λ. (An integer is zero mod $\alpha - 1$ if and only if it is divisible by λ, in which case it is zero mod $(\alpha - 1)^{\lambda - 1}$.)

2. Prove that if λ is a regular prime greater than 3 and if $x, y,$ and z are relatively prime cyclotomic integers then $x^\lambda + y^\lambda = z^\lambda$ is impossible. (The case in which any one of them is divisible by $\alpha - 1$ is covered by the proof of Case II that was given in the text, so that only Case I remains. By replacing x by $\alpha^r x$ if necessary one can assume at the outset that x is congruent to an integer mod $(\alpha - 1)^2$ and similarly for y and z, say $x \equiv a, y \equiv b, z \equiv c \bmod (\alpha - 1)^2$ where a, b, c are integers. The proof in the text shows that for every j there is an r such that $x + \alpha^{-j}y \equiv \alpha^{-r}(\bar{x} + \alpha^j \bar{y})$. Then, using $\alpha^j = [1 + (\alpha - 1)]^j$ and computing mod $(\alpha - 1)^2$, one finds $r(a + b) \equiv 2jb \bmod \lambda$. This shows that $r \equiv vj \bmod \lambda$ where v is an integer independent of j. In fact, one can assume that v is even so that the equation takes the symmetrical form $\alpha^{kj}(x + \alpha^{-j}y) \equiv \alpha^{-kj}(\bar{x} + \alpha^j \bar{y}) \bmod \lambda$. In particular $x + y \equiv \bar{x} + \bar{y} \bmod \lambda$. If the original equation is put in the symmetric form $x^\lambda + y^\lambda + z^\lambda = 0$ the same argument applies to the pairs x, z and y, z and $x \equiv \bar{x}, y \equiv \bar{y}, z \equiv \bar{z} \bmod \lambda$ results. Therefore $\alpha^{kj}(x + \alpha^{-j}y)$ is unchanged mod λ when j is changed to $-j$. Use the cases $j = 1$ and $j = 2$ to conclude that, since y contains no factors $(\alpha - 1)$, $(\alpha^{k-1} - \alpha^{-(k-1)}) \cdot (\alpha^{-k} - \alpha^{k-1})(1 - \alpha)$ is divisible by $(\alpha - 1)^{\lambda - 1}$. Since $\lambda \geqslant 5$ this implies either $\alpha^{k-1} = \alpha^{-(k-1)}$ or $\alpha^{-k} = \alpha^{k-1}$, that is, either $k \equiv 1 \bmod \lambda$ or $2k \equiv 1 \bmod \lambda$. In the first case $a \equiv 0$ contrary to assumption. In the second case $a \equiv b$. Thus $a \equiv c$ as well and $0 = x^\lambda + y^\lambda + z^\lambda \equiv a^\lambda + b^\lambda + c^\lambda \equiv 3a$ and again $a \equiv 0$ contrary to assumption.)

5.6 Quadratic reciprocity

The surprising fact that Kummer never developed in full the relation between his theory and Gauss's theory of binary quadratic forms is all the more surprising because the famous theorem of quadratic reciprocity is a simple consequence of Kummer's theory. Gauss considered this theorem so important that he called it the "fundamental theorem" and he published a number of proofs of it, two in the *Disquisitiones Arithmeticae* and four others subsequently. (See Gauss [G5].) Kummer himself, as has already been mentioned in Section 4.1, considered the quadratic reciprocity law and its generalizations to higher powers as the most important subject of number theory. It is almost inconceivable that he could have overlooked the following simple deduction of quadratic reciprocity from his theory of ideal factorization;* a more plausible hypothesis is that he preferred not to publish anything on the subject until he had worked out the generalizations to higher powers, something which he did several years later. The theory of binary quadratic forms is related to quadratic reciprocity but not

*According to Hecke ([H2], p. 113) this proof was first given by Kronecker, who was Kummer's student and close friend. This reference probably comes from Hilbert [H3, §122]. However, Smith [53, addition to Section 22] made this remark before Kronecker did, and, as Smith observes, the proof is very closely related to Gauss's 6th proof (see Exercises 12 and 13).

to higher reciprocity laws, and Kummer's later publications on reciprocity laws therefore bypassed the theory of quadratic forms.

Consider again the example $\lambda = 23$ which was studied in such detail in Section 5.2. A very natural question to ask in connection with this example is the question "For which primes p does the divisor of p have a factor of the form (p, u)?" In fact, a full solution of the problem of Section 5.2 probably should include an answer to this question. It was avoided in Section 5.2 only because it was not relevant to the main goal there of motivating the definition of equivalence of divisors, and because it would lead into a sidetrack. A careful study of this question leads almost inevitably to the law of quadratic reciprocity.

It was already observed in Section 5.2 that there is a divisor (p, u) *if and only if p has an even number of prime divisors.* (Here, and in the entire discussion which follows, it is assumed that $p \neq 23$.) On the one hand, if the divisor of p can be factored $(p, u)(p, -1 - u)$ then, because $\alpha \mapsto \alpha^{-2}$ interchanges the prime divisors of (p, u) with those of $(p, -1 - u)$, these two divisors must split the prime divisors of p into two sets of equal size, and there must be an even number of them. On the other hand, if there are an even number of them, then 2 divides $e = 22/f$, from which it follows that f divides 11. Therefore, because periods of length f are congruent to integers modulo any prime divisor of p, θ_0 is congruent to an integer modulo any prime divisor of p, say $\theta_0 \equiv u$ modulo a particular prime divisor of p. Then exactly half of the prime divisors of p divide $\theta_0 - u$ and their product is a divisor of the form (p, u), as was to be shown.

This necessary and sufficient condition for the existence of (p, u) can be reformulated as follows. In the first place, e is even if and only if f divides 11. Since f is by definition the exponent of p mod 23, this is true if and only if $p^{11} \equiv 1$ mod 23. Now it is a well-known fact of elementary number theory, called *Euler's criterion*, that $x^{(\lambda - 1)/2} \equiv 1$ mod λ if and only if x is a nonzero square mod λ, that is, if and only if there is a $y \not\equiv 0$ mod λ such that $x \equiv y^2$ mod λ. This is easy to prove in a number of ways (see Exercise 7) and it implies, in the case at hand, that, if $p \neq 23$, *there is a divisor of the form (p, u) if and only if p is a square* mod 23.

The squares mod 23 are 1, 4, 9, $16 \equiv -7$, $25 \equiv 2$, $36 \equiv -10$, $49 \equiv 3$, $64 \equiv -5$, $81 \equiv -11$, $100 \equiv 8$, $121 \equiv 6$ or, in a more easily legible order, 1, 2, 3, 4, -5, 6, -7, 8, 9, -10, -11. In the sequence of primes 2, 3, 5, 7, 11, ... those for which there are divisors (p, u) can then be listed very easily: 2, 3, $13 \equiv -10$, $29 \equiv 6$, $31 \equiv 8$, $41 \equiv -5$, ... This list is of course in accord with Table 5.2.2.

This argument does not, however, provide the value of u to go with each of these values of p; it is in fact the quest for the value of u which leads to the other criterion for the existence of a divisor (p, u). The basic property of u is that half of the prime divisors of p divide $\theta_0 - u$. Then all prime divisors of p—and hence p itself—divide $(\theta_0 - u)(\theta_1 - u) = u^2 + u + 6$.

Therefore u is a solution of the congruence $u^2 + u + 6 \equiv 0 \bmod p$. The connection of this congruence with $\lambda = 23$ is made clearer when it is written in the form $\frac{1}{4}[(2u + 1)^2 + 23] \equiv 0 \bmod p$. This shows that if there is a divisor of the form (p, u) then $(2u + 1)^2 \equiv -23 \bmod p$ and, in particular, -23 *is a square* mod p. The converse is also true, provided one excludes the prime $p = 2$. If $p \neq 2$ and $x^2 \equiv -23 \bmod p$ then $(x + p)^2 \equiv -23 \bmod p$ and either x or $x + p$ is odd. Then there is a u such that $(2u + 1)^2 \equiv -23 \bmod p$. This implies that p divides $(2u + 1)^2 + 23 = 4(\theta_0 - u)(\theta_1 - u)$ and, since $p \neq 2$, that every prime divisor of p divides either $\theta_0 - u$ or $\theta_1 - u$, and that there is therefore a divisor of the form (p, u). In short, if $p \neq 23$ and $p \neq 2$ then *there is a divisor of the form (p, u) if and only if -23 is a square* mod p.

Comparison of these two criteria for the existence of a divisor (p, u) reveals the surprising fact that *a prime $p \neq 2$ or 23 is a square* mod 23 *if and only if -23 is a square* mod p. This is one case of the law of quadratic reciprocity. The other cases can easily be obtained by generalizing the above arguments to other values of λ than $\lambda = 23$.

To say that there is a divisor of the form (p, u) in cyclotomic integers for a given λ means that for this λ exactly half of the prime divisors of p divide $\theta_0 - u$ (where θ_0 is the period of length $\frac{1}{2}(\lambda - 1)$ which contains α). Exactly the same argument as in the case $\lambda = 23$ proves that *if $p \neq \lambda$ then there is a divisor of the form (p, u) if and only if p is a square* mod λ. The general case of the formula $(\theta_0 - u)(\theta_1 - u) = \frac{1}{4}[(2u + 1)^2 + 23]$ in the case $\lambda = 23$ is, as will be shown below, $(\theta_0 - u)(\theta_1 - u) = \frac{1}{4}[(2u + 1)^2 \pm \lambda]$ where the sign in front of λ is determined by the condition that 4 must divide $(2u + 1)^2 \pm \lambda$—that is, the sign is $+$ if $\lambda \equiv 3 \bmod 4$ and $-$ if $\lambda \equiv 1 \bmod 4$. The same argument then proves that for $p \neq 2$ and $p \neq \lambda$, there is a divisor of the form (p, u) if and only if $-\lambda$ is a square mod p when $\lambda \equiv 3 \bmod 4$, and if and only if λ is a square mod p when $\lambda \equiv 1 \bmod 4$. Thus:

Let p and λ be distinct odd primes ($p \neq 2$, $\lambda \neq 2$, $p \neq \lambda$). If $\lambda \equiv 1 \bmod 4$ *then p is a square* mod λ *if and only if λ is a square* mod p. *If $\lambda \equiv 3 \bmod 4$ then p is a square* mod λ *if and only if $-\lambda$ is a square* mod p.

This is precisely the form in which Gauss stated the *law of quadratic reciprocity* or, as he called it, the fundamental theorem [*Disquisitiones Arithmeticae*, Art. 131]. Note that the above argument does much more than prove the theorem; it also provides a simple, though completely unhistorical, way of discovering the *statement* of the theorem. All that remains to be done is to prove the formula $(\theta_0 - u) \cdot (\theta_1 - u) = \frac{1}{4}[(2u + 1)^2 \pm \lambda]$ that was stated above. This can be done as follows. (See Exercises 3 and 4 of Section 4.5.)

Of course $(\theta_0 - u)(\theta_1 - u) = u^2 + u + \theta_0\theta_1 = \frac{1}{4}[(2u + 1)^2 + 4\theta_0\theta_1 - 1]$. The problem, therefore, is to evaluate $4\theta_0\theta_1 - 1$. As was noted in Section 4.9 (for periods of length f, where here $f = \frac{1}{2}(\lambda - 1)$), $\theta_0\theta_1$ has the form

$a \cdot \frac{1}{2}(\lambda - 1) + b\theta_0 + c\theta_1$ where $a + b + c = \frac{1}{2}(\lambda - 1)$ and a is 0 or 1. By symmetry, $b = c$. If $a = 0$ then $a + b + c = \frac{1}{2}(\lambda - 1)$ becomes $2b = \frac{1}{2}(\lambda - 1)$, $\lambda = 4b + 1$, $\lambda \equiv 1 \bmod 4$, $\theta_0\theta_1 = a + b\theta_0 + c\theta_1 = 0 + b(\theta_0 + \theta_1) = -b$, $4\theta_0\theta_1 - 1 = -4b - 1 = -\lambda$. If $a = 1$ then $a + b + c = \frac{1}{2}(\lambda - 1)$ becomes $1 + 2b = \frac{1}{2}(\lambda - 1)$, $4b + 3 = \lambda$, $\lambda \equiv 3 \bmod 4$, $\theta_0\theta_1 = \frac{1}{2}(\lambda - 1) + b\theta_0 + b\theta_1 = 1 + 2b - b = b + 1$, $4\theta_0\theta_1 - 1 = 4b + 4 - 1 = \lambda$. Thus either $\lambda \equiv 1 \bmod 4$ and $4\theta_0\theta_1 - 1 = -\lambda$ or $\lambda \equiv 3 \bmod 4$ and $4\theta_0\theta_1 - 1 = \lambda$, as was to be shown.

The reason that the law of quadratic reciprocity has held such fascination for so many great mathematicians should be apparent. On the face of it there is absolutely no relation between the questions "is p a square mod λ?" and "is λ a square mod p?" yet here is a theorem which shows that they are practically the same question. Surely the most fascinating theorems in mathematics are those in which the premises bear the least obvious relation to the conclusions, and the law of quadratic reciprocity is an example *par excellence* of such a theorem. Not only Gauss and Kummer but also Euler, Lagrange, Legendre, Dirichlet, Jacobi, Eisenstein, Liouville, Hilbert, Artin, and many, many other great mathematicians have taken up the challenge presented by this theorem to find a natural proof or to find a more comprehensive "reciprocity" phenomenon of which this theorem is a special case. Gauss called the quadratic reciprocity law the *fundamental theorem*, however, not because of its aesthetic value but because of its very great usefulness in the theory of congruences of degree 2 and in the theory of binary quadratic forms, the topics of the fourth and fifth sections of the *Disquisitiones Arithmeticae*. It is also, as will be seen in Chapter 7, very useful in the theory of ideal factorization of quadratic integers. Specifically, in all these theories it is frequently necessary to determine, given a prime p and an integer $n \not\equiv 0 \bmod p$, whether the congruence $n \equiv x^2 \bmod p$ has a solution. This is of course a finite problem which can be solved simply by computing the squares of all integers less than $\frac{1}{2}p$ and seeing whether one of them is congruent to $n \bmod p$. However, even when some elementary simplifications are used (see Exercises 1 and 2) this is a long process of trial-and-error for large primes p, and the method using the quadratic reciprocity law is immeasurably simpler. This method is as follows.

Suppose, for example, that the problem is to determine whether the congruence $x^2 \equiv 31 \bmod 79$ has a solution. The so-called *Legendre symbol* $(\frac{n}{p})$, in which p is a (positive) odd prime and n is an integer $n \not\equiv 0 \bmod p$, is defined to be ± 1, being $+1$ if the congruence $x^2 \equiv n \bmod p$ has a solution and being -1 otherwise. The problem is to evaluate the Legendre symbol $(\frac{31}{79})$. Now because $79 \equiv 3 \bmod 4$ the law of quadratic reciprocity states $(\frac{-79}{31}) = (\frac{31}{79})$. But $-79 = n \equiv 14 \bmod 31$, so $(\frac{31}{79}) = (\frac{14}{31})$ and one is confronted with the much smaller problem of determining whether $x^2 \equiv 14 \bmod 31$ has a solution. As was noted in the proof above, Euler's criterion states that $(\frac{n}{p}) = +1$ if and only if $n^{(p-1)/2} \equiv 1 \bmod p$. On the other hand, the square of $n^{(p-1)/2}$ is $1 \bmod p$ and $n^{(p-1)/2}$ can therefore

177

have only the values 1 or $-1 \bmod p$. Therefore $\left(\frac{n}{p}\right) = -1$ if and only if $n^{(p-1)/2} \equiv -1 \bmod p$. These observations imply $n^{(p-1)/2} \equiv \left(\frac{n}{p}\right) \bmod p$, from which the fundamental relation

$$\left(\frac{n}{p}\right)\left(\frac{m}{p}\right) = \left(\frac{nm}{p}\right),$$

satisfied by the Legendre symbol whenever n, m and p are such that $\left(\frac{n}{p}\right)$ and $\left(\frac{m}{p}\right)$ are defined, follows. Therefore, in order to evaluate $\left(\frac{14}{31}\right)$ it will suffice to evaluate $\left(\frac{2}{31}\right)$ and $\left(\frac{7}{31}\right)$. By quadratic reciprocity the latter is $\left(\frac{-31}{7}\right) = \left(\frac{4}{7}\right) = \left(\frac{2}{7}\right)\left(\frac{2}{7}\right) = (\pm 1)^2 = +1$ and it follows that $\left(\frac{31}{79}\right) = \left(\frac{2}{31}\right)$. This can be further reduced $\left(\frac{2}{31}\right) = \left(\frac{33}{31}\right) = \left(\frac{3}{31}\right)\left(\frac{11}{31}\right) = \left(\frac{-31}{3}\right)\left(\frac{-31}{11}\right) = \left(\frac{2}{3}\right)\left(\frac{2}{11}\right) = -\left(\frac{2}{11}\right)$. This is a problem of small magnitude: the squares mod 11 are $1, 4, 9 \equiv -2, 16 \equiv 5, 25 \equiv 3, 49 \equiv 5, 64 \equiv -2, 81 \equiv 4, 100 \equiv 1, 121 \equiv 0$. Since this list does not include 2, it follows that $\left(\frac{2}{11}\right) = -1$ and finally that $\left(\frac{31}{79}\right) = +1$ and 31 *is a square mod* 79.

In applying this process it is helpful to use the so-called *supplementary laws of quadratic reciprocity* (from which the element of reciprocity is absent) to evaluate the Legendre symbols $\left(\frac{-1}{p}\right)$, $\left(\frac{2}{p}\right)$. These state:

$$\left(\frac{-1}{p}\right) = \begin{cases} +1 & \text{if } p \equiv 1 \bmod 4, \\ -1 & \text{if } p \equiv 3 \bmod 4. \end{cases} \qquad (1)$$

Otherwise stated, $\left(\frac{-1}{p}\right) = \left(\frac{-1}{p+4n}\right)$ whenever both are defined (that is, whenever $p + 4n$ is a positive prime) and this remains true when $\left(\frac{-1}{1}\right)$ is defined to be $+1$. One has $\left(\frac{-1}{3}\right) = -1$ by inspection.

$$\left(\frac{2}{p}\right) = \begin{cases} +1 & \text{if } p \equiv 1 \text{ or } 7 \bmod 8, \\ -1 & \text{if } p \equiv 3 \text{ or } 5 \bmod 8. \end{cases} \qquad (2)$$

Otherwise stated, $\left(\frac{2}{p}\right) = \left(\frac{2}{p+8n}\right)$ and this remains true if $\left(\frac{2}{1}\right)$ is defined to be $+1$. One has $\left(\frac{2}{3}\right) = -1$, $\left(\frac{2}{5}\right) = -1$, and $\left(\frac{2}{7}\right) = +1$ by inspection.

These two statements follow from theorems on representations $p = a^2 + b^2$ and $p = a^2 + 2b^2$ proved in Chapter 2 (see Exercises 5 and 6 below). Alternatively, (1) can be deduced by applying quadratic reciprocity to p and 3 as follows. If $p \equiv 1 \bmod 4$ then $\left(\frac{-3}{p}\right) = \left(\frac{p}{3}\right) = \left(\frac{3}{p}\right) = \left(\frac{-1}{p}\right)\left(\frac{-3}{p}\right)$ from which $\left(\frac{-1}{p}\right) = +1$, whereas if $p \equiv 3 \bmod 4$ then $\left(\frac{-3}{p}\right) = \left(\frac{p}{3}\right) = \left(\frac{-1}{3}\right)\left(\frac{-p}{3}\right) = \left(\frac{-1}{3}\right)\left(\frac{3}{p}\right) = \left(\frac{-1}{3}\right)\left(\frac{-1}{p}\right)\left(\frac{-3}{p}\right)$ from which $\left(\frac{-1}{3}\right)\left(\frac{-1}{p}\right) = +1$ and $\left(\frac{-1}{p}\right) = \left(\frac{-1}{3}\right) = -1$. (2) can be proved by a straightforward extension of the proof of the reciprocity law above to the excluded case $p = 2$: It is still true that 2 is a square mod λ if and only if its divisor factors as required and, on the other hand, that its divisor factors if and only if $u^2 + u + \theta_0 \theta_1 \equiv 0 \bmod 2$ is possible. This is equivalent to $(2u + 1)^2 \pm \lambda \equiv 0 \bmod 8$ which depends only on the class of $\lambda \bmod 8$. If $\lambda \equiv 1 \bmod 8$ then the needed congruence $(2u - 1)^2 - 1 \equiv 0 \bmod 8$ is not only possible but true for all integers u.

The rule (1) and the Legendre symbol can be used to put the law of quadratic reciprocity in a simple and symmetrical form: *If p and q are distinct odd primes and if either is 1 mod 4 then $\left(\frac{p}{q}\right) = \left(\frac{q}{p}\right)$, but if both are*

3 mod 4 *then* $\left(\frac{p}{q}\right) = -\left(\frac{q}{p}\right)$ (because $\left(\frac{p}{q}\right) = \left(\frac{-1}{q}\right)\left(\frac{-p}{q}\right) = -\left(\frac{q}{p}\right)$). Thus the evaluation of $\left(\frac{31}{79}\right)$ can be given simply as $\left(\frac{31}{79}\right) = -\left(\frac{79}{31}\right) = -\left(\frac{17}{31}\right) = -\left(\frac{31}{17}\right) = -\left(\frac{14}{17}\right) = -\left(\frac{2}{17}\right)\left(\frac{7}{17}\right) = -\left(\frac{2}{1}\right)\left(\frac{17}{7}\right) = -\left(\frac{3}{7}\right) = \left(\frac{7}{3}\right) = \left(\frac{1}{3}\right) = +1$.

EXERCISES

1. Solve the congruence $x^2 \equiv 31$ mod 79. (Do this by trial-and-error. Rather than computing squares one can add successive odd integers $2^2 = 1^2 + 3$, $3^2 = 2^2 + 5$, $4^2 = 3^2 + 7$, . . . and reduce mod 79.)

2. Gauss proposes the following method of solving congruences $x^2 \equiv A$ mod m (*Disquisitiones Arithmeticae*, Art. 319–322). Write $x^2 = A + my$ and consider y to be the unknown. One can assume $-A/m \leqslant y < \frac{1}{4}m - (A/m)$. For any modulus E (which Gauss calls the *excludent*) one must have $A + my \equiv$ square mod E. Only about half the possible classes mod E are squares and one can in this way exclude about half the possible values of y. Show that in Exercise 1, this method with $E = 3$, 4, and 5 excludes all but four possible values of y, and find them. Show that use of $E = 8$ instead of $E = 4$ excludes one of these, and use of $E = 7$ excludes a second. Try the two which remain.

3. Show that $\left(\frac{22}{97}\right) = +1$ and then solve $x^2 \equiv 22$ mod 97. ($E = 8, 9, 5$ excludes all but three values, the smallest two of which are easily ruled out.)

4. Evaluate the following Legendre symbols: $\left(\frac{79}{101}\right)$, $\left(\frac{97}{139}\right)$, $\left(\frac{91}{139}\right)$. (Beware: The evaluation $\left(\frac{91}{139}\right) = -\left(\frac{139}{91}\right) = -\left(\frac{48}{91}\right) = -\left(\frac{3}{91}\right) = -\left(\frac{91}{3}\right) = \left(\frac{91}{3}\right) = +1$ has not been proved to be valid even though it gives the correct answer.)

5. Deduce the rule (1) for evaluating $\left(\frac{-1}{p}\right)$ from (4) and (5) of Section 2.4.

6. Prove the rule (2) for the evaluation of $\left(\frac{2}{p}\right)$ as follows. Exercise 4 of Section 2.4 shows that if $\left(\frac{2}{p}\right) = +1$ then $p = c^2 - 2d^2$ and therefore $p \equiv \pm 1$ mod 8. The converse is proved in Exercises 6 and 7 of Section 2.4.

7. Prove that the formula $\left(\frac{n}{p}\right) \equiv n^{(p-1)/2}$ mod p is valid whenever the Legendre symbol is defined. (This is easy using the existence of a primitive root mod p, but the use of a primitive root can be avoided by *counting* arguments: Since squaring is a two-to-one function, just half the nonzero classes mod p are squares. For other reasons the product of a square and a nonsquare is a nonsquare and, therefore, the product of two nonsquares is a square, etc.) Conclude that $\left(\frac{nm}{p}\right) = \left(\frac{n}{p}\right)\left(\frac{m}{p}\right)$.

8. Assume the second statement of the quadratic reciprocity law "If p and q are odd primes and $p \equiv 1$ or $q \equiv 1$ mod 4 then $\left(\frac{p}{q}\right) = \left(\frac{q}{p}\right)$ but if $p \equiv 3$ mod 4 and $q \equiv 3$ mod 4 then $\left(\frac{p}{q}\right) = -\left(\frac{q}{p}\right)$" and deduce the first "If $p \equiv 3$ mod 4 then $\left(\frac{-p}{q}\right) = \left(\frac{q}{p}\right)$ and if $p \equiv 1$ mod 4 then $\left(\frac{p}{q}\right) = \left(\frac{q}{p}\right)$." Do not use rule (1) without proving it first.

9. Show that the quadratic reciprocity law can be stated $\left(\frac{p}{q}\right)\left(\frac{q}{p}\right) = (-1)^{(p-1)(q-1)/4}$.

10. Determine whether the congruences
$$3u^2 - 2u \equiv 7 \text{ mod } 23$$
$$7u^2 + 6u \equiv 2 \text{ mod } 37$$
have solutions. [Complete the square and use quadratic reciprocity.]

11. Lagrange observed [L1, Art. 54] that if $p \equiv 3 \bmod 4$, if p is prime, and if $x^2 \equiv B \bmod p$ has a solution, then $x \equiv B^{(p+1)/4} \bmod p$ satisfies $x^2 \equiv B$. Prove this fact and use it to solve the congruence of Exercise 1. Gauss's method of the excludent is not very practical. For a better method see Shanks [S1].

12. Streamline the proof of quadratic reciprocity given in this section as follows. First prove: *Theorem.* Let $ef = \lambda - 1$, let $\eta_1, \eta_2, \ldots \eta_e$ be the periods of length f, and let $\phi(X)$ be the polynomial $(X - \eta_1)(X - \eta_2) \cdots (X - \eta_e)$. ($\phi$ has integer coefficients.) Consider the congruence $\phi(u) \equiv 0 \bmod q$ where q is a prime integer. This congruence has an integer solution u if and only if $q^f \equiv 1$ mod λ. [If f' is the exponent of q mod λ then $q^f \equiv 1$ mod λ implies f' divides f. The periods of length f' are congruent to integers mod any ideal prime factor of q and therefore so are $\eta_1, \eta_2, \ldots, \eta_e$. If $\eta_1 \equiv u$ mod an ideal prime factor of q then $\phi(u) \equiv 0 \bmod q$. For the converse, note that if q divides $(u - \eta_1)$ $(u - \eta_2) \cdots (u - \eta_e)$ then the ideal prime divisors of q are partitioned into e subsets of equal size.] If $e = 2$ then $\phi(X) = [(2X + 1)^2 - \lambda^*]/4$ where $\lambda^* = \pm\lambda$, $\lambda^* \equiv 1$ mod 4. Thus $r^2 \equiv \lambda^*$ mod q has a solution if and only if $q^{(\lambda-1)/2} \equiv 1$ mod λ. In short, $\left(\frac{\lambda^*}{q}\right) = \left(\frac{q}{\lambda}\right)$.

13. Gauss's 6th proof of quadratic reciprocity is in essence the following: Let $\xi = \theta_0 - \theta_1$. Then $\xi^2 = \lambda^*$ where λ^* is $\pm\lambda$ and $\lambda^* \equiv 1$ mod 4. On the other hand, $\xi = \Sigma(\frac{j}{\lambda})\alpha^j$ as j ranges over $1, 2, \ldots, \lambda - 1$. This gives $\xi^p \equiv \Sigma(\frac{j}{\lambda})^p\alpha^{jp} = (\frac{p}{\lambda}) \xi$ mod p. Thus $(\frac{p}{\lambda}) \equiv \xi^{p-1} = (\lambda^*)^{(p-1)/2} \equiv (\frac{\lambda^*}{p})$ mod p (by Euler's criterion). Therefore $(\frac{p}{\lambda}) = (\frac{\lambda^*}{p})$. Fill in the steps of this proof.

6.1 Introduction

Kummer published his proof that conditions* (A) and (B) imply Fermat's Last Theorem for the exponent λ in April of 1847. In order to apply this theorem to prove Fermat's Last Theorem for any particular prime exponent λ it is necessary to have some method of *testing*, for a given λ, whether conditions (A) and (B) are satisfied. Kummer said that he had proved rigorously that they were satisfied when $\lambda = 3, 5, 7$ and that they were probably satisfied not for all primes λ but for an infinite number of them. The natural way to approach the problem of testing (A) is to attempt to *compute the class number* for a given value of λ. This problem of the determination of the class number is one which Kummer mentioned in his very first announcement of the theory of ideal complex numbers in 1847; he mentioned it at that time only to say, however, that he had not pursued the problem because he had learned by word of mouth that Dirichlet had already succeeded in finding a formula for the class number. Kummer alluded to this fact again when he communicated his paper of April 1847 to Dirichlet, saying that, while he himself was unable to say which primes λ satisfied conditions (A) and (B), "for you it will probably be easy." In the remarks which Dirichlet appended to this communication, however, he said only that he had a formula for the class number that made it possible to test in particular cases whether condition (A) was satisfied; he did not give any values of λ for which he had tested the condition and he admitted both that his method for testing (B) was impractical for $\lambda > 7$ and that he was unable to say whether Kummer's conjecture that (A) implied (B) was

*These conditions are that (A) the class number is not divisible by λ and (B) any unit which is congruent to an integer mod λ is a λth power.

true. Finally, in September 1847, Kummer sent to Dirichlet and the Berlin Academy a paper [K10] in which he gave a very satisfactory theorem on the conditions (A) and (B).

Theorem. (A) *does imply* (B). *Condition* (A) *is equivalent to the statement that* λ *does not divide the numerators of any of the Bernoulli numbers* $B_2, B_4, \ldots, B_{\lambda-3}$ *where the Bernoulli numbers are defined* by the equation* $x/(e^x - 1) = \Sigma B_n x^n / n!$.

This theorem and the known values of the Bernoulli numbers enabled Kummer to say that (A) and (B) hold for all primes less than 37 but that they *fail* for $\lambda = 37$, because 37 divides the numerator of B_{32}. The values of the Bernoulli numbers are not at all simple to compute, but the values up to B_{60} had been tabulated in Kummer's time and this made it possible to test all primes λ up to $\lambda = 61$. Kummer does not appear to have carried out such tests before he sent his paper to Berlin because he does not announce the fact that the only other prime $\lambda \leqslant 61$ for which (A) and (B) fail is $\lambda = 59$.

Notwithstanding Kummer's deference to Dirichlet and his acknowledgement that he had received some helpful information from Dirichlet personally, his success in working out over the summer of 1847 both the formula for the class number and the above theorem on the conditions (A) and (B) must be regarded as an extraordinary *tour de force*. The reader can judge for himself the degree of computational skill and perseverance needed to achieve the results of this chapter—all of which are Kummer's—between May and September of the same year.

6.2 The Euler product formula

The basis of virtually all of analytic number theory, of which the class number formula is an important part, is the *Euler product formula*

$$\sum_n \frac{1}{n^s} = \prod_p \frac{1}{1 - \frac{1}{p^s}} \qquad (s > 1) \tag{1}$$

in which the sum on the left is a sum over all positive integers $n = 1, 2, 3, \ldots$ and the product on the right is a product over all primes $p = 2, 3, 5, 7, 11, \ldots$. Both the product and the sum are easily shown to converge for positive real numbers s greater than 1. The proof that they are equal is a simple exercise in the use of absolute convergence and the fundamental theorem of arithmetic (every positive integer n can be written in exactly one way as a product of prime powers $n = p_1^{\mu_1} p_2^{\mu_2} \ldots p_k^{\mu_k}$) to

---

*Kummer used a different definition of the Bernoulli numbers and his B_j is here denoted by B_{2j}.

justify the formal manipulations

$$\prod_p \left(1 - \frac{1}{p^s}\right)^{-1} = \prod_p \left(1 + \frac{1}{p^s} + \frac{1}{p^{2s}} + \cdots\right)$$

$$= \sum_{\substack{p \\ \mu_1, \mu_2, \ldots}} \frac{1}{p_1^{\mu_1 s} p_2^{\mu_2 s} \cdots p_k^{\mu_k s}} = \sum_n \frac{1}{n^s}.$$

This formula has the infinitude of primes as an immediate consequence because $\sum n^{-s} > \int_1^\infty x^{-s}\, dx = (s-1)^{-1}$ implies that the left side of (1) approaches ∞ as $s \downarrow 1$ and this could not be the case if the product on the right side of (1) contained only a finite number of factors.

In a nutshell, the idea behind the derivation of the class number formula is to find the analogous formula when the fundamental theorem of arithmetic for positive integers is replaced with the obvious fact that a divisor can be written in exactly one way as a product of prime divisors. Since the norm of a product is the product of the norms this gives

$$\sum_A \frac{1}{N(A)^s} = \prod_P \frac{1}{1 - \frac{1}{N(P)^s}} \tag{2}$$

where A ranges over all divisors and P over all prime divisors. The function of s in this equation approaches ∞ as $s \downarrow 1$ like a constant times $(s-1)^{-1}$. The constant can be computed by two methods, one for each side of the equation, and the answer on the left involves the class number whereas the answer on the right does not. The equality of the two ways of computing this constant then gives the *class number formula*. It is a bit lengthy to write the class number formula in closed form (see Section 6.14) but it is simply a working out of this one basic idea.

The Euler product formula also lay at the base of much of Dirichlet's work, including both his famous theorem on the infinitude of primes in an arithmetic progression and his equally important class number formula for the number of classes of binary quadratic forms with a given determinant. In fact, both of these theorems depend on a study of the way in which functions analogous to $\sum n^{-s}$ approach ∞ as $s \downarrow 1$. Thus Kummer's idea had ample precedents, and a historical approach such as the one that is taken in this book would seem to call for an examination of the precedents in order to discover the basic ideas in their simplest form. However, a detailed study of Dirichlet's work hinders rather than helps in the effort to show the ideas in their simplest form. The reason for this is the fact that Dirichlet, following Gauss, formulated the definition of the class number in terms of binary quadratic forms. It was not until Kummer showed that this theory could be formulated much more elegantly in terms of divisors (ideal complex numbers) that it became possible to see the *direct* link —Dirichlet was well aware of a less direct link—between the Euler product formula and Dirichlet's class number formula. In terms of Kummer's theory, Dirichlet's formula is simply the limiting case as $s{\downarrow}1$ of the generalization (2) of the Euler product formula (1) when the divisors and prime divisors that occur in (2) are those of the arithmetic of quadratic integers $x + y\sqrt{D}$ (see Chapter 9).

For this reason it is natural to skip over Dirichlet's work and to go directly from the Euler product formula to Kummer's generalization (2) in the case of cyclotomic integers. Suffice it to say that Dirichlet's work provided many useful indications of the direction to take, showing not only the importance of the limit $s{\downarrow}1$ but also the fact that if the sum $\Sigma N(A)^{-s}$ is partitioned into a finite number of sums over *equivalence classes* of divisors A then in the limit as $s{\downarrow}1$ each of these sums will be the same, and so forth. In addition, Kummer acknowledged having received helpful suggestions from Dirichlet in conversations about the computation of the class number.

Kummer did say, however, in a letter to Jacobi dated 27 September 1847 that "My derivation of this formula is apparently quite different from Dirichlet's because I never encountered the difficulties of which Dirichlet spoke. The difficulties which I had to overcome were of a more subjective nature, and lay only in the trifling computations of the sort that I encountered in applying the techniques which Dirichlet has already propounded in the study of quadratic forms." He goes on to say that he hesitates to publish his derivation because to do so might deprive the mathematical world of a work of Dirichlet for which he, Kummer, could in no way give an equivalent.

In the end, Dirichlet never did publish any generalization of his class number formula beyond the quadratic case. It would of course be interesting to know what form this generalization might have taken, but interest in it would stem more from curiosity than from real mathematical substance. It seems most likely that, lacking the simplifying concepts of Kummer's theory, Dirichlet's results were more difficult and less general than Kummer's and that it was for this reason that Dirichlet did not publish them.

6.3 First steps

As was outlined in the preceding section, the class number formula is derived by studying the limit as $s{\downarrow}1$ of the Euler product formula

$$\sum_A N(A)^{-s} = \prod_P \left(1 - N(P)^{-s}\right)^{-1} \qquad (s > 1) \tag{1}$$

in which A ranges over all divisors of cyclotomic integers and P ranges over all *prime* divisors (the cyclotomic integers being those corresponding to some fixed prime $\lambda > 2$ for which the class number is to be computed).

This section is devoted to the simple proof of the Euler product formula (1) and to a basic fact about the Riemann zeta function.

The first step is to prove that the product on the right side of (1) converges for all real $s > 1$. Because $N(P)^{-s} > 0$ (it is p^{-fs} where p is the prime integer which P divides and f is the exponent of p mod λ), it follows from a basic theorem on the convergence of infinite products that this product converges if and only if the series $\Sigma N(P)^{-s}$ converges. For each prime integer p this series contains $e \leqslant \lambda - 1$ terms equal to $p^{-fs} \leqslant p^{-s}$ where f is the exponent of p mod λ (or $f = 1$ if $p = \lambda$) and $e = (\lambda - 1)/f$ (or $e = 1$ if $p = \lambda$). Therefore

$$\sum N(P)^{-s} \leqslant (\lambda - 1) \sum_p \frac{1}{p^s} < (\lambda - 1) \sum_n \frac{1}{n^s}$$

is convergent, and the convergence of the product follows.

Any one of the finite products of which this product is the limit is equal to a sum of *some* of the terms of $\Sigma N(A)^{-s}$, namely, the sun of those terms in which all prime divisors of A are included in the finite product; this follows from multiplication of the absolutely convergent series $(1 - N(P)^{-s})^{-1} = 1 + N(P)^{-s} + N(P^2)^{-s} + N(P^3)^{-s} + \cdots$ and rearrangement of the terms. This shows that any finite sum $\Sigma N(A)^{-s}$ is less than some partial product $\Pi(1 - N(P)^{-s})^{-1}$, for example less than the product over all prime divisors which occur in the given finite sum. Therefore, the infinite sum $\Sigma N(A)^{-s}$ converges and is less than or equal to the infinite product. On the other hand, $\Sigma N(A)^{-s}$ is greater than any sum in which some terms are omitted and therefore is greater than any partial product. Therefore $\Sigma N(A)^{-s}$ is greater than or equal to the infinite product. This proves the Euler product formula (1).

In the course of the derivation of the class number formula it will be necessary to use the simplest fact about the Euler product formula

$$\sum \frac{1}{n^s} = \prod \frac{1}{1 - \frac{1}{p^s}} \qquad (s > 1) \tag{2}$$

for ordinary integers, namely, the fact that the function of s defined by these expressions approaches ∞ like $(s - 1)^{-1}$ as $s{\downarrow}1$. This can be proved as follows. The function of s defined by (2) for $s > 1$ is called* the *Riemann zeta function* and is denoted $\zeta(s)$. The expression of $\zeta(s)$ on the left side of (2) gives

$$\int_1^\infty \frac{1}{x^s}\, dx < \sum_{n=1}^\infty \frac{1}{n^s} < 1 + \int_1^\infty \frac{1}{x^s}\, dx$$

$$\frac{1}{s-1} < \zeta(s) < 1 + \frac{1}{s-1}$$

---

*It was not so called by Dirichlet and Kummer, of course, because their work preceded that of Riemann by many years.

from which it follows immediately that $\lim_{s\downarrow 1}(s-1)\zeta(s)=1$, that is, $\zeta(s)$ approaches ∞ like $(s-1)^{-1}$ as $s\downarrow 1$, as was to be shown.

EXERCISES

1. The function of s defined by the Euler product formula (1) is called the zeta function of the cyclotomic integers. Show that it approaches ∞ as $s\downarrow 1$ if and only if the sum of all reciprocal primes $1/p$ in which $p \equiv 1 \bmod \lambda$ diverges. [In $\Sigma N(P)^{-s}$ only primes with exponent one can produce divergence.] This statement is true for all primes λ, as Dirichlet proved.

2. Estimate the value of $(s-1)\zeta(s)$ for $s = 1.001$.

3. Prove that the sums of the reciprocals of the prime integers $\Sigma(1/p)$ is a divergent series. This famous theorem of Euler was the beginning of analytic number theory.

6.4 Reformulation of the right side

The class number formula results from showing that the function of s in the Euler product formula approaches ∞ like a constant times $(s-1)^{-1}$ as $s\downarrow 1$ and from evaluating the constant in different ways for the two sides. In other words, the formula results from multiplying the Euler product formula by $s-1$ and taking the limit as $s\downarrow 1$. The first step will be to rearrange and reformulate the product $\Pi(1-N(P)^{-s})^{-1}$ on the right side of the formula.

The product in question can be written explicitly as

$$\left(1-\frac{1}{\lambda^s}\right)^{-1} \prod_{p\neq\lambda}\left(1-\frac{1}{p^{fs}}\right)^{-e}$$

where in the infinite product p ranges over all primes other than λ and where, for each p, f is by definition the least positive integer such that $p^f \equiv 1 \bmod \lambda$ and $e = (\lambda-1)/f$. For example, when $\lambda = 5$ the product is $(1-5^{-s})^{-1}$ times

$$\prod_{p\equiv1}\left(1-\frac{1}{p^s}\right)^{-4} \prod_{p\equiv2}\left(1-\frac{1}{p^{4s}}\right)^{-1} \prod_{p\equiv3}\left(1-\frac{1}{p^{4s}}\right)^{-1} \prod_{p\equiv4}\left(1-\frac{1}{p^{2s}}\right)^{-2}$$

where the congruences are mod 5. It will be convenient to factor and rearrange this product to put it in the form

$$= \prod_{p \equiv 1} \left(1 - \frac{1}{p^s}\right)^{-1} \left(1 - \frac{1}{p^s}\right)^{-1} \left(1 - \frac{1}{p^s}\right)^{-1} \left(1 - \frac{1}{p^s}\right)^{-1}$$

$$\times \prod_{p \equiv 2} \left(1 - \frac{1}{p^s}\right)^{-1} \left(1 - \frac{i}{p^s}\right)^{-1} \left(1 + \frac{1}{p^s}\right)^{-1} \left(1 + \frac{i}{p^s}\right)^{-1}$$

$$\times \prod_{p \equiv 3} \left(1 - \frac{1}{p^s}\right)^{-1} \left(1 - \frac{i}{p^s}\right)^{-1} \left(1 + \frac{1}{p^s}\right)^{-1} \left(1 + \frac{i}{p^s}\right)^{-1}$$

$$\times \prod_{p \equiv 4} \left(1 - \frac{1}{p^s}\right)^{-1} \left(1 + \frac{1}{p^s}\right)^{-1} \left(1 - \frac{1}{p^s}\right)^{-1} \left(1 + \frac{1}{p^s}\right)^{-1}$$

$$= \prod_{p \neq 0} \left(1 - \frac{1}{p^s}\right)^{-1} \times \left[\prod_{p \equiv 1, 4} \left(1 - \frac{1}{p^s}\right)^{-1} \prod_{p \equiv 2, 3} \left(1 + \frac{1}{p^s}\right)^{-1} \right]$$

$$\times \left[\prod_{p \equiv 1} \left(1 - \frac{1}{p^s}\right)^{-1} \prod_{p \equiv 2} \left(1 + \frac{i}{p^s}\right)^{-1} \prod_{p \equiv 3} \left(1 - \frac{i}{p^s}\right)^{-1} \prod_{p \equiv 4} \left(1 + \frac{1}{p^s}\right)^{-1} \right]$$

$$\times \left[\prod_{p \equiv 1} \left(1 - \frac{1}{p^s}\right)^{-1} \prod_{p \equiv 2} \left(1 - \frac{i}{p^s}\right)^{-1} \prod_{p \equiv 3} \left(1 + \frac{i}{p^s}\right)^{-1} \prod_{p \equiv 4} \left(1 + \frac{1}{p^s}\right)^{-1} \right]$$

$$= \prod_{k=0}^{3} \left[\prod_{p \equiv 2} \left(1 - \frac{i^k}{p^s}\right)^{-1} \prod_{p \equiv 2^2} \left(1 - \frac{i^{2k}}{p^s}\right)^{-1} \prod_{p \equiv 2^3} \left(1 - \frac{i^{3k}}{p^s}\right)^{-1} \prod_{p \equiv 2^4} \left(1 - \frac{i^{4k}}{p^s}\right)^{-1} \right]$$

where $i = \sqrt{-1}$. In exactly the same way, when $\lambda = 7$ each term other than $(1 - 7^{-s})^{-1}$ has one of the four forms $(1 - p^{-s})^{-6}$, $(1 - p^{-2s})^{-3} = (1 - p^{-s})^{-3}(1 + p^{-s})^{-3}$, $(1 - p^{-3s})^{-2} = (1 - p^{-s})^{-2}(1 - \omega p^{-s})^{-2}(1 - \omega^2 p^{-s})^{-2}$ where ω is a primitive cube root of unity, or $(1 - p^{-6s})^{-1} = (1 - p^{-s})^{-1}(1 - \beta p^{-s})^{-1}(1 - \beta^2 p^{-s})^{-1}(1 - \beta^3 p^{-s})^{-1}(1 - \beta^4 p^{-s})^{-1}(1 - \beta^5 p^{-s})^{-1}$ where β is a primitive sixth root of unity. [That is, $\beta^6 = 1$ but no lesser power of β is 1. Then ω is β^2 or β^4 and $x^6 - y^6 = (x - y) \cdot (x - \beta y) \cdots (x - \beta^5 y)$—see Exercise 1.] Each of these factorizations is of the form $(1 - p^{-s})^{-1}(1 - \beta^j p^{-s})^{-1}(1 - \beta^{2j} p^{-s})^{-1} \cdots (1 - \beta^{5j} p^{-s})^{-1}$ for an appropriate choice of $j = 0, 1, 2, 3, 4, 5$. ($\beta^3 = -1$. The factorizations with $j = 1$ and $j = 5$ coincide, as do those with $j = 2$ and $j = 4$.) Moreover, j must be 0 when $p \equiv 1$ mod 7, 3 when $p \equiv 6$ mod 7, 2 or 4 when $p \equiv 2$ or 4 mod 7, and 1 or 5 when $p \equiv 3$ or 5 mod 7. This can be accomplished by choosing j according to the rule $p \equiv 3^j$ mod 7. Then

$$\prod (1 - N(P)^{-s})^{-1} = \left(1 - \frac{1}{7^s}\right)^{-1} \prod_{k=0}^{5} \prod_{p \equiv 3^j} \left(1 - \frac{\beta^{jk}}{p^s}\right)^{-1}.$$

187

In the general case the formula is (see Exercise 2)

$$\prod (1 - N(P)^{-s})^{-1} = \left(1 - \frac{1}{\lambda^s}\right)^{-1} \prod_{k=0}^{\lambda-2} \prod_{p \equiv \gamma^j} \left(1 - \frac{\beta^{jk}}{p^s}\right)^{-1} \qquad (1)$$

where β is a primitive $(\lambda - 1)$st root of unity, where γ is a primitive root mod λ, and where the second product is a product over all primes p, for each of which j is determined by $p \equiv \gamma^j$ mod λ.

The products $\prod_{p \equiv \gamma^j} (1 - \beta^{jk} p^{-s})^{-1}$ which occur in this formula had already been studied and named by Dirichlet in connection with his work on primes in arithmetic progressions. A *character mod* λ is a complex-valued function χ of an integer variable n with the properties $\chi(n + \lambda) = \chi(n)$, $\chi(nm) = \chi(n)\chi(m)$, $\chi(0) = 0$, $\chi(1) \neq 0$. It is a simple exercise to prove that *there are precisely* $\lambda - 1$ *characters* mod λ and that they can be defined as follows. Let γ be a primitive root mod λ (in the examples above, 2 when $\lambda = 5$, 3 when $\lambda = 7$), let β be a primitive $(\lambda - 1)$st root of unity (in the examples i when $\lambda = 5$, β when $\lambda = 7$) and for each integer k define a character χ_k by setting $\chi_k(\gamma) = \beta^k$, after which the remaining values are determined by $\chi_k(n) = \chi_k(\gamma^j) = \chi_k(\gamma)^j = \beta^{jk}$ when $n \equiv \gamma^j$ mod λ, $\chi_k(n) = \chi_k(0) = 0$ if $n \equiv 0$ mod λ. If χ is any character mod λ then $\chi(\gamma)^{\lambda-1} = \chi(\gamma^{\lambda-1}) = \chi(1)$ because $\gamma^{\lambda-1} \equiv 1$ mod λ. Since $\chi(1) \neq 0$ and $\chi(1)^2 = \chi(1^2) = \chi(1)$ it follows that $\chi(1) = 1$. Therefore $\chi(\gamma)^{\lambda-1} = 1$ and $\chi(\gamma)$ must be a power of β. Since β has exactly $\lambda - 1$ distinct powers and since $\chi(\gamma)$ determines all other values of χ, it follows that there are exactly $\lambda - 1$ characters mod λ, as claimed.

With this notation formula (1) can be rewritten

$$\prod (1 - N(P)^{-s})^{-1} = \left(1 - \frac{1}{\lambda^s}\right)^{-1} \prod_{\chi} \prod_{p} \left(1 - \frac{\chi(p)}{p^s}\right)^{-1}$$

where χ ranges over all $\lambda - 1$ characters mod λ and where p ranges over all primes. (Note that $\chi(\lambda) = 0$ for all χ.) Dirichlet dealt specifically with these products over p which he denoted

$$L(s, \chi) = \prod_{p} \left(1 - \frac{\chi(p)}{p^s}\right)^{-1} \qquad (s > 1)$$

and he observed that the method used to derive the Euler product formula gives

$$L(s, \chi) = \sum_{n} \frac{\chi(n)}{n^s} \qquad (s > 1)$$

because, as before, $(1 - \chi(p)p^{-s})^{-1} = \sum_{\mu=0}^{\infty} [\chi(p)p^{-s}]^{\mu}$ and the product of these expressions over all primes p gives $\sum \chi(p_1)^{\mu_1} \chi(p_2)^{\mu_2} \cdots \chi(p_\nu)^{\mu_\nu} p_1^{-\mu_1 s} p_2^{-\mu_2 s} \cdots p_\nu^{-\mu_\nu s} = \sum \chi(n) n^{-s}$ because χ is a character and because every positive integer n can be written in just one way in the form

$n = p_1^{\mu_1} p_2^{\mu_2} \ldots p_\nu^{\mu_\nu}$. Therefore, in Dirichlet's notation, the right side of the Euler product formula can be written

$$\prod \left(1 - \frac{1}{N(P)^s}\right)^{-1} = \left(1 - \frac{1}{\lambda^s}\right)^{-1} L(s, \chi_0) L(s, \chi_1) \cdots L(s, \chi_{\lambda-2})$$

$$(s > 1)$$

where $\chi_0, \chi_1, \ldots, \chi_{\lambda-2}$ are the characters mod λ defined by $\chi_j(\gamma) = \beta^j$.

Now since $\chi_0(p) = 1$ for all primes p except λ, and $\chi_0(\lambda) = 0$, the first two terms combine to give $(1 - \lambda^{-s})^{-1} L(s, \chi_0) = \zeta(s)$ where $\zeta(s) = \prod(1 - p^{-s})^{-1} = \sum n^{-s}$ is the Riemann zeta function. Therefore, since $\lim (s - 1)\zeta(s) = 1$ as $s \downarrow 1$,

$$\lim_{s \downarrow 1} (s - 1) \prod \left(1 - \frac{1}{N(P)^s}\right)^{-1} = \lim_{s \downarrow 1} L(s, \chi_1) L(s, \chi_2) \cdots L(s, \chi_{\lambda-2})$$

and the problem of finding the way in which $\prod(1 - N(P)^{-s})^{-1}$ approaches ∞ as $s \downarrow 1$ turns on evaluating the limit of the product of Dirichlet L-functions on the right. This problem had already been solved completely by Dirichlet, and Kummer needed only to use Dirichlet's solution for this part of the problem. This solution is presented in the next section.

EXERCISES

1. Prove that if β is a primitive nth root of unity then $(x - \beta y)(x - \beta^2 y) \cdots (x - \beta^n y) = x^n - y^n$. What form does this formula take if β is an nth root of unity but not a primitive one?

2. Prove formula (1).

3. For $\lambda = 13$ find the two characters χ mod λ which have the property that all their values are real. Write the first 15 terms of the corresponding L-series in each of the two cases. Write the first 10 nonzero terms of the series expansion of $(1 - \chi(2)2^{-s})^{-1} \cdot (1 - \chi(3)3^{-s})^{-1}$ in each case.

4. The definition of "character mod λ" in no way depends on the assumption that λ is prime. Therefore the meaning of "character mod 8" is clear. Find 4 real-valued characters mod 8 and show that there are only these 4 characters mod 8.

6.5 Dirichlet's evaluation of $L(1, \chi)$

Let λ be a given prime, $\lambda > 2$, let χ be a character mod λ, and let $L(s, \chi)$ denote the function

$$\sum_n \frac{\chi(n)}{n^s} = \prod_p \frac{1}{1 - \dfrac{\chi(p)}{p^s}} \qquad (s > 1).$$

The character χ defined by $\chi(n) = 1$ for $n \not\equiv 0$ mod λ and $\chi(n) = 0$ for $n \equiv 0$ mod λ is called the *principal* character mod λ. For reasons explained in the preceding section, a part of the derivation of Kummer's formula for the class number is the evaluation of the limit $\lim L(s, \chi)$ as $s \downarrow 1$ for all $\lambda - 2$ characters χ *other than* the principal character.

The first step in the evaluation of this limit will be to show that the series $L(s, \chi) = \Sigma \chi(n) n^{-s}$ converges *conditionally* for $s > 0$ (provided χ is not the principal character) and that $L(s, \chi)$ is a well-defined continuous (in fact analytic) function for $s > 0$. Therefore the desired limit can be written $L(1, \chi) = \Sigma \chi(n) n^{-1}$ where the series converges conditionally.

The method of proving that $\Sigma \chi(n) n^{-s}$ converges conditionally for $s > 0$ is that of *summation by parts*. For this, let $S(n)$ denote the sum function of $\chi(n)$, that is, $S(0) = 0$, $S(n) = S(n - 1) + \chi(n)$, so that $S(n) = \chi(1) + \chi(2) + \cdots + \chi(n)$. Then, provided χ is not the principal character, $S(\lambda) = 0$ because $S(\lambda)$ is the sum of $\lambda - 1$ nonzero values of χ in some order, which is $1 + \beta^k + \beta^{2k} + \cdots + \beta^{(\lambda-2)k} = (1 - \beta^{(\lambda-1)k})/(1 - \beta^k) = 0/(1 - \beta^k) = 0$ provided $\beta^k \neq 1$ (but if $\beta^k = 1$, χ is the principal character, and $S(\lambda) = \lambda - 1$). Therefore $S(\lambda + 1) = S(1)$, $S(\lambda + 2) = S(2), \ldots, S(\lambda + n) = S(n)$, from which it follows that $S(n)$ *is bounded*. Then summation by parts gives

$$\sum_{n=1}^{N} \frac{\chi(n)}{n^s} = \sum_{n=1}^{N} \frac{S(n) - S(n-1)}{n^s} = \sum_{n=1}^{N} \frac{S(n)}{n^s} - \sum_{n=2}^{N} \frac{S(n-1)}{n^s}$$

$$= \frac{S(N)}{N^s} + \sum_{n=1}^{N-1} S(n) \left[\frac{1}{n^s} - \frac{1}{(n+1)^s} \right]$$

because $S(0) = 0$. As $N \to \infty$ the first term goes to 0 when $s > 0$ because the numerator is bounded and the denominator goes to ∞. The second term—that is, the series—converges as $N \to \infty$ as one sees by comparing it with the series

$$\sum_{n=1}^{N} \left[\frac{1}{n^s} - \frac{1}{(n+1)^s} \right] = 1 - \frac{1}{2^s} + \frac{1}{2^s} - \frac{1}{3^s} + \cdots = 1 - \frac{1}{(N+1)^s}$$

which is a series with positive terms which converges for $s > 0$, uniformly for $s \geqslant \delta > 0$. This proves not only that the series $L(s, \chi) = \Sigma \chi(n) n^{-s}$ converges conditionally for $s > 0$ but also that $L(s, \chi)$ is a continuous and analytic function of s for $s > 0$ (χ not the principal character).

Therefore, to evaluate $\lim L(s, \chi)$ as $s \downarrow 1$ it suffices to sum the conditionally convergent series $\Sigma \chi(n)/n$. This can be done by writing it as a superposition of the conditionally convergent series.*

$$\log \frac{1}{1 - x} = x + \frac{x^2}{2} + \frac{x^3}{3} + \frac{x^4}{4} + \cdots$$

*Dirichlet in his derivation uses the description of this function $-\log(1 - x)$ as a definite integral rather than as a conditionally convergent series.

for $x = \alpha, \alpha^2, \ldots, \alpha^{\lambda-1}$ where, as before, α is a primitive λth root of unity. For example, let $\lambda = 5$ and let χ be the character mod 5 determined by $\chi(2) = -1$. Then the problem is to sum

$$1 - \frac{1}{2} - \frac{1}{3} + \frac{1}{4} + \frac{1}{6} - \frac{1}{7} - \frac{1}{8} + \frac{1}{9} + \frac{1}{11} - \cdots.$$

The method is to write it as a superposition of the series

$$\alpha + \frac{\alpha^2}{2} + \frac{\alpha^3}{3} + \frac{\alpha^4}{4} + \frac{\alpha^5}{5} + \cdots = \log \frac{1}{1-\alpha}$$

$$\alpha^2 + \frac{\alpha^4}{2} + \frac{\alpha^6}{3} + \frac{\alpha^8}{4} + \frac{\alpha^{10}}{5} + \cdots = \log \frac{1}{1-\alpha^2}$$

$$\alpha^3 + \frac{\alpha^6}{2} + \frac{\alpha^9}{3} + \frac{\alpha^{12}}{4} + \frac{\alpha^{15}}{5} + \cdots = \log \frac{1}{1-\alpha^3}$$

$$\alpha^4 + \frac{\alpha^8}{2} + \frac{\alpha^{12}}{3} + \frac{\alpha^{16}}{4} + \frac{\alpha^{20}}{5} + \cdots = \log \frac{1}{1-\alpha^4}$$

and, as a computational convenience, of the divergent series

$$1 + \frac{1}{2} + \frac{1}{3} + \frac{1}{4} + \frac{1}{5} + \cdots = \log \frac{1}{1-1}.$$

What is to be found, then, is a set of constants c_0, c_1, c_2, c_3, c_4 such that

$$1 - \frac{1}{2} - \frac{1}{3} + \frac{1}{4} + \frac{1}{6} - \cdots$$

$$= c_0\left(1 + \frac{1}{2} + \cdots\right) + c_1\left(\alpha + \frac{\alpha^2}{2} + \cdots\right) + c_2\left(\alpha^2 + \frac{\alpha^4}{2} + \cdots\right)$$

$$+ c_3\left(\alpha^3 + \frac{\alpha^6}{2} + \cdots\right) + c_4\left(\alpha^4 + \frac{\alpha^8}{2} + \cdots\right).$$

It will suffice for the c's to satisfy

$$1 = c_0 + c_1\alpha + c_2\alpha^2 + c_3\alpha^3 + c_4\alpha^4$$
$$-1 = c_0 + c_1\alpha^2 + c_2\alpha^4 + c_3\alpha^6 + c_4\alpha^8$$
$$-1 = c_0 + c_1\alpha^3 + c_2\alpha^6 + c_3\alpha^9 + c_4\alpha^{12}$$
$$1 = c_0 + c_1\alpha^4 + c_2\alpha^8 + c_3\alpha^{12} + c_4\alpha^{16}$$
$$0 = c_0 + c_1 + c_2 + c_3 + c_4$$

(because $\alpha^5 = 1$). The sum of these equations gives $0 = 1 - 1 - 1 + 1 = 5c_0$ because $1 + \alpha + \alpha^2 + \alpha^3 + \alpha^4 = 1 + \alpha^2 + \alpha^4 + \alpha^6 + \alpha^8 = \cdots = 0$. Similarly, if the first equation is multiplied by α^{-1}, the second by α^{-2}, etc., and the five equations are added the result is $\alpha^{-1} - \alpha^{-2} - \alpha^{-3} + \alpha^{-4} = 5c_1$. In this way the 5×5 system is very easily solved to find

$$c_0 = 0 \qquad c_1 = \frac{\alpha^4 - \alpha^3 - \alpha^2 + \alpha}{5}$$

$$c_2 = \frac{\alpha^3 - \alpha - \alpha^4 + \alpha^2}{5} = -c_1 \qquad c_3 = -c_1 \qquad c_4 = c_1$$

and

$$1 - \frac{1}{2} - \frac{1}{3} + \frac{1}{4} + \frac{1}{6} - \cdots = c_1 \log \frac{1}{1-\alpha} - c_1 \log \frac{1}{1-\alpha^2}$$

$$- c_1 \log \frac{1}{1-\alpha^3} + c_1 \log \frac{1}{1-\alpha^4}$$

where $c_1 = (\alpha - \alpha^2 - \alpha^3 + \alpha^4)/5$ and where $\log (1 - \alpha^j)^{-1}$ stands for the conditionally convergent series $\alpha^j + \frac{1}{2}\alpha^{2j} + \frac{1}{3}\alpha^{3j} + \cdots$. If it can be shown that these 4 conditionally convergent series actually converge to the numbers $\log (1 - \alpha^j)^{-1}$ it will follow that the number $1 - \frac{1}{2} - \frac{1}{3} + \frac{1}{4} + \frac{1}{6} - \cdots$ is equal to the number on the right side of the above equation.

The validity of the formula

$$\log \frac{1}{1-x} = x + \frac{x^2}{2} + \frac{x^3}{3} + \frac{x^4}{4} + \cdots \tag{1}$$

for all complex numbers x in the unit disk $|x| \leqslant 1$ other than $x = 1$ can be proved by summation by parts as follows. Let $S(0) = 0$, $S(n) = S(n-1) + x^n$. Then $S(n) = x + x^2 + \cdots + x^n = (x - x^{n+1})/(1 - x)$ and $|S(n)|$ is at most $2/|1 - x|$. Since

$$\sum_{n=1}^{N} \frac{x^n}{n} = \sum_{n=1}^{N} \frac{S(n) - S(n-1)}{n}$$

$$= \sum_{n=1}^{N} \frac{S(n)}{n} - \sum_{n=2}^{N} \frac{S(n-1)}{n}$$

$$= \frac{S(N)}{N} + \sum_{n=1}^{N-1} S(n) \left[\frac{1}{n} - \frac{1}{n+1} \right]$$

it follows that as $N \to \infty$ the series on the left not only converges but converges *uniformly* in x for x in the set $\{|x| \leqslant 1, |x - 1| \geqslant \epsilon\}$ by comparison with the convergent series

$$\sum \left[(1/n) - (1/(n+1)) \right] = 1 - \frac{1}{2} + \frac{1}{2} - \frac{1}{3} + \frac{1}{3} - \frac{1}{4} + \cdots.$$

Therefore $\sum x^n/n$ defines a continuous function of x for $|x| \leqslant 1$, $x \neq 1$. Since for $|x| < 1$ this function is $\log (1/(1 - x))$ by elementary complex analysis, and since $\log (1/(1 - x))$ is defined and continuous for $|x| \leqslant 1$, $x \neq 1$, the desired formula (1) follows.

Therefore

$$1 - \frac{1}{2} - \frac{1}{3} + \frac{1}{4} + \frac{1}{6} - \cdots = \frac{\alpha - \alpha^2 - \alpha^3 + \alpha^4}{5} \times$$

$$\left[\log \frac{1}{1-\alpha} - \log \frac{1}{1-\alpha^2} - \log \frac{1}{1-\alpha^3} + \log \frac{1}{1-\alpha^4} \right]$$

$$= \frac{\theta_0 - \theta_1}{5} \left[\log \frac{(1-\alpha^2)(1-\alpha^3)}{(1-\alpha)(1-\alpha^4)} + 2\pi n i \right]$$

$$= \frac{\theta_0 - \theta_1}{5} \left[\log (1 - \theta_1) + 2\pi n i \right]$$

where, as before, $\theta_0 = \alpha + \alpha^4$, $\theta_1 = \alpha^2 + \alpha^3$, so that $(1 - \alpha^2)(1 - \alpha^3) = 2 - \theta_1$, $(1 - \alpha)(1 - \alpha^4) = 2 - \theta_0$, $(2 - \theta_1)/(2 - \theta_0) = (2 - \theta_1)^2/(2 - \theta_0)(2 - \theta_1)$ $= 1 - \theta_1$ as an easy computation shows. Note that since θ_0, θ_1 are real numbers less than 1 the integer n must be zero when the real branch of $\log (1 - \theta_1)$ is used. Moreover, $(\theta_0 - \theta_1)^2 = 5$. By interchanging θ_0 and θ_1 if necessary (change α to α^2), one can assume $\theta_0 - \theta_1 > 0$. Then $\theta_0 - \theta_1 = \sqrt{5}$, $-1 - 2\theta_1 = \sqrt{5}$, $1 - \theta_1 = \frac{1}{2}(3 + \sqrt{5})$ and the final formula

$$L(1, \chi) = \frac{\sqrt{5}}{5} \log \frac{3 + \sqrt{5}}{2}$$

has been proved in the case $\lambda = 5$, $\chi(2) = -1$.

The general formula for $L(1, \chi)$ can be derived by exactly the same sequence of steps. Let χ be a nonprincipal character mod λ and let α be a λth root of unity $\alpha \neq 1$. Set

$$1 + \frac{\chi(2)}{2} + \frac{\chi(3)}{3} + \frac{\chi(4)}{4} + \cdots = c_0 \left(1 + \frac{1}{2} + \frac{1}{3} + \frac{1}{4} + \cdots \right)$$

$$+ c_1 \left(\alpha + \frac{\alpha^2}{2} + \frac{\alpha^3}{3} + \cdots \right) + \cdots$$

$$+ c_{\lambda-1} \left(\alpha^{\lambda-1} + \frac{\alpha^{2\lambda-2}}{2} + \frac{\alpha^{3\lambda-3}}{3} + \cdots \right).$$

For this equation to hold, it suffices for the coefficients c_k to satisfy

$$\chi(j) = \sum_{k=0}^{\lambda-1} c_k \alpha^{jk}.$$

These equations can be solved by multiplying the jth equation by $\alpha^{-j\nu}$ and adding to find

$$\sum_{j=1}^{\lambda-1} \chi(j) \alpha^{-j\nu} = \lambda c_\nu.$$

In particular $c_0 = 0$ because the sum of the values of χ is zero (χ is not the principal character). Therefore

$$L(1, \chi) = \sum_{\nu=1}^{\lambda-1} c_\nu \log \frac{1}{1-\alpha^\nu} \qquad (2)$$

where

$$c_\nu = \frac{1}{\lambda} \sum_{j=1}^{\lambda-1} \chi(j)\alpha^{-j\nu}. \qquad (3)$$

The expression for $L(1, \chi)$ given by (2) and (3) can be further simplified as follows. Let γ be a primitive root mod λ and let $\sigma : \alpha \mapsto \alpha^\gamma$ be the corresponding conjugation of cyclotomic integers. Then it is natural to write the $\lambda - 1$ terms of (2) with the powers of α in the order $\alpha, \sigma\alpha, \sigma^2\alpha, \ldots$ rather than in the order $\alpha, \alpha^2, \alpha^3, \ldots$. (Note that $c_0 = 0$.) When this is done, (2) takes the form

$$L(1, \chi) = \sum_{k=0}^{\lambda-2} b_k \log \frac{1}{1-\sigma^k\alpha}$$

where the b_k are the c_ν in a different order. In fact, $b_k = c_\nu$ where $\nu \equiv \gamma^k$ mod λ, from which

$$b_k = c_\nu = \frac{1}{\lambda} \sum_{j=1}^{\lambda-1} \chi(j)\sigma^k\alpha^{-j} = \sigma^k(b_0)$$

when σ is applied in the obvious way to a linear combination of powers of α with complex coefficients. Let ρ denote $\chi(\gamma)$. Then

$$b_0 = \frac{1}{\lambda}\left[\chi(1)\alpha^{-1} + \chi(2)\alpha^{-2} + \chi(3)\alpha^{-3} + \cdots \right]$$

$$= \frac{1}{\lambda}\left[\alpha^{-1} + \chi(\gamma)\alpha^{-\gamma} + \cdots + \chi(\gamma^j)\sigma^j\alpha^{-1} + \cdots \right]$$

$$= \frac{1}{\lambda}\left[\sigma^\mu\alpha + \rho\sigma^{\mu+1}\alpha + \cdots + \rho^j\sigma^{\mu+j}\alpha + \cdots \right]$$

$$= \frac{\rho^{-\mu}}{\lambda}\left[\alpha + \rho\sigma\alpha + \cdots + \rho^k\sigma^k\alpha + \cdots \right],$$

where $\mu = \frac{1}{2}(\lambda - 1)$ (see Exercise 3) and where each sum is a sum of $\lambda - 1$ terms. Therefore

$$b_k = \sigma^k b_0 = \frac{\rho^{-\mu}}{\lambda}\left[\sigma^k\alpha + \rho\sigma^{k+1}\alpha + \cdots \right]$$

$$= \frac{\rho^{-\mu-k}}{\lambda}\left[\rho^k\sigma^k\alpha + \rho^{k+1}\sigma^{k+1}\alpha + \cdots \right]$$

$$= \rho^{-k}b_0,$$

and

$$L(1, \chi) = b_0 \sum_{k=0}^{\lambda-2} \rho^{-k} \log \frac{1}{1 - \sigma^k \alpha}.$$

The factor $\rho^{-\mu}$ which occurs in b_0 is ± 1 because $(\rho^\mu)^2 = \rho^{\lambda-1} = \chi(\gamma^{\lambda-1}) = \chi(1) = 1$. Moreover, $\rho = \beta^j$ for some primitive $(\lambda - 1)$st root of unity β and $\beta^\mu = -1$. Therefore $\rho^{-\mu} = 1$ if j is even and -1 if j is odd. Let χ_j denote the character mod λ for which $\chi(\gamma) = \beta^j$. Then the formulas above give

$$L(1, \chi_j) = (-1)^j m_j \sum_{k=0}^{\lambda-2} \beta^{-jk} \log \frac{1}{1 - \sigma^k \alpha} \tag{4}$$

where

$$m_j = \frac{1}{\lambda} \sum_{k=0}^{\lambda-2} \beta^{jk} \sigma^k \alpha. \tag{5}$$

When j is even, the two terms

$$\beta^{-jk} \log \frac{1}{1 - \sigma^k \alpha} + \beta^{-j(k+\mu)} \log \frac{1}{1 - \sigma^{k+\mu} \alpha}$$

in the formula for $L(1, \chi_j)$ combine, because $\beta^{-j\mu} = 1$ and $\sigma^{k+\mu} \alpha$ is the complex conjugate of $\sigma^k \alpha$, to give

$$\beta^{-jk} 2 \operatorname{Re} \log \frac{1}{1 - \sigma^k \alpha} = -2\beta^{-jk} \log |1 - \sigma^k \alpha|.$$

Therefore

$$L(1, \chi_{2\nu}) = -2m_{2\nu} \sum_{k=0}^{\mu-1} \beta^{-2\nu k} \log |1 - \sigma^k \alpha|. \tag{6}$$

On the other hand, when j is odd the real parts cancel and

$$L(1, \chi_{2\nu+1}) = -m_{2\nu+1} \sum_{k=0}^{\lambda-2} \beta^{-(2\nu+1)k} i \operatorname{Im} \log \frac{1}{1 - \sigma^k \alpha}$$

where the imaginary part of $\log (1 - x)^{-1}$ is determined on the circle $|x| = 1$ by the condition that it be the continuous extension of the value inside the circle that is zero for real x. If x is inside this circle then $\operatorname{Re} (1 - x)^{-1}$ is easily seen to be greater than $\frac{1}{2}$, from which it follows that the imaginary part of its logarithm is between $-\pi/2$ and $\pi/2$. Therefore the same is true of $\operatorname{Im} \log (1 - x)^{-1}$ when $|x| = 1$ $(x \neq 1)$. Now if $\sigma^k \alpha = e^{i\theta}$

then

$$\frac{1}{1-\sigma^k\alpha} = \frac{1}{1-e^{i\theta}} = \frac{e^{-i\theta/2}}{e^{-i\theta/2}-e^{i\theta/2}} = \frac{e^{-i\theta/2}}{-2i\sin\dfrac{\theta}{2}} = \frac{e^{i(\pi-\theta)/2}}{2\sin\dfrac{\theta}{2}}$$

$$\log\frac{1}{1-\sigma^k\alpha} = -\log\left(2\sin\frac{\theta}{2}\right) + \frac{i}{2}(\pi-\theta) \tag{7}$$

where $\frac{1}{2}(\pi-\theta)$ lies between $-\pi/2$ and $\pi/2$, so that θ lies between 0 and 2π. If α is chosen to be $e^{2\pi i/\lambda}$ then $\sigma^k\alpha$ is $e^{2\pi i/\lambda}$ to the power γ^k and the corresponding θ is $2\pi\gamma^k/\lambda$ reduced mod 2π so that it lies between 0 and 2π. Therefore $\theta = 2\pi\gamma_k/\lambda$ where γ_k is the integer defined by $0 < \gamma_k < \lambda$, $\gamma_k \equiv \gamma^k$ mod λ. Thus when j is odd the formula becomes

$$L(1,\chi_{2\nu+1}) = -m_{2\nu+1}\sum_{k=0}^{\lambda-2}\beta^{-(2\nu+1)k}\,i\left(\frac{\pi-\theta}{2}\right)$$

$$= \text{const.}\sum_{k=0}^{\lambda-2}\beta^{-(2\nu+1)k} - m_{2\nu+1}\sum_{k=0}^{\lambda-2}\beta^{-(2\nu+1)k}\,i\left(-\frac{2\pi\gamma_k}{2\lambda}\right)$$

and finally

$$L(1,\chi_{2\nu+1}) = \frac{i\pi m_{2\nu+1}}{\lambda}\sum_{k=0}^{\lambda-2}\gamma_k\beta^{-(2\nu+1)k} \tag{8}$$

where $m_{2\nu+1}$ is defined by (5), where $\chi_{2\nu+1}$ is the character defined by $\chi_{2\nu+1}(\gamma) = \beta^{2\nu+1}$, and where the integers γ_k are defined by $0 < \gamma_k < \lambda$, $\gamma_k \equiv \gamma^k$ mod λ.

EXERCISES

1. The formula for $L(1,\chi)$ in the case $\lambda = 5$, $\chi(2) = -1$ was found in the text with the assumption $\theta_0 - \theta_1 > 0$. If $\theta_0 - \theta_1 < 0$ one obtains an apparently different formula. Show that the two formulas represent the same number.

2. In the case $\lambda = 7$, $\chi(3) = -1$, find $L(1,\chi)$ to 3 significant places.

3. Show that if $\mu = \frac{1}{2}(\lambda-1)$ then $\sigma^\mu\alpha = \alpha^{-1}$.

4. Find $L(1,\chi)$ explicitly for all nonprincipal characters χ in all cases with $\lambda = 3, 5, 7$.

6.6 The limit of the right side

Let λ be a given prime, let γ be a primitive root mod λ, let β be a primitive $(\lambda-1)$st root of unity (that is, $\beta^{\lambda-1} = 1$, $\beta^j \neq 1$ for $0 < j < \lambda - 1$), and let χ_j denote the character mod λ determined by $\chi_j(\gamma) = \beta^j$. Then $\chi_0, \chi_1, \ldots, \chi_{\lambda-2}$ is a complete list of the characters mod λ and it was shown in Section 6.4 that

$$\lim_{s \downarrow 1} (s - 1) \prod \left(1 - \frac{1}{N(P)^s}\right)^{-1} = L(1, \chi_1)L(1, \chi_2) \cdots L(1, \chi_{\lambda-2}) \quad (1)$$

where the product on the left is over all prime divisors P of cyclotomic integers built up from a primitive λth root of unity. In Section 6.5 it was shown that each factor $L(1, \chi_j)$ on the right can be given explicitly. Specifically, if

$$m_j = \frac{1}{\lambda} \sum_{k=0}^{\lambda-2} \beta^{jk} \sigma^k \alpha \quad (2)$$

(as usual, $\sigma : \alpha \mapsto \alpha^\gamma$ for the chosen primitive root γ mod λ) then

$$L(1, \chi_{2\nu}) = -2m_{2\nu} \sum_{k=0}^{\mu-1} \beta^{-2\nu k} \log |1 - \sigma^k \alpha|$$

(where $\mu = \frac{1}{2}(\lambda - 1)$) and

$$L(1, \chi_{2\nu+1}) = \frac{i\pi}{\lambda} m_{2\nu+1} \sum_{k=0}^{\lambda-2} \gamma_k \beta^{-(2\nu+1)k}$$

when $\alpha = e^{2\pi i/\lambda}$ is used in the determination of $m_{2\nu+1}$ and when γ_k is the integer determined by $0 < \gamma_k < \lambda$, $\gamma_k \equiv \gamma^k$ mod λ.

Let

$$C_\nu = \sum_{k=0}^{\mu-1} \beta^{-2\nu k} \log |1 - \sigma^k \alpha|$$

and let $\phi(X)$ denote the polynomial

$$\phi(X) = \sum_{k=0}^{\lambda-2} \gamma_k X^k.$$

Then, because $\beta^{-1} = \beta^{\lambda-2}$, $\beta^{-3} = \beta^{\lambda-4}, \ldots$, it follows that the limit in (1) is equal to

$$(-2)^{\mu-1} \left(\frac{i\pi}{\lambda}\right)^\mu m_1 m_2 \cdots m_{\lambda-2} C_1 C_2 \cdots C_{\mu-1} \phi(\beta) \phi(\beta^3) \cdots \phi(\beta^{\lambda-2}).$$

The other half of the derivation of the class number formula consists of an evaluation of the limit as $s \downarrow 1$ of $(s - 1) \sum N(A)^{-s}$. Before beginning this second half of the derivation it is useful to notice that the above formulas have as a simple consequence *the nonvanishing of $L(1, \chi)$* for nonprincipal characters χ. The next section is devoted to the proof that $L(1, \chi) \neq 0$ and the derivation of the class number formula resumes in the following one.

6.7 The nonvanishing of L-series

The main difficulty in the proof of Dirichlet's theorem on primes in an arithmetic progression is the proof that if χ is a nonprincipal character mod λ (in Dirichlet's theorem λ need not be prime) then $L(1, \chi) \neq 0$. For different reasons, the same fact is needed in the derivation of Kummer's class number formula, except that in this case one can confine oneself to the simplest case $\lambda = $ prime. In this case $L(1, \chi) \neq 0$ can be proved very simply as follows.

Let χ be a nonprincipal character mod λ. It is to be shown that $L(1, \chi) \neq 0$. Let $\rho = \chi(\gamma)$ where γ is a primitive root mod λ. By assumption $\rho \neq 1$. If $\rho \neq -1$ then ρ is not real and $\overline{\chi(\gamma)} = \bar{\rho} \neq \chi(\gamma)$. Thus, in the case $\rho \neq -1$, the characters χ and $\bar{\chi}$ (the complex conjugate of χ) are *distinct*. Since $L(1, \bar{\chi}) = \Sigma \bar{\chi}(n)/n = \overline{L(1, \chi)}$ it follows that in this case $L(1, \chi) = 0$ would imply the existence of *two distinct* nonprincipal characters χ for which $L(1, \chi) = 0$. The theorem will be proved, then, if it is shown that $L(1, \chi) = 0$ for *at most one* nonprincipal character χ, and if it is shown that $L(1, \chi) \neq 0$ for the specific character determined by $\chi(\gamma) = -1$.

Recall that

$$\prod \left(1 - \frac{1}{N(P)^s}\right)^{-1} = \zeta(s)L(s, \chi_1)L(s, \chi_2) \ldots L(s, \chi_{\lambda-2})$$

where $\chi_1, \chi_2, \ldots, \chi_{\lambda-1}$ are the nonprincipal characters mod λ. Moreover, $L(s, \chi_j)$ is a uniform limit of analytic functions for $s \geqslant \epsilon > 0$ and is therefore analytic for $s > 0$, that is, it can be expanded as a power series in the neighborhood of any s. In particular it is differentiable at $s = 1$ and if $L(1, \chi_j) = 0$ then

$$\lim_{s \to 1} \frac{L(s, \chi_j)}{s - 1} = L'(1, \chi_j)$$

exists. Thus if $L(1, \chi_j) = 0$ for two distinct values of j

$$\lim_{s \downarrow 1} \prod \left(1 - \frac{1}{N(P)^s}\right)^{-1}$$

$$= \lim_{s \downarrow 1} [s-1] \cdot [(s-1)\zeta(s)] \left[\frac{L(s, \chi_1) \ldots L(s, \chi_{\lambda-2})}{(s-1)^2}\right]$$

$$= 0 \cdot 1 \cdot \text{finite} = 0$$

which is absurd because the product on the left is at least 1 for each $s > 1$ and therefore cannot have the limit 0 as $s \downarrow 1$. Thus $L(1, \chi_j) = 0$ can occur for at most one nonprincipal character and consequently can occur only for the character χ with $\chi(\gamma) = -1$, which is the character χ_μ (where

$\mu = \frac{1}{2}(\lambda - 1)$). For this character $L(1, \chi_\mu) \neq 0$ can be proved by using the explicit formulas for this number.

Since $\chi_\mu(\gamma) = \rho = -1$, formulas (4) and (5) of Section 6.5 give

$$L(1, \chi_\mu) = \pm m_\mu \sum_{k=0}^{\lambda-2} (-1)^k \log \frac{1}{1 - \sigma^k \alpha}$$

$$= \pm \frac{1}{\lambda} (\alpha - \sigma\alpha + \sigma^2\alpha - \cdots) \left(\log \frac{1}{1 - \alpha} - \log \frac{1}{1 - \sigma\alpha} \right.$$

$$\left. + \log \frac{1}{1 - \sigma^2\alpha} - \cdots \right)$$

$$= \pm \frac{\theta_0 - \theta_1}{\lambda} \left[\log \frac{(1 - \sigma\alpha)(1 - \sigma^3\alpha) \cdots}{(1 - \alpha)(1 - \sigma^2\alpha) \cdots} + 2\pi i n \right]$$

where θ_0 is the period of length μ containing α, where $\theta_1 = \sigma\theta_0$, and where n is an integer. The first factor can be zero only if $\theta_0 = \theta_1$. This is absurd because $\theta_0 = \theta_1 = \sigma\theta_0$ would imply that θ_0 was an ordinary integer at the same time that $0 = 1 + \theta_0 + \theta_1 = 1 + 2\theta_0$, which is impossible. The second factor can be zero only if the complex number whose log it contains is 1, that is, only if the numerator $\prod(1 - \sigma^{2\nu+1}\alpha)$ is equal to the denominator $\prod(1 - \sigma^{2\nu}\alpha)$; this is absurd because their common value would be an integer (because it is invariant under σ) whose square was $\prod(1 - \sigma^\nu\alpha) = N(1 - \alpha) = \lambda$ contrary to the assumption that λ is prime. This completes the proof that $L(1, \chi) \neq 0$ for all non-principal characters χ mod λ.

6.8 Reformulation of the left side

The second half of the derivation of the class number formula depends on the evaluation of the left side of the Euler product formula

$$\sum \frac{1}{N(A)^s} = \prod \left(1 - \frac{1}{N(P)^s} \right)^{-1}.$$

More specifically, it depends on the estimation of the way in which the left side grows as $s \downarrow 1$. The basic technique, which is a very obvious adaptation of the technique Dirichlet used to derive his class number formula, is to *break the sum over all divisors into sums over divisor classes and show that in the limit as $s \downarrow 1$ all these sums are equal*. That is, let $A_1, A_2, \ldots, A_h$ be a representative set of divisors (see Section 5.3) and let the sum be rewritten

$$\sum N(A)^{-s} = \sum_{A \sim A_1} N(A)^{-s} + \sum_{A \sim A_2} N(A)^{-s} + \cdots + \sum_{A \sim A_h} N(A)^{-s}.$$

(Rearrangement of the sum is justified by absolute convergence.) What is suggested by Dirichlet's work—and what is relatively easy to show—is that these terms all approach ∞ in the same way as $s{\downarrow}1$, specifically that

$$\lim_{s{\downarrow}1} (s - 1) \sum_{A \sim A_j} N(A)^{-s}$$

exists and is the same for all j. Thus if L denotes this common limit

$$\lim_{s{\downarrow}1} (s - 1) \sum N(A)^{-s} = hL.$$

Since this limit was evaluated by another method in the preceding sections —a method which did not involve h at all—one need only find L in order to arrive at the desired formula for h.

The main part of the remaining work lies in the evaluation of the limit L. One can do this by evaluating it for any one equivalence class of divisors, and it is natural to use the *principal* class, that is, to evaluate

$$\lim_{s{\downarrow}1} (s - 1) \sum_{A \sim I} N(A)^{-s} = L.$$

Since $A \sim I$ means A is the divisor of some cyclotomic integer $g(\alpha)$, the sum in this limit can be written as a sum over cyclotomic integers $\sum Ng(\alpha)^{-s}$ if one can devise a rule for selecting, from among all cyclotomic integers having a given divisor A, one specific $g(\alpha)$ with divisor A. To find such a selection rule it is of the essence to study the structure of the *units* of the arithmetic because if $g(\alpha)$ is any cyclotomic integer then the set of all cyclotomic integers with the same divisor as $g(\alpha)$ is $\{e(\alpha)g(\alpha) : e(\alpha)$ a unit$\}$ and the task is to pick out a particular element of this set $\{e(\alpha)g(\alpha)\}$.

Once such a selection rule is found, the desired limit

$$L = \lim_{s{\downarrow}1} (s - 1) \sum_{\substack{g(\alpha) \text{ selected} \\ \text{by the rule}}} Ng(\alpha)^{-s}$$

can be evaluated relatively easily by showing that the *sum* differs by a bounded amount from an analogous *integral* as $s{\downarrow}1$ so that the desired limit can be expressed as an integral which can be evaluated by the techniques of integral calculus. This last step is analogous to the evaluation of

$$\lim_{s{\downarrow}1} (s - 1) \sum_{n=N}^{\infty} n^{-s}$$

by observing that as $s{\downarrow}1$ the sum differs by a bounded amount from the integral $\int_c^\infty x^{-s}\, dx$ (because, except on finite intervals where both are bounded, the summand differs from the integrand by a quantity whose order of magnitude $n^{-s} - (n + 1)^{-s} \sim (-s)n^{-s-1}$ has a finite sum for

$s \geqslant 1$) from which

$$\lim_{s \downarrow 1} (s-1) \sum_{n=N}^{\infty} n^{-s} = \lim_{s \downarrow 1} (s-1) \left[\int_{c}^{\infty} x^{-s}\, dx + \text{bounded} \right]$$

$$= \lim_{s \downarrow 1} (s-1) \frac{c^{1-s}}{s-1} + 0$$

$$= c^{0} = 1.$$

In summary, the order of topics in the next few sections is the following. First, the cyclotomic units will be studied in order to arrive at a selection rule for choosing, for each principal divisor A, a cyclotomic integer $g(\alpha)$ with divisor A. (In actual fact the selection rule will do just a bit less than that.) Then the limit L for principal divisors will be evaluated by comparing the sum to an integral. Following this, it will be shown that L is the same for all other divisor classes as it is for the principal class. Finally, hL will be equated to the other expression for this limit obtained in the preceding section and the formula for h will be deduced.

6.9 Units: The first few cases

For reasons explained in the preceding section, the evaluation of the sum of $N(A)^{-s}$ over all principal divisors A is closely related to a knowledge of the cyclotomic units. In this section the units are studied in the cases $\lambda = 3$, 5, 7, 11, and 17. The general case, which is a simple generalization of these, is treated in the next section.

It was shown in Section 5.5 that a cyclotomic unit divided by its complex conjugate $e(\alpha)/e(\alpha^{-1})$ is a power of α. In this section it will be useful to reformulate this statement as the statement that *every cyclotomic unit is a power of α times a real unit*. To prove this form of the statement, note that in the equation $e(\alpha)/e(\alpha^{-1}) = \alpha^{k}$ one can assume without loss of generality that k is even, because if k is odd it can be replaced by $k + \lambda$. With $k = 2j$, then, the equation can be put in the form $e(\alpha)\alpha^{-j} = e(\alpha^{-1})\alpha^{j}$. Thus $E(\alpha) = e(\alpha)\alpha^{-j}$ is a unit for which $E(\alpha) = E(\alpha^{-1})$, that is, $E(\alpha)$ is real. Since $e(\alpha) = \alpha^{j}E(\alpha)$ this proves the theorem.

Thus, in order to find all units it suffices to find all *real* units. Consider first the case $\lambda = 3$. In this case a cyclotomic integer is real if and only if it is invariant under the single conjugation $\alpha \mapsto \alpha^{2}$, which is true if and only if it is an ordinary integer. The norm of an ordinary integer n is just its square n^{2}. Thus the only real units are ± 1 and there are *precisely six units* ± 1, $\pm \alpha$, $\pm \alpha^{2}$. In this case each principal divisor A is the divisor of precisely six cyclotomic integers $g(\alpha)$ and

$$\sum_{\substack{A \text{ principal}}} N(A)^{-s} = \frac{1}{6} \sum Ng(\alpha)^{-s} \qquad (\lambda = 3)$$

where the sum on the right is the sum over all cyclotomic integers.

In all other cases, however, the number of cyclotomic units is infinite. Consider the next case $\lambda = 5$. In that case $a + b\alpha + c\alpha^2 + d\alpha^3 + e\alpha^4$ is real if and only if it is invariant under $\alpha \mapsto \alpha^4$, which is true if and only if it can be written in the form $a\theta_0 + b\theta_1$ where, as usual, $\theta_0 = \alpha + \alpha^4$, $\theta_1 = \alpha^2 + \alpha^3$. The norm of $a\theta_0 + b\theta_1$ is the square of $(a\theta_0 + b\theta_1)(a\theta_1 + b\theta_0) = (a^2 + b^2) \cdot \theta_0\theta_1 + ab(\theta_0^2 + \theta_1^2) = (a^2 + b^2)\frac{1}{4}[(\theta_0 + \theta_1)^2 - (\theta_0 - \theta_1)^2] + ab\frac{1}{2}[(\theta_0 + \theta_1)^2 + (\theta_0 - \theta_1)^2] = (a^2 + b^2)\frac{1}{4}[1 - 5] + ab\frac{1}{2}[1 + 5] = -a^2 - b^2 + 3ab$. Thus $a\theta_0 + b\theta_1$ is a real unit if and only if a and b are integers for which $a^2 + b^2 - 3ab = \pm 1$. In particular, θ_0 is a real unit. Therefore ± 1, $\pm \theta_0$, $\pm \theta_0^2 = \pm (2 + \theta_1)$, $\pm \theta_0^3 = \pm (-1 + 2\theta_0)$, . . . are real units. The units in this list are distinct and the list contains all real units; this can be proved in various ways (see Exercise 1), for example by specializing the argument given below in the case $\lambda = 7$.

Even with this complete description of the units, however, it is not entirely obvious how to compute $\sum N(A)^{-s}$ over all principal divisors A. For this it is helpful to think of cyclotomic integers as points in the complex plane, say by setting $\alpha = e^{2\pi i/5}$. Then there are ten units $\pm \alpha^j$ on the unit circle, ten units $\pm \alpha^j\theta_0$ on the circle with radius $|\theta_0| = 2\cos 2\pi/5$, ten units $\pm \alpha^j\theta_0^2$ on the circle with radius $2^2 \cos^2 2\pi/5$, and so forth. It is then clear geometrically that for any given nonzero cyclotomic integer $g(\alpha)$ there is a unit $e(\alpha)$ such that $1 \le |e(\alpha)g(\alpha)| < |\theta_0|$ where $|\ |$ denotes the modulus of the complex number; for this one need only choose n so that $|\theta_0|^n \le |g(\alpha)| < |\theta_0|^{n+1}$ and set $e(\alpha) = \theta_0^{-n}$. If $e_1(\alpha)$ and $e_2(\alpha)$ are two units which have this property then $|\theta_0|^{-1} < |e_1(\alpha)^{-1}g(\alpha)^{-1}e_2(\alpha)g(\alpha)| < |\theta_0|$, $|\theta_0|^{-1} < |e_1(\alpha)^{-1}e_2(\alpha)| < |\theta_0|$. Because every unit is of the form $\pm \theta_0^n\alpha^j$, this implies that $e_1(\alpha)^{-1}e_2(\alpha)$ is one of the ten units ± 1, $\pm \alpha$, $\pm \alpha^2$, $\pm \alpha^3$, $\pm \alpha^4$. Therefore there are precisely ten units $e(\alpha)$ for which $1 \le |e(\alpha)g(\alpha)| < |\theta_0|$. In other words, given a nonzero cyclotomic integer $g(\alpha)$, there are precisely ten cyclotomic integers $h(\alpha)$ with the same divisor as $g(\alpha)$ which lie in the range $1 \le |h(\alpha)| < |\theta_0|$. Then

$$\sum_{A \text{ principal}} N(A)^{-s} = \frac{1}{10} \sum_{1 \le |g(\alpha)| < |\theta_0|} Ng(\alpha)^{-s}$$

gives the sum over all principal divisors in terms of a sum over cyclotomic integers.

In the next case $\lambda = 7$ there are 14 units of the form $\pm \alpha^j$. In following the above procedure, the first step is a study of the real units. A cyclotomic integer in the case $\lambda = 7$ is real if and only if it is of the form $a\eta_0 + b\eta_1 + c\eta_2$ where $\eta_0 = \alpha + \alpha^{-1}$, $\eta_1 = \alpha^3 + \alpha^{-3}$, $\eta_2 = \alpha^2 + \alpha^{-2}$ are the periods of length 2. A medium long computation (Exercise 2) shows that the norm of $a\eta_0 + b\eta_1 + c\eta_2$ is

$$\left[(a + b + c)^3 - 7(a^2b + b^2c + c^2a + abc) \right]^2.$$

Therefore the real units are those $a\eta_0 + b\eta_1 + c\eta_2$ for which $(a + b + c)^3 -$

$7(a^2b + b^2c + c^2a + abc) = \pm 1$. The problem, then, is to find all triples of integers a, b, c for which this equation is satisfied. This problem is not at all simple, but it can be solved as follows.

Note first that η_0 is a unit. This is also clear from the fact that $\eta_0 = \alpha + \alpha^6 = \alpha^{-1}(\alpha^2 + 1) = \alpha^{-1}(\alpha^4 - 1)/(\alpha^2 - 1)$ because $\alpha^4 - 1$ and $\alpha^2 - 1$, since they have the same divisor $(\alpha - 1)$, differ by a unit multiple. Therefore η_1 and η_2 are also units and $\pm \eta_0^l \eta_1^m \eta_2^n$ (l, m, n integers) is a formula which gives many real units. Since $\eta_0 \eta_1 \eta_2 = 1$ (an easy computation) every unit of the form $\pm \eta_0^l \eta_1^m \eta_2^n$ can be written in the form $\pm \eta_0^r \eta_1^s$ (r, s integers). It is natural to ask, then, whether the real units $\pm \eta_0^r \eta_1^s$ are distinct and whether they include all the real units.

That the units $\pm \eta_0^r \eta_1^s$ are all distinct can be proved as follows. Since $\pm \eta_0^r \eta_1^s = \pm \eta_0^R \eta_1^S$ is equivalent to $\pm \eta_0^{r-R} \eta_1^{s-S} = 1$, what is to be shown is that $\eta_0^r \eta_1^s = \pm 1$ is possible only when $r = 0, s = 0$. Let the cyclotomic integers be regarded as complex numbers by setting $\alpha = e^{2\pi i/7}$. Then $\log |g(\alpha)|$ is defined for all cyclotomic integers $g(\alpha)$, and $\eta_0^r \eta_1^s = \pm 1$ implies

$$r \log |\eta_0| + s \log |\eta_1| = 0,$$

a relation between r and s. Another relation between r and s can be obtained by taking the conjugate of this one under $\sigma : \alpha \mapsto \alpha^3$. This gives

$$r \log |\eta_1| + s \log |\eta_2| = 0.$$

To conclude that $r = s = 0$ it suffices to prove that the determinant of the coefficients

$$\begin{vmatrix} \log |\eta_0| & \log |\eta_1| \\ \log |\eta_1| & \log |\eta_2| \end{vmatrix}$$

is not zero. Because $\log |\eta_0| = \log (2 \cos 2\pi/7) > 0$, $\log |\eta_1| = \log (-2 \cos 6\pi/7) > 0$, $\log |\eta_2| = -\log |\eta_0| - \log |\eta_1| < 0$ this is immediate. Therefore the real units $\pm \eta_0^r \eta_1^s$ are all distinct.

Now let $E(\alpha) = a\eta_0 + b\eta_1 + c\eta_2$ be a given real unit. The problem is to find, if possible, a pair of integers (r, s) such that $\eta_0^r \eta_1^s = \pm E(\alpha)$. Using the function $\log |g(\alpha)|$ and its conjugate $\log |g(\alpha^3)|$ as above shows that the desired relation implies

$$r \log |\eta_0| + s \log |\eta_1| = \log |E(\alpha)|,$$
$$r \log |\eta_1| + s \log |\eta_2| = \log |E(\alpha^3)|. \tag{1}$$

Since the matrix of coefficients on the left has nonzero determinant, these equations can be solved for *real* numbers r and s when $E(\alpha)$ is given. It will suffice to show that r and s are necessarily *integers* because then $E(\alpha)\eta_0^{-r}\eta_1^{-s}$ is a real unit the log of whose modulus is 0, that is, $E(\alpha)\eta_0^{-r}\eta_1^{-s} = \pm 1$, as is to be shown.

Therefore let $\Phi : E(\alpha) \mapsto (r, s)$ be the mapping from real cyclotomic units to the real rs-plane defined implicitly by (1). The problem is to prove that the image of Φ consists exclusively of points with integer coefficients. Note that $\Phi(1) = (0, 0)$, $\Phi(\eta_0) = (1, 0)$, $\Phi(\eta_1) = (0, 1)$, and $\Phi(E_1 E_2) = \Phi(E_1) + \Phi(E_2)$ when points in the plane are added componentwise $(r_1, s_1) + (r_2, s_2) = (r_1 + r_2, s_1 + s_2)$. Therefore, if $\Phi(E)$ fails to have integer coordinates for some real unit E, one can find a real unit E_1 for which $\Phi(E_1)$ is a point in the square $\{|r| \leqslant \frac{1}{2}, s \leqslant \frac{1}{2}\}$ and fails to have integer coordinates—for this one need only set $E_1 = E\eta_0^{-\rho}\eta_1^{-\sigma}$ where (ρ, σ) is the point with integer coordinates nearest to (r, s) so that $\rho - \frac{1}{2} \leqslant r \leqslant \rho + \frac{1}{2}$, $\sigma - \frac{1}{2} \leqslant s \leqslant \sigma + \frac{1}{2}$. Therefore it will suffice to prove that $\Phi(E)$ can lie in the square $\{|r| \leqslant \frac{1}{2}, |s| \leqslant \frac{1}{2}\}$ only when $\Phi(E) = (0, 0)$.

Let $E(\alpha)$ be a real unit for which $\Phi(E) = (r, s)$ lies in the square $\{|r| \leqslant \frac{1}{2}, |s| \leqslant \frac{1}{2}\}$. Then

$$-\tfrac{1}{2} \log |\eta_0| - \tfrac{1}{2} \log |\eta_1| \leqslant \log |E(\alpha)| \leqslant \tfrac{1}{2} \log |\eta_0| + \tfrac{1}{2} \log |\eta_1|$$
$$-\tfrac{1}{2} \log |\eta_1| + \tfrac{1}{2} \log |\eta_2| \leqslant \log |E(\alpha^3)| \leqslant \tfrac{1}{2} \log |\eta_1| - \tfrac{1}{2} \log |\eta_2|$$

(recall that $\log |\eta_i|$ is positive for $i = 0, 1$, negative for $i = 2$). Therefore

$$|\eta_0\eta_1|^{-1/2} \leqslant |E(\alpha)| \leqslant |\eta_0\eta_1|^{1/2}$$
$$|\eta_1/\eta_2|^{-1/2} \leqslant |E(\alpha^3)| \leqslant |\eta_1/\eta_2|^{1/2}.$$

These bounds can be evaluated numerically to find

$$0.667 \leqslant |E(\alpha)| \leqslant 1.499$$
$$0.497 \leqslant |E(\alpha^3)| \leqslant 2.012.$$

Moreover, because $E(\alpha)E(\alpha^3)E(\alpha^2) = \pm\sqrt{NE(\alpha)} = \pm 1$,

$$0.332 \leqslant |E(\alpha^2)| \leqslant 3.017.$$

What is to be shown is that no real unit other than ± 1 can lie within these bounds.

The remaining idea of the proof is to write $E(\alpha) = a\eta_0 + b\eta_1 + c\eta_2$ and to *solve* this equation and its conjugates $E(\alpha^3) = a\eta_1 + b\eta_2 + c\eta_0$, $E(\alpha^2) = a\eta_2 + b\eta_0 + c\eta_1$ for a, b, c in terms of $E(\alpha)$, $E(\alpha^3)$, $E(\alpha^2)$. This can be done by using the relations

$$2 + \eta_0^2 + \eta_1^2 + \eta_2^2 = 7, \qquad 2 + \eta_0\eta_1 + \eta_1\eta_2 + \eta_2\eta_0 = 0$$

of Section 4.9 to find

$$\eta_0 E(\alpha) + \eta_1 E(\alpha^3) + \eta_2 E(\alpha^2) = 7a - 2(a + b + c)$$
$$\eta_1 E(\alpha) + \eta_2 E(\alpha^3) + \eta_0 E(\alpha^2) = 7b - 2(a + b + c)$$
$$\eta_2 E(\alpha) + \eta_0 E(\alpha^3) + \eta_1 E(\alpha^2) = 7c - 2(a + b + c).$$

Since $E(\alpha) + E(\alpha^3) + E(\alpha^2) = -a - b - c$ it follows that

$$7a = (\eta_0 - 2)E(\alpha) + (\eta_1 - 2)E(\alpha^3) + (\eta_2 - 2)E(\alpha^2)$$

$$7b = (\eta_1 - 2)E(\alpha) + (\eta_2 - 2)E(\alpha^3) + (\eta_0 - 2)E(\alpha^2)$$

$$7c = (\eta_2 - 2)E(\alpha) + (\eta_0 - 2)E(\alpha^3) + (\eta_1 - 2)E(\alpha^2).$$

The real numbers η_0, η_1, η_2 are all known and bounds on $|E(\alpha)|$, $|E(\alpha^2)|$, $|E(\alpha^2)|$ were found above. These observations combine to give rather narrow bounds on the possible values of a, b, c. Since a, b, c are integers, this limits $E(\alpha)$ to a *finite number* of possibilities, each of which can be checked to see that $\Phi(E(\alpha))$ is not a unit which lies in $\{|r| < \frac{1}{2}, |s| < \frac{1}{2}\}$ unless $E(\alpha) = \pm 1$.

Explicitly, the above bounds on $|E(\alpha^j)|$ and the numerical values of $\eta_j - 2$ combine to show that $7a$, $7b$, and $7c$ are all comfortably less than 21 in absolute value. Therefore a, b, and c can only have the values 0, ± 1, ± 2 and the question is whether any one of these 125 cyclotomic integers is a unit of the desired type. Since $E(\alpha)$ is a unit of the desired type if and only if $-E(\alpha)$ is, one can assume $a \geqslant 0$. Since $(a + b + c)^3 - 7[a^2b + b^2c + c^2a + abc] = \pm 1$ many of the 75 remaining possibilities can be rejected. (i) If $a = 0$ there are 25 possibilities, of which the 9 in which b or c is 0 can be rejected immediately. The cases $b = -c$ can also be rejected, which leaves just 12 cases. Since b can be assumed to be positive, only 6 cases remain. Of these only 2 are units, namely, $2\eta_1 + \eta_2$ and $\eta_1 + \eta_2$. Since $\eta_1 + \eta_2 = \eta_0\eta_1$, $\Phi(\eta_1 + \eta_2) = (1, 1)$ does not lie in the square $\{|r| < \frac{1}{2}, |s| < \frac{1}{2}\}$. Similarly, $2\eta_1 + \eta_2 = -\eta_0\eta_1^2$ (trial-and-error) and $\Phi(2\eta_1 + \eta_2)$ does not lie in the square. (ii) If $a = 1$ there are 25 possibilities. Those in which b or c are zero were covered in case (i). Of the 16 remaining cases, 10 are not units. The 6 that are units are $\eta_0 - \eta_1 - \eta_2$, $\eta_0 - \eta_1 + \eta_2$, $\eta_0 + \eta_1 - \eta_2$, $\eta_0 + \eta_1 + \eta_2$, $\eta_0 + \eta_1 + 2\eta_2$, $\eta_0 + 2\eta_1 + \eta_2$. The first three are variations of $\eta_0 - \eta_1 - \eta_2 = 1 + 2\eta_0$. By computing Φ of this unit it is possible to find a representation of it in the form $\pm \eta_0^r \eta_1^s$ and to conclude that it does not have the desired property (Exercise 4). The fourth is $\eta_0 + \eta_1 + \eta_2 = -1$ and the fifth and sixth are essentially the same as $-1 + \eta_2 = \eta_1^2\eta_2$ (trial-and-error). Therefore these are not units with the desired property. (iii) If $a = 2$ then $a\eta_0 + b\eta_1 + c\eta_2$ is essentially the same as one of those already tried in cases (i) and (ii) unless $b = \pm 2$ and $c = \pm 2$. But in this case $a\eta_0 + b\eta_1 + c\eta_2$ is divisible by 2 and is therefore not a unit. This completes the proof that $\Phi(E(\alpha))$ always assumes integer values or, what is the same, every real unit is of the form $\pm \eta_0^r \eta_1^s$.

The sum $\sum N(A)^{-s}$ over all principal divisors A can now be written as a sum over cyclotomic integers as follows. The function Φ can be defined for all nonzero cyclotomic integers $g(\alpha)$ as well as for real units $E(\alpha)$. One

simply defines $\Phi(g(\alpha)) = (r, s)$ to mean that
$$r \log |\eta_0| + s \log |\eta_1| = \log |g(\alpha)|$$
$$r \log |\eta_1| + s \log |\eta_2| = \log |g(\alpha^3)|$$
where η_0, η_1, η_2, $g(\alpha)$, $g(\alpha^3)$ are the complex numbers obtained* by setting $\alpha = e^{2\pi i/7}$. Then $\Phi(e(\alpha)g(\alpha)) = \Phi(e(\alpha)) + \Phi(g(\alpha))$ and as $e(\alpha)$ ranges over all units $\Phi(e(\alpha))$ ranges over all points of the rs-plane with integer coordinates. Let $g(\alpha)$ be given and let $e(\alpha)$ be chosen in such a way that $\Phi(e(\alpha)) = (\rho, \sigma)$ is near $\Phi(g(\alpha)) = (R, S)$; specifically, let ρ, σ be chosen so that $\rho \leqslant R < \rho + 1, \sigma \leqslant S < \sigma + 1$. Then the image under Φ of $e(\alpha)^{-1}g(\alpha)$ lies in the square $\{0 \leqslant r < 1, 0 \leqslant s < 1\}$. Let $\mathcal{S}$ denote this square. Then every principal divisor A is the divisor of some $g(\alpha)$ for which $\Phi(g(\alpha))$ lies in $\mathcal{S}$. If $g_1(\alpha)$, $g_2(\alpha)$ have the same divisor and if $\Phi(g_1(\alpha))$, $\Phi(g_2(\alpha))$ both lie in $\mathcal{S}$ then $g_1(\alpha) = e(\alpha)g_2(\alpha) = \alpha^k E(\alpha)g_2(\alpha)$, where $e(\alpha)$ is a unit and $E(\alpha)$ a real unit, and $\Phi(g_1(\alpha)) - \Phi(g_2(\alpha)) = \Phi(\alpha^k) + \Phi(E(\alpha)) = \Phi(E(\alpha))$ lies in the square $\{|r| < 1, |s| < 1\}$. Since this implies $\Phi(E(\alpha)) = (0, 0)$, $E(\alpha) = \pm 1$, it follows that there are precisely 14 possible values for $g_2(\alpha)$, namely, $g_2(\alpha) = \pm g_1(\alpha), \pm \alpha g_1(\alpha), \ldots, \pm \alpha^6 g_1(\alpha)$. Therefore
$$\sum_{A \text{ principal}} N(A)^{-s} = \frac{1}{14} \sum_{\Phi(g(\alpha)) \text{ in } \mathcal{S}} Ng(\alpha)^{-s}$$
gives the desired representation of $\sum N(A)^{-s}$ as a sum over cyclotomic integers.

In the next case $\lambda = 11$ a similar procedure can be followed, but the explicit computations become very long. However, for Kummer's purposes —in which one needs to check conditions (A) and (B) and need not necessarily compute the class number itself—these computations need not be carried out and can remain implicit.

To begin the study of the case $\lambda = 11$, note first that $\eta_0 = \alpha + \alpha^{-1} = \alpha + \alpha^{10}$ is a unit for essentially the same reason as before, namely, $\eta_0 = \alpha \cdot (1 + \alpha^9) = \alpha(\alpha^{18} - 1)/(\alpha^9 - 1)$ is a unit times a quotient of two cyclotomic integers $\alpha^7 - 1$, $\alpha^9 - 1$ with the same divisor $(\alpha - 1)$. Therefore the conjugates of η_0 are also units. With the primitive root $\gamma = 2$ these conjugates are named $\eta_1 = \alpha^2 + \alpha^{-2}$, $\eta_2 = \alpha^4 + \alpha^{-4}$, $\eta_3 = \alpha^3 + \alpha^{-3}$, $\eta_4 = \alpha^5 + \alpha^{-5}$. Since $\eta_0\eta_1\eta_2\eta_3\eta_4 = \pm\sqrt{N\eta_0} = \pm 1$, every real unit of the form $\pm \eta_0^a\eta_1^b\eta_2^c\eta_3^d\eta_4^e$ can also be written in the form $\pm \eta_0^r\eta_1^s\eta_2^t\eta_3^u$. In analogy with the previous cases it is natural to ask whether every real unit $E(\alpha)$ is, up to sign, a unit of this form. If so then r, s, t, u satisfy
$$r \log |\eta_0| + s \log |\eta_1| + t \log |\eta_2| + u \log |\eta_3| = \log |E(\alpha)|$$
$$r \log |\eta_1| + s \log |\eta_2| + t \log |\eta_3| + u \log |\eta_4| = \log |E(\alpha^2)|$$
$$r \log |\eta_2| + s \log |\eta_3| + t \log |\eta_4| + u \log |\eta_0| = \log |E(\alpha^4)|$$
$$r \log |\eta_3| + s \log |\eta_4| + t \log |\eta_0| + u \log |\eta_1| = \log |E(\alpha^8)|.$$
(2)

*If $g(\alpha) \neq 0$ then the integer $Ng(\alpha)$ is nonzero, from which it follows that the complex number $g(\alpha)$ is nonzero and $\log |g(\alpha)|$ is defined.

The analog of the function Φ would assign to each real unit $E(\alpha)$ the quadruple of real numbers (r, s, t, u) obtained by solving this 4×4 system of equations. Its definition depends, of course, on proving that the matrix of coefficients is nonzero.

The analysis of this 4×4 system is simplified by symmetrizing it. This can be done by adding the fifth equation $r \log |\eta_4| + \cdots = \log |E(\alpha^{16})|$ which it implies and by adding a fifth variable v with the missing coefficient in each equation, that is, adding a term $v \log |\eta_4|$ in the first equation, $v \log |\eta_0|$ in the second, and so forth. The equations then take the form

$$M \begin{bmatrix} r \\ s \\ t \\ u \\ v \end{bmatrix} = \begin{bmatrix} \log |E(\alpha)| \\ \log |E(\alpha^2)| \\ \log |E(\alpha^4)| \\ \log |E(\alpha^8)| \\ \log |E(\alpha^{16})| \end{bmatrix} \qquad (3)$$

where M is the 5×5 matrix

$$M = \begin{bmatrix} \log |\eta_0| & \log |\eta_1| & \log |\eta_2| & \log |\eta_3| & \log |\eta_4| \\ \log |\eta_1| & \log |\eta_2| & \log |\eta_3| & \log |\eta_4| & \log |\eta_0| \\ \log |\eta_2| & \log |\eta_3| & \log |\eta_4| & \log |\eta_0| & \log |\eta_1| \\ \log |\eta_3| & \log |\eta_4| & \log |\eta_0| & \log |\eta_1| & \log |\eta_2| \\ \log |\eta_4| & \log |\eta_0| & \log |\eta_1| & \log |\eta_2| & \log |\eta_3| \end{bmatrix}.$$

This matrix might be said to be *anti-circulatory* in that the entry in the ijth location depends only on $i + j \mod 5$. (An $n \times n$ matrix (a_{ij}) is said to be *circulatory* if a_{ij} depends only on $i - j \mod n$. Another way of saying the same thing—see Exercise 5—is to say that a linear map $(x_1, x_2, \ldots, x_n) \to (x_1, x_2, \ldots, x_n)$ is circulatory if it is a polynomial in the "shift" map which carries $(x_1, x_2, \ldots, x_n)$ to $(x_2, x_3, \ldots, x_n, x_1)$.) In fact an interchange of rows (interchange the 1st and 5th and interchange the 2nd and 4th) converts M into a circulatory matrix. The advantage of this observation is that the techniques of Fourier analysis can be applied to the study of circulatory matrices as follows.

Finite dimensional Fourier analysis is essentially the observation that if ρ is a primitive nth root of unity then the n vectors $(1, \rho^j, \rho^{2j}, \ldots, \rho^{-j})$ for $j = 0, 1, \ldots, n - 1$ form a basis with respect to which all circulatory matrices are diagonal. This observation, once made, is very obvious and can be proved in various ways (Exercises 6 and 7). (In infinite dimensional Fourier analysis—Fourier series and Fourier integrals—the difficulties are always in proving *convergence*, not in proving algebraic properties.) In the case of the circulatory matrix obtained from M by interchanging rows, say $\tilde{M}$, the diagonal form with respect to the new basis has on the diagonal the complex numbers c_j for which $\tilde{M}(1, \rho^j, \rho^{2j}, \rho^{3j}, \rho^{4j}) = c_j(1, \rho^j, \rho^{2j}, \rho^{3j}, \rho^{4j})$ where ρ is a primitive 5th root of unity. Therefore c_j is the first component of $\tilde{M}(1, \rho^j, \rho^{2j}, \rho^{3j}, \rho^{4j})$ which is

$$c_j = \log |\eta_4| + \rho^j \log |\eta_0| + \rho^{2j} \log |\eta_1| + \rho^{3j} \log |\eta_2| + \rho^{4j} \log |\eta_3|.$$

In particular $c_0 = \log |\eta_0 \eta_1 \eta_2 \eta_3 \eta_4| = \log 1 = 0$ and $\det \tilde{M} = 0$. Thus $\det M = 0$ and the system of equations (3) cannot be solved by inverting the matrix M. However, the remaining c's are all nonzero, as will be shown below. This implies that two vectors can have the same image under $\tilde{M}$ only if their difference is a multiple of $(1, 1, 1, 1, 1)$. The same is then true of two vectors with the same image under M. Therefore two vectors of the form $(r, s, t, u, 0)$ can have the same image under M only if they are identical. Since the 4 equations of (2) imply the 5th equation of (3), this shows that the solution of (2), if it exists, is unique. This shows that the 4×4 system of equations (2) is nonsingular and that for any 4 values on the right side there exist unique real numbers r, s, t, u satisfying (2). Therefore the mapping Φ is well defined.

It remains to show that c_1, c_2, c_3, c_4 are nonzero. Kummer in effect *recognized* these numbers as being essentially of the form $L(1, \chi)$ so that their nonvanishing was a consequence of Dirichlet's theorem that $L(1, \chi) \neq 0$. To find the connection between c_j and $L(1, \chi)$, note first that $\eta_0 = \alpha^{-1}(\alpha^2 + 1) = \alpha^{-1}(\alpha^4 - 1)(\alpha^2 - 1)^{-1}$, $\log|\eta_0| = \log|\alpha^4 - 1| - \log|\alpha^2 - 1| = \log |1 - \sigma^2 \alpha| - \log |1 - \sigma \alpha|$. Therefore $\log |\eta_k| = \log |1 - \sigma^{2+k} \alpha| - \log |1 - \sigma^{1+k} \alpha|$. Substitution of these expressions for $\log |\eta_k|$ in the formula for c_j gives

$$c_j = \log |1 - \sigma^6 \alpha| - \log |1 - \sigma^5 \alpha| + \rho^j \log |1 - \sigma^2 \alpha|$$
$$- \rho^j \log |1 - \sigma \alpha| + \rho^{2j} \log |1 - \sigma^3 \alpha| - \cdots .$$

Now $\sigma^5 \alpha = \alpha^{-1}$ is the complex conjugate of α. Therefore $\log |1 - \sigma^5 \alpha| = \log |1 - \alpha|$, $\log |1 - \sigma^6 \alpha| = \log |1 - \sigma \alpha|$, and so forth. This gives

$$c_j = \log |1 - \sigma \alpha| + \rho^j \log |1 - \sigma^2 \alpha| + \cdots$$
$$+ \rho^{4j} \log |1 - \sigma^5 \alpha| - \log |1 - \alpha| - \rho^j \log |1 - \sigma \alpha|$$
$$- \cdots - \rho^{4j} \log |1 - \sigma^4 \alpha|$$
$$= (\rho^{-j} - 1) \big[\log |1 - \alpha| + \rho^j \log |1 - \sigma \alpha| + \cdots$$
$$+ \rho^{4j} \log |1 - \sigma^4 \alpha| \big].$$

The factor $\rho^{-j} - 1$ is of course nonzero. If $\rho = \beta^2$ where β is a primitive 10th root of unity then the other factor of c_j is precisely equal to the number that was denoted C_j in Section 6.6. Since $L(1, \chi_{2j})$ is a multiple of C_j and $L(1, \chi_{2j}) \neq 0$, this proves that $c_j \neq 0$.

This completes the proof that the system (2) does define implicitly a function from real cyclotomic units $E(\alpha)$ to quadruples (r, s, t, u) of real numbers. Clearly $\Phi(1) = (0, 0, 0, 0)$, $\Phi(\eta_0) = (1, 0, 0, 0)$, $\Phi(\eta_1) = (0, 1, 0, 0)$, $\Phi(\eta_2) = (0, 0, 1, 0)$, $\Phi(\eta_3) = (0, 0, 0, 1)$, $\Phi(\eta_4) = (-1, -1, -1, -1)$, and $\Phi(E_1E_2) = \Phi(E_1) + \Phi(E_2)$. It follows that every quadruple of *integers* (r, s, t, u) is Φ of some real unit E, and in fact it is Φ of precisely the two real units $\pm \eta_0^r \eta_1^s \eta_2^t \eta_3^u$ because $\Phi(E) = (0, 0, 0, 0)$ implies $\log|E(\alpha)| = 0$, from which $E(\alpha) = \pm 1$ because $E(\alpha)$ is real. The question whether there are any real units other than those of the form $\pm \eta_0^r \eta_1^s \eta_2^t \eta_3^u$ is therefore the same as the question whether $\Phi(E(\alpha))$ can ever assume values that are not integers.

As in the previous case, the search for real units $E(\alpha) = a\eta_0 + b\eta_1 + c\eta_2 + d\eta_3 + e\eta_4$ that are not of the form $\pm \eta_0^r \eta_1^s \eta_2^t \eta_3^u$ can be restricted to units $E(\alpha)$ for which $\Phi(E(\alpha))$ lies in the range $\{|r| \leqslant \frac{1}{2}, |s| \leqslant \frac{1}{2}, |t| \leqslant \frac{1}{2}, |u| \leqslant \frac{1}{2}\}$. As before, the assumption that $\Phi(E(\alpha))$ lies in this range implies bounds on $\log|E(\alpha^j)|$ and therefore on $E(\alpha^j)$ for $j = 1, 2, \ldots, 5$. These bounds, combined with the equations

$$11a = (\eta_0 - 2)E(\alpha) + (\eta_1 - 2)E(\alpha^2) + (\eta_2 - 2)E(\alpha^4)$$
$$+ (\eta_3 - 2)E(\alpha^3) + (\eta_4 - 2)E(\alpha^5)$$
$$11b = \text{etc.},$$

(derived in exactly the same way as in the case $\lambda = 7$) then imply bounds on a, b, c, d, e. Since these are *integers*, only a finite number of possibilities for E remain. For each of these, one can compute the norm and eliminate from consideration those which are not units. Let $E_1, E_2, \ldots, E_N$ be the real units which remain. One can then compute $\Phi(E_1), \Phi(E_2), \ldots, \Phi(E_N)$ and eliminate those E_i for which $\Phi(E_i)$ is not in the range $\{|r| \leqslant \frac{1}{2}, |s| \leqslant \frac{1}{2}, |t| \leqslant \frac{1}{2}, |u| \leqslant \frac{1}{2}\}$.

The fact is that if this is done, then, as in the case $\lambda = 7$, it will be found that all the E's are eliminated and therefore it will follow that all real units are of the form $\pm \eta_0^r \eta_1^s \eta_2^t \eta_3^u$. However, Kummer's proof of Fermat's Last Theorem for $\lambda = 11$ does not depend on this fact; it is not necessary to carry out the very lengthy computation described above as an explicit computation, and one can instead proceed implicitly as follows.

If one encounters a unit E_i in the list which is not of the form $\pm \eta_0^r \eta_1^s \eta_2^t \eta_3^u$, one can renumber so that it is E_1 and for the remaining E_i's in the list one can test whether it is of the form $\pm \eta_0^r \eta_1^s \eta_2^t \eta_3^u$ or of the form $\pm E_1 \eta_0^r \eta_1^s \eta_2^t \eta_3^u$. If so, eliminate it from the list. If not, renumber it E_2 and proceed. This process terminates with a list $E_1, E_2, \ldots, E_n$ of real units with the property that *each real unit E can be written in exactly one way in the form $E = \pm E_i \eta_0^r \eta_1^s \eta_2^t \eta_3^u$* for some choice of the sign, for some integers r, s, t, u, and for some choice of $i = 0, 1, 2, \ldots, n$ where E_0 is by definition 1. To see this it suffices to note that r, s, t, u can be chosen in such a

209

way that Φ of $E\eta_0^{-r}\eta_1^{-s}\eta_2^{-t}\eta_3^{-u}$ lies in $\{|r| \leq \frac{1}{2}, |s| \leq \frac{1}{2}, |t| \leq \frac{1}{2}, |u| \leq \frac{1}{2}\}$ and therefore is included in the original list $E_1, E_2, \ldots, E_N$; therefore, by the way the final list was constructed by eliminating elements from the original list, $E\eta_0^{-r}\eta_1^{-s}\eta_2^{-t}\eta_3^{-u}$ is either ± 1 or it is $\pm E_i\eta_0^\rho\eta_1^\sigma\eta_2^\tau\eta_3^\nu$ for exactly one E_i in the final list.

For reasons which will be explained in Section 6.14, the number $n + 1$ of real units in the final list when it is extended to include $E_0 = 1$ is called the *second factor of the class number* and it is denoted h_2. As was stated above, $h_2 = 1$ in the case $\lambda = 11$, but this fact will not be needed.

The sum of $N(A)^{-s}$ over all principal divisors can now be written as a sum over cyclotomic integers by first extending Φ to all cyclotomic integers as was done in the case $\lambda = 7$ above. As it stands, the definition of Φ implicitly by (2) applies just as well to arbitrary nonzero cyclotomic integers $g(\alpha)$ as to real cyclotomic units $E(\alpha)$. In this way Φ is defined as a function from cyclotomic integers to quadruples of real numbers and $\Phi(g_1(\alpha)g_2(\alpha)) = \Phi(g_1(\alpha)) + \Phi(g_2(\alpha))$. Now if A is a given principal divisor, say it is the divisor of $g(\alpha)$, then every cyclotomic integer with divisor A can be written in just one way in the form $\pm \alpha^k E_i(\alpha) \cdot \eta_0^\rho\eta_1^\sigma\eta_2^\tau\eta_3^\nu g(\alpha)$. Therefore the images under Φ of the cyclotomic integers with divisor A consist precisely of the quadruples of the form $\Phi(E_i(\alpha)) + \Phi(g(\alpha)) + (\rho, \sigma, \tau, \nu)$ and precisely 22 cyclotomic integers are mapped to each of these points. Given any one of the h_2 units $E_i(\alpha)$, it is obviously possible to find one and only one quadruple $(\rho, \sigma, \tau, \nu)$ such that this point lies in the region $\mathbb{S} = \{(r, s, t, u) : 0 \leq r < 1, 0 \leq s < 1, 0 \leq t < 1, 0 \leq u < 1\}$. Therefore precisely $22h_2$ cyclotomic integers $g(\alpha)$ with divisor A have the property that $\Phi(g(\alpha))$ lies in $\mathbb{S}$ and

$$\sum_{A \text{ principal}} N(A)^{-s} = \frac{1}{22h_2} \sum_{\Phi(g(\alpha)) \text{ in } \mathbb{S}} Ng(\alpha)^{-s}.$$

This writes the desired sum as a sum over cyclotomic integers.

The procedure in the general case is a straightforward generalization of the case $\lambda = 11$ except for one aspect, namely, the choice of the basic units $\eta_0, \eta_1, \eta_2, \ldots$. It is true that the periods of length 2 are always units, but one can give in a simple way a formula for units which is, in general, more inclusive than the formula $\pm \eta_0^r\eta_1^s\eta_2^t \ldots$ that was used above. This formula is

$$\pm \alpha^k (1 - \sigma\alpha)^{x_1}(1 - \sigma^2\alpha)^{x_2} \ldots (1 - \sigma^\mu\alpha)^{x_\mu} \tag{4}$$

where $\mu = \frac{1}{2}(\lambda - 1)$ and where $x_1, x_2, \ldots, x_\mu$ are integers for which $x_1 + x_2 + \cdots + x_\mu = 0$; then the expression is a *unit* because each term of the product is of the form unit $\cdot (1 - \alpha)^{x_j}$ and the product of the $(1 - \alpha)^{x_j}$ cancels out. In the case $\lambda = 11$ this formula gives no more units than the formula $\pm \alpha^k\eta_0^r\eta_1^s\eta_2^t\eta_3^u$ because $\eta_0 = \alpha + \alpha^{-1} = \alpha^{-1}(\alpha^2 + 1) = \alpha^{-1}(\alpha^4 - 1) \cdot (\alpha^2 - 1)^{-1} = \alpha^{-1}(1 - \sigma^2\alpha)(1 - \sigma\alpha)^{-1}$ and it is always possible to solve

$$(1 - \sigma\alpha)^{x_1}(1 - \sigma^2\alpha)^{x_2} \cdots (1 - \sigma^5\alpha)^{x_5}$$

$$= \pm \alpha^k \eta_0^r \eta_1^s \eta_2^t \eta_3^u$$

$$= \pm \alpha^{k-r-2s-4t-8u} (1 - \sigma^2\alpha)^r (1 - \sigma\alpha)^{-r}$$

$$\times (1 - \sigma^3\alpha)^s (1 - \sigma^2\alpha)^{-s} \cdots (1 - \sigma^4\alpha)^{-u}$$

$$= \pm \alpha^{k-r-2s-4t-8u} (1 - \sigma\alpha)^{-r} (1 - \sigma^2\alpha)^{r-s}$$

$$\times (1 - \sigma^3\alpha)^{s-t} (1 - \sigma^4\alpha)^{t-u} (1 - \sigma^5\alpha)^u$$

for k, r, s, t, u when x_1, x_2, x_3, x_4, x_5 are given satisfying $x_1 + x_2 + x_3 + x_4 + x_5 = 0$ ($k \equiv r + 2s + 4t + 8u$ mod 11, $r = -x_1, s = -x_1 - x_2, t = -x_1 - x_2 - x_3, u = -x_1 - x_2 - x_3 - x_4 = x_5$). However, in the case $\lambda = 17$ there are units of the form (4) which are not of the form $\eta_0^r \eta_1^s \eta_2^t \eta_3^u \eta_4^v \eta_5^w \eta_6^x \eta_7^y$ (see Exercise 8). Therefore, in the general case the attempt will be made to write units in the form (4) rather than to write them as products of $\pm \alpha^k$ and periods of length 2. This is done in the following section.

EXERCISES

1. Prove that in the case $\lambda = 5$ the real units $\pm \theta^j$ ($\theta = \alpha + \alpha^4$, $j = $ integer) are distinct [easy] and include *all* real units. [Solve Pell's equation or use the method suggested in the text.]

2. With $\lambda = 7$ and η_0, η_1, η_2 the periods of length 2, compute the product of the 3 conjugates of $a\eta_0 + b\eta_1 + c\eta_2$ (a, b, c integers).

3. Prove that $E(\alpha) = a\eta_0 + b\eta_1 + \cdots + c\eta_{\mu-1}$ implies $\lambda a = (\eta_0 - 2)E(\alpha) + (\eta_1 - 2)\sigma E(\alpha) + \cdots + (\eta_{\mu-1} - 2)\sigma^{\mu-1}E(\alpha)$.

4. In the case $\lambda = 7$, write $1 + 2\eta_0$ in the form $\pm \eta_0^r \eta_1^s$. [Use either trial-and-error or the method suggested in the text of evaluating $\Phi(1 + 2\eta_0)$.]

5. Show that a linear map of the form $y_i = \sum_{j=1}^{n} a_{ij}x_j$ whose matrix of coefficients is *circulatory* ($a_{i+1,j+1} = a_{i,j}$ for all i, j when subscripts are interpreted mod n) can be written as a polynomial in the linear map $y_1 = x_2, y_2 = x_3, \ldots, y_{n-1} = x_n, y_n = x_1$. Here linear maps are regarded as mappings from the set $\mathbf{R}^n$ of n real variables to itself, so that the *product* (composition) and the *sum* (pointwise addition) of linear maps is meaningful.

6. Prove that if ρ is a primitive nth root of unity then the n vectors $(1, \rho^j, \rho^{2j}, \ldots, \rho^{-j})$ for $j = 0, 1, \ldots, n - 1$ are linearly independent. Otherwise stated, show that the system of equations $y_i = \sum_{j=0}^{n-1} \rho^{ij}x_j$ has a unique solution $(x_0, x_1, \ldots, x_{n-1})$ for each given $(y_0, y_1, \ldots, y_{n-1})$. The simplest way to prove this is by giving the *explicit solution* as in the special case $n = 5$ that was considered in Section 6.5. This is the *Fourier inversion formula* in this finite-dimensional case. It is essentially just the statement that the vectors $(1, \rho^j, \ldots, \rho^{-j})$ are orthogonal.

7. Show that if M is a circulatory matrix then M applied to any one of the vectors $(1, \rho^j, \rho^{2j}, \ldots, \rho^{-j})$ carries it to a multiple of itself. [By virtue of Exercise 5 it suffices to prove this just for the shift map $y_i = x_{i+1}, y_n = x_1$, which is trivial.] Conclude that with respect to the basis $(1, \rho^j, \rho^{2j}, \ldots, \rho^{-j})$ the matrix M becomes diagonal. That is, if $y_i = \Sigma \rho^{ij} x_j$ then application of M to the vector y merely multiplies each component x_j of x by a constant c_j.

8. In the case $\lambda = 17$ find a unit of the form (4) which is not $\pm \alpha^k$ times a product of periods of length 2. Under what conditions on λ can you say that all units of the form (4) are $\pm \alpha^k$ times products of periods of length 2? Your conditions should include the cases $\lambda < 17$.

6.10 Units: The general case

As usual, let λ be a prime greater than 2 and let $\mu = \frac{1}{2}(\lambda - 1)$. Then for any integers $x_1, x_2, \ldots, x_\mu$ satisfying $x_1 + x_2 + \cdots + x_\mu = 0$ the cyclotomic integer

$$\pm \alpha^k (1 - \sigma\alpha)^{x_1} (1 - \sigma^2\alpha)^{x_2} \cdots (1 - \sigma^\mu\alpha)^{x_\mu} \tag{1}$$

(where γ is a primitive root mod λ and σ is the conjugation which carries α to α^γ) is well defined (despite the negative exponents) and is a unit. This follows from the fact that $(1 - \sigma^j\alpha)/(1 - \alpha)$ is a unit for each j and from the observation that this implies that the product (1) is a product of units times $(1 - \alpha)^{x_1+x_2+ \cdots +x_\mu} = (1 - \alpha)^0 = 1$.

For a given cyclotomic integer $g(\alpha)$, let $Lg(\alpha)$ denote the vector whose μ components are $(\log|g(\alpha)|, \log|\sigma g(\alpha)|, \ldots, \log|\sigma^{\mu-1}g(\alpha)|)$ where $|\ |$ denotes the modulus of the complex number obtained by setting $\alpha = e^{2\pi i/\lambda}$. If a given cyclotomic unit $e(\alpha)$ can be written in the form (1) then obviously

$$Le(\alpha) = x_1 L(1 - \sigma\alpha) + x_2 L(1 - \sigma^2\alpha) + \cdots + x_\mu L(1 - \sigma^\mu\alpha). \tag{2}$$

This equation uniquely determines $x_1, x_2, \ldots, x_\mu$, as can be proved as follows.

Consider the $\mu \times \mu$ matrix M whose columns are $L(1 - \sigma\alpha), L(1 - \sigma^2\alpha), \ldots, L(1 - \sigma^\mu\alpha)$. Then (2) is the statement that M times the column vector $(x_1, x_2, \ldots, x_\mu)$ is the column vector $Le(\alpha)$. The matrix M is *anti-circulatory* in the sense defined in the preceding section in that the entry in the ith row of the jth column is $\log|1 - \sigma^{i+j-1}\alpha|$ and, since $\sigma^\mu\alpha = \alpha^{-1} = \bar{\alpha}$ and $\log|\bar{z}| = \log|z|$, this depends only on $i + j$ mod μ. Let $\tilde{M}$ be the matrix obtained from M by interchanging the first and last rows, the second and next-to-last rows, and so forth. Then $\tilde{M}$ is circulatory and is diagonalized by the basis $(1, \rho^j, \rho^{2j}, \ldots, \rho^{-j})$, where $j = 0, 1, \ldots, \mu - 1$ and ρ is a primitive μth root of unity. Therefore the determinant of $\tilde{M}$ is the product of the c_j defined by $\tilde{M}(1, \rho^j, \ldots, \rho^{-j}) = c_j(1, \rho^j, \ldots, \rho^{-j})$. Thus

$$c_j = \log|1 - \sigma^\mu\alpha| + \rho^j \log|1 - \sigma\alpha| + \cdots + \rho^{(\mu-1)j} \log|1 - \sigma^{\mu-1}\alpha|.$$

For $j = 0$ this gives $c_0 = \log|1 - \sigma\alpha| + \log|1 - \sigma^2\alpha| + \cdots + \log|1 - \sigma^\mu\alpha|$
$= \frac{1}{2}[\log|1 - \sigma\alpha| + \log|1 - \sigma^2\alpha| + \cdots + \log|1 - \sigma^{\lambda-1}\alpha|] = \frac{1}{2}\log N(1 - \alpha) =$
$\log \sqrt{\lambda}$. For $j = 1, 2, \ldots, \mu - 1$ it gives $c_j = C_j$ where C_j is defined as in
Section 6.6 when $\rho = \beta^{-2}$. Therefore $\det \tilde{M} = \log \sqrt{\lambda} \cdot C_1 C_2 \cdots C_{u-1}$
and since none of the factors is zero ($L(1, \chi_{2j}) \neq 0$ is a multiple of C_j), $\tilde{M}$ is
invertible. Since $\tilde{M}$ is the matrix of coefficients of the system (2) when the
equations of (2) are rearranged, it follows that the system (2) is invertible,
as was to be shown.

Therefore the equations (2) define a function from certain cyclotomic
units to μ-tuples $(x_1, x_2, \ldots, x_\mu)$ of integers. In fact, this definition can be
extended to all cyclotomic integers $g(\alpha)$. Let Φ denote this function; that
is, let $\Phi(g(\alpha))$ denote the μ-tuple of real numbers $(x_1, x_2, \ldots, x_\mu)$ such
that

$$Lg(\alpha) = x_1 L(1 - \sigma\alpha) + x_2 L(1 - \sigma^2\alpha) + \cdots + x_\mu L(1 - \sigma^\mu\alpha). \quad (3)$$

Then Φ carries cyclotomic units of the special form (1) to μ-tuples of
integers $(x_1, x_2, \ldots, x_\mu)$ for which $x_1 + x_2 + \cdots + x_\mu = 0$. More generally, $g(\alpha)$ is a unit if and only if $\Phi(g(\alpha)) = (x_1, x_2, \ldots, x_\mu)$ has the
property that $x_1 + x_2 + \cdots + x_\mu = 0$. This can be proved as follows.

The sum of the components of the vector on the left side of (3) is
$\log|g(\alpha)| + \log|\sigma g(\alpha)| + \cdots + \log|\sigma^{\mu-1}g(\alpha)|$. Since $\log|\sigma^{\mu+k}g(\alpha)| =$
$\log|\sigma^k g(\alpha)|$, this sum is $\frac{1}{2}\log Ng(\alpha)$. On the other hand it is x_1 times the
sum of the components of $L(1 - \sigma\alpha)$ plus x_2 times the sum of the
components of $L(1 - \sigma^2\alpha)$ and so forth. Because the sum of the compo-
nents of $L(1 - \sigma^k\alpha)$ is $\frac{1}{2}\log N(1 - \sigma^k\alpha) = \frac{1}{2}\log \lambda$ for all k, it follows that
(3) implies

$$\tfrac{1}{2}\log Ng(\alpha) = \tfrac{1}{2}(\log \lambda)(x_1 + x_2 + \cdots + x_\mu)$$

$$x_1 + x_2 + \cdots + x_\mu = \frac{\log Ng(\alpha)}{\log \lambda} = \log_\lambda Ng(\alpha),$$

an identity which will be useful later. In particular, $g(\alpha)$ is a unit if and
only if $x_1 + x_2 + \cdots + x_\mu = 0$, as was to be shown.

Let two units be called *equivalent* if their quotient is a unit of the form
(1). This is the same as saying that $e_1(\alpha)$ and $e_2(\alpha)$ are equivalent if
$\Phi(e_1(\alpha)) - \Phi(e_2(\alpha))$ has integer entries. (Exercise 2). Then it is clear that
every unit $e(\alpha)$ is equivalent to one whose image under Φ is of the form
$(x_1, x_2, \ldots, x_\mu)$ where $|x_1| \leqslant \frac{1}{2}, |x_2| \leqslant \frac{1}{2}, \ldots, |x_{\mu-1}| \leqslant \frac{1}{2}$. There are only a
finite number of units which satisfy this condition, and all of them can be
found as follows. The bounds $|x_1| \leqslant \frac{1}{2}, |x_2| \leqslant \frac{1}{2}, \ldots, |x_{\mu-1}| \leqslant \frac{1}{2}$ imply
$|x_\mu| \leqslant \frac{1}{2}(\mu - 1)$ (because the sum is 0 for units), and the definition (2) of Φ
then implies that all μ components of $Le(\alpha)$ are bounded by explicit
bounds. Therefore $|e(\alpha^j)|$ is bounded for all $j = 1, 2, \ldots, \lambda - 1$. Since
$e(\alpha) = \alpha^k E(\alpha)$ where $E(\alpha)$ is a real unit, say $E(\alpha) = a\eta_0 + b\eta_1 + \cdots +$
$c\eta_{\mu-1}$ where $a, b, \ldots, c$ are integers and $\eta_0, \eta_1, \ldots, \eta_{\mu-1}$ are periods of

213

length 2, the bounds on $|e(\alpha^j)|$ are bounds on $|E(\alpha^j)|$ and these bounds can be used in the relations

$$\lambda a = (\eta_0 - 2)E(\alpha) + (\eta_1 - 2)\sigma E(\alpha) + \cdots + (\eta_{\mu-1} - 2)\sigma^{\mu-1}E(\alpha)$$
$$\lambda b = \text{etc.}$$

to give bounds on $a, b, \ldots, c$. Since these are *integers* there are only finitely many possible values, and therefore only finitely many possibilities for $E(\alpha)$ and for $e(\alpha) = \alpha^k E(\alpha)$.

From this finite list of possibilities, one can eliminate all those which are not units. One obtains in this way a finite list $E_1(\alpha), E_2(\alpha), \ldots, E_N(\alpha)$ of real units with the property that every unit is equivalent to one of the $E_i(\alpha)$. One can assume without loss of generality that $E_1(\alpha) = 1$, and one can go through the list eliminating duplicates—that is, eliminating $E_i(\alpha)$ from the list if it is equivalent to one of the units which precedes it in the list—to obtain a list with the property that every unit is equivalent to one and only one unit in the list. Let h_2 be the number of units in this final list. Another way of saying that every unit $e(\alpha)$ is equivalent to one and only one of the units E_i $(i = 1, 2, \ldots, h_2)$ is to say that every unit can be written in one and only one way in the form $\pm \alpha^k E_i(1 - \sigma\alpha)^{x_1}(1 - \sigma^2\alpha)^{x_2} \cdots (1 - \sigma^\mu\alpha)^{x_\mu}$. This gives a complete description of the cyclotomic units.

Now let A be a principal divisor and let $g(\alpha)$ be a cyclotomic integer whose divisor is A. For each of the h_2 units $E_i(\alpha)$ constructed above there is a cyclotomic integer $g_i(\alpha)$ of the form $g_i(\alpha) = \pm \alpha^k E_i(\alpha)(1 - \sigma\alpha)^{x_1}(1 - \sigma^2\alpha)^{x_2} \cdots (1 - \sigma^\mu\alpha)^{x_\mu}g(\alpha)$ whose image under Φ has its first $\mu - 1$ components ≥ 0 and < 1; in fact, the factor $\pm \alpha^k$ is arbitrary (2λ choices) but the integers $x_1, x_2, \ldots, x_\mu$ are uniquely determined by the condition that x_j plus the jth component of $\Phi(E_i(\alpha)) + \Phi(g(\alpha))$ is ≥ 0 and < 1 for $j = 1, 2, \ldots, \mu - 1$ and by the condition that $x_1 + x_2 + \cdots + x_\mu = 0$. Therefore there are exactly 2λ cyclotomic integers $g_i(\alpha)$ satisfying these conditions. Since i can have any one of h_2 values, it follows that there are exactly $2\lambda h_2$ cyclotomic integers of the form $e(\alpha)g(\alpha)$—that is, cyclotomic integers whose divisor is A—whose images under Φ have their first $\mu - 1$ components ≥ 0 and < 1. Therefore if $\mathcal{S}$ denotes $\{(x_1, x_2, \ldots, x_\mu): 0 \leq x_1 < 1, 0 \leq x_2 < 1, \ldots, 0 \leq x_{\mu-1} < 1, x_\mu \text{ unrestricted}\}$ then

$$\sum_{A \text{ principal}} N(A)^{-s} = \frac{1}{2\lambda h_2} \sum_{\Phi(g(\alpha)) \text{ in } \mathcal{S}} Ng(\alpha)^{-s}.$$

This is the desired representation of the left side in terms of a sum over cyclotomic integers.

EXERCISES

1. Prove that the equivalence relation for units that was described in the text is reflexive, symmetric, transitive, and consistent with multiplication.

2. Clearly if $e_1(\alpha)$ and $e_2(\alpha)$ are equivalent then $\Phi(e_1(\alpha)) - \Phi(e_2(\alpha))$ has integer entries. Prove the converse.

6.11   Evaluation of the integral

The way in which Σn^{-s} approaches ∞ as $s\downarrow 1$ was found in Section 6.3 by showing in the first place that it differs by a bounded amount from the corresponding integral $\int_1^{\infty} x^{-s}dx$ and by then evaluating this integral using the fundamental theorem of calculus. In the same way, there is an integral which differs by a bounded amount from the sum $\Sigma Ng(\alpha)^{-s}$ over all $g(\alpha)$ for which $\Phi(g(\alpha))$ is in S that was encountered in the previous section. This integral is

$$\int_{\mathcal{D}} N\big(u_1\sigma\alpha + u_2\sigma^2\alpha + \cdots + u_{\lambda-1}\sigma^{\lambda-1}\alpha\big)^{-s} du_1\,du_2\cdots du_{\lambda-1} \quad (1)$$

where $u_1, u_2, \ldots, u_{\lambda-1}$ are real variables, where the functions N and Φ are extended from cyclotomic integers $u_1\sigma\alpha + u_2\sigma^2\alpha + \cdots + u_{\lambda-1}\sigma^{\lambda-1}\alpha$ (u_j integers) to arbitrary real values of the u_j in the obvious ways, and where the domain of integration $\mathcal{D}$ is the set of all points whose images under Φ lie in $S = \{(x_1, x_2, \ldots, x_\mu) : 0 \leqslant x_1 < 1, 0 \leqslant x_2 < 1, \ldots, 0 \leqslant x_{\mu-1} < 1, x_\mu$ unrestricted$\}$. More precisely, let $g(\alpha)$ denote $u_1\sigma\alpha + u_2\sigma^2\alpha + \cdots + u_{\lambda-1}\alpha$ where the u_j are not necessarily integers, let $Lg(\alpha)$ denote the column vector $(\log|g(\alpha)|, \log|\sigma g(\alpha)|, \ldots, \log|\sigma^{\mu-1}g(\alpha)|)$ in which $|g(\alpha)|$ denotes the modulus of the complex number obtained by setting $\alpha = e^{2\pi i/\lambda}$ in $g(\alpha)$, let $Ng(\alpha)$ denote the real number $g(\alpha)\cdot\sigma g(\alpha)\cdot\cdots\cdot\sigma^{\lambda-1}g(\alpha)$, and let Φ be the function defined implicitly by $\Phi(g(\alpha)) = (x_1, x_2, \ldots, x_\mu)$ where

$$Lg(\alpha) = x_1 L(1 - \sigma\alpha) + x_2 L(1 - \sigma^2\alpha) + \cdots + x_\mu L(1 - \sigma^\mu\alpha).$$

The functions $Ng(\alpha)^{-s}$ and $Lg(\alpha)$ are defined only if $g(\alpha)$, considered as a complex number, is nonzero. This is the case if and only if $Ng(\alpha) \neq 0$. Therefore points at which $Ng(\alpha) = 0$ must be excluded from the domain of integration $\mathcal{D}$ at the outset. In fact, in order to avoid points where $Ng(\alpha)^{-s}$ is large, it is natural to exclude points at which $Ng(\alpha) < 1$ from $\mathcal{D}$. Note that this does not exclude any cyclotomic integers other than $g(\alpha) = 0$. Thus $\mathcal{D}$ will denote the set of all $g(\alpha)$ in $u_1u_2 \cdots u_{\lambda-1}$-space for which $Ng(\alpha) \geqslant 1$ and for which $\Phi(g(\alpha))$ (which is defined because $Ng(\alpha) \neq 0$) lies in S.

It will be shown in the next section that the improper integral (1) converges for $s > 1$ and that its value differs by a bounded amount from $\Sigma Ng(\alpha)^{-s}$ ($g(\alpha)$ a cyclotomic integer in $\mathcal{D}$) as $s\downarrow 1$. This section is devoted to the explicit evaluation of the integral by means of changes of variable and the fundamental theorem of calculus.

As the first step in this evaluation, consider the complex change of variable $z_k = u_1\sigma^k\alpha + u_2\sigma^{k+1}\alpha + \cdots + u_{\lambda-1}\sigma^{k-1}\alpha$ for $k = 1, 2, \ldots, \lambda-1$. Because $\sigma^{k+\mu}\alpha$ is the complex conjugate of $\sigma^k\alpha$, $z_{k+\mu}$ is the complex conjugate of z_k for real values of the u_j. Therefore when real numbers v_k, w_k are defined by $z_k = v_k + iw_k$, the $\lambda - 1$ quantities $v_1, v_2, \ldots,$ $v_\mu, w_1, w_2, \ldots, w_\mu$ are real-valued linear functions of $u_1, u_2, \ldots, u_{\lambda-1}$ and $z_k = v_k + iw_k$, $z_{\mu+k} = v_k - iw_k$. The integral (1) can be expressed in terms of the variables v_k, w_k as follows. If A is the $(\lambda - 1) \times (\lambda - 1)$ complex matrix which defines the transformation $z = Au$ then, by the rules for calculating with differential forms (see [E1]), $dz_1 \, dz_2 \cdots dz_{\lambda-1} = \det A \, du_1 \, du_2 \cdots du_{\lambda-1}$. On the other hand

$$dz_1 \, dz_2 \cdots dz_{\lambda-1} = (dv_1 + idw_1)(dv_2 + idw_2) \cdots (dv_\mu + idw_\mu) \cdot$$
$$(dv_1 - idw_1)(dv_2 - idw_2) \cdots (dv_\mu - idw_\mu)$$
$$= \pm \prod_{k=1}^{\mu} (dv_k + idw_k)(dv_k - idw_k)$$
$$= \pm \prod_{k=1}^{\mu} (-2i dv_k \, dw_k)$$
$$= \pm (2i)^\mu \, dv_1 \, dw_1 \, dv_2 \, dw_2 \cdots dv_\mu \, dw_\mu$$

where the sign is neglected because the sign of the integral to be computed is known to be $+$. The integrand is $(\Pi z_i)^{-s} = (v_1^2 + w_1^2)^{-s}(v_2^2 + w_2^2)^{-s} \cdots (v_\mu^2 + w_\mu^2)^{-s}$ and the domain of integration is the set, call it $\mathcal{D}'$, for which $(v_1^2 + w_1^2)(v_2^2 + w_2^2) \cdots (v_\mu^2 + w_\mu^2) \geqslant 1$ and for which $(\log|z_1|, \log|z_2|, \ldots, \log|z_\mu|) = (\frac{1}{2} \log|v_1^2 + w_1^2|, \frac{1}{2} \log|v_2^2 + w_2^2|, \ldots, \frac{1}{2} \log|v_\mu^2 + w_\mu^2|)$ is of the form $x_1 L(1 - \sigma\alpha) + x_2 L(1 - \sigma^2\alpha) + \cdots + x_\mu L(1 - \sigma^\mu\alpha)$ with $0 \leqslant x_j < 1$ (for $j = 1, 2, \ldots, \mu - 1$ but not necessarily for $j = \mu$). Thus the integral becomes

$$\frac{\pm (2i)^\mu}{\det A} \int_{\mathcal{D}'} (v_1^2 + w_1^2)^{-s} \cdots (v_\mu^2 + w_\mu^2)^{-s} dv_1 \, dw_1 \cdots dv_\mu \, dw_\mu.$$

The determinant which occurs in this formula can be evaluated as follows.

The matrix A is (a_{ij}) where $a_{ij} = \sigma^{i+j-1}\alpha$, which depends only on $i + j \mod(\lambda - 1)$. Therefore the $(\lambda - 1) \times (\lambda - 1)$ matrix A is *anticirculatory* in the sense defined in the previous section. Therefore its determinant can be evaluated in the same way that the determinant of M was evaluated in the preceding section, namely, by interchanging rows (or columns) until it becomes a circulatory matrix, diagonalizing the circulatory matrix using the basis $(1, \rho^j, \rho^{2j}, \ldots, \rho^{-j})$ (ρ a primitive $(\lambda - 1)$st root of unity, $j = 0, 1, \ldots, \lambda - 2$), and noting that the determinant is then the product of the diagonal elements. Explicitly, μ interchanges of rows (first and last, second and next-to-last, etc.) convert A to a circulatory matrix $\tilde{A}$ and multiply its determinant by $(-1)^\mu$. The determinant of $\tilde{A}$ is the product of the $\lambda - 1$ numbers $\sigma^{\lambda-1}\alpha + \rho^j\sigma\alpha + \cdots + \rho^{-j}\sigma^{\lambda-2}\alpha = \sum_{k=0}^{\lambda-2} \rho^{jk}\sigma^k\alpha$. If $j = 0$ this number is $\sigma\alpha + \sigma^2\alpha + \cdots + \sigma^{\lambda-1}\alpha = \alpha + \alpha^2 + \cdots + \alpha^{\lambda-1} = -1$. If

$j = 1, 2, \ldots, \lambda - 2$ then it is λ times the number that was called m_j in (2) of Section 6.6. Therefore the determinant of A is $\pm m_1 m_2 \cdots m_{\lambda-2} \cdot \lambda^{\lambda-2}$.

It will also be useful to make the following observation about det A. If the anti-circulatory matrix A is *squared* then it simplifies considerably, becoming the matrix (b_{ij}) in which

$$b_{ij} = \sum_{k=1}^{\lambda-1} \sigma^{i+k-1} \alpha \cdot \sigma^{k+j-1} \alpha = \sum_{k=1}^{\lambda-1} \sigma^{i+k-1} [\alpha \cdot \sigma^{j-i} \alpha]$$

which is the sum of the conjugates of $\alpha \cdot \sigma^{j-i} \alpha$ and which is therefore $\lambda - 1$ if $j - i \equiv \mu \bmod (\lambda - 1)$, -1 otherwise. Thus A^2 is circulatory and its determinant is the product of its effects on $(1, \rho^j, \rho^{2j}, \ldots, \rho^{-j})$ for $j = 0, 1, \ldots, \lambda - 1$. If $j = 0$ then A^2 multiplies this vector by $(-1) + (-1) + \cdots + (\lambda - 1) + \cdots + (-1) = \lambda - (\lambda - 1) = 1$, whereas if $j = 1, 2, \ldots, \lambda - 1$ it multiplies it by $-1 - \rho^j - \rho^{2j} - \cdots + (\lambda - 1)\rho^{\mu j} - \cdots - \rho^{-j} = \lambda \rho^{\mu j}$. Since $\rho^\mu = -1$ this is λ if j is even and $-\lambda$ if j is odd. Therefore det $A^2 = \lambda(-\lambda)\lambda(-\lambda) \ldots (-\lambda)\lambda = \pm \lambda^{\lambda-2}$. This gives

$$\frac{1}{\det A} = \frac{\det A}{\det A^2} = \frac{\pm m_1 m_2 \ldots m_{\lambda-2} \lambda^{\lambda-2}}{\pm \lambda^{\lambda-2}} = \pm m_1 m_2 \ldots m_{\lambda-2}.$$

(As is easy to show, the sign in this formula is in fact $+$ in all cases.)

Now consider the change of variables $v_k + i w_k = e^{q_k + i r_k}$, where $0 < q_k < \infty$, $0 \leqslant r_k < 2\pi$. Then $-2i \, dv \, dw = (dv + i \, dw)(dv - i \, dw) = e^{q+ir}(dq + i \, dr)e^{q-ir}(dq - i \, dr) = -2ie^{2q} \, dq \, dr$. Therefore $dv_1 \, dw_1 \, dv_2 \, dw_2 \ldots dv_\mu \, dw_\mu = \exp(2q_1 + 2q_2 + \cdots + 2q_\mu) \, dq_1 \, dr_1 \, dq_2 \, dr_2 \ldots dq_\mu \, dr_\mu$. The integrand becomes the product of the $(v^2 + w^2)^{-s} = e^{-2qs}$ and the domain of integration becomes the set $\mathcal{D}''$ for which $q_1 + q_2 + \cdots + q_\mu \geqslant 0$, for which the vector $(q_1, q_2, \ldots, q_\mu)$ is of the form $\Sigma x_j L(1 - \sigma^j \alpha)$ with $0 \leqslant x_j < 1$ for $j = 1, 2, \ldots, \mu - 1$, and for which $0 \leqslant r_k < 2\pi$ $(k = 1, 2, \ldots, \mu)$. Thus the integral becomes

$$\pm (2i)^\mu m_1 m_2 \ldots m_{\lambda-2} \int_{\mathcal{D}''} e^{(2-2s)(q_1 + q_2 + \cdots + q_\mu)} \, dq_1 \, dr_1 \, dq_2 \, dr_2 \cdots dq_\mu \, dr_\mu.$$

Let the r's be moved to the back of the list of variables and integrated out to give

$$\pm (2i)^\mu (2\pi)^\mu m_1 m_2 \ldots m_{\lambda-2} \int_{\mathcal{D}'''} e^{(2-2s)(q_1 + q_2 + \cdots + q_\mu)} \, dq_1 \, dq_2 \cdots dq_\mu$$

where $\mathcal{D}'''$ is the projection of $\mathcal{D}''$ onto its q-coordinates.

Finally, consider the change of coordinates defined implicitly by $q = Mx$, where q is the column vector $(q_1, q_2, \ldots, q_\mu)$, x the column vector $(x_1, x_2, \ldots, x_\mu)$, and M the $\mu \times \mu$ anti-circulatory matrix $(\log|1 - \sigma^{i+j-1}\alpha|)$ of the preceding section. Then, by the definition of $\mathcal{D}'''$, the domain of integration with respect to the variables x is simply those x in $\mathcal{S}$ which correspond to points $(q_1, q_2, \ldots, q_\mu)$ with $q_1 + q_2 + \cdots + q_\mu \geqslant 0$.

217

It was noted in the preceding section that, since the column sums of M are all $\frac{1}{2} \log \lambda$, $q_1 + q_2 + \cdots + q_\mu = \frac{1}{2} (\log \lambda)(x_1 + x_2 + \cdots + x_\mu)$. Therefore the integral becomes

$$\pm (2i)^\mu (2\pi)^\mu m_1 m_2 \ldots m_{\lambda-2} \det M$$

$$\times \int_{\substack{x \text{ in } \mathcal{S} \\ x_1 + x_2 + \cdots + x_\mu \geqslant 0}} e^{(1-s)(\log \lambda)(x_1 + x_2 + \cdots + x_\mu)} \, dx_1 \, dx_2 \cdots dx_\mu.$$

With $t = x_1 + x_2 + \cdots + x_\mu$ the variables $x_1, x_2, \ldots, x_{\mu-1}$ can be moved to the back of the list of variables and integrated out. This leaves only a factor times

$$\int_{t>0} \exp \left[(1-s)(\log \lambda) t \right] \, dt = \left[(s-1) \log \lambda \right]^{-1}.$$

Therefore the value of (1) is

$$\pm (4\pi i)^\mu m_1 m_2 \ldots m_{\lambda-2} \frac{\det M}{(s-1) \log \lambda}.$$

It was shown in the preceding section that $\det \tilde{M} = (\log \sqrt{\lambda}\,)$ $C_1 C_2 \cdots C_{\mu-1}$. Since M is $\tilde{M}$ with some rows interchanged, $\det (M) = \pm \det (\tilde{M})$. Therefore (1) is the absolute value of

$$\pm 2^{2\mu-1} (\pi i)^\mu m_1 m_2 \ldots m_{\lambda-2} C_1 C_2 \cdots C_{\mu-1} (s-1)^{-1}.$$

This completes the evaluation of the integral.

6.12 Comparison of the integral and the sum

Let $I(s)$ and $S(s)$ denote the integral and the sum under consideration, namely

$$I(s) = \int_{\mathcal{D}} N(u_1 \sigma \alpha + \cdots + u_{\lambda-1} \sigma^{\lambda-1} \alpha)^{-s} \, du_1 \, du_2 \cdots du_{\lambda-1},$$

$$S(s) = \sum_{\mathcal{D}} Ng(\alpha)^{-s}.$$

It has been shown that the sum converges for all $s > 1$ and that $\lim (s-1)S(s)$ exists as $s \downarrow 1$. What is to be shown is that the integral also converges for all $s > 1$, and that $(s-1)I(s)$ has a limit as $s \downarrow 1$ which is equal to the limit of $(s-1)S(s)$.

In order to prove these facts it is helpful to alter the set $\mathcal{S} = \{(x_1, \ldots, x_\mu) : 0 \leqslant x_1 < 1, \ldots, 0 \leqslant x_{\mu-1} < 1, x_\mu \text{ unrestricted}\}$ in terms of which the domain $\mathcal{D}$ is defined. The reason for this is that in estimating $I(s)$ and $S(s)$ it is useful to consider scale changes $(u_1, u_2, \ldots, u_{\lambda-1}) \mapsto (cu_1, cu_2, \ldots, cu_{\lambda-1})$ in $u_1 u_2 \ldots u_{\lambda-1}$-space and $\mathcal{D}$, as it was defined above, is not invariant under scale changes. The first step of the proof will be to define a new domain $\mathcal{D}'$ over which the integral and sum are the same and for which scale changes are more easily handled.

The property of $\mathbb{S}$ which led to its selection is the property that for any given point $(x_1, x_2, \ldots, x_\mu)$ in $x_1 x_2 \ldots x_\mu$-space there is exactly one way to choose integers $n_1, n_2, \ldots, n_\mu$ such that $n_1 + n_2 + \cdots + n_\mu = 0$ and such that $(x_1 + n_1, x_2 + n_2, \ldots, x_\mu + n_\mu)$ lies in $\mathbb{S}$. Scale changes in $u_1 u_2 \ldots u_{\lambda-1}$-space correspond under Φ to the addition of multiples of $(1, 1, \ldots, 1)$ in $x_1 x_2 \ldots x_\mu$-space. This motivates the introduction of a set $\mathbb{S}'$ with the same property as $\mathbb{S}$ but which is invariant under translation by multiples of $(1, 1, \ldots, 1)$. Specifically, let $\mathbb{S}'$ be the set of all points of $x_1 x_2 \ldots x_\mu$-space for which $0 \leqslant x_j - \mu^{-1}(x_1 + x_2 + \cdots + x_\mu) < 1$ for $j = 1, 2, \ldots, \mu - 1$. Then for points with $x_1 + x_2 + \cdots + x_\mu = 0$ the sets $\mathbb{S}$ and $\mathbb{S}'$ coincide, but $\mathbb{S}'$ is invariant under addition of multiples of $(1, 1, \ldots, 1)$ (addition of $c(1, 1, \ldots, 1)$ to $(x_1, x_2, \ldots, x_\mu)$ adds c to x_j and to $\mu^{-1}(x_1 + x_2 + \cdots + x_\mu)$ and therefore leaves $x_j - \mu^{-1}(x_1 + x_2 + \cdots + x_\mu)$ unchanged) whereas $\mathbb{S}$ is invariant under addition of multiples of $(0, 0, \ldots, 0, 1)$.

Let $\mathcal{D}'$ be the set of points in $u_1 u_2 \ldots u_{\lambda-1}$-space at which $Ng(\alpha) \geqslant 1$ and for which the image under Φ lies in $\mathbb{S}'$. It is simple to prove that $\mathbb{S}'$ has the same property as $\mathbb{S}$ described above; since this implies that for every cyclotomic integer $g(\alpha)$ there is exactly one way to choose integers $n_1, n_2, \ldots, n_\mu$ with $n_1 + n_2 + \cdots + n_\mu = 0$ and with $(1 - \sigma\alpha)^{n_1}(1 - \sigma^2\alpha)^{n_2} \ldots (1 - \sigma^\mu\alpha)^{n_\mu} g(\alpha)$ in $\mathcal{D}'$, it implies that

$$S(s) = \sum_{g(\alpha) \in \mathcal{D}'} Ng(\alpha)^{-s}$$

because both are equal to $2\lambda h_2 \sum N(A)^{-s}$ where A ranges over all principal divisors. At the same time

$$I(s) = \int_{\mathcal{D}'} N\left(u_1 \sigma\alpha + \cdots + u_{\lambda-1}\sigma^{\lambda-1}\alpha\right)^{-s} du_1 \, du_2 \cdots du_{\lambda-1}.$$

This is perhaps most easily seen by retracing the evaluation of $I(s)$ in the preceding section and noting that the entire argument applies to $\mathcal{D}'$ as well as to $\mathcal{D}$ except that in the last step it is a matter of evaluating the integral of

$$e^{(1-s)(\log \lambda)(x_1 + x_2 + \cdots + x_\mu)} \, dx_1 \, dx_2 \cdots dx_\mu$$

over $\mathbb{S}'$ instead of $\mathbb{S}$. The value of this integral is $[(s - 1) \log \lambda]^{-1}$ in either case, as can be seen by taking the change of variables $y_1 = x_1, \ldots, y_{\mu-1} = x_{\mu-1}, t = x_1 + x_2 + \cdots + x_\mu$ in the case of $\mathbb{S}'$ and the change of variables $y_1 = x_1 - \mu^{-1}t, y_2 - \mu^{-1}t, \ldots, y_{\mu-1} = x_{\mu-1} - \mu^{-1}t, t = x_1 + x_2 + \cdots + x_\mu$ in the case of $\mathbb{S}'$.

Let the domain $\mathcal{D}'$ be broken into the domains $\mathcal{D}_0 = \{(u) \in \mathcal{D}' : 1 \leqslant Nu < 2^{\lambda-1}\}$, $\mathcal{D}_1 = \{(u) \in \mathcal{D}' : 2^{\lambda-1} \leqslant Nu < 4^{\lambda-1}\}, \ldots, \mathcal{D}_k = \{(u) \in \mathcal{D}' : 2^{k(\lambda-1)} \leqslant Nu < 2^{(k+1)(\lambda-1)}\}, \ldots$ or, what is the same (because $\mathcal{D}'$ is invariant under scale changes) $\mathcal{D}_k = \{(u) \in \mathcal{D}' : (2^{-k}u) \in \mathcal{D}_0\}$. Here (u) denotes $(u_1, u_2, \ldots, u_{\lambda-1})$, $(2^{-k}u)$ denotes $(2^{-k}u_1, 2^{-k}u_2, \ldots, 2^{-k}u_{\lambda-1})$ and Nu denotes the product of the $\lambda - 1$ conjugates of $u_1\sigma\alpha + u_2\sigma^2\alpha + \cdots + u_{\lambda-1}\sigma^{\lambda-1}\alpha$. Let $S(s) = S_0(s) + S_1(s) + S_2(s) + \cdots$ and $I(s) = I_0(s) + I_1(s) + I_2(s) + \cdots$ be the corresponding decompositions of the sum $S(s)$ and the integral $I(s)$. In other words, let $S_k(s)$ be the sum of $Ng(\alpha)^{-s}$ over all cyclotomic integers in $\mathcal{D}_k$ and let $I_k(s)$ be the integral of $N(u_1\sigma\alpha + \cdots + u_{\lambda-1}\sigma^{\lambda-1}\alpha)^{-s} \, du_1 \, du_2 \cdots du_{\lambda-1}$ over $\mathcal{D}_k$. Because $S(s)$ converges absolutely (it is a series of positive terms) it can be summed in the order $S(s) = S_0(s) + S_1(s) + \cdots$; to prove that the improper integral $I(s)$ converges for $s > 1$ it will suffice to prove that there is a convergent series $\delta_0(s) + \delta_1(s) + \delta_2(s) + \cdots$ such that $|S_k(s) - I_k(s)| < \delta_k(s)$ because this implies $I_m(s) + I_{m+1}(s) + \cdots + I_n(s) \leq S_m(s) + |\delta_m(s)| + S_{m+1}(s) + |\delta_{m+1}(s)| + \cdots + S_n(s) + |\delta_n(s)|$, which implies that $I(s)$ satisfies the Cauchy criterion. If it is shown, moreover, that as $s \downarrow 1$ the terms $\delta_k(s)$ remain bounded by a convergent series of constants $\delta_0 + \delta_1 + \delta_2 + \cdots$ it will follow that $|S(s) - I(s)|$ remains bounded as $s \downarrow 1$, which implies that $\lim (s-1)I(s) = \lim [(s-1)S(s) + (s-1)(I(s) - S(s))]$ exists and is equal to $\lim (s-1)S(s) + 0$ as desired. Thus the desired theorem will be proved if estimates $|S_k(s) - I_k(s)| < \delta_k(s)$ can be given in such a way that $\Sigma \delta_k(s) < \infty$ and in such a way that for for $1 < s \leq 2$ one has $\delta_k(s) < \delta_k$ where $\Sigma \delta_k < \infty$. In proving these statements one can of course ignore any finite number of terms at the beginning of the series (both $S_k(s)$ and $I_k(s)$ converge for $s > 1$ and both are bounded as $s \downarrow 1$). Therefore one can focus on the estimation of $|S_k(s) - I_k(s)|$ for large integers k.

Let $u_1 u_2 \ldots u_{\lambda-1}$-space be subdivided into "cubes" of the form $\{(u) : |u_j - n_j| \leq \frac{1}{2}\}$ which are 1 on a side and have their centers at points $(n_1, n_2, \ldots, n_{\lambda-1})$ all of whose coordinates are integers. For a given integer k the difference $S_k(s) - I_k(s)$ can be written as the sum over all cubes of the difference between the term, if any, of $S_k(s)$ which is the value of $Ng(\alpha)^{-s}$ at the center of that cube and the portion, if any, of $I_k(s)$ which is an integral over that cube. There are three types of cubes:

(i) cubes which contain no points of $\mathcal{D}_k$
(ii) cubes which are contained entirely inside $\mathcal{D}_k$
(iii) cubes which lie on the boundary of $\mathcal{D}_k$, that is, cubes which contain points inside $\mathcal{D}_k$ and points outside $\mathcal{D}_k$.

For cubes of type (i) of course the corresponding parts of $S_k(s)$ and $I_k(s)$ are both zero and their difference is therefore zero. It remains to estimate the total difference between $S_k(s)$ and $I_k(s)$ over cubes of types (ii) and (iii). The estimates for the two types of cubes are quite different and it is natural to separate them. The estimate for cubes of type (iii) is the easier of the two and it will be given first.

The term of $S_k(s)$ corresponding to a cube of type (iii) is equal to 0 or to N^{-s} and is therefore at most $[2^{k(\lambda-1)}]^{-s} \leq 2^{-k(\lambda-1)}$. Similarly the portion of $I_k(s)$ corresponding to such a cube is at most $2^{-k(\lambda-1)}$. Since the difference of two positive numbers has absolute value at most the maximum of the two numbers, it follows that the portion of $|S_k(s) - I_k(s)|$ which corresponds to cubes of type (iii) is at most $2^{-k(\lambda-1)}$ times the number of such cubes (for all $s > 1$). The number of such cubes is equal to the number of cubes which lie on the boundary of $\mathcal{D}_0$ when $u_1 u_2 \ldots u_{\lambda-1}$-space is divided into cubes that are 2^{-k} on a side. The problem, therefore, is to estimate the number of these cubes. Such an estimate is basic to the theory of integration over domains like $\mathcal{D}_0$; using the fact that the boundary of $\mathcal{D}_0$ is nonsingular and compact, it is simple (see Exercise 1) to prove that the number of such cubes is at most a constant times $(2^k)^{\lambda-2}$. Therefore the portion of $|S_k(s) - I_k(s)|$ corresponding to cubes of type (iii) is at most a constant times $2^{k(\lambda-2)}2^{-k(\lambda-1)} = 2^{-k}$ for all $s \geqslant 1$. Since $\Sigma 2^{-k} < \infty$, this proves the desired estimate for cubes of type (iii).

Consider now a cube of type (ii). Let N_c^{-s} denote the value of N^{-s} at the center of the cube. Then, since the volume of the cube is 1, the term of $S_k(s)$ corresponding to this cube can be written as the integral of the constant function N_c^{-s} over the cube and the portion of $|S_k(s) - I_k(s)|$ corresponding to this cube is at most the integral of $|N^{-s} - N_c^{-s}|$ over the cube. This in turn is at most the maximum value of $|N^{-s} - N_c^{-s}|$ on the cube. Therefore it is natural to seek an estimate of $|N^{-s} - N_c^{-s}|$. By the fundamental theorem of calculus $N^{-s} - N_c^{-s}$ is equal to the integral, along a path in the cube from the center to a point of the cube, of $d(N^{-s}) = (-s)N^{-s}(dN/N)$. The value of N^{-s} is roughly equal to N_c^{-s} and in fact the ratio of N^{-s} to N_c^{-s} is at most $[2^{k(\lambda-1)}]^s / [2^{(k+1)(\lambda-1)}]^s = 2^{-s(\lambda-1)} < 2^{-(\lambda-1)}$. Thus $|(-s)N^{-s}|$ is less than a constant times N_c^{-s}, where the constant is independent of s. The differential form dN/N is invariant under scale changes (N is a homogeneous polynomial in $u_1, u_2, \ldots, u_{\lambda-1}$) and from this it follows easily that its integral between the center of the cube and another point of the cube is bounded by a bound valid for all cubes of type (ii). (In fact, as will be shown in a moment, this bound goes to zero as $k \to \infty$.) Thus the portion of $S_k(s) - I_k(s)$ corresponding to this cube is at most a constant times N_c^{-s} and the total difference is at most a constant times $S_k(s)$. This estimate suffices to prove that $I(s)$ converges because $\Sigma S_k(s) < \infty$. However, as $s \downarrow 1$ this does not give a uniform bound for $|S(s) - I(s)|$ because the series $\Sigma S_k(s) = S(s)$ approaches ΣN^{-1}, which diverges. Therefore a finer estimate, one which takes into account the fact that the integral of dN/N is small, is required.

As was noted above, the integral of dN/N between two points in a cube of type (ii) is equal to the integral of dN/N between the images of these two points under the scale change $(u) \mapsto (2^{-k}u)$. (Note that $dN/N = d \log N$ is a closed differential form so that this integral is independent of the path.) These two points lie in $\mathcal{D}_0$ and their coordinates differ by at most 2^{-k} in all $\lambda - 1$ coordinate directions. Since $\mathcal{D}_0$ is bounded and $\log N$

is continuously differentiable on the closure of $\mathcal{D}_0$, the partial derivatives of $\log N$ are bounded on $\mathcal{D}_0$ and it follows easily that this integral is at most a constant times 2^{-k}. Since $N_c \leqslant 2^{(k+1)(\lambda-1)}$, $N_c^{1/(\lambda-1)} \leqslant 2^{k+1}$, $2N_c^{-1/(\lambda-1)} \geqslant 2^{-k}$, the total over all such cubes is at most a constant times ΣN_c^{-t} where $t = 1 + (\lambda - 1)^{-1}$. This estimate holds for $s \geqslant 1$ and, since ΣN_c^{-t} converges, this completes the proof.

EXERCISES

1. Prove that there is a constant B such that when $u_1 u_2 \ldots u_{\lambda-1}$-space is divided into "cubes" that are 2^{-k} on a side the total number of cubes on the boundary of $\mathcal{D}_0$ is at most $B \cdot 2^{k(\lambda-2)}$ for $k = 1, 2, \ldots$. [Prove first that the boundary of $\mathcal{D}_0$ is compact and nonsingular in the sense that it can be written as the image of a finite number of (possibly overlapping) continuously differentiable nonsingular mappings of a $(\lambda - 2)$-dimensional cube. In fact, the image of $\mathcal{D}_0$ under Φ is a cube in $x_1 x_2 \cdots x_\mu$-space and it is possible to use the mappings of the preceding sections to give quite explicit parameterizations of the boundary of $\mathcal{D}_0$. By the Implicit Function Theorem it follows that locally the boundary of $\mathcal{D}_0$ can be parameterized by $\lambda - 2$ of the $\lambda - 1$ coordinate functions in $u_1 u_2 \cdots u_{\lambda-1}$-space. For each such parameterized piece, the number of cubes of side 2^{-k} it can cut through is a constant times the number of cubes it lies over in the $(\lambda - 2)$-dimensional coordinate plane, the constant arising from an upper bound on the partial derivatives of the parameterizing function. The number of cubes in the coordinate plane is of course less than a constant times $2^{k(\lambda-2)}$. Since the entire boundary is covered by a finite number of these pieces the desired estimate follows.]

6.13 The sum over other divisor classes

It was mentioned in Section 6.2 that the sum of $N(A)^{-s}$ over any two divisor classes is, in the limit as $s \downarrow 1$, the same. In other words

$$\lim_{s \downarrow 1} \left[\sum_{A \sim B} N(A)^{-s} / \sum_{A \sim I} N(A)^{-s} \right] = 1 \tag{1}$$

where B is a given divisor, where the sum in the numerator is over all divisors equivalent to B, and where the sum in the denominator is over all principal divisors. This fact, which was clearly presaged by Dirichlet's formula for the class number in the quadratic case, can be proved as follows.

Let B be a given divisor and let C be a divisor such that $BC \sim I$. Then $A \sim B$ implies $AC \sim I$, that is, $A' = AC$ is a principal divisor divisible by C. Conversely, if A' is a principal divisor divisible by C, say $A' = AC$, then the quotient A satisfies $A \sim ABC = A'B \sim IB \sim B$. That is, $A \mapsto AC$ establishes a one-to-one correspondence between divisors A equivalent to B and principal divisors A' divisible by C. Since $N(A') = N(AC) = N(A)N(C)$ it follows that

$$\sum_{A \sim B} N(A)^{-s} = \sum_{A \sim B} N(C)^s N(AC)^{-s} = N(C)^s \sum_{\substack{A' \sim I \\ C|A'}} N(A')^{-s}.$$

Therefore the statement to be proved can be reformulated as

$$\lim_{s \downarrow 1} \left[N(C)^s \sum_{\substack{A \sim I \\ C|A}} N(A)^{-s} / \sum_{A \sim I} N(A)^{-s} \right] = 1.$$

Since $\lim N(C)^s = N(C)$, this is the statement that in the limit as $s \downarrow 1$ the sum of $\Sigma N(A)^{-s}$ over *all* principal divisors is $N(C)$ times as great as the sum over just those principal divisors which are divisible by C. Now $N(C)$ is a positive integer and, roughly speaking, the statement to be proved is simply the statement that *one out of every $N(C)$ cyclotomic integers is divisible by C.*

More precisely, consider the problem of evaluating, in the limit as $s \downarrow 1$, the sum of $N(A)^{-s}$ over all principal divisors A that are divisible by a given divisor C. If L denotes the set of all cyclotomic integers that are divisible by C then clearly, as in the case $C = I$ already covered,

$$\sum_{\substack{A \text{ principal} \\ C|A}} N(A)^{-s} = \frac{1}{2\lambda h_2} \sum_{\substack{\Phi(g(\alpha)) \text{ in } \mathbb{S} \\ g(\alpha) \text{ in } L}} Ng(\alpha)^{-s}$$

for $s > 1$. In other words, for each principal divisor A divisible by C there are exactly $2\lambda h_2$ cyclotomic integers $g(\alpha)$ whose divisor is A and whose images under Φ lie in $\mathbb{S}$.

Now let $u_1 u_2 \cdots u_{\lambda-1}$-space be divided into "cubes" with vertices at the points whose coordinates are all integers divisible by $N(C)$. The vertices of these cubes all lie in L because divisibility by $N(C)$ implies divisibility by C. Then by the arguments of the preceding section

$$\lim_{s \downarrow 1} (s - 1) \int_{\mathcal{D}} N(u_1 \sigma \alpha + \cdots + u_{\lambda-1} \sigma^{\lambda-1} \alpha) \, du_1 \cdots du_{\lambda-1}$$

$$= \lim_{s \downarrow 1} (s - 1) \sum_{\text{cubes in } \mathcal{D}} N(\text{point of the cube})^{-s}(\text{volume of the cube})$$

where the variation of the value of N^{-s} over any one cube is insignificant and where the sum over all cubes partly inside $\mathcal{D}$ and partly outside $\mathcal{D}$ is also insignificant.

Let each point $(u_1, u_2, \ldots, u_{\lambda-1})$ of L be associated to the cube which contains $(u_1 + \varepsilon, u_2 + \varepsilon, \ldots, u_{\lambda-1} + \varepsilon)$ for all sufficiently small ε. Then to each cube there is associated the same number of points of L, say v, because $(u_1, u_2, \ldots, u_{\lambda-1})$ lies in L if and only if $(u_1 + n_1 N(C), u_2 + n_2 N(C), \ldots, u_{\lambda-1} + n_{\lambda-1} N(C))$ lies in L for all integers $n_1, n_2, \ldots, n_{\lambda-1}$. In the above sum over cubes one can use as the value of N^{-s} the *average* of the values of $Ng(\alpha)^{-s}$ over the ν points of L associated to the cube. Then each cube inside $\mathcal{D}$ contributes a term

$$\frac{Ng(\alpha)^{-s} + Ng_2(\alpha)^{-s} + \cdots + Ng_\nu(\alpha)^{-s}}{\nu} N(C)^{\lambda-1}$$

to the above sum, where $g_1(\alpha), g_2(\alpha), \ldots, g_\nu(\alpha)$ are the ν points of L associated to the cube. This shows that the limit of the above sum is equal to

$$\lim_{s\downarrow 1} (s-1) \frac{N(C)^{\lambda-1}}{\nu} \sum_{\substack{g(\alpha) \text{ in } \mathcal{D} \\ g(\alpha) \text{ in } L}} Ng(\alpha)^{-s}.$$

On the other hand, the limit of the integral on the left side of the above equation is also equal to

$$\lim_{s\downarrow 1} (s-1) \sum_{\mathcal{D}} Ng(\alpha)^{-s}$$

which is $2\lambda h_2$ times the sum of $N(A)^{-s}$ over all principal divisors. Therefore

$$\frac{N(C)^{\lambda-1}}{\nu} \lim_{s\downarrow 1} (s-1) \sum_{\substack{A \sim I \\ C \mid A}} N(A)^{-s} = \lim_{s\downarrow 1} (s-1) \sum_{A \sim I} N(A)^{-s}$$

and the statement to be proved is simply

$$\frac{N(C)^{\lambda-1}}{\nu} = N(C).$$

This is a purely arithmetical statement, from which limits and infinite sums have been eliminated. It can be proved very simply as follows.

As was shown in Section 4.14, $N(C)$ is equal to the number of congruence classes mod C. Since $N(C)^{\lambda-1}$ is equal to the number of cyclotomic integers in a cube of side $N(C)$, the desired equation $N(C)^{\lambda-1} = \nu \cdot N(C)$ will be proved if it is shown that each of the $N(C)$ congruence classes contains exactly ν cyclotomic integers in each cube. If $g(\alpha)$ is a cyclotomic integer then the cyclotomic integers congruent to $g(\alpha) \bmod C$ are those of the form $g(\alpha) + h(\alpha)$ where $h(\alpha)$ is divisible by C. In other words, the points in $u_1 u_2 \ldots u_{\lambda-1}$-space which correspond to cyclotomic integers in a given congruence class $\{ g(\alpha) + h(\alpha) \} \bmod C$ are a *translation* of the set of points L corresponding to cyclotomic integers divisible by C. It will suffice to prove, therefore, that L has the property that any translation of L has the same number of points in a given cube of side $N(C)$ as L itself. This is obviously true for a translation by 1 in any coordinate direction because each point of L which leaves the cube at one end is matched by one which enters at the opposite end. Since every translation is a composition of translations by 1 in coordinate directions, this completes the proof of (1).

6.14 The class number formula

If h is the number of distinct classes of inequivalent divisors then, by the fact proved in the preceding section that $\lim (s - 1)\Sigma N(A)^{-s}$ is the same when the sum is extended over any class, $\lim (s - 1)\Sigma N(A)^{-s}$ when the sum is extended over *all* classes is equal to h times $\lim (s - 1)\Sigma N(A)^{-s}$ when the sum is extended over the principal class. Therefore, when the Euler product formula

$$\sum_{\text{all divisors } A} N(A)^{-s} = \prod_{\substack{\text{all prime} \\ \text{divisors } P}} \left(1 - N(P)^{-s}\right)^{-1}$$

is multiplied by $(s - 1)$, the limits as $s\downarrow 1$ have been found above to be

$$\pm h \frac{1}{2\lambda h_2} 2^{2\mu - 1}(\pi i)^\mu m_1 m_2 \cdots m_{\lambda - 2} C_1 C_2 \cdots C_{\mu - 1}$$

$$= (-2)^{\mu - 1}\left(\frac{i\pi}{\lambda}\right)^\mu m_1 m_2 \cdots m_{\lambda - 2} C_1 C_2$$

$$\cdots C_{\mu - 1}\phi(\beta)\phi(\beta^3) \cdots \phi(\beta^{\lambda - 2})$$

which reduces to

$$h = \pm \frac{\phi(\beta)\phi(\beta^3) \cdots \phi(\beta^{\lambda - 2})}{(2\lambda)^{\mu - 1}} \cdot h_2.$$

This is the *class number formula* in essentially the form in which Kummer gave it.

The class number is here given as a product of two factors which are · traditionally known as the *first factor* and the *second factor* of the class number. The second factor h_2 is the positive integer that was defined in Section 6.10. Briefly put, h_2 is the number of units E_j that one must have in order to be able to write all cyclotomic units in the form

$$\pm \alpha^k E_j (1 - \sigma\alpha)^{x_1}(1 - \sigma^2\alpha)^{x_2} \cdots (1 - \sigma^\mu\alpha)^{x_\mu}$$

for some choice of the sign, of $k = 1, 2, \ldots, \lambda$, of $j = 1, 2, \ldots, h_2$, and of integers $x_1, x_2, \ldots, x_\mu$ satisfying $x_1 + x_2 + \cdots + x_\mu = 0$. (In terms of group theory, h_2 is the index of the group of units of the form $\pm \alpha^k \Pi(1 - \sigma^\nu\alpha)^{x_\nu}$ in the group of all units.) The first factor, which is also denoted

$$h_1 = \pm \frac{\phi(\beta)\phi(\beta^3) \cdots \phi(\beta^{\lambda - 2})}{(2\lambda)^{\mu - 1}},$$

is in fact an integer. The sign is of course chosen so that it is *positive*. Recall that β is a primitive $(\lambda - 1)$st root of unity and that $\phi(X)$ is by definition the polynomial $\phi(X) = 1 + \gamma X + \gamma_2 X^2 + \cdots + \gamma_{\lambda - 2}X^{\lambda - 2}$

where $0 < \gamma_j < \lambda$, $\gamma = \gamma_1$ is a primitive root mod λ, and $\gamma_j \equiv \gamma^j \bmod \lambda$. Thus, unlike h_2, h_1 is given by an explicit formula all of whose terms depend in a simple way on λ. It will be proved in the following sections that $2^{\mu-1}h_1$ is an integer. It is not difficult to prove that h_1 itself is an integer (see Exercise 2 of Section 6.15)—as Kummer stated it was—but this fact will not be needed.

Kummer used the letter P to denote $\phi(\beta)\phi(\beta^3) \ldots \phi(\beta^{\lambda-2})$ so that the class number formula takes the form

$$h = \frac{|P|}{(2\lambda)^{\mu-1}} \cdot h_2.$$

He also described the integer h_2 as a quotient of two determinants $h_2 = D/\Delta$, but the numerator D cannot in fact be found without first determining enough about the structure of the group of units to find h_2 itself. Therefore the direct description of h_2 seems preferable to this quotient of determinants. In any event, the computation of h_2 is much more difficult than the computation of h_1. Fortunately, as will be seen in the following sections, it is not necessary to evaluate *either* h_1 or h_2 in order to prove Fermat's Last Theorem for a given λ, that is, in order to prove that (A) and (B) hold. This is the great convenience of Kummer's criterion in terms of numerators of Bernoulli numbers.

EXERCISE

1. Prove by direct computations that the first factor of the class number is 1 for $\lambda = 3, 5, 7, 11, 13$. Determine the sign of $\phi(\beta)\phi(\beta^3) \ldots \phi(\beta^{\lambda-2})$ in each of these cases.

6.15 Proof that 37 is irregular

The formula $\pm P/(2\lambda)^{\mu-1}$ for the first factor of the class number prescribes a quite explicit algebraic computation for this number which, for small values of λ, can be carried through to find the number (see Exercise 1 of the preceding section). The case $\lambda = 37$ is not beyond the feasible range of this computation, and Kummer ultimately computed* not only this factor—finding it to be 37—but also the first factor of the class number for all primes $\lambda < 100$. How many of these computations were completed in the summer of 1847 is not known, but from what is known about Kummer's methods of work it seems safe to guess that his computations of particular class numbers were already rather extensive by that time. In any case, whether by virtue of long computational experience or

---

* [K14, p. 473]. Kummer later said [K15, p. 199] that these computations were made "not without great effort." He also corrected the answer in the case $\lambda = 71$. In [K17] Kummer carried these computations clear to $\lambda = 163$ and made a few elementary remarks on his method of computation.

direct inspiration, he succeeded in discovering in 1847 some ingenious techniques for resolving the particular question about $P/(2\lambda)^{\mu-1}$ which was of greatest interest to him, namely, the question of whether it is divisible by λ, without actually doing the computation. This section is devoted to proving that in the special case $\lambda = 37$ the answer is that λ *does* divide the first factor of the class number. In the next section it will be shown how the same techniques can be used to determine a necessary and sufficient condition for divisibility of $P/(2\lambda)^{\mu-1}$ by λ, a condition which is entirely independent of the actual computation of $P/(2\lambda)^{\mu-1}$.

Computation of the powers of 2 mod 37 gives 1, 2, 4, 8, 16, 32, 27, . . . , 14, 28, 19, 1, 2, 4, ... Since this sequence goes the full 36 terms before beginning to repeat, 2 is a primitive root mod 37. Therefore P is by definition the integer obtained by setting

$$\phi(x) = 1 + 2x + 4x^2 + 8x^3 + 16x^4 + 32x^5 + 27x^6 + 17x^7$$
$$+ \cdots + 14x^{33} + 28x^{34} + 19x^{35}$$
$$\beta = \text{a primitive 36th root of unity}$$
$$P = \phi(\beta)\phi(\beta^3)\phi(\beta^5) \ldots \phi(\beta^{33})\phi(\beta^{35})$$

and what was shown in the preceding section is that

$$h = \pm \frac{P}{(2 \cdot 37)^{17}} \cdot h_2.$$

What is to be shown in this section is that the integer h is divisible by 37.

It is not entirely obvious that $P = \phi(\beta)\phi(\beta^3) \ldots \phi(\beta^{35})$ is even an *integer*, much less that it is divisible by $2^{17}37^{17}$ as it must be in order for the first factor to be, as Kummer stated, an integer. The proof that P is an integer is an immediate consequence of the general fact that if $f(\beta) = a_0 + a_1\beta + a_2\beta^2 + \cdots + a_n\beta^n$ is a number which can be expressed as a polynomial in a primitive kth root of unity β and if $f(\beta)$ has the property that substitution of any other primitive kth root of unity β' for β gives the same number $f(\beta) = f(\beta')$ then the number $f(\beta)$ must be an integer. [In the language of Galois theory, any integer of the kth cyclotomic field which is invariant under the Galois group is a rational integer.] Since in the case $k = 36$ a primitive root β' must be an odd power of β, it is easy to see that the substitution $\beta \mapsto \beta'$ merely permutes the odd powers of β and therefore leaves P invariant. Thus the fact that P is an integer is an immediate consequence of the theorem. For the proof of the theorem, see Exercise 1.

The main step in proving that P is divisible by 37^{17} is an algebraic identity which results from the observation that $\phi(x)$ is by definition $1 + 2x + (2x)^2 + (2x)^3 + \cdots + (2x)^{35} = [(2x)^{36} - 1]/(2x - 1)$ except that the integer coefficients of the powers of x are reduced mod 37. This motivates multiplying $\phi(x)$ by $2x - 1$ to find

$$(2x - 1)\phi(x) = 0 \cdot x + 0 \cdot x^2 + 0 \cdot x^3 + 0 \cdot x^4 + 0 \cdot x^5 + 37 \cdot x^6 + 37 \cdot x^7$$
$$+ 0 \cdot x^8 + \cdots + 0 \cdot x^{34} + 37 \cdot x^{35} + 38 \cdot x^{36} - 1.$$

If one sets $x = \beta, \beta^3, \beta^5, \ldots, \beta^{35}$ in this identity and multiplies all the resulting identities one finds

$$(2\beta - 1)(2\beta^3 - 1) \cdots (2\beta^{35} - 1)P = 37^{18} \psi(\beta)\psi(\beta^3) \cdots \psi(\beta^{35})$$

where ψ is defined to be

$$\psi(x) = 0 \cdot x + 0 \cdot x^2 + 0 \cdot x^3 + 0 \cdot x^4 + 0 \cdot x^5 + 1 \cdot x^6 + 1 \cdot x^7 + 0 \cdot x^8$$
$$+ \cdots + 0 \cdot x^{34} + 1 \cdot x^{35} + 1 \cdot x^{36}.$$

(Note that $38\beta^{36} - 1 = 37 = 37\beta^{36}$.) Now $(2\beta - 1)(2\beta^3 - 1) \cdots (2\beta^{35} - 1)$ is an integer for the same reason that P is, and this integer can be found explicitly by means of the formula

$$(X - \beta)(X - \beta^3) \cdots (X - \beta^{35}) = X^{18} + 1$$

(the two sides are polynomials with the 18 distinct roots $\beta, \beta^3, \ldots, \beta^{35}$) from which

$$(x - \beta y)(x - \beta^3 y) \cdots (x - \beta^{35}y) = x^{18} + y^{18}$$
$$(2\beta - 1)(2\beta^3 - 1) \cdots (2\beta^{35} - 1) = 1 + 2^{18}.$$

Because 2 is a primitive root mod 37, Fermat's theorem implies $2^{18} \equiv -1 \bmod 37$ and 37 must divide $2^{18} + 1$. In fact, $2^{18} + 1 = 262145 = 37 \cdot 7085$. Therefore

$$7085 P = 37^{17}\psi(\beta)\psi(\beta^3) \cdots \psi(\beta^{35}).$$

Because 37 does not divide 7085 and because $\psi(\beta)\psi(\beta^3) \cdots \psi(\beta^{35})$ is an integer, this shows that P is divisible by 37^{17}.

A similar type of argument can be used to prove that P is also divisible by 2^{17}, so that the first factor $P/(2 \cdot 37)^{17}$ of the class number is indeed an integer. However, this fact will not be needed in what follows and will therefore be left to the excercises (Exercise 2).

The question of determining whether the integer h is divisible by 37 is, by virtue of the equation in integers $2^{17}h = \pm(P/37^{17}) \cdot h_2$, the same as the question of determining whether either of the integers $P/37^{17}$ or h_2 is divisible by 37. In order to prove that h *is* divisible by 37 it will suffice to prove that $P/37^{17}$ is divisible by 37, for which it will suffice to prove that

$$\psi(\beta)\psi(\beta^3) \cdots \psi(\beta^{35}) \equiv 0 \bmod 37$$

because this implies that $7085P$ is divisible by 37^{18}. Now in computations mod 37, 2 has the same properties as β, namely, $2^{36} \equiv 1 \bmod 37$ but $2^j \not\equiv 0 \bmod 37$ for $0 < j < 36$. Therefore

$$\psi(\beta)\psi(\beta^3) \cdots \psi(\beta^{35}) \equiv \psi(2)\psi(2^3) \cdots \psi(2^{35}) \bmod 37$$

because all the simplifications* used to eliminate β from the expression of the integer on the left can be applied to 2 mod 37 on the right. Therefore, *in order to prove that 37 divides the first factor of its class number it suffices to prove that one of the integers* $\psi(2)$, $\psi(2^3)$, $\psi(2^5)$, ..., $\psi(2^{35})$ *is zero* mod 37.

Because the powers of 2 mod 37 and the polynomials $\psi(x)$ can be written down explicitly with no trouble, the computation of these 18 integers mod 37 is very simple. If one carries it out one finds that $\psi(2^{31}) \equiv 0$ mod 37, and this proves that 37 is irregular. Kummer, however, was able to simplify the problem even further and to avoid even this simple computation as follows. Let $b_1, b_2, \ldots, b_{36}$ represent the coefficients of $\psi(x) = b_1 x + b_2 x^2 + \cdots + b_{36} x^{36}$, and let $\gamma_0, \gamma_1, \ldots, \gamma_{35}$ represent the coefficients of $\phi(x) = \gamma_0 + \gamma_1 x + \gamma_2 x^2 + \cdots + \gamma_{35} x^{35}$ so that, by definition, $0 < \gamma_j < 37$, $\gamma_j \equiv 2^j$ mod 37, $(2x - 1)\phi(x) = 37 \cdot \psi(x) + (x^{36} - 1)$, and, consequently, $2\gamma_{j-1} - \gamma_j = 37 b_j$ for $j = 1, 2, \ldots, 35$, and $2\gamma_{35} - 1 = 37 b_{36}$. Then

$$\psi(2^n) = b_1 2^n + b_2 2^{2n} + \cdots + b_{36} 2^{36n}$$

$$\equiv b_1 \gamma_1^n + b_2 \gamma_2^n + \cdots + b_{36} \gamma_{36}^n \text{ mod } 37.$$

But it is clear that the b_j are always 0 or 1 and, more specifically, that b_j is 0 if $2\gamma_{j-1} < 37$ and 1 if $2\gamma_{j-1} > 37$. (This also holds for $j = 36$.) Therefore $\psi(2^n) = \sum \gamma_j^n$ where j ranges over all indices for which $\gamma_{j-1} \geqslant 19$. This can be simplified by noting that $\gamma_j^n \equiv (2\gamma_{j-1})^n \equiv \gamma_n \gamma_{j-1}^n$, which gives

$$\psi(2^n) \equiv \gamma_n \left[b_1 \gamma_0^n + b_2 \gamma_1^n + \cdots + b_{36} \gamma_{35}^n \right] \text{ mod } 37$$

$$= \gamma_n \left[19^n + 20^n + \cdots + 36^n \right]$$

because each integer m from 1 to 36 occurs just once as a γ_j, and the corresponding b_{j+1} is 1 if and only if $m \geqslant 19$. Therefore, in order to determine whether $\psi(2^n)$ is divisible by 37 it is necessary and sufficient to determine whether $19^n + 20^n + \cdots + 36^n$ is divisible by 37.

The formula for the sum of consecutive nth powers was developed by Jakob Bernoulli in the 18th century. Briefly, the formula can be described in modern notation as follows. For each $n = 0, 1, 2, \ldots$ let $B_n(x)$ be the polynomial for which

$$\int_x^{x+1} B_n(t) \, dt = x^n.$$

It is easily seen that this identity determines a unique nth degree polynomial. (Exercise 3). Then clearly $M^n + (M + 1)^n + \cdots + N^n = \int_M^{M+1} B_n(t) \, dt + \int_{M+1}^{M+2} B_n(t) \, dt + \cdots + \int_N^{N+1} B_n(t) \, dt = \int_M^{N+1} B_n(t) \, dt$, and to find a formula for the sum of consecutive nth powers is essentially the

---

* The congruence in question is best viewed as a congruence modulo the prime divisor common to $\beta - 2$ and 37. However, to state it in this way requires a theory of divisors for numbers built up from a 36th root of unity β, and this theory was not developed in Chapter 4 because, of course, 36 is not prime.

same as to find a formula for $B_n(x)$. Now differentiation of the defining equation with respect to the limits of integration gives

$$B_n(x+1) - B_n(x) = nx^{n-1}$$

$$\int_x^{x+1} \frac{d}{dt} B_n(t) \, dt = n \int_x^{x+1} B_{n-1}(t) \, dt$$

from which it follows that $dB_n(t)/dt = nB_{n-1}(t)$. Therefore, in order to find the polynomial $B_n(t)$ it suffices to know $B_{n-1}(t)$ and $B_n(0)$. Explicitly, denoting the constant $B_n(0)$ by B_n, one has

$$B_0(x) = 1 = B_0$$

$$B_1(x) = \int B_0(x) \, dx + \text{const} = x + B_1$$

$$B_2(x) = \int 2B_1(x) \, dx + \text{const} = x^2 + 2B_1 x + B_2$$

$$B_3(x) = \int 3B_2(x) \, dx + \text{const} = x^3 + 3B_1 x^2 + 3B_2 x + B_3$$

$$B_4(x) = x^4 + 4B_1 x^3 + 6B_2 x^2 + 4B_3 x + B_4$$

and in general

$$B_n(x) = x^n + nB_1 x^{n-1} + \frac{n(n-1)}{2} B_2 x^{n-2} + \cdots$$

$$+ \binom{n}{k} B_k x^{n-k} + \cdots + B_n$$

where $\binom{n}{k}$ denotes the binomial coefficient $n!/(n-k)!k!$. Therefore the task is to find the so-called *Bernoulli numbers* $B_0, B_1, B_2, \ldots$. The computation of these numbers is facilitated by the observation that for $n \geqslant 1$

$$0^n = \int_0^1 B_n(t) \, dt = \frac{B_{n+1}(1) - B_{n+1}(0)}{n+1} \qquad (n \geqslant 1)$$

which implies

$$B_{n+1}(1) = B_{n+1}(0) \qquad (n \geqslant 1)$$

$$B_0 + (n+1)B_1 + \cdots + \binom{n+1}{k} B_k + \cdots + (n+1)B_n + B_{n+1} = B_{n+1}$$

$$B_n = -\frac{1}{n+1} \left[B_0 + (n+1)B_1 + \cdots + \frac{n(n+1)}{2} B_{n-1} \right]$$

for $n \geqslant 1$. Thus*

$$B_1 = -\frac{1}{2}[1] = -\frac{1}{2}$$

$$B_2 = -\frac{1}{3}\left[1 + 3\left(-\frac{1}{2}\right)\right] = \frac{1}{6}$$

$$B_3 = -\frac{1}{4}\left[1 + 4\left(-\frac{1}{2}\right) + 6\left(\frac{1}{6}\right)\right] = 0$$

$$B_4 = -\frac{1}{5}\left[1 + 5\left(-\frac{1}{2}\right) + 10\left(\frac{1}{6}\right) + 10(0)\right] = -\frac{1}{30}$$

$$B_5 = -\frac{1}{6}\left[1 + 6\left(-\frac{1}{2}\right) + 15\left(\frac{1}{6}\right) + 20(0) + 15\left(-\frac{1}{30}\right)\right] = 0$$

etc. Euler computed the values of B_n for $n \leqslant 30$. For odd n greater than 1, $B_n = 0$ (Exercise 5). For even n, the B's alternate in sign (Exercise 6) and grow very rapidly in absolute value as $n \to \infty$. For example, the last one computed by Euler was

$$B_{30} = \frac{8615841276005}{14322}.$$

In 1842, M. Ohm published in Crelle's famous Journal [O1] the values of all the Bernoulli numbers up to and including B_{62}, so that these values were certainly at Kummer's disposal in 1847.

Thus the sums in question are

$$19^n + 20^n + \cdots + 36^n = \int_{19}^{37} B_n(t)\, dt = \frac{B_{n+1}(37) - B_{n+1}(19)}{n+1}$$

where, of course, $B_{n+1}(37)$ and $B_{n+1}(19)$ denote values of the Bernoulli polynomials and not the Bernoulli number B_{n+1} times 37 and 19. Since $n + 1 = 2, 4, \ldots 36$ is always invertible mod 37, the question of whether this sum is divisible by 37 is the same as the question of whether $B_{n+1}(37) - B_{n+1}(19) \equiv 0 \bmod 37$. One would like to say that since $37 \equiv 0 \bmod 37$ all terms of $B_{n+1}(37)$ other than the constant term can be disregarded. However, the polynomial $B_{n+1}(x)$ has *rational* coefficients, not integer coefficients, and its use in a congruence requires special attention.

The fact is that $B_{n+1}(37) - B_{n+1}(19)$ is an integer, even though the several terms of these polynomials are not integers and even though, for that matter, $B_{n+1}(37)$ and $B_{n+1}(19)$ are not integers (see Exercise 7). The evaluation of this integer calls for *divisions* as well as additions, subtractions, and multiplications, namely, divisions by the denominators of the Bernoulli numbers which occur as coefficients in $B_{n+1}(x)$, the point being that in the end one will arrive at a rational number whose denominator is 1. However, as is clear from the above formula for the Bernoulli numbers

---

*The description of the Bernoulli numbers in terms of the expansion of $x/(e^x - 1)$ that is given in the theorem on p. 182 follows easily from this formula.

and from the fact that $n < 35$, the computation never calls for division by a multiple of 37 except in the case of B_{36}—which involves a division by 37—but B_{36} occurs only as the constant term in $B_{36}(x)$ and drops out of $B_{36}(37) - B_{36}(19)$. Therefore, since division by any integer not a multiple of 37 *is* possible mod 37, the entire computation can be carried out mod 37 to determine not the integer $B_{n+1}(37) - B_{n+1}(19)$ itself but merely to determine it mod 37, which is all that is required.

[This is exactly analogous to the arguments above, in which the integer $\psi(\beta)\psi(\beta^3)\ldots\psi(\beta^{35})$ was computed mod 37 by using 2 instead of β because 2 is a primitive 36th root of unity mod 37. Similarly, $B_{n+1}(19) - B_{n+1}$ will be evaluated below by setting $19 \equiv \frac{1}{2}$ mod 37. Although this seems quite bizarre at first, it is a very natural thing to do in view of the *meaning* of the number $\frac{1}{2}$—it is the number x for which $2x = 1$, and mod 37 this number is 19.]

Thus the Bernoulli numbers can be found mod 37 (up to but not including B_{36}) and the congruence $B_{n+1}(37) - B_{n+1}(19) \equiv 0$ mod 37 becomes perfectly meaningful. Because, of course, $37 \equiv 0$ mod 37, one has immediately that all terms other than the constant term of $B_{n+1}(37)$ are zero and the problem is to determine whether $B_{n+1} - B_{n+1}(19)$ is zero mod 37. Kummer achieved this by the ingenious device of noting that $19 \equiv \frac{1}{2}$ mod 37 and by using a special formula for $B_n(\frac{1}{2})$.

The needed identity can be derived as follows.* From the defining equation

$$2^m x^m = \int_{2x}^{2x+1} B_m(t)\, dt = 2\int_x^{x+\frac{1}{2}} B_m(2u)\, du;$$

on the other hand

$$2^m x^m = 2^m \int_x^{x+1} B_m(t)\, dt = 2^m \int_x^{x+\frac{1}{2}} \left[B_m(t) + B_m\left(t + \frac{1}{2}\right) \right] dt.$$

It follows that

$$2^m \left[B_m(x) + B_m\left(x + \frac{1}{2}\right) \right] = 2 B_m(2x)$$

(because these two polynomials have the same integrals over all intervals of length $\frac{1}{2}$). Thus, with $x = 0$ and $m = n + 1$,

$$B_{n+1} + B_{n+1}\left(\tfrac{1}{2}\right) = 2^{-n} B_{n+1}$$

$$B_{n+1}\left(\tfrac{1}{2}\right) - B_{n+1} = (2^{-n} - 2) B_{n+1} = \left(\tfrac{1}{2}\right)^n (1 - 2^{n+1}) B_{n+1}.$$

---

*This derivation differs from Kummer's. Kummer believed this identity to be original with him ([K14], p. 478) and H. J. S. Smith ([S3], p. 117) echoes this opinion. It is surprising that this basic formula in difference calculus (see Norlund [N3]) was discovered in the course of research on Fermat's Last Theorem.

This equation of rational numbers can be regarded as saying that the additions, subtractions, multiplications, and divisions called for by $B_{n+1}(\frac{1}{2}) - B_{n+1}$ produce the same result as those called for by $(\frac{1}{2})^n(1 - 2^{n+1})B_{n+1}$. (Both can be simplified to the form $\pm p/q$, that is, simplified until they call for one multiplication by an integer and one division by a positive integer.) The same is true of these same operations viewed as operations on integers mod 37, provided that they never call for division by a multiple of 37. Since B_n never calls for division by 37 when $n < 36$, and since division by 2 is the same as multiplication by 19 mod 37, this proves that

$$B_{n+1}(19) - B_{n+1} \equiv 19^n(1 - 2^{n+1})B_{n+1} \bmod 37$$

for $n = 1, 3, 5, \ldots, 33$, provided that B_{n+1} is interpreted as the integer mod 37 obtained by dividing the numerator of B_{n+1} by the denominator of B_{n+1}, treating both as integers mod 37. Since the right side cannot be zero mod 37 unless $2^{n+1} - 1$ or B_{n+1} is zero, and since $2^{n+1} \not\equiv 1 \bmod 37$ for $n = 1, 3, \ldots, 33$, this shows that *except in the case $n = 35$, the congruence $19^n + 20^n + \cdots + 36^n \equiv 0 \bmod 37$ is equivalent to the statement that the numerator of B_{n+1} is divisible by* 37. Therefore the problem of determining whether $\psi(2^n)$ is divisible by 37 for $n = 1, 3, \ldots, 33$ can be resolved merely by consulting the table of values of the Bernoulli numbers and seeing whether the numerators of $B_2, B_4, \ldots, B_{34}$ are divisible by 37. The fact is that the numerator of B_{32}, which was given in Ohm's tabulation as 7709321041217, *is* divisible by 37. (Division by 37 is greatly facilitated by the observation that 37 divides 111. Thus the fact that the numerator of B_{32} is $111 \cdot 69453342713 + 74$ shows that it is divisible by 37. This also makes it possible to decide whether $\psi(2^n) \equiv 0 \bmod 37$ for the other values of $n = 1, 3, 5, \ldots, 33$. The answer is that *only* $\psi(2^{31})$ is divisible by 37). Thus 37^{18} divides P, 37 divides $h = \pm P \cdot 37^{-17} \cdot 2^{-17} \cdot h_2$, and $\lambda = 37$ does not satisfy Kummer's condition (A), that is, 37 *is an irregular prime*. This completes the proof of this section.

EXERCISES

1. Prove that if β is a primitive nth root of unity and if $g(\beta) = a_0 + a_1\beta + \cdots + a_{n-1}\beta^{n-1}$ is a polynomial in β with integer coefficients with the property that $g(\beta') = g(\beta)$ for all primitive nth roots of unity β' then $g(\beta)$ is an integer.

2. Prove that $2^{\mu-1}$ divides P. (See [B2], Section 5.5.3).

3. Show that if $p(x) = a_N x^N + a_{N-1} x^{N-1} + \cdots + a_0$ then the equation $\int_x^{x+1} p(t)\, dt = x^n$ uniquely determines rational numbers $a_N, a_{N-1}, \ldots, a_0$ provided $N \geqslant n$.

4. Prove by mathematical induction that $B_n(x) = \sum_{k=0}^{n} \binom{n}{k} B_k x^{n-k}$.

5. Prove that $B_n(1 - t) = (-1)^n B_n(t)$. [Work directly from the defining equation of $B_n(x)$.] Conclude that $B_{2n+1} = 0$ for $n > 0$.

6. Find the pattern of the signs of $B_n(t)$ for $0 \leqslant t \leqslant 1$ and, in particular, prove that the signs of B_{2n} and B_{2n+2} are opposite for $n > 0$.

7. Prove that if B_n is not an integer then $B_n(x)$ is never an integer when x is an integer.

8. Compute $\psi(2^n)$ mod 37 directly from the definition for $n = 1, 2, \ldots, 35$.

9. Compute $B_0, B_1, \ldots, B_{10}$ mod 13 by first constructing Pascal's triangle mod 13. Evaluate $B_n(\frac{1}{2})$ mod 13 two ways, first by using $\frac{1}{2} \equiv 7$ mod 13 and second by using the identity for $B_n(\frac{1}{2})$.

10. Prove that $\psi(2^{35})$ is not divisible by 37. [One method for doing this is direct computation of $\psi(2^{35})$ mod 37. Another is to apply the method used in the general case in the next section. A third is the following: $B_{36}(\frac{1}{2}) - B_{36} = -(\frac{1}{2})^{36}(2^{18} - 1)(2^{18} + 1)B_{36}$. Thus in order to show that $B_{36}(19) - B_{36}$ is not divisible by 37 it will suffice to show that $37B_{36}$ is a rational number in which neither numerator nor denominator is divisible by 37. Kummer calls this a "known property" of Bernoulli numbers. (It is a special case of *von Staudt's theorem*.) Since $37B_{36} = -[1 + 37B_1 + \binom{37}{2}B_2 + \cdots + \binom{37}{2}B_{35}]$ it is clear that $37B_{36}$ involves no division by 37. Divide $37(1^{36} + 2^{36} + 3^{36} + \cdots + 36^{36}) = B_{37}(37) - B_{37}$ by 37 and consider it as a congruence mod 37 to find $-1 \equiv 37B_{36}$.]

6.16 Divisibility of the first factor by λ

Only slight modifications of the arguments of the preceding section are needed in order to answer the same question "Is $P/\lambda^{\mu-1}$ divisible by λ?" in the case of any prime λ modulo which 2 is a primitive root. The result is that $P/\lambda^{\mu-1}$ is divisible by λ if and only if the numerator of one of the Bernoulli numbers $B_2, B_4, \ldots, B_{\lambda-3}$ is divisible by λ. If 2 is not a primitive root mod λ, then a less immediate modification of the argument is necessary, but the final theorem is the same.

Consider, for example, the case $\lambda = 31$. Then 2 is not a primitive root because $2^5 \equiv 1$ mod 31. However, 3 is a primitive root, as was noted in Section 4.7. With $\gamma = 3$ one finds $\phi(x) = 1 + 3x + 9x^2 + 27x^3 + 19x^4 + \cdots + 21x^{29}$. Therefore $(3x - 1)\phi(x) = 0 \cdot x + 0 \cdot x^2 + 0 \cdot x^3 + 62 \cdot x^4 + \cdots + 63x^{30} - 1 = 31(b_1x + b_2x^2 + \cdots + b_{30}x^{30}) + (x^{30} - 1)$ where b_j is defined by $31b_j = 3\gamma_{j-1} - \gamma_j$, $\gamma_j \equiv 3^j$ mod 31, $0 < \gamma_j < 31$. Define $\psi(x)$ by this equation $(3x - 1)\phi(x) = 31\psi(x) + (x^{30} - 1)$, let β be a primitive 30th root of unity, and multiply the 15 equations found by setting $x = \beta^j$ where $j = 1, 3, \ldots, 29$ to find, by the definition of P,

$$(3^{15} + 1)P = 31^{15}\psi(\beta)\psi(\beta^3) \cdots \psi(\beta^{29}).$$

Now $3^{15} + 1$ is divisible by 31 because $(3^{15})^2 \equiv 1$ mod 31 (Fermat's theorem) but $3^{15} \not\equiv 1$ mod 31 (3 is a primitive root). However, $3^{15} + 1$ is not divisible by 31^2 because, as a brief computation shows, $3^{15} + 1 \equiv 7 \cdot 31$ mod 31^2. Thus the equation just found shows that P is divisible by 31^{14} and that the quotient is divisible by 31 if and only if $\psi(\beta)\psi(\beta^3) \cdots \psi(\beta^{29}) \equiv 0 \mod 31$. Since $\psi(\beta)\psi(\beta^3) \cdots \psi(\beta^{29}) \equiv \psi(3)\psi(3^3) \cdots \psi(3^{29}) \mod 29$ (because mod 29 the integer 3 satisfies all the relations satisfied by β) this shows that to determine whether $P/31^{14} \equiv 0 \mod 31$ is equivalent to determining whether any of the integers $\psi(3), \psi(3^3), \ldots, \psi(3^{29})$ are zero mod 31.

Now $\psi(3^n) = b_1 3^n + b_2 3^{2n} + \cdots + b_{30} 3^{30n} \equiv b_1 \gamma_1^n + b_2 \gamma_2^n + \cdots + b_{30} \gamma_{30}^n \equiv b_1 (3\gamma_0)^n + \cdots + b_{30}(3\gamma_{29})^n \equiv \gamma_n[b_1\gamma_0^n + b_2\gamma_1^n + \cdots + b_{30}\gamma_{29}^n]$ where b_j is defined by $31 b_j = 3\gamma_{j-1} - \gamma_j$. Since $-31 < -\gamma_j < 31$, $b_j < 3\gamma_{j-1} < 3 \cdot 31$, one has $-1 < b_j < 3$, that is, $b_j = 0, 1,$ or 2. Moreover, if $b_j = 0$ then $3\gamma_{j-1} = \gamma_j$ lies between 0 and 31, if $b_j = 1$ then $3\gamma_{j-1} = \gamma_j + 31$ lies between 31 and $2 \cdot 31$, and if $b_j = 2$ then $3\gamma_{j-1} = \gamma_j + 2 \cdot 31$ lies between $2 \cdot 31$ and $3 \cdot 31$. In other words, $b_j = 0$ for $\gamma_{j-1} = 1, 2, \ldots, 10$; $b_j = 1$ for $\gamma_{j-1} = 11, 12, \ldots, 20$; and $b_j = 2$ for $\gamma_{j-1} = 21, 22, \ldots, 30$. Since each integer from 1 to 30 occurs exactly once as a γ_j, this shows that

$$\psi(3^n) \equiv \gamma_n[11^n + 12^n + \cdots + 20^n + 2 \cdot 21^n + 2 \cdot 22^n + \cdots + 2 \cdot 30^n]\mod 31.$$

Since $\gamma_n \not\equiv 0 \mod 31$, the question of the divisibility of this integer by 31 is the same as the question of the divisibility of the integer

$$(11^n + 12^n + \cdots + 30^n) + (21^n + 22^n + \cdots + 30^n)$$

$$= \int_{11}^{31} B_n(t)\, dt + \int_{21}^{31} B_n(t)\, dt$$

$$= \frac{B_{n+1}(31) - B_{n+1}(11) + B_{n+1}(31) - B_{n+1}(21)}{n+1}$$

by 31. Except in the case $n = 29$, the Bernoulli numbers which occur in this equation can all be regarded as integers mod 31—that is, they do not call for division by any multiple of 31—and, because $B_{n+1}(31) \equiv B_{n+1}(0) = B_{n+1} \mod 31$, it follows that $\psi(3^n) \equiv 0 \mod 31$ if and only if $B_{n+1}(11) + B_{n+1}(21) \equiv 2B_{n+1}$ (for $n = 1, 3, \ldots, 27$, but not for $n = 29$).

Now $3 \cdot 11 \equiv 2 \mod 31$ and $3 \cdot 21 \equiv 1 \mod 31$, so that the congruence in question can be rewritten as $B_{n+1}(\frac{2}{3}) + B_{n+1}(\frac{1}{3}) \equiv 2B_{n+1}$. The method that was used to derive the identity for $B_n(\frac{1}{2})$ in the preceding section generalizes immediately to give

$$3^m x^m = \int_{3x}^{3x+1} B_m(t)\, dt = 3\int_x^{x+\frac{1}{3}} B_m(3u)\, du$$

$$3^m x^m = 3^m \int_x^{x+1} B_m(t)\, dt$$

$$= 3^m \int_x^{x+\frac{1}{3}} \left[B_m(t) + B_m\left(t + \frac{1}{3}\right) + B_m\left(t + \frac{2}{3}\right) \right] dt$$

$$3^{1-m} B_m(3x) = B_m(x) + B_m\left(x + \frac{1}{3}\right) + B_m\left(x + \frac{2}{3}\right)$$

$$B_{n+1} + B_{n+1}\left(\frac{1}{3}\right) + B_{n+1}\left(\frac{2}{3}\right) = 3^{-n} B_{n+1}$$

235

and the congruence in question is

$$3^{-n}B_{n+1} \equiv 3B_{n+1}$$

$$\left(\tfrac{1}{3}\right)^{n}(3^{n+1} - 1)B_{n+1} \equiv 0 \bmod 31$$

which for $n = 1, 3, \ldots, 27$ is obviously satisfied if and only if the numerator of B_{n+1} is divisible by 31. A table of Bernoulli numbers can be consulted and it can be verified that this is not true in any of the cases $n = 1, 3, 5, \ldots, 27$, that is, $\psi(3^{n}) \not\equiv 0$ for $n = 1, 3, \ldots, 27$.

The case $n = 29$ remains. In this case one can prove directly that $\psi(3^{29}) \not\equiv 0 \bmod 31$ as follows. Let $3^{30} - 1 = (3^{15} - 1)(3^{15} + 1) = 31\nu$ where, as was noted above, ν is not divisible by 31. Then the defining equation of ψ, namely, $(3x - 1)\phi(x) = 31\psi(x) + (x^{30} - 1)$ with $x = 3^{29}$ gives $(3^{30} - 1) \cdot \phi(3^{29}) = 31\psi(3^{29}) + (3^{30})^{29} - 1$, $\nu\phi(3^{29}) = \psi(3^{29}) + [(1 + 31\nu)^{29} - 1] \cdot 31^{-1}$. By the definition of ϕ, one sees immediately that $\phi(3^{29}) \equiv 1 + 3^{30} + 3^{60} + \cdots \equiv 1 + 1 + \cdots + 1 = 30 \equiv -1 \bmod 31$. Therefore $\psi(3^{29}) = \nu\phi(3^{29}) - 29\nu - \binom{29}{2}\nu^2 31 - \cdots \equiv -\nu - 29\nu \equiv \nu \not\equiv 0 \bmod 31$, as was to be shown.

A similar technique can be used to determine the divisibility of P by powers of λ in the general case. Let λ be an odd prime, and let γ be a primitive root mod λ. Let $\phi(x) = 1 + \gamma_1 x + \gamma_2 x^2 + \cdots + \gamma_{\lambda-2}x^{\lambda-2}$ where γ_j is defined by $\gamma_j \equiv \gamma^j \bmod \lambda$, $0 < \gamma_j < \lambda$. Then, by definition, $P = \phi(\beta)\phi(\beta^3) \ldots \phi(\beta^{\lambda-2})$ where β is a primitive $(\lambda - 1)$st root of unity. Define $\psi(x)$ by $(\gamma x - 1)\phi(x) = \lambda\psi(x) + (x^{\lambda-1} - 1)$ where $\psi(x) = b_1 x + b_2 x^2 + \cdots + b_{\lambda-1}x^{\lambda-1}$, $\lambda b_j = \gamma\gamma_{j-1} - \gamma_j$. Then the product of the equations $(\gamma\beta^j - 1)\phi(\beta^j) = \lambda\psi(\beta^j)$ for $j = 1, 3, \ldots, \lambda - 2$ gives

$$(\gamma^{\mu} + 1)P = \lambda^{\mu}\,\psi(\beta)\psi(\beta^3) \ldots \psi(\beta^{\lambda-2}).$$

Assume for the moment that γ can be chosen in such a way that $\gamma^{\mu} + 1$—which is necessarily divisible by λ—is not divisible by λ^2. Then this equation shows that P is divisible by $\lambda^{\mu-1}$ and that the quotient $P/\lambda^{\mu-1}$ is divisible by λ if and only if the integer $\psi(\beta)\psi(\beta^3) \ldots \psi(\beta^{\lambda-2})$ is divisible by λ. This integer coincides with the integer $\psi(\gamma)\psi(\gamma^3) \ldots \psi(\gamma^{\lambda-2}) \bmod \lambda$ (because the integer γ satisfies mod λ all the relations satisfied by the complex number β) and this integer is zero mod λ if and only if one of the factors $\psi(\gamma)$, $\psi(\gamma^3)$, $\ldots$, $\psi(\gamma^{\lambda-2})$ is zero mod λ. Therefore the quotient $P/\lambda^{\mu-1}$ is divisible by λ if and only if $\psi(\gamma)$ or $\psi(\gamma^3)$ or $\ldots$ or $\psi(\gamma^{\lambda-2})$ is divisible by λ (still assuming $\gamma^{\mu} + 1 \not\equiv 0 \bmod \lambda^2$).

Since $\lambda b_j = \gamma\gamma_{j-1} - \gamma_j$ lies between $-\lambda$ and $\gamma\lambda$, b_j lies between -1 and γ, that is, $b_j = 0, 1, \ldots, \gamma - 1$. Moreover, $b_j = k$ if and only if $\gamma\gamma_{j-1} = \lambda k + \gamma_j$ lies between $k\lambda$ and $k\lambda + \lambda$, that is, γ_{j-1} lies between $k(\lambda/\gamma)$ and $k(\lambda/\gamma) + (\lambda/\gamma)$. Since

$$\psi(\gamma^{n}) \equiv b_1\gamma^{n} + b_2\gamma_2^{n} + \cdots + b_{\lambda-1}\gamma_{\lambda-1}^{n}$$

$$\equiv \gamma_n[b_1\gamma_0^{n} + b_2\gamma_1^{n} + \cdots + b_{\lambda-1}\gamma_{\lambda-2}^{n}]$$

and since every integer between 0 and λ is γ_j for exactly one value of j, it follows that $\psi(\gamma^n)$ is congruent mod λ to the integer γ_n times

$$0 \cdot 1^n + \cdots + 0 \cdot (t_1 - 1)^n + 1 \cdot t_1^n + \cdots + 1 \cdot (t_2 - 1)^n + 2 \cdot t_2^n$$

$$+ \cdots + (\gamma - 2)(t_{\gamma-1} - 1)^n + (\gamma - 1)t_{\gamma-1}^n + \cdots + (\gamma - 1) \cdot (\lambda - 1)^n$$

$$= t_1^n + (t_1 + 1)^n + \cdots + (\lambda - 1)^n + t_2^n + (t_2 + 1)^n + \cdots + (\lambda - 1)^n$$

$$+ \cdots + t_{\gamma-1}^n + (t_{\gamma-1} + 1)^n + \cdots + (\lambda - 1)^n$$

where t_k denotes the least integer greater than $k(\lambda/\gamma)$. Therefore

$$\psi(\gamma^n) \equiv \frac{\gamma_n}{n+1} \Big[B_{n+1}(\lambda) - B_{n+1}(t_1) + B_{n+1}(\lambda) - B_{n+1}(t_2)$$

$$+ \cdots + B_{n+1}(\lambda) - B_{n+1}(t_{\gamma-1}) \Big].$$

Except in the case $n = \lambda - 2$, the Bernoulli numbers in this congruence never call for division by multiples of λ and the individual terms can be regarded as integers mod λ. The integers $t_1, t_2, \ldots, t_{\gamma-1}$ are congruent to $(1/\gamma), (2/\gamma), \ldots, ((\gamma - 1)/\gamma)$ mod λ (possibly in a different order) for the following reasons. For every integer r in the range $0 < r < \gamma$ there is an integer t in the range $0 < t < \lambda$ such that $t\gamma \equiv r$ mod λ (γ is invertible mod λ). Thus $t\gamma - r = k\lambda$ for some k greater than 0 and less than γ (because $t\gamma - r > 0$ and $t\gamma - r < t\gamma < \lambda\gamma$). Therefore $t = k(\lambda/\gamma) + (r/\gamma)$ is the least integer greater than $k(\lambda/\gamma)$, $t = t_k$, and $t_k \equiv r/\gamma$ mod λ. Now

$$B_{n+1} + B_{n+1}\left(\frac{1}{\gamma}\right) + B_{n+1}\left(\frac{2}{\gamma}\right) + \cdots + B_{n+1}\left(\frac{\gamma - 1}{\gamma}\right) = \gamma^{-n}B_{n+1}.$$

Therefore, except in the case $n = \lambda - 2$,

$$\psi(\gamma^n) \equiv \frac{\gamma_n}{n+1} \left[(\gamma - 1)B_{n+1} - B_{n+1}\left(\frac{1}{\gamma}\right) - B_{n+1}\left(\frac{2}{\gamma}\right) \right.$$

$$\left. - \cdots - B_{n+1}\left(\frac{\gamma - 1}{\gamma}\right) \right]$$

$$\equiv \frac{\gamma_n}{n+1} \left[\gamma B_{n+1} - \gamma^{-n}B_{n+1} \right] \equiv \frac{1}{n+1}\left(\gamma^{n+1} - 1\right)B_{n+1} \bmod \lambda.$$

Since $\gamma^{n+1} - 1 \not\equiv 0$ mod λ (again excepting $n = \lambda - 2$) this shows that $\psi(\gamma^n) \equiv 0$ mod λ if and only if $B_{n+1} \equiv 0$ mod λ, that is, if and only if the numerator of the Bernoulli number B_{n+1} is divisible by λ.

In the case $n = \lambda - 2$, $\psi(\gamma^n)$ is not zero mod λ in any case. This can be proved exactly as in the case $\gamma = 3$, $\lambda = 31$. With ν defined by $\gamma^{\lambda-1} - 1 = \lambda\nu$ one finds, by setting $x = \gamma^{\lambda-2}$ in $(\gamma x - 1)\phi(x) = \lambda\psi(x) + x^{\lambda-1} - 1$, dividing by λ, and reducing mod λ, that $\nu\phi(\gamma^{\lambda-2}) \equiv \psi(\gamma^{\lambda-2}) + (\lambda - 2)\nu$ mod λ. Since $\phi(\gamma^{\lambda-2}) \equiv -1$ mod λ it follows that $\psi(\gamma^{\lambda-2}) \equiv \nu$ mod λ.

Since $\gamma^{\lambda-1} - 1 \equiv (\gamma^\mu - 1)(\gamma^\mu + 1)$ is, by the assumption on γ, not divisible by λ^2, $\nu \not\equiv 0 \bmod \lambda$ and $\psi(\gamma^{\lambda-2}) \not\equiv 0 \bmod \lambda$, as was to be shown.

Thus, provided there is a primitive root $\gamma \bmod \lambda$ for which $\gamma^\mu + 1 \not\equiv 0 \bmod \lambda^2$, the proof of the theorem is complete.

Theorem. *P is divisible by $\lambda^{\mu-1}$. The quotient is divisible by λ if and only if the numerator of one of the Bernoulli numbers $B_2, B_4, \ldots, B_{\lambda-3}$ is divisible by λ.*

Consider finally the problem of finding a primitive root $\gamma \bmod \lambda$ for which $\gamma^\mu + 1 \not\equiv 0 \bmod \lambda^2$. Let γ be any primitive root in the range $0 < \gamma < \lambda$. If $\gamma^\mu + 1 \equiv 0 \bmod \lambda^2$ then $(\gamma + \lambda)^\mu + 1 \equiv \gamma^\mu + \mu\gamma^{\mu-1}\lambda + 1 \equiv \mu\gamma^{\mu-1}\lambda \not\equiv 0 \bmod \lambda^2$ and $\gamma + \lambda$ is a primitive root of the type needed for the proof. In this proof it was tacitly assumed that γ lay in the range $0 < \gamma < \lambda$ natural for integers mod λ, but a careful analysis of the proof will show that it requires only the assumption that $\gamma > 0$ *provided** the i's are counted with the appropriate multiplicities. Alternatively, one can retain the assumption $0 < \gamma < \lambda$ and use the equation $[(\gamma + \lambda)x - 1]\phi(x) = \lambda\psi(x) + \lambda x\phi(x) + x^{\lambda-1} - 1$ with $x = \beta, \beta^3, \ldots, \beta^{\lambda-2}$ to find

$$\left[(\gamma + \lambda)^\mu + 1\right]P = \lambda^\mu \prod_n \left[\psi(\beta^n) + \beta^n \phi(\beta^n)\right]$$

where, in the product on the right, n ranges over $1, 3, \ldots, \lambda - 2$. In the case $\gamma^\mu + 1 \equiv 0 \bmod \lambda^2$ under consideration, this shows that P is divisible by $\lambda^{\mu-1}$ and that the quotient is divisible by λ if and only if one of the integers $\psi(\gamma^n) + \gamma^n\phi(\gamma^n)$ is $0 \bmod \lambda$. Except in the case $n = \lambda - 2$ the equation $(\gamma^{n+1} - 1)\phi(\gamma^n) = \lambda\psi(\gamma^n) + \gamma^{(\lambda-1)n} - 1$ shows that $\phi(\gamma^n) \equiv 0 \bmod \lambda$ and $\psi(\gamma^n) + \gamma^n\phi(\gamma^n) \equiv \psi(\gamma^n) \equiv (n + 1)^{-1}(\gamma^{n+1} - 1)B_{n+1} \bmod \lambda$; therefore the criterion for divisibility is, as before, the divisibility of the numerator of B_{n+1}. In the case $n = \lambda - 2$, the previous argument gives $\psi(\gamma^{\lambda-2}) \equiv \nu$ mod λ where ν is defined by $\gamma^{\lambda-1} - 1 = \lambda\nu$ and where, consequently, $\nu \equiv 0$ mod λ. Therefore $\psi(\gamma^{\lambda-2}) + \gamma^{\lambda-2}\phi(\gamma^{\lambda-2}) \equiv \gamma^{\lambda-2}\phi(\gamma^{\lambda-2}) \equiv \gamma^{-1} \cdot (-1) \not\equiv 0$ mod λ and the theorem is proved.

EXERCISES

1. Prove that $\psi(\gamma^n) \equiv 0 \bmod \lambda$ for $n = 2, 4, \ldots, \lambda - 3$.

2. Evaluate $\psi(\gamma^n) \bmod \lambda$ in all cases with $\lambda \leqslant 13$ (λ prime) and $0 < n < \lambda$.

6.17 Divisibility of the second factor by λ

Kummer's condition (A) is the condition that h not be divisible by λ. The equation $2^{\mu-1}h = (P/\lambda^{\mu-1}) \cdot h_2$ shows that this is true if and only if $P/\lambda^{\mu-1}$ is not divisible by λ *and* h_2 is not divisible by λ. Simple criteria for

---

*Kummer appears to have been a bit lax on this point.

the divisibility of $P/\lambda^{\mu-1}$ by λ were found in the preceding section. In order to be able to test condition (A) one must also establish criteria for the divisibility of h_2 by λ. Kummer remarked that it was fortunate that this can be done without a knowledge of h_2 itself because the computation of h_2 in any given case is extremely difficult. What he showed was that $\lambda|h_2$ implies $\lambda|(P/\lambda^{\mu-1})$. Therefore if λ does not divide $P/\lambda^{\mu-1}$ it does not divide h_2 either and condition (A) is satisfied.

Theorem. *A prime $\lambda > 2$ satisfies Kummer's condition* (A) *if and only if it does not divide the numerators of* $B_2, B_4, \ldots, B_{\lambda-3}$.

What must be shown in order to prove this theorem is that $\lambda|h_2$ implies the Bernoulli number criterion or, what is the same, that it implies $\lambda|(P/\lambda^{\mu-1})$.

If $\lambda|h_2$ then there is a unit not of the form $\pm\alpha^k(1-\sigma\alpha)^{x_1}(1-\sigma^2\alpha)^{x_2}\cdots(1-\sigma^\mu\alpha)^{x_\mu}$ whose λth power is of this form. In terms of group theory, this is a consequence of the fact that an Abelian group whose order is divisible by λ must contain an element of order λ (the group being the group of all units modulo those of the special form $\pm\alpha^k\|(1-\sigma^k\alpha)^{x_k})$. Alternatively, this can be proved by examining the determinant of the change of coordinates involved in transforming the lattice of images under Φ of the units of the special form into the lattice of images under Φ of all units (see Exercise 3).

Let such a unit be chosen, say $e_0(\alpha)$, and let $e(\alpha)$ denote its λth power. Then $e(\alpha)$ can be written in the special form $e(\alpha) = \pm\alpha^k(1-\sigma\alpha)^{x_1}(1-\sigma^2\alpha)^{x_2}\cdots(1-\sigma^\mu\alpha)^{x_\mu}$ where $x_1 + x_2 + \cdots + x_\mu = 0$. The x's are not all divisible by λ because if they were then $e(\alpha)$ would have the form $e(\alpha) = \pm\alpha^k e_1(\alpha)^\lambda$ where $e_1(\alpha)$ has the special form $e_1(\alpha) = (1-\sigma\alpha)^{y_1}(1-\sigma^2\alpha)^{y_2}\cdots(1-\sigma^\mu\alpha)^{y_\mu}$; then $e_0(\alpha)e_1(\alpha)^{-1}$ would be a unit whose λth power was $\pm\alpha^k$, from which it would follow (Exercise 2) that $e_0(\alpha)e_1(\alpha)^{-1} = \pm\alpha^j$, contrary to the assumption that $e_0(\alpha)$ does not have the form $\pm\alpha^j(1-\sigma\alpha)^{y_1}(1-\sigma^2\alpha)^{y_2}\cdots(1-\sigma^\mu\alpha)^{y_\mu}$. Since $e_0(\alpha)^\lambda = (a_0 + a_1\alpha + \cdots + a_{\lambda-1}\alpha^{\lambda-1})^\lambda \equiv a_0^\lambda + (a_1\alpha)^\lambda + \cdots + (a_{\lambda-1}\alpha^{\lambda-1})^\lambda \equiv a_0 + a_1 + \cdots + a_{\lambda-1} \equiv c \bmod \lambda$, where c is an integer, this shows that there exist integers $x_1, \ldots, x_\mu, c$ such that $x_1 + x_2 + \cdots + x_\mu = 0$ and

$$\alpha^k(1-\sigma\alpha)^{x_1}(1-\sigma^2\alpha)^{x_2}\cdots(1-\sigma^\mu\alpha)^{x_\mu} \equiv c \bmod \lambda \qquad (1)$$

but such that the x's are not all zero mod λ. What is to be shown is that this implies $\lambda|(P/\lambda^{\mu-1})$.

It would be natural to want to take the logarithm of the relation (1) so that it could become a linear relation in the x's—in fact several linear relations, one for each power of α. Although the notion of taking a logarithm seems remote from congruences among cyclotomic integers, the notion of *logarithmic derivative* is more nearly algebraic and can in fact be applied to the relation (1) as follows. First write the relation (1) as a rational relation of polynomials by changing α to a variable X and writing it as

$$X^k(1-\sigma X)^{x_1}(1-\sigma^2 X)^{x_2}\ldots(1-\sigma^\mu X)^{x_\mu}$$

$$= c + \lambda\Phi(X) + (1 + X + \cdots + X^{\lambda-1})\Psi(X)$$

where $\Phi(X)$, $\Psi(X)$ are polynomials with integer coefficients and where $\sigma^j X$ denotes the monomial X to the power γ^j (γ a fixed primitive root mod λ). The logarithmic derivative is' a formal algebraic operation that can be applied to both sides to give

$$\frac{k}{X} - x_1 \frac{\gamma\sigma X}{(1-\sigma X)X} - x_2 \frac{\gamma^2\sigma^2 X}{(1-\sigma^2 X)X} - \cdots - x_\mu \frac{\gamma^\mu\sigma^\mu X}{(1-\sigma^\mu X)X}$$

$$= \frac{\lambda\Phi'(X) + (1 + X + \cdots + X^{\lambda-1})\Psi'(X) + (1 + 2X + \cdots + (\lambda-1)X^{\lambda-2})\Psi(X)}{c + \lambda\Phi(X) + (1 + X + \cdots + X^{\lambda-1})\Psi(X)}.$$

Now set $X = \alpha$ in this identity and multiply by α to find

$$k - x_1 \frac{\gamma\sigma\alpha}{1-\sigma\alpha} - x_2 \frac{\gamma^2\sigma^2\alpha}{1-\sigma^2\alpha} - \cdots - x_\mu \frac{\gamma^\mu\sigma^\mu\alpha}{1-\sigma^\mu\alpha}$$

$$= \frac{\lambda\alpha\Phi'(\alpha) + \alpha(1 + 2\alpha + \cdots + (\lambda-1)\alpha^{\lambda-2})\Psi(\alpha)}{e(\alpha)}.$$

The right side of this equation is a well-defined cyclotomic integer because $e(\alpha)$ is a unit. Because the original relation (1) was a congruence mod λ and because one can not get more information out of the relation than it contained at the outset, it is logical to reduce the equation mod λ and therefore to disregard the first term of the numerator. The left side of the equation can be written in the form

$$k - \left(x_1\gamma\sigma + x_2\gamma^2\sigma^2 + \cdots + x_\mu\gamma^\mu\sigma^\mu\right)\left(\frac{\alpha}{1-\alpha}\right)$$

where the polynomial $x_1\gamma\sigma + x_2\gamma^2\sigma^2 + \cdots + x_\mu\gamma^\mu\sigma^\mu$ in σ with integer coefficients is regarded in the obvious way as a mapping from rational expressions in α with integer coefficients to other rational expressions in α with integer coefficients. The terms on the left can all be written with the denominator λ so that the left side becomes a cyclotomic integer divided by λ; since the right side is a cyclotomic integer, the cyclotomic integer in the numerator must in fact be divisible by λ.

The expression of the left side as a cyclotomic integer divided by λ can be found* very quickly by differentiating $(X-1)(X^{\lambda-1} + X^{\lambda-2} + \cdots +$

*See Exercise 11, Section 4.4, for an alternative derivation.

$X + 1) = X^\lambda - 1$ and setting $X = \alpha$ to find

$$(\alpha - 1)\big[(\lambda - 1)\alpha^{\lambda-2} + (\lambda - 2)\alpha^{\lambda-3} + \cdots + 2\alpha + 1\big] = \lambda\alpha^{\lambda-1}$$

$$(\lambda - 1)\alpha^{\lambda-1} + (\lambda - 2)\alpha^{\lambda-2} + \cdots + 2\alpha^2 + \alpha = \frac{\lambda}{\alpha - 1}.$$

If the powers of α are arranged in the order $\alpha, \sigma\alpha, \sigma^2\alpha, \ldots$ this becomes

$$\alpha + \gamma\sigma\alpha + \gamma_2\sigma^2\alpha + \cdots + \gamma_{\lambda-2}\sigma^{\lambda-2}\alpha = \frac{\lambda}{\alpha - 1}$$

where, as before, $0 < \gamma_j < \lambda$, $\gamma_j \equiv \gamma^j \bmod \lambda$. Then

$$\frac{\alpha}{\alpha - 1} = \frac{1}{1 - \alpha^{-1}} = -\sigma^\mu\left(\frac{1}{\alpha - 1}\right)$$

$$= -\frac{\sigma^\mu}{\lambda}(\alpha + \gamma\sigma\alpha + \cdots + \gamma_{\lambda-2}\sigma^{\lambda-2}\alpha).$$

The term in parentheses can be rewritten as $\phi(\sigma)(\alpha)$ where, as before, $\phi(X) = 1 + \gamma X + \gamma_2 X + \cdots + \gamma_{\lambda-2}X^{\lambda-2}$, and where $\phi(\sigma)$ is a mapping from cyclotomic integers to cyclotomic integers. Thus the relation takes the form

$$k + \left(x_1\gamma\sigma + x_2\gamma^2\sigma^2 + \cdots + x_\mu\gamma^\mu\sigma^\mu\right)\frac{\sigma^\mu}{\lambda}\phi(\sigma)(\alpha)$$

$$\equiv e(\alpha)^{-1}\Psi(\alpha)\big[\lambda/(\alpha - 1)\big] \bmod \lambda.$$

Division of polynomials can be used to put the cyclotomic integer $e(\alpha)^{-1}\Psi(\alpha)$ in the form $q(\alpha)(\alpha - 1) + a$ for some integer a. The right side then becomes $\lambda q(\alpha) + a[\lambda/(\alpha - 1)] \equiv a[\lambda/(\alpha - 1)] \equiv a\phi(\sigma)(\alpha) \bmod \lambda$. Thus

$$k + \frac{\sigma^\mu}{\lambda}\left(x_1\gamma\sigma + x_2\gamma^2\sigma^2 + \cdots + x_\mu\gamma^\mu\sigma^\mu\right)\phi(\sigma)(\alpha) \equiv a\phi(\sigma)(\alpha) \bmod \lambda.$$

To cancel the denominator of λ from the left side one can apply the operator $\gamma\sigma - 1$ to both sides and use the defining relation $(\gamma X - 1)\phi(X) = \lambda\psi(X) + (X^{\lambda-1} - 1)$ of $\psi(X)$ to find

$$\gamma k - k + \sigma^\mu\left(x_1\gamma\sigma + x_2\gamma^2\sigma^2 + \cdots + x_\mu\gamma^\mu\sigma^\mu\right)\psi(\sigma)(\alpha)$$

$$\equiv a\lambda\psi(\sigma)(\alpha) \equiv 0 \bmod \lambda$$

because $\sigma^{\lambda-1} - 1$ carries everything to 0. This shows that the operator

$$\left(x_1\gamma\sigma + x_2\gamma^2\sigma^2 + \cdots + x_\mu\gamma^\mu\sigma^\mu\right)\psi(\sigma) \qquad (2)$$

applied to α gives an integer mod λ, say the integer K. Then the same operator applied to $\sigma^j\alpha$ gives $\sigma^j K = K \bmod \lambda$ for all j. It is to be shown that if this is true and if the x's are not all zero mod λ then $\lambda | (P/\lambda^{\mu-1})$.

The method for doing this is finite dimensional Fourier analysis (see Section 6.9) mod λ. For this, note that there are $\lambda^{\lambda-1}$ distinct cyclotomic integers mod λ and that each of them can be written in just one way as a linear combination (with integer coefficients mod λ) of the $\lambda - 1$ cyclotomic integers

$$\alpha + \gamma^j \sigma \alpha + \gamma^{2j} \sigma^2 \alpha + \cdots + \gamma^{-j} \sigma^{-1} \alpha$$

mod λ ($j = 0, 1, \ldots, \lambda - 2$). This is proved exactly as before using Fourier inversion $(1 + \gamma^{j-k} + \gamma^{2(j-k)} + \cdots + \gamma^{-(j-k)} \equiv 0 \mod \lambda$ unless $j \equiv k$ mod $\lambda - 1$, but if $j \equiv k \mod \lambda - 1$ then it is $\equiv (\lambda - 1) \not\equiv 0 \mod \lambda$).

Application of the operator σ to the cyclotomic integer $\alpha + \gamma^j \sigma \alpha + \cdots + \gamma^{-j} \sigma^{-1} \alpha$ multiplies it by $\gamma^{-j} \mod \lambda$. Therefore application of the operator (2) to this cyclotomic integer multiplies it by

$$\left(x_1 \gamma^{1-j} + x_2 \gamma^{2(1-j)} + \cdots + x_\mu \gamma^{\mu(1-j)} \right) \psi(\gamma^{-j}). \tag{3}$$

On the other hand, application of the operator (2) to $\alpha + \gamma^j \sigma \alpha + \cdots + \gamma^{-j} \sigma^{-1} \alpha$ carries it to $K + \gamma^j K + \cdots + \gamma^{-j} K = K(1 + \gamma^j + \cdots + \gamma^{-j}) \equiv 0 \mod \lambda$ for $j = 1, 2, \ldots, \lambda - 2$. Therefore the integer (3) must be zero mod λ for $j = 1, 2, \ldots, \lambda - 2$.

It was shown in the preceding section that λ divides the numerator B_{n+1} if and only if $\psi(\gamma^n) \equiv 0 \mod \lambda$. Therefore if λ does not divide the numerators of $B_2, B_4, \ldots, B_{\lambda-3}$ the second factor $\psi(\gamma^{-j})$ of (3) is nonzero mod λ for the $\mu - 1$ cases $j = -1, -3, -5, \ldots, -\lambda + 4$. Moreover, as was shown in the preceding section, $\psi(\gamma^{\lambda-2}) \not\equiv 0 \mod \lambda$. This gives μ congruences

$$x_1 \gamma^{j+1} + x_2 \gamma^{2j+2} + \cdots + x_\mu \gamma^{\mu j + \mu} \equiv 0 \mod \lambda$$

for $j = 1, 3, 5, \ldots, \lambda - 2$. These can be solved by Fourier inversion to find $x_i \equiv 0 \mod \lambda$ for $i = 1, 2, \ldots, \mu$. Thus, if $x_i \not\equiv 0 \mod \lambda$ for some i, then λ must divide the numerator of at least one of the Bernoulli numbers $B_2, B_4, \ldots, B_{\lambda-3}$ and the theorem is proved.

EXERCISES

1. Show that if the number of elements of a group is not divisible by λ then every element of the group (written multiplicatively) is a λth power.

2. Show that if $e_2(\alpha)$ is a unit such that $e_2(\alpha)^\lambda = \pm \alpha^k$ then $e_2(\alpha) = \pm \alpha^j$.

3. Kummer's proof that if $\lambda | h_2$ then there is a unit not of the form $\pm \alpha^k (1 - \sigma \alpha)^{x_1} (1 - \sigma^2 \alpha)^{x_2} \ldots (1 - \sigma^\mu \alpha)^{x_\mu}$ whose λth power is of this form is roughly the following. Fill in the details.

   Let Λ_1 denote the lattice of all points of the form $\Phi(e(\alpha))$ in $x_1 x_2 \ldots x_\mu$-space as $e(\alpha)$ ranges over all units, where Φ is as in Section 6.10. Let $\Lambda_2 \subseteq \Lambda_1$ be the sublattice consisting of those points of Λ_1 all of whose coordinates are integers or, what is the same, the images under Φ of units that are equivalent to 1. The points $v_1 = (1, -1, 0, \ldots, 0), v_2 = (0, 1, -1, 0, \ldots, 0), \ldots, v_{\mu-1} = (0, 0, \ldots, 0, 1, -1)$ are a basis of Λ_2. By elementary linear algebra there is a basis $w_1, w_2, \ldots, w_{\mu-1}$ of Λ_1 and each w_i can be written in just one way in the form $w_i = \Sigma c_{ij} v_j$ where the

c_{ij} are rational numbers. Since one in every h_2 units is equivalent to 1, $\det(c_{ij}) = 1/h_2$. Let n_i be the least common denominator of c_{ij} for $j = 1, 2, \ldots, \mu - 1$. Then $\det(c_{ij})$ is an integer divided by the product of the n_i. If this is $1/h_2$ and if $\lambda | h_2$ it follows that $\lambda | n_i$ for at least one i. If $e(\alpha)$ is a unit for which $\Phi e(\alpha) = w_i$ then $e(\alpha)^{n_i}$ is a unit equivalent to 1. On the other hand $\lambda | n_i$ and the unit $e(\alpha)^{n_i/\lambda}$ is not equivalent to 1, since otherwise n_i/λ would be a common denominator of the c_{ij}.

4. Prove that if G is an abelian group whose order is divisible by a prime p then it contains an element of order p.

6.18 Kummer's lemma

Kummer's condition (A) has been shown to be equivalent to the condition that λ does not divide the numerators of the Bernoulli numbers $B_2, B_4, \ldots, B_{\lambda-3}$. In order to prove Fermat's Last Theorem for all primes λ which satisfy this condition it remains only to prove that, as Kummer had guessed, (A) implies (B). This is known as *Kummer's lemma*. The proof is a brief addition to the argument of the preceding section.

In the preceding section it was shown that if $e(\alpha)$ is a unit of the form $\pm \alpha^k (1 - \sigma\alpha)^{x_1}(1 - \sigma^2\alpha)^{x_2} \cdots (1 - \sigma^\mu\alpha)^{x_\mu}$, and if $e(\alpha) \equiv c \bmod \lambda$ where c is an integer, then either condition (A) fails or the integers $x_1, x_2, \ldots, x_\mu$ are all divisible by λ. Now if $e(\alpha)$ is any unit then $e(\alpha)^{h_2}$ is of the form $\pm \alpha^k (1 - \sigma\alpha)^{x_1}(1 - \sigma^2\alpha)^{x_2} \cdots (1 - \sigma^\mu\alpha)^{x_\mu}$. This follows from the argument used to prove Fermat's theorem and several of its analogs in this book (see Exercise 2) because h_2 is by definition the number of distinct units modulo units of the special form. Suppose now that condition (A) holds and that $e(\alpha) \equiv c \bmod \lambda$. Then $e(\alpha)^{h_2} = \pm \alpha^k (1 - \sigma\alpha)^{x_1}(1 - \sigma^2\alpha)^{x_2} \ldots (1 - \sigma^\mu\alpha)^{x_\mu}$ and $e(\alpha)^{h_2} \equiv c^{h_2} \bmod \lambda$, and it follows that $x_1, x_2, \ldots, x_\mu$ are all divisible by λ. Therefore $e(\alpha)^{h_2} = \pm \alpha^k e_0(\alpha)^\lambda$ for some unit $e_0(\alpha)$. Moreover, h_2 is not divisible by λ (by condition (A) and the fact that $2^{\mu-1}h_1$ is an integer) and there exist integers a and b such that $ah_2 + b\lambda = 1$. Thus $e(\alpha) = e(\alpha)^{ah_2}e(\alpha)^{b\lambda} = (\pm \alpha^k)^a e_0(\alpha)^{a\lambda}e(\alpha)^{b\lambda}$. Mod λ this shows that α^{ak} is congruent to an integer, from which it follows that $ak \equiv 0 \bmod \lambda$, $\alpha^{ak} = 1$ (Exercise 1). Thus $e(\alpha) = [e_0(\alpha)^a e(\alpha)^b]^\lambda$ or $e(\alpha) = [-e_0(\alpha)^a e(\alpha)^b]^\lambda$. That is, condition (A) and $e(\alpha) \equiv c \bmod \lambda$ imply that $e(\alpha)$ is a λth power. This proves Kummer's lemma.

EXERCISES

1. Prove that if $\alpha^j \equiv$ integer mod λ then $j \equiv 0 \bmod \lambda$. [Write $\alpha = 1 - (1 - \alpha)$.]

2. Let a group with h elements be given and let b be an element of the group. Show that if d is the number of distinct powers of b in the group then the elements of the group can be partitioned into sets containing d each. Conclude that d divides h and that b^h is the identity of the group.

6.19 Summary

It has now been proved in this chapter that a prime λ satisfies Kummer's conditions (A) and (B)—that is, λ is regular—if and only if λ does not divide the numerator of any of the Bernoulli numbers $B_2, B_4, \ldots, B_{\lambda-3}$. For all such primes Fermat's Last Theorem is true, as was proved in Chapter 5. With this theorem it becomes a matter of routine calculation to prove, say, that Fermat's Last Theorem is true for all prime exponents less than 100 except possibly for 37, 59, and 67. (From this it follows that it is true for all exponents, prime or not, less than 100, except possibly for these three and, in addition, the exponent 74.)

Kummer at first jumped to the conclusion that the number of regular primes is *infinite*, but he later realized that he could not prove this. Although it is clear on both empirical and theoretical grounds that over 60% of all primes are regular (see [J4]) it is unproved to this day that the number of regular primes is infinite. (Ironically, it is relatively easy to prove that the number of irregular primes is infinite—see [B2].)

Kummer's theorem does not imply, of course, that Fermat's Last Theorem is false for 37, 59, and 67, but only that its proof requires more powerful techniques and an understanding of even more subtle properties of the arithmetic of cyclotomic integers built up from a 37th, a 59th, or a 67th root of unity. Such techniques have been developed, beginning with Kummer himself and continuing with Mirimanoff, Wieferich, Furtwängler, and Vandiver, among others. These continuations of Kummer's work are the subject of the projected second volume of this book. Suffice it for now to say that Fermat's Last Theorem has been proved for all exponents well up into the thousands (see [W1]) but that the many criteria for the truth of the theorem still leave much to be desired. For example, it has not yet been proved that Fermat's Last Theorem is true for an infinite set of prime exponents.

7.1 The prime divisors

This chapter is devoted to an investigation of what Kummer might have had in mind when he spoke of ideal complex numbers of the form $x + y\sqrt{D}$ (see Section 5.1). The main problem in establishing the theory is to define the notion of a *prime divisor* of numbers of this form, or, as Kummer would have put it, the notion of an ideal prime number of this form. As in Chapter 4, the method will be to assume that such a notion exists and to deduce from that assumption what the notion must be. However, it will turn out that in order to make the theory work it will be necessary to revise somewhat the notion of "complex numbers of the form $x + y\sqrt{D}$ " as well.

Let D be a fixed integer and consider the set of all numbers of the form $x + y\sqrt{D}$ in which x and y are integers. It is natural to exclude the cases $D = 0$ and $D = 1$ because in these cases $x + y\sqrt{D}$ is an integer. Moreover, if D has any square factor, say $D = t^2 D'$ then $x + y\sqrt{D} = x + ty\sqrt{D'}$ and the numbers $x + y\sqrt{D}$ are *included* among the numbers $x + y\sqrt{D'}$. As will be seen in Section 8.1, this does not imply that the theory of divisors of numbers of the form $x + y\sqrt{D}$ reduces entirely to the theory of divisors of numbers of the form $x + y\sqrt{D'}$. However it does mean that there is a very strong relation between the two problems and that it is reasonable to consider first the case in which D is *squarefree*, which is to say that D is not 0 or 1 and is not divisible by any square other than 1. This chapter is devoted entirely to the case in which D is squarefree.

It is clear how numbers of the form $x + y\sqrt{D}$ should be added and multiplied. These two operations satisfy the commutative, associative, and distributive laws, and they define an arithmetic (a *ring* in the terminology of modern algebra) in which the usual properties hold. Specifically, *subtraction* is possible; that is, $a + X = b$ has a unique solution X for any pair of numbers of the form $a = x + y\sqrt{D}$, $b = x' + y'\sqrt{D}$. Very simply, $X = (x' - x) + (y' - y)\sqrt{D}$. *Division* is not normally possible, but there is a multiplicative unit 1 ($1 \cdot a = a$ for all a) and nonzero factors can be cancelled, that is, $ab = ac$ and $a \neq 0$ implies $b = c$. To prove this last fact it is necessary and sufficient to prove that if $(x + y\sqrt{D})(x' + y'\sqrt{D}) = 0 + 0\sqrt{D}$ and if $x + y\sqrt{D} \neq 0 + 0\sqrt{D}$ then $x' + y'\sqrt{D} = 0 + 0\sqrt{D}$. For this, multiply by $x - y\sqrt{D}$ to find $(x^2 - Dy^2)(x' + y'\sqrt{D}) = 0 + 0\sqrt{D}$. This implies $(x^2 - Dy^2)x' = 0$, $(x^2 - Dy^2)y' = 0$ and the desired conclusion follows provided $x^2 - Dy^2 \neq 0$. Since $x^2 = Dy^2$ implies* D is a square unless $y^2 = 0$, the desired conclusion follows from the assumption that D is squarefree.

There is one *conjugation* of numbers of the form $x + y\sqrt{D}$, namely, the one which carries $x + y\sqrt{D}$ to $x - y\sqrt{D}$. The product of all the conjugates of a given element $x + y\sqrt{D}$ is invariant under conjugation and is therefore an ordinary integer. Explicitly, $(x + y\sqrt{D})(x - y\sqrt{D}) = x^2 - Dy^2$. This integer is called the *norm* of $x + y\sqrt{D}$. The use of the norm makes it possible, as it did for cyclotomic integers in Section 4.2, to reduce division by numbers of the form $x + y\sqrt{D}$ to division by ordinary integers because to say that $x + y\sqrt{D}$ divides $u + v\sqrt{D}$ with quotient $r + s\sqrt{D}$ is the same as to say that the integer $x^2 - Dy^2$ divides $(u + v\sqrt{D}) \cdot (x - y\sqrt{D})$ with quotient $r + s\sqrt{D}$. (This is what is called "rationalizing the denominator" in elementary algebra.) Because D is not a square, norms $x^2 - Dy^2$ cannot be zero unless the element $x + y\sqrt{D}$ itself is zero. However, unlike the situation for cyclotomic integers, *norms may be negative* in the case $D > 0$.

In order to develop a theory of divisors for this arithmetic of numbers of the form $x + y\sqrt{D}$ the technique of *analysis* of Sections 4.3 and 4.6 will be adopted. That is, it will be assumed that it is possible to define prime divisors in such a way that the expected properties hold. From this assumption, *necessary* conditions about the divisor theory will be deduced. In fact, enough necessary conditions will be deduced to determine precisely what the prime divisors must be and how divisibility by a prime divisor (with multiplicities) must be determined. This will then be followed in Section 7.2 by a *synthesis* in which it will be shown that the necessary conditions that have been found are also *sufficient*; that is, when they are used to define a divisor theory the resultant theory is free of contradictions and the expected properties hold.

As the first step in the analysis, note that for any prime divisor A there is a unique positive integer p with the property that integers (things of the form $x + 0\sqrt{D}$) are divisible by A if and only if they are divisible by p.

*See the proposition of Section 1.3.

This is true because A must divide some $x + y\sqrt{D}$ and therefore must divide its *norm* $N(x + y\sqrt{D}) = (x + y\sqrt{D})(x - y\sqrt{D}) = x^2 - Dy^2$, which is an integer. Therefore A divides $|x^2 - Dy^2|$ which, being a positive integer, can be written as a product of positive primes. (The special case $|x^2 - Dy^2| = 1$ cannot occur because this would imply that A divided 1 and therefore divided all quadratic integers.) By the assumption that A is a *prime* divisor (if it divides a product then it divides one of the factors) it then follows that A divides at least one prime positive integer, say $A|p$. An arbitrary integer x can be written in the form $x = qp + r$ where $0 \leqslant r < p$. If $r = 0$ then of course A divides x. Conversely, if $r \neq 0$ then r and p are relatively prime and there is an integer y which is 0 mod p and 1 mod r; if A divided x then it would divide both r and $y = nr + 1$, which would imply that A divided 1 and is therefore impossible. This proves that A divides x if and only if p divides x. In particular, p is the only prime positive integer divisible by A.

If $\sqrt{D}$ $(= 0 + 1\sqrt{D})$ were congruent to an integer mod A (i.e. if $r - \sqrt{D}$ were divisible by A for some integer r) then every $x + y\sqrt{D}$ would be congruent to an integer mod A; since it is possible to determine when two integers are congruent mod A it would then be possible to tell when two numbers of the form $x + y\sqrt{D}$ were congruent mod A. Therefore it is natural to try to determine whether $\sqrt{D}$ is congruent to an integer mod A. Because A is prime, this is the case if and only if $(\sqrt{D} - 1)(\sqrt{D} - 2) \ldots (\sqrt{D} - p)$ is divisible by A. Now by the generalization of Fermat's theorem proved in Section 4.6, $(X - 1)(X - 2) \cdots (X - p) \equiv X^p - X$ mod p. Therefore $(\sqrt{D} - 1)(\sqrt{D} - 2) \cdots (\sqrt{D} - p) \equiv \sqrt{D}(\sqrt{D}^{\,p-1} - 1)$ mod A. This shows that $\sqrt{D}$ *is congruent to an integer* mod A *if and only if* $\sqrt{D} \equiv 0$ mod A *or* $\sqrt{D}^{\,p-1} \equiv 1$ mod A.

Consider first the case in which $\sqrt{D} \equiv 0$ mod A. If one assumes that divisors have *norms*—that is, there is an integer whose divisor is the product of the conjugates of A—then the norm of A must divide both $D = -N(\sqrt{D})$ and $p^2 = N(p)$. The norm of A is therefore either 1 or p (because D is squarefree) and, since a norm of 1 would imply that A divided 1, it follows that the norm of A must be p. Now A divides $\sqrt{D}$, and $D = p \cdot k$ where k is an integer not divisible by p. Thus A^2 divides $(\sqrt{D})^2 = pk$ and, since A does not divide k at all, A^2 divides p. On the other hand, $N(A^2) = N(A)^2 = p^2 = N(p)$. Therefore A^2 *is the divisor of p, A divides p with multiplicity exactly 2, and A divides $\sqrt{D}$ with multiplicity exactly 1.* Thus A divides $x + y\sqrt{D}$ with multiplicity μ if and only if it divides $(x + y\sqrt{D})(\sqrt{D})^\mu$ with multiplicity 2μ, and this is true if and only if p^μ divides $(x + y\sqrt{D})(\sqrt{D})^\mu$. Since this latter condition is meaningful within the arithmetic of numbers of the form $x + y\sqrt{D}$ without any assumptions about divisors, it can be taken as the *definition* of divisibility by the prime divisor A of p.

Definition. If p divides D then p has one prime divisor A, and A divides $x + y\sqrt{D}$ with multiplicity μ if p^μ divides $(x + y\sqrt{D})(\sqrt{D})^\mu$.

Consider next the case in which $(\sqrt{D})^{p-1} \equiv 1 \bmod A$. The case $p = 2$ will be considered later. For primes p other than 2, the expression $(\sqrt{D})^{p-1}$ describes an *integer* because $p - 1$ is even. If p is a prime for which this integer is 1 mod p then, as was shown above, for any prime divisor A of p there is an integer u such that $\sqrt{D} \equiv u \bmod A$. Given such a divisor A, its conjugate, which will be denoted $\bar{A}$, is also a prime divisor of p and, since A divides $u - \sqrt{D}$, $\bar{A}$ divides $u + \sqrt{D}$, that is, $\sqrt{D} \equiv -u \bmod \bar{A}$. This implies that $\bar{A}$ is distinct from A, because $A = \bar{A}$ would imply $u \equiv -u \bmod A$, $2u \equiv 0 \bmod p$, which implies either $2 = p$ or $(\sqrt{D})^{p-1} \equiv u^{p-1} \equiv 0 \bmod p$, both of which are contrary to assumption. Therefore A and $\bar{A}$ are distinct and both divide p. Since p has norm p^2 and since both A and $\bar{A}$ have norms equal to a power of p it follows that both A and $\bar{A}$ must have norm p, each of them must divide p with multiplicity 1, and $A\bar{A}$ must be the divisor of p. Since A does not divide $u + \sqrt{D}$, A divides $x + y\sqrt{D}$ with multiplicity μ if and only if it divides $(x + y\sqrt{D}) \cdot (u + \sqrt{D})^\mu$ with multiplicity μ, and this is true, by the fundamental theorem, if and only if p^μ divides $(x + y\sqrt{D})(u + \sqrt{D})^\mu$. Similarly, $\bar{A}$ divides $x + y\sqrt{D}$ with multiplicity μ if and only if p^μ divides $(x + y\sqrt{D}) \cdot (u - \sqrt{D})^\mu$. This can be made the *definition* of the divisors A and $\bar{A}$ once the integer u is found.

It was shown above that if p has a prime divisor A then $\sqrt{D} \equiv u \bmod A$ for some integer u, and this implies that the congruence $u^2 \equiv D \bmod p$ has a solution for this p. The same conclusion can be reached without assuming any divisor theory by setting $X = \sqrt{D}$ in $(X - 1) \cdot (X - 2) \cdots (X - p + 1) \equiv X^{p-1} - 1$ and taking norms on both sides to find $(1^2 - D)(2^2 - D) \ldots ((p - 1)^2 - D) \equiv 0 \bmod p$ (because, by assumption, $\sqrt{D}^{p-1} \equiv 1 \bmod p$). Thus $D \equiv u^2 \bmod p$ for at least one value $u = 1, 2, \ldots, p - 1$. If $D \equiv u^2$ and $D \equiv v^2 \bmod p$ then $(u - v)(u + v) = u^2 - v^2 \equiv 0 \bmod p$, from which $u \equiv \pm v \bmod p$. Since $u \not\equiv -u \bmod p$ ($p \neq 2$ and $u^2 \not\equiv 0 \bmod p$), this proves that *for odd primes p which satisfy $\sqrt{D}^{p-1} \equiv 1 \bmod p$ there are precisely two distinct solutions* mod p of the congruence $u^2 \equiv D \bmod p$.

Definition. If $p \neq 2$ and $D^{(p-1)/2} \equiv 1 \bmod p$ then there are precisely two solutions $u \bmod p$ of the congruence $u^2 \equiv D \bmod p$. For each of these solutions u define a prime divisor A of p by the condition that A divides $x + y\sqrt{D}$ with multiplicity μ if p^μ divides $(x + y\sqrt{D})(\sqrt{D} + u)^\mu$.

The prime divisors of other primes $p \neq 2$ are even simpler because if $p \neq 2$, $D \not\equiv 0 \bmod p$, and $D^{(p-1)/2} \not\equiv 1 \bmod p$ then p is itself prime. It is easy to see that this must be the case, because p must have some prime

248

divisor A (otherwise the condition for divisibility by p would be vacuous and p would have to divide 1) and if this prime divisor is not p itself then it divides something of the form $x + y\sqrt{D}$ which is *not* divisible by p. If p divided y then, because A divides $x + y\sqrt{D}$ and consequently p divides $x^2 - Dy^2$, it would follow that p divided x and that p divided $x + y\sqrt{D}$, contrary to assumption. Therefore if p is not prime it has a prime divisor which divides something of the form $x + y\sqrt{D}$ in which $y \not\equiv 0 \mod p$. Then $-x \equiv y\sqrt{D} \mod A$ and there is an integer z such that $yz \equiv 1 \mod p$, from which it follows that $\sqrt{D}$ must be congruent to an integer $\mod A$, namely, $-zx$. This implies, as was shown above, that $D \equiv 0 \mod p$ or $D^{(p-1)/2} \equiv 1 \mod p$. If p satisfies neither of these conditions and if a divisor theory is possible, it then must be the case that p is itself prime, that is, that a product $(x + y\sqrt{D})(u + v\sqrt{D})$ is divisible by p only if one of the factors is.

In order to prove that p is prime it is both necessary and sufficient (Exercise 1) to prove that if p divides the norm $x^2 - Dy^2$ of $x + y\sqrt{D}$ then p divides $x + y\sqrt{D}$. Let $n = D^{(p-1)/2}$ (assuming still that $p \neq 2$ so that n is an integer). Then $n^2 = D^{p-1}$ and by Fermat's theorem and the assumption $D \not\equiv 0 \mod p$ it follows that $n^2 \equiv 1 \mod p$. Thus $(n + 1)(n - 1) \equiv 0 \mod p$ and the assumption $n \not\equiv 1 \mod p$ implies $n \equiv -1 \mod p$. If p divides $x^2 - Dy^2$ then $x^2 \equiv Dy^2 \mod p$ and raising both sides to the power $\frac{1}{2}(p - 1)$ gives $x^{p-1} \equiv ny^{p-1}$, $x^{p-1} + y^{p-1} \equiv 0 \mod p$. By Fermat's theorem x^{p-1} and y^{p-1} can only be 0 or 1 $\mod p$ and since $p \neq 2$ the congruence $x^{p-1} + y^{p-1} \equiv 0 \mod p$ can hold only when $x^{p-1} \equiv y^{p-1} \equiv 0$. This implies that $x \equiv 0$, $y \equiv 0 \mod p$, that is, p divides $x + y\sqrt{D}$, as was to be shown.

Definition. If $p \neq 2$, $D^{(p-1)/2} \equiv -1 \mod p$, then the only prime divisor of p is the one defined by "A divides $x + y\sqrt{D}$ with multiplicity μ if p^μ divides $x + y\sqrt{D}$."

All that remains is to determine the prime divisors of 2 in the case $D \not\equiv 0 \mod p$, that is, in the case where D is odd. In this case $1 - D = (1 - \sqrt{D})(1 + \sqrt{D})$ is divisible by 2 and it follows that any prime divisor of 2 must divide either $1 - \sqrt{D}$ or $1 + \sqrt{D}$. That is, if A is any prime divisor of 2 then $\sqrt{D} \equiv \pm 1 \mod A$. Since $1 \equiv -1 \mod 2$ these two cases are in fact the same and $\sqrt{D}$ is necessarily congruent to $1 \mod A$. Therefore A divides $1 - D = (1 - \sqrt{D})(1 + \sqrt{D})$ with multiplicity at least 2. If $1 - D = 2k$ where k is odd—that is, if $D \equiv 3 \mod 4$—then A divides 2 with multiplicity at least 2. On the other hand, $N(A)$ is 2 or 4 because it divides $N(2) = 4$ and is not 1. If A^2 divides 2 then not only is $N(A) = 2$ but also A^2 is the divisor of 2 because the divisor of 2 divided by A^2 has norm $N(2)/N(A^2) = 4/N(A)^2 \leqslant 1$. Thus, when $D \equiv 3 \mod 4$, A divides both $1 - \sqrt{D}$ and $1 + \sqrt{D}$ with multiplicity exactly 1

249

and A divides $x + y\sqrt{D}$ with multiplicity μ if and only if 2^μ divides $(x + y\sqrt{D})(1 - \sqrt{D})^\mu$.

Definition. If $D \equiv 3 \bmod 4$ then 2 has a single prime divisor A and A divides $x + y\sqrt{D}$ with multiplicity μ if 2^μ divides $(x + y\sqrt{D}) \cdot (1 - \sqrt{D})^\mu$.

The only case which remains is the case of the prime divisors of 2 when $D \equiv 1 \bmod 4$. In this case, however, one encounters an absurdity. If A is any prime divisor of 2 then $N(A)$ divides $N(2) = 4$ and is therefore 2 or 4. If it were 4 then A would be the divisor of 2 and 2 would be prime, contrary to the fact that 2 divides the product $1 - D = (1 - \sqrt{D})(1 + \sqrt{D})$ without dividing either factor. Therefore $N(A)$ must be 2; that is, the divisor of 2 must be $A\overline{A}$. On the other hand, $A = \overline{A}$ because the test for divisibility of $x + y\sqrt{D}$ by A, which is $x + y \equiv 0 \bmod 2$, coincides with the test for divisibility of $x - y\sqrt{D}$ by A. Therefore A^2 must be the divisor of 2 and anything divisible by A with multiplicity 2 must be divisible by 2. But this too contradicts the factorization $1 - D = (1 - \sqrt{D})(1 + \sqrt{D})$ because it shows that neither factor on the right is divisible by A^2 despite the fact that A^4 divides the integer on the left. Thus a theory of divisors for numbers of the form $x + y\sqrt{D}$ is *impossible* when $D \equiv 1 \bmod 4$.

Bourbaki [B3, p. 127] is of the opinion that it was this apparent contradiction which prevented Kummer from pursuing the theory of "ideal complex numbers of the form $x + y\sqrt{D}$." Strangely enough, however, it is *only* the case $D \equiv 1 \bmod 4$ for which Kummer did develop the theory. For example, when $D = -3$ the numbers of the form $x + y\sqrt{D}$ are *included* in the cyclotomic integers with $\lambda = 3$ because in this case one can take the cube root α of 1 to be $\frac{1}{2}(-1 + \sqrt{-3})$ so that $x + y\sqrt{-3} = x + y + 2\alpha y$ is a cyclotomic integer. More generally, as Kummer observed (see [K8, p. 366] and [K11, p. 114]), when $D \equiv 1 \bmod 4$ and $|D| =$ prime, say $|D| = \lambda$, the numbers of the form $x + y\sqrt{D}$ are *included* among the cyclotomic integers for this λ and a theory of divisors for these numbers can be *deduced* from the theory of divisors of cyclotomic integers. (See Section 5.6.)

How then is the contradiction which was found above resolved? If one follows through the proof above that a theory of divisors is impossible in the case $D = -3$ one finds that it fails at the point where it asserts that 2 does not divide $1 - \sqrt{D}$, because when the numbers of the form $x + y\sqrt{-3}$ are considered as being included in the cyclotomic integers $a + b\alpha$ ($\alpha^3 = 1$) then 2 *does* divide $1 - \sqrt{-3}$ and the quotient is the cyclotomic integer $-\alpha$. More generally, if one declares that 2 *shall* divide $1 - \sqrt{D}$ in the case $D \equiv 1 \bmod 4$ then the contradiction above disappears and, as will be shown below, one can proceed to develop a theory of divisors.

Definition. When D is a squarefree integer and $D \not\equiv 1$ mod 4 a *quadratic integer* for the determinant D is a number of the form $x + y\sqrt{D}$ in which x and y are integers. When D is squarefree and $D \equiv 1$ mod 4 a quadratic integer for the determinant D is a number of the form $x + y \cdot \frac{1}{2}(1 - \sqrt{D})$ in which x and y are integers. Otherwise stated, a quadratic integer for the determinant $D \equiv 1$ mod 4 is a number of the form $u + v\sqrt{D}$ in which u and v are both integers *or* u and v are both halfintegers, i.e. $2u$, $2v$ are both odd integers. (For another, more natural, formulation of the definition of quadratic integers with determinant D see Exercise 3.)

It is clear that the quadratic integers defined in this way are closed under addition—the sum of two quadratic integers is a quadratic integer—but it is not entirely obvious that in the case $D \equiv 1$ mod 4 the quadratic integers are closed under multiplication. That they are closed under multiplication follows simply from the observation that if $\omega = \frac{1}{2}(1 - \sqrt{D})$ then $\omega^2 = \frac{1}{4}(1 + D) - \frac{1}{2}\sqrt{D} = \omega + \frac{1}{4}(D - 1)$, which is a quadratic integer.

In the case of primes $p \neq 2$, the definitions above of the prime divisors of p, together with their derivations, apply equally well to quadratic integers for determinants $D \equiv 1$ mod 4, the only difference being that $x + y\sqrt{D}$ may be a number in which x and y are halfintegers. It remains only to determine the prime divisors of 2 in the case $D \equiv 1$ mod 4. This can be done as follows.

Let A be a prime divisor of 2. Since every quadratic integer can be written in the form $x + y\omega$ ($\omega = \frac{1}{2}(1 - \sqrt{D})$) and since congruence mod A for integers coincides with congruence mod 2, it is natural to ask whether ω is congruent to an integer mod A, that is, to ask whether $\omega(\omega - 1)$ is divisible by A. Since $\omega(\omega - 1) = \frac{1}{2}(1 - \sqrt{D})\frac{1}{2}(-1 - \sqrt{D}) = (D - 1)/4$ is an integer, it is divisible by A if and only if it is even, and this is true if and only if $D \equiv 1$ mod 8. Thus, when $D \equiv 1$ mod 8, A must divide either ω or $\omega - 1$. It cannot divide both because then it would divide $1 = \omega - (\omega - 1)$. If it divides ω then its conjugate divides $\bar{\omega} = \frac{1}{2}(1 + \sqrt{D}) = 1 - \omega$ and if it divides $\omega - 1$ then its conjugate divides $\bar{\omega} - 1 = -\omega$. In either case there are two *distinct* prime divisors of 2. Since the norm of 2 is 2^2 it follows that $A\bar{A}$ is the divisor of 2. Thus the divisor of 2 is the product of two prime divisors. The one which divides ω divides $x + y\omega$ with multiplicity μ if and only if it divides $(x + y\omega)(\omega - 1)^{\mu}$ with multiplicity μ, and this is true if and only if 2^{μ} divides $(x + y\omega)(\omega - 1)^{\mu}$. Similarly the prime divisor of 2 which divides $\omega - 1$ divides $x + y\omega$ with multiplicity μ if and only if 2^{μ} divides $(x + y\omega)\omega^{\mu}$. In the remaining case $p = 2$, $D \equiv 1$ mod 4, $D \not\equiv 1$ mod 8 —that is, $D \equiv 5$ mod 8—the argument used above shows that ω is not congruent to an integer mod A and consequently, by an argument given earlier, that there is no quadratic integer $x + y\omega$ which is divisible by A but not by 2. In this case it is to be expected, therefore, that divisibility

251

by A coincides with divisibility by 2 and that 2 is prime. To prove this it is necessary and sufficient to prove that 2 divides the norm $(x + y\omega)(x + y\bar{\omega})$ of $x + y\omega$ if and only if it divides $x + y\omega$ (Exercise 1). Now $(x + y\omega)(x + y\bar{\omega}) = x^2 + (\omega + \bar{\omega})xy + \omega\bar{\omega}y^2 = x^2 + xy - \frac{1}{4}(D - 1)y^2$. By assumption $\frac{1}{4}(D - 1)$ is an odd integer and it follows immediately that $x^2 + xy - \frac{1}{4}(D - 1)y^2 \equiv x^2 + xy + y^2$ mod 2 is even only if x and y are both even, that is, 2 divides the norm of $x + y\omega$ only if it divides $x + y\omega$.

Definition. If $D \equiv 1$ mod 8 then 2 has two prime divisors, one of which divides $x + y\omega$ with multiplicity μ if 2^μ divides $(x + y\omega)(\omega - 1)^\mu$ and the other of which divides $x + y\omega$ with multiplicity μ if 2^μ divides $(x + y\omega)\omega^\mu$. (Here $\omega = \frac{1}{2}(1 - \sqrt{D})$.) If $D \equiv 5$ mod 8 then the only prime divisor of 2 is the one defined by "A divides $x + y\omega$ with multiplicity μ if 2^μ divides $x + y\omega$."

This completes the analysis, the deduction of the way in which prime divisors must be defined. The synthesis, the proof that these definitions lead to a consistent theory in which the expected properties hold, is given in the next section.

Summary. There are three different ways in which a prime positive integer p can "factor" when it is considered as a quadratic integer for the determinant D. In the first place, p may *remain prime*; that is, the divisor of p may be a prime divisor. In the second place p may *split*, that is, its divisor may be the product of two distinct prime divisors. Finally, the divisor of p may be the square of a prime divisor. In this case, because of an analogy with the theory of Riemann surfaces, it is said that p *ramifies*. If $p \neq 2$ then p remains prime if $D^{(p-1)/2} \equiv -1$ mod p, splits if $D^{(p-1)/2} \equiv 1$ mod p, and ramifies if $D^{(p-1)/2} \equiv 0$ mod p. Another way to state this criterion is to say that p remains prime if $u^2 \equiv D$ mod p has no solution, splits if $u^2 \equiv D$ mod p has two distinct solutions u mod p, and ramifies if $u^2 \equiv D$ mod p has the one solution $u \equiv 0$ mod p. The prime 2 remains prime if $D \equiv 5$ mod 8, splits if $D \equiv 1$ mod 8, and ramifies if $D \equiv 2$ or 3 mod 4. (Because D is squarefree, the case $D \equiv 0$ mod 4 does not occur.)

Notation. It will be useful in what follows to have a notation for the prime divisors. If p is a prime integer which remains prime then (p) will denote its divisor. If p is a prime which ramifies then $(p, *)$ will denote its unique prime divisor. Then $(p, *)^2$ is the divisor of p. If p is a prime which splits, if $p \neq 2$, and if u is a solution of $u^2 \equiv D$ mod p then (p, u) will denote the prime divisor of p which divides $u - \sqrt{D}$, that is, the prime divisor of p modulo which $\sqrt{D} \equiv u$. Then $(p, u)(p, -u)$ is the divisor of p. If 2 splits (that is, if $D \equiv 1$ mod 8) then $(2, 0)$ and $(2, 1)$ will denote the prime divisors of 2 modulo which $\frac{1}{2}(1 - \sqrt{D}) \equiv 0$ and 1 respectively. In this case the divisor of 2 is $(2, 0)(2, 1)$. Otherwise 2 either remains prime or ramifies and has divisor (2) or $(2, *)^2$.

EXERCISES

1. Show that p, considered as a quadratic integer for the determinant D, is prime if and only if $p|(x^2 - Dy^2)$ implies $p|(x + y\sqrt{D})$.

2. Show that if $D \equiv 1 \bmod 4$ then the norm $x^2 - Dy^2$ of any quadratic integer is an *integer*.

3. Let $a = x + y\sqrt{D}$ where x and y are arbitrary real numbers. Show that a is a quadratic integer for the determinant D if and only if the second degree polynomial $(X - a)(X - \bar{a})$ has integer coefficients. Another way of stating the definition of a quadratic integer is therefore to say that it satisfies an equation of the form $X^2 + bX + c = 0$ in which b and c are integers.

4. Show that if $p|D$ then $(p, *)$ divides $x + y\sqrt{D}$ if and only if $x \equiv 0 \bmod p$. Show also that p divides $x + y\sqrt{D}$ if and only if $(p, *)$ divides $x + y\sqrt{D}$ with multiplicity at least 2.

5. List all squarefree integers D in the range $|D| \leqslant 10$.

6. For $D = -1$ use Girard's theorem on sums of two squares (see Section 1.7) to say exactly which primes p split, which ramify, and which remain prime.

7. Determine how primes p factor (that is, whether they split, ramify, or remain prime) when $D = -2$. (Use the theorems of Chapter 2 on numbers of the form $x^2 + 2y^2$.)

8. Determine how primes p factor when $D = 2$.

9. Using facts about factorization of primes in cyclotomic integers in the case $\lambda = 3$, determine how primes p factor when $D = -3$.

10. Using cyclotomic integers for $\lambda = 5$, determine how primes factor when $D = 5$.

11. Determine by trial-and-error which primes less than 20 split when $D = -5$. Use quadratic reciprocity to solve this problem in general.

12. For each squarefree D in the range $|D| \leqslant 5$ find all divisors with norm 36.

13. Prove that if $D \equiv 1 \bmod 4$ and if p is a prime which splits then every quadratic integer is congruent to an integer mod (p, u) (where (p, u) is one of the prime divisors of p).

7.2 The divisor theory

Let the quadratic integers for the determinant D be defined as in the preceding section (that is, when $D \equiv 1 \bmod 4$ denominators of 2 are allowed) and let the prime divisors (p), $(p, *)$, (p, u) also be defined as in the preceding section. What is to be shown is that these definitions give a consistent theory of divisors in which the expected properties hold.

Proposition 1. *If p remains prime then p is divisible by one prime divisor with multiplicity exactly 1 and by no other prime divisors. If p ramifies then it is*

253

divisible by one prime divisor with multiplicity exactly 2 and by no other prime divisors. If p splits then it is divisible by two prime divisors with multiplicity exactly 1 and by no other prime divisors. In all cases the fundamental theorem holds for division by a prime p in the sense that if $x + y\sqrt{D}$ is divisible by all prime divisors of p with multiplicity at least as great as that with which they divide p then $x + y\sqrt{D}$ is divisible by p.

PROOF. It follows easily from the definition that each prime divisor divides a single prime integer and no other. Therefore, in considering the prime divisors of p one can ignore all those except the ones defined in connection with this particular prime p. If p remains prime then divisibility by (p) means ordinary divisibility by p and there is nothing to prove. If p ramifies then divisibility of $x + y\sqrt{D}$ by $(p, *)$ with multiplicity μ means divisibility of $(x + y\sqrt{D})\tau^\mu$ by p^μ where $\tau = \sqrt{D}$ if $p|D$ and $\tau = 1 + \sqrt{D}$ if $p = 2$ and D is odd. Thus p is divisible with multiplicity 2 if and only if τ^2 is divisible by p, which is immediately seen to be true in both cases. It is divisible with multiplicity exactly 2 if p does not divide $(\tau^2/p)\tau$. When $\tau = \sqrt{D}$ this follows immediately from the fact that D is squarefree. When $\tau = 1 + \sqrt{D}$ it follows from $(\tau^2/p)\tau = \frac{1}{2}[(1 + 3D) + (3 + D)\sqrt{D}]$ because in this case $D \equiv 3 \bmod 4$. If $x + y\sqrt{D}$ is divisible by $(p, *)$ with multiplicity 2 then p divides $(\tau^2/p)(x + y\sqrt{D})$. When $\tau = \sqrt{D}$ this clearly implies p divides $x + y\sqrt{D}$, as desired, because in this case τ^2/p is an integer relatively prime to p. When $\tau = 1 + \sqrt{D}$ (and therefore $p = 2$, $D \equiv 3 \bmod 4$) the same conclusion follows from $(\tau^2/p)(x + y\sqrt{D}) \equiv \sqrt{D}(x + y\sqrt{D}) \equiv y + x\sqrt{D} \bmod 2$, which shows that $x + y\sqrt{D}$ is divisible with multiplicity 2 only if $y \equiv x \equiv 0 \bmod 2$. Finally, suppose that p splits. Obviously p is divisible with multiplicity 1 by both of its prime divisors. Such a prime divisor divides $x + y\sqrt{D}$ with multiplicity μ if and only if p^μ divides $(x + y\sqrt{D})\tau^\mu$ where $\tau = u - \sqrt{D}$ when $p \neq 2$ ($u^2 \equiv D \bmod p$) and $\tau = \omega$ or $\omega - 1$ when $p = 2$ (and therefore $D \equiv 1 \bmod 8$), where $\omega = \frac{1}{2}(1 - \sqrt{D})$. To prove that such a prime divisor divides p with multiplicity exactly 1 it is necessary and sufficient to prove that p does not divide τ^2. When $p \neq 2$ this follows from $\tau^2 = u^2 + D - 2\sqrt{D} \equiv 2(D - \sqrt{D}) \not\equiv 0 \bmod p$. When $p = 2$ it follows from $\tau^2 = \omega^2 = \frac{1}{4}(D + 1 - 2\sqrt{D}) = \frac{1}{4}(D - 1) + \omega \equiv \omega \not\equiv 0 \bmod 2$ or from $\tau^2 = (\omega - 1)^2 \equiv \omega^2 - 1 \equiv \omega - 1 \not\equiv 0 \bmod 2$. It remains only to show that if $x + y\sqrt{D}$ is divisible by both prime divisors of p then $x + y\sqrt{D}$ is divisible by p. If $p \neq 2$ this follows from the observation that if p divides both $(x + y\sqrt{D})(u - \sqrt{D})$ and $(x + y\sqrt{D})(-u - \sqrt{D})$ then it divides $2u(x + y\sqrt{D})$ and $2u$ is relatively prime to p. If $p = 2$ and 2 divides both $(x + y\sqrt{D})\omega$ and $(x + y\sqrt{D})(\omega - 1)$ then 2 divides $x + y\sqrt{D}$.

Proposition 2. *Let A be a prime divisor. If A divides both $x + y\sqrt{D}$ and $r + s\sqrt{D}$ with multiplicity μ then it also divides $(x + y\sqrt{D}) + (r + s\sqrt{D})$*

with multiplicity μ. If A divides $x + y\sqrt{D}$ with multiplicity exactly μ (that is, divides with multiplicity μ but not with multiplicity $\mu + 1$) and divides $r + s\sqrt{D}$ with multiplicity exactly ν then it divides $(x + y\sqrt{D})(r + s\sqrt{D})$ with multiplicity exactly $\mu + \nu$. Finally, if $x + y\sqrt{D} \neq 0$ then there is a unique integer $\mu \geqslant 0$ such that A divides $x + y\sqrt{D}$ with multiplicity exactly μ.

PROOF. The first of these statements is immediate in all cases. For the proof of the second statement it is useful to consider first the case $\mu = \nu = 0$, in which it states that *a prime divisor is prime*, that is, if A divides neither factor then it cannot divide the product. If $A = (p)$ where p remains prime this was proved in the preceding section. If $A = (p, *)$ where p ramifies or $A = (p, u)$ where p splits then every quadratic integer is congruent to an integer mod A and two integers are congruent mod A if and only if they are congruent mod p. Thus $(x + yu)(r + su) \equiv (x + y\sqrt{D})(r + s\sqrt{D}) \equiv 0$ mod A implies $x + yu$ or $r + su \equiv 0$ mod p and therefore $x + y\sqrt{D}$ or $r + s\sqrt{D} \equiv 0$ mod A. Next observe that A divides $x + y\sqrt{D}$ with multiplicity exactly μ if and only if p^μ divides $(x + y\sqrt{D})\tau^\mu$ but the quotient is not divisible by A (where $\tau = 1$ if $A = (p), \tau = \sqrt{D}$ if A divides D, $\tau = \sqrt{D} + 1$ if $A = (2, *)$ and D is odd, $\tau = u - \sqrt{D}$ if $A = (p, u), p \neq 2$, and $\tau = \omega$ or $\omega - 1$ if $A = (2, 0)$ or $(2, 1)$). This is simply the observation that A divides $(x + y\sqrt{D})\tau^\mu/p^\mu$ if and only if p divides $(x + y\sqrt{D})\tau^{\mu+1}/p^\mu$, which is true if and only if $p^{\mu+1}$ divides $(x + y\sqrt{D})\tau^{\mu+1}$ (given that p^μ divides $(x + y\sqrt{D})\tau^\mu$). Thus A divides neither $(x + y\sqrt{D})\tau^\mu/p^\mu$ nor $(r + s\sqrt{D})\tau^\nu/p^\nu$ and, by what was just shown, it therefore does not divide $(x + y\sqrt{D})(r + s\sqrt{D})\tau^{\mu+\nu}/p^{\mu+\nu}$, which shows that it divides the product with multiplicity exactly $\mu + \nu$, as was to be shown. Now if A divides $x + y\sqrt{D}$ with multiplicity exactly μ and with multiplicity exactly $\mu + j$ where $j > 0$ then $(x + y\sqrt{D})\tau^\mu/p^\mu$ is not divisible by A but when it is multiplied by τ^j it becomes divisible by p^j, and therefore by A. This is obviously impossible except in cases where A divides τ, that is, cases where $A = (p, *)$. In these cases $\tau^2 = p\sigma$ where σ is not divisible by A (because, as was shown above, p does not divide $\sigma\tau = (\tau^2/p)\tau$). Then j cannot be even, for, if $j = 2k$, then $(x + y\sqrt{D})\tau^{\mu+j} = p^{\mu+k}[(x + y\sqrt{D})\tau^\mu/p^\mu]\sigma^k$ is divisible by $p^{\mu+k}$ but the quotient is not divisible by A and therefore not divisible by p^k. Nor can j be odd because if $j = 2k + 1$ ($k \geqslant 0$) then $(x + y\sqrt{D})\tau^{\mu+j} = p^{\mu+k}[(x + y\sqrt{D})\tau^\mu/p^\mu]\sigma^k\tau$ can be divisible by $p^{\mu+k+1}$ only if A divides $[(x + y\sqrt{D})\tau^\mu/p^\mu]\sigma^k$, which it does not. It remains only to show that if $x + y\sqrt{D} \neq 0$ then $x + y\sqrt{D}$ is divisible with multiplicity exactly μ for at least one μ. By the nature of the definition, it will suffice to show that there is at least one integer $\nu > 0$ such that $x + y\sqrt{D}$ is not divisible ν times by A. For this, let $N(x + y\sqrt{D}) = p^n k$ where k is relatively prime to p. Since p is divisible at most 2 times by A and k is not divisible at all, $p^n k$ is not divisible $2n + 1$ times by

A, and therefore $x + y\sqrt{D}$ is not divisible $2n + 1$ times by A. This completes the proof.

Fundamental theorem. *A quadratic integer $x + y\sqrt{D}$ divides a quadratic integer $u + v\sqrt{D}$ if and only if every prime divisor which divides $x + y\sqrt{D}$ divides $u + v\sqrt{D}$ with multiplicity at least as great.*

PROOF. This follows easily from Propositions 1 and 2 by the argument given in Section 4.11.

Corollary. *If two quadratic integers are divisible by exactly the same prime divisors with exactly the same multiplicities then their quotient is a unit, that is, a quadratic integer which divides 1.*

PROOF. Their quotient is not divisible by any prime divisor. Therefore by the fundamental theorem it must divide all quadratic integers and in particular it must divide 1.

The *divisor of a quadratic integer* is simply a list of all the prime divisors which divide it, counted with multiplicities. A *divisor* is a (finite) list of prime divisors in which some prime divisors may occur more than once. The *empty* divisor, which is the divisor of· 1, will be denoted by I. The *product* of two divisors is the list obtained by combining the two lists. Products will be written in the usual way by writing the factors side-by-side. Thus $(2, *)(3, 1)^2(17, 5)$ denotes (for some value of D which must be inferred from the context) the divisor which contains the divisors $(2, *)$, $(3, 1)$ twice, and $(17, 5)$. With these definitions, the mapping

$$\left\{\begin{array}{c} \text{nonzero} \\ \text{quadratic} \\ \text{integers} \end{array}\right\} \rightarrow \{\text{divisors}\}$$

is defined and has the properties (ii)–(v) of Section 4.15. (For the proof of (iv) see Exercise 4.)

The notion of *equivalence* of divisors can be defined in the same way—and for the same reasons—as in Chapter 5. A divisor which is the divisor of a quadratic integer is called a *principal* divisor. (For the origin of this terminology see Section 8.5.) Two divisors A and B are said to be *equivalent*, denoted $A \sim B$, if it is true that a divisor of the form AC is principal when and only when BC is principal. In other words, in any divisor divisible by A one can replace A by B and the new divisor will be principal if and only if the original one was. This relation is obviously reflexive, symmetric, transitive, and consistent with multiplication of divisors. Moreover, properties (1)–(8) of Section 5.3 all hold. For example, A is principal if and only if $A \sim I$. Kummer's definition of $A \sim B$, which is to say that there is a third divisor C for which AC and BC are both principal, is equivalent to the definition given above.

As in the cyclotomic case, it is possible to find a finite *representative set* of divisors, a finite set of divisors with the property that every divisor is equivalent to exactly one divisor in the set. This will be proved in Section 7.4 and, in addition, an actual method of *constructing* a representative set will be derived, something which is much more difficult in the cyclotomic case and which was not even attempted in Chapter 5.

Every divisor A has a conjugate $\bar{A}$ defined by the condition that A divides $x + y\sqrt{D}$ with multiplicity μ if and only if $\bar{A}$ divides $x - y\sqrt{D}$ with multiplicity μ. [In other words, $\bar{A}$ is obtained from A by leaving divisors (p) and $(p, *)$ unchanged while changing divisors (p, u) to their conjugates $(p, -u)$ for $p \neq 2$ and interchanging $(2, 0) \leftrightarrow (2, 1)$ when $p = 2$, $D \equiv 1 \bmod 8$.] It is natural, then, to consider the divisor $A\bar{A}$ to be the *norm* of A. As in the cyclotomic case, it is often useful to be able to consider the norm of a divisor to be an *integer*, not a divisor. It is easy to show, as in the cyclotomic case, that there is a unique *positive* integer whose divisor is $A\bar{A}$, and one might be tempted to define this to be the norm of A. It is natural to do this in the case $D < 0$ because then norms $x^2 - Dy^2$ are always positive. When $D > 0$ norms may be negative and the norm of a divisor needs to be treated in a different way. This is the subject of the following very brief section.

EXERCISES

1. Prove that if AB is a principal divisor and A is a principal divisor then B is a principal divisor.

2. Find the divisors of the following quadratic integers. (The value of D in each case is the integer under the radical sign.)
   (a) $4 + 7\sqrt{3}$
   (b) $5 - 9\sqrt{-2}$
   (c) $3\sqrt{-5}$
   (d) $\sqrt{21}$
   (e) $55 + 12\sqrt{21}$
   (f) $20 + 5\sqrt{14}$

3. Prove that the divisor of the product of two quadratic integers is the product of their divisors.

4. Prove that if A and B are divisors and if every quadratic integer divisible by A is also divisible by B then A is divisible by B. [Imitate the proof of the theorem of Section 4.13.]

7.3 The sign of the norm

In the divisor theory for cyclotomic integers it was useful to define the norm of a divisor to be the positive integer whose divisor is equal to the product of the conjugates of the given divisor. Among other useful properties which then followed was the fact that if a divisor is principal and is the divisor of a particular cyclotomic integer $f(\alpha)$ then its norm is equal to $Nf(\alpha)$. (Indeed, Kummer's terminology did not allow for a distinction between the integer $Nf(\alpha)$ and the ideal complex number $Nf(\alpha)$.) When $D < 0$ norms $x^2 - Dy^2$ are always positive and the norm of a divisor A can be defined in the same way for quadratic integers—namely, $N(A)$ is the positive integer whose divisor is $A\overline{A}$. When $D > 0$, however, this definition is unsatisfactory because the norm of the divisor of a quadratic integer will not always be the same as the norm of the quadratic integer.

This anomaly can be remedied by introducing a new divisor in the case $D > 0$ which has norm -1. Since there is to be just one such divisor, it will be its own conjugate and the notation $(-1, *)$ for this divisor is a natural one. Then *the divisor of a quadratic integer*, which is a list of all prime divisors which divide it, counted with multiplicities, should be modified so that it *includes* $(-1, *)$ *if the norm is negative and otherwise does not.* In order for it to be true that the divisor of a product is the product of the divisors, it is necessary and sufficient to define $(-1, *)^2 = I$. With these definitions all the usual properties hold. (From $(-1, *)^2 = I$ it follows that $(-1, *)$ divides any quadratic integer and therefore that this divisor does not enter into the condition of the fundamental theorem.) In addition, the norm of the divisor is a signed integer which, when the divisor is principal, is equal to the norm of any quadratic integer of which it is the divisor.

The divisor $(-1, *)$ is principal if and only if there is a quadratic integer with norm -1. As examples show (see Exercises) this may or may not be the case for any given $D > 0$.

In the remainder of the book the divisor $(-1, *)$ will always be included in the case $D > 0$ (but never in the case $D < 0$). Special notice should be taken of the fact that with this new divisor $(-1, *)$ in the case $D > 0$ *it is no longer true that a divisor is determined by the set of quadratic integers that it divides*—in other words property (iv) of Section 4.15 fails—because $(-1, *)$, like I, divides everything, but $(-1, *) \neq I$. More generally, $(-1, *)A$ divides the same quadratic integers that A does. The theorem of Section 4.13 is still true in the new theory, but its corollary is false because $A|B$ and $B|A$ no longer imply $A = B$, but only $A = B$ or $A = (-1, *)B$.

As was explained in Section 4.13, Dedekind identified ideal complex numbers with the set of all things they divide, and he *defined* an ideal (or, what is the same in these cases, a divisor) to *be* the set of all things that it divides. The above observation implies that *from Dedekind's point of view it is impossible to distinguish A and $(-1, *)A$* and, indeed, in most modern treatments this distinction is not made. Nonetheless, it is a useful distinction—one which Gauss, in a different context, invariably made—and it is made here both because it seems natural to do so and because it is easier to ignore the distinction in situations where it is inappropriate than it is to introduce the distinction after the entire theory has been developed without it. Moreover, it appears that the distinction is essential to Gauss's proof of quadratic reciprocity in Section 7.11.

1. Find a quadratic integer for the determinant $D = 2$ with divisor $(-1, *)$. Find its square and its cube and their divisors.

2. Show that $(-1, *)$ is not principal when $D = 3$.

3. Show that $(-1, *)$ is principal when $D = 5$.

4. Prove the theorem of Section 4.13 for the new theory.

7.4 Quadratic integers with given divisors

Unique factorization holds for divisors by their very definition—a divisor *is* a product of prime divisors. The fundamental theorem shows that a factorization of a quadratic integer implies a factorization of its divisor. Thus, since the factorizations of its divisor can easily be enumerated, the problem of factoring a quadratic integer leads to the problem of determining, for a given divisor, whether it is the divisor of a quadratic integer, and, if so, of determining all quadratic integers of which it is the divisor. This section, as its title indicates, is devoted to the solution of this problem.

Let A be a given divisor. The first part of the problem is to determine whether A is principal, that is, whether $A \sim I$. A natural approach to this problem—which is precisely the approach that was used in Sections 4.4 and 4.7 in attempting to find cyclotomic integers with given divisors—is to attempt to find quadratic integers $x + y\sqrt{D}$ *divisible* by A in which x and y are in some way small. In order to do this it is natural to try to find a simple way of testing whether a given quadratic integer is divisible by A.

Note first that if A is divisible by an integer n or, more precisely, if A is divisible by the divisor (n) of an integer n, say $A = (n)A'$, then division by n gives a one-to-one correspondence between quadratic integers (if any) with divisor A and quadratic integers with divisor A'. Therefore the problem can be simplified, with no loss of generality, to the case in which the given divisor A is divisible by the divisor of no integer greater than 1.

If A is divisible by no integer greater than 1 then every quadratic integer is congruent mod A to an ordinary integer, and two ordinary integers are congruent mod A if and only if they are congruent mod the norm of A. This can be proved as follows. Since A cannot be divisible by (p) or by $(p, *)^2$ or by $(p, u)(p, -u)$ (where p remains prime, ramifies, or splits,* respectively) it must have the form

$$A = (p_1, u_1)^{\mu_1}(p_2, u_2)^{\mu_2} \cdots (p_\sigma, u_\sigma)^{\mu_\sigma}(p'_1, *)(p'_2, *) \cdots (p'_\tau, *)$$

where $p_1, p_2, \ldots, p_\sigma$ are distinct primes which split, where $\mu_1 \geqslant 1, \mu_2 \geqslant 1, \ldots, \mu_\sigma \geqslant 1$, and where $p'_1, p'_2, \ldots, p'_\tau$ are distinct primes which ramify. (Of course σ or τ can be 0.) If $D > 0$ then p'_i might be -1 for some i. In

---

*When 2 splits—that is, when $D \equiv 1 \bmod 8$—one should take $(p, u)(p, -u)$ to mean $(2, 0)(2, 1)$.

this case, however, $A' = (-1, *)A$ does not contain the divisor $(-1, *)$ and, since congruence mod A coincides with congruence mod A', it will suffice to prove the theorem for A'. In other words, one can assume without loss of generality that $p_i' = -1$ does not occur. The norm of A is $p_1^{\mu_1} p_2^{\mu_2} \cdots p_\sigma^{\mu_\sigma} p_1' p_2' \cdots p_\tau'$. Let a denote this integer. If an *integer* is divisible by A then it is divisible by $\overline{A}$, as is clear directly from the definition. For the factors of the form $(p, u)^\mu$ of A this implies that the integer is divisible by $(p, u)^\mu (p, -u)^\mu$ and therefore that it is divisible by p^μ. On the other hand, a prime divisor of the form $(p, *)$ divides an integer only if p divides that integer. Therefore, because the p_i and p_i' are distinct, an integer is divisible by A if and only if it is divisible by a. It remains to show that every quadratic integer is congruent to an ordinary integer mod A.

Consider first the case $D \equiv 2$ or $3 \bmod 4$. Then every quadratic integer is of the form $x + y\sqrt{D}$ where x and y are integers and it will suffice to prove that there is an integer r such that $\sqrt{D} \equiv r \bmod A$. For each factor $(p, u)^\mu$ of A there is an integer, namely u, such that $u \equiv \sqrt{D} \bmod (p, u)$. Then $(u - \sqrt{D})^\mu \equiv 0 \bmod (p, u)^\mu$. Let $(u - \sqrt{D})^\mu = b + c\sqrt{D}$. The integer c is not divisible by p because if it were then $c\sqrt{D} \equiv 0 \bmod (p, u)$, $b \equiv 0 \bmod (p, u)$, $b \equiv 0 \bmod p$, $b + c\sqrt{D} \equiv 0 \bmod p$, $b + c\sqrt{D} \equiv 0 \bmod (p, -u)$, $u - \sqrt{D} \equiv 0 \bmod (p, -u)$ would follow, and $u - \sqrt{D} \not\equiv 0 \bmod (p, -u)$. Therefore c and p^μ are relatively prime and there is an integer d such that $cd \equiv 1 \bmod p^\mu$, from which $db + dc\sqrt{D} \equiv db + \sqrt{D} \bmod (p, u)^\mu$, $\sqrt{D} \equiv -db \bmod (p, u)^\mu$. Thus $\sqrt{D}$ is congruent to an integer mod $(p, u)^\mu$. Similarly $\sqrt{D} \equiv$ integer mod $(p', *)$, namely, $\sqrt{D} \equiv 0 \bmod (p', *)$ except in the case $p = 2$, $D \equiv 3 \bmod 4$, in which case $\sqrt{D} \equiv 1 \bmod (p, *)$. Thus for each factor $(p, u)^\mu$ or $(p', *)$ of A there is an integer r_i such that $\sqrt{D} \equiv r \bmod (p, u)^\mu$ or $\bmod (p', *)$ if and only if $r \equiv r_i \bmod p^\mu$ or $\bmod p'$. By the Chinese remainder theorem there is an integer r which satisfies all these congruences and it follows that $\sqrt{D} \equiv r \bmod A$.

If $D \equiv 1 \bmod 4$ then every quadratic integer is of the form $x + y\omega$ where x and y are integers and $\omega = \frac{1}{2}(1 - \sqrt{D})$. It will suffice, therefore, to prove that ω is congruent to an integer mod A. For factors of A of the form $(p, u)^\mu$ or $(p', *)$ in which $p \neq 2$ the argument above proves that $\sqrt{D} \equiv r_i$ for some integer r_i. Then $2\omega = 1 - \sqrt{D} \equiv 1 - r_i$ and inversion of $2 \bmod p^\mu$ or $\bmod p'$ gives an integer congruent to $\omega \bmod (p, u)^\mu$ or $\bmod (p', *)$. If $p = 2$ then $D \equiv 1 \bmod 8$ and the corresponding factor of A is either $(2, 0)^\mu$ or $(2, 1)^\mu$. By definition $\omega \equiv 0$ or $1 \bmod (2, 0)$ or $(2, 1)$. Then ω^μ or $(1 - \omega)^\mu$ is divisible by $(2, 0)^\mu$ or $(2, 1)^\mu$ and is of the form $b + c\omega$ where c is odd because otherwise 2 would divide ω or $1 - \omega$. Thus c is invertible mod 2^μ and there is an integer congruent to $\omega \bmod (2, 0)^\mu$ or $\bmod (2, 1)^\mu$. The Chinese remainder theorem then gives an integer congruent to $\omega \bmod A$, as required.

Thus, for a given divisor A, provided it is divisible by no integer greater than 1, there is an integer r such that $r - \sqrt{D}$ is divisible by A. Moreover,

r can be reduced mod a where a is the norm of A. These observations suffice to prove that *the class number is finite*; that is, there exists a finite set of divisors with the property that every divisor is equivalent to an element of the set. For this, note that the divisor of $r - \sqrt{D}$ is of the form AB where the norm b of B satisfies $|ab| = |(r - \sqrt{D})(r + \sqrt{D})| \leqslant r^2 + |D|$ $\leqslant \frac{1}{4} a^2 + |D|$. Thus $|b| < |a|$ unless $\frac{1}{4} a^2 + |D| \geqslant |ab| \geqslant a^2, |D| \geqslant 3a^2/4, |a| \leqslant 2\sqrt{|D|/3}$. Since AB and $\bar{B}B$ are both principal, $A \sim \bar{B}$. The norm of $\bar{B}$ is b and is less than a in absolute value unless $|a| \leqslant 2\sqrt{|D|/3}$. If the original A had been divisible by an integer greater than 1 then it could have been replaced by an equivalent divisor with smaller norm which was not divisible by an integer greater than 1. Thus *a divisor A is always equivalent to a divisor with a norm smaller in absolute value except possibly when its norm is $\leqslant 2\sqrt{|D|/3}$ in absolute value.* By repeated applications of these reductions one can, by the principle of infinite descent, find a divisor equivalent to A which has norm $|a| \leqslant 2\sqrt{|D|/3}$. Since the number of divisors with norm less than a given bound in absolute value is finite, this proves that the class number is finite.

Only a slight refinement of this reduction process is necessary to determine whether a given divisor is principal. Consider first the case where $D < 0$ and $D \equiv 2$ or $3 \bmod 4$. Let A_0 be the given divisor and assume, without loss of generality, that A_0 is not divisible by an integer greater than 1. Then the reduction above gives $A_0 \sim A_1$ where $A_0 \bar{A}_1$ is the divisor of $r_0 - \sqrt{D}$, r_0 is an integer congruent to $\sqrt{D} \bmod A_0$, and r_0 is reduced* modulo the norm a_0 of A_0. Since A_1 divides $r_0 + \sqrt{D}$ and $r_0 + \sqrt{D}$ is divisible by no integer, A_1 is divisible by no integer. The process can therefore be repeated to give a sequence of equivalent divisors $A_0 \sim A_1 \sim A_2 \sim \cdots$. Moreover, once a_0, r_0 are known $a_1 = (r_0^2 - D)/a_0$ is known and r_1 is easily found because A_1 divides $(-r_0) - \sqrt{D}$, which shows that r_1 is $-r_0$ reduced mod a_1. Similarly, if a_i, r_i are known then a_{i+1}, r_{i+1} can be found by $a_{i+1} = (r_i^2 - D)/a_i$ and $r_{i+1} + r_i \equiv 0 \bmod a_{i+1}$. If there is an i such that $a_i = 1$ then $A_i = I$ and, in particular, A_i is principal. That is, a *sufficient* condition for the given divisor to be principal is that the sequence of integers $a_0, a_1, a_2, \ldots$ contain the integer 1.

It will now be shown that this condition is *necessary* as well (still under the assumption that $D < 0$, $D \equiv 2$ or $3 \bmod 4$). By the principle of infinite descent there must be an i such that $a_{i+1} \geqslant a_i$ and, as was shown above, this implies $a_i \leqslant 2\sqrt{|D|/3}$. This implies $a_i \leqslant |D|$ with strict inequality unless $D = -1$. ($|D| \leqslant 2\sqrt{|D|/3}$ implies $3|D|^2 \leqslant 4|D|, |D| \leqslant 4/3$.) If $A \sim I$ then $A_i \sim I$ and there is a quadratic integer $u + v\sqrt{D}$ with divisor A_i. Then $u^2 - Dv^2 = a_i \leqslant |D|$ with strict inequality except in the case

---

*The reduction is to make $|r_0|$ as small as possible, so that $-\frac{1}{2} a_0 < r_0 \leqslant \frac{1}{2} a_0$. It may happen that *two* values of r_0 are possible, $-\frac{1}{2} a_0$ and $\frac{1}{2} a_0$. For the sake of definiteness, let the positive value $\frac{1}{2} a_0$ be chosen in such cases.

$D = -1$. Thus $v \neq 0$ implies $D = -1$, from which it follows that $a_i \leqslant 1$, that is, $a_i = 1$, as was to be shown. Otherwise $v = 0$ and A_i is the divisor of u, which contradicts the fact that A_i is divisible by no integer greater than 1 unless $u = \pm 1$ and $a_i = 1$. Therefore $A_0 \sim I$ implies not only that $a_i = 1$ is reached but also that the sequence of a's decreases until $a_i = 1$ is reached.

This gives a very simple algorithm for determining, in the case $D < 0$, $D \equiv 2$ or $3 \bmod 4$, whether a given divisor is principal. Only a slight extension of the algorithm is required to find the set of all quadratic integers with divisor A when $A \sim I$. Let $A \sim A_0 \sim A_1 \sim \cdots \sim A_i = I$ where $A = (n)A_0$ and where $A_j \overline{A}_{j+1}$ is the divisor of $r_j - \sqrt{D}$ for $j = 0, 1, \ldots, i - 1$. Given a quadratic integer with divisor A_{j+1} one can find a quadratic integer with divisor A_j simply by multiplying by $r_j - \sqrt{D}$ with divisor $A_j \overline{A}_{j+1}$ and dividing by a_{j+1} with divisor $A_{j+1} \overline{A}_{j+1}$. Starting with 1, which has divisor $A_i = I$, the composition of these operations gives a quadratic integer with divisor A_0; multiplication by n then gives a quadratic integer with divisor A. The most general quadratic integer with divisor A is, by the fundamental theorem, a unit times this one. Now a unit $u + v\sqrt{D}$ satisfies $u^2 - Dv^2 = 1$ which implies $u = \pm 1$, $v = 0$ or $D = -1$ and $u = 0$, $v = \pm 1$. That is, ± 1 are the only units except when $D = -1$, in which case there are 4 units ± 1, $\pm \sqrt{-1}$. This completes the solution of the given problem in the case $D < 0$, $D \equiv 2$ or $3 \bmod 4$: in a finite number of simple steps one can determine whether the given divisor is principal and, if so, one can find all quadratic integers (there will be just two or, when $D = -1$, four) which have that divisor.

The modifications required in the case $D < 0$, $D \equiv 1 \bmod 4$ are minor. Given A, let $A = (n)A_0$ where A_0 is divisible by no integer greater than 1. There is a halfinteger (half an odd integer) r such that $r - \frac{1}{2}\sqrt{D} \equiv 0 \bmod A_0$. Let a_0 be the norm of A_0, let r_0 be the halfinteger obtained by reducing* $r \bmod a_0$, and let A_1 be the divisor such that $A_0 \overline{A}_1$ is the divisor of $r_0 - \frac{1}{2}\sqrt{D}$. If a_1 is the norm of A_1 then $a_0 a_1 = r_0^2 - \frac{1}{4}D \leqslant \frac{1}{4}a_0^2 - \frac{1}{4}D$. If a_0 is not greater than a_1 then $a_0^2 \leqslant a_0 a_1 \leqslant (a_0^2 - D)/4$, $a_0 \leqslant \sqrt{|D|/3}$. Therefore iteration of the procedure leads eventually to a divisor A_i with norm $a_i \leqslant \sqrt{|D|/3}$. This proves that the class number is finite. If the sequence $a_0, a_1, a_2, \ldots$ contains the value 1 then $A_i = I$ and it follows not only that A is principal but also that all quadratic integers with divisor A can be found once all the units are known. If $u^2 - Dv^2 = 1$ and if $u + v\sqrt{D} \neq \pm 1$ then either $u = 0$—which is impossible because then v is an integer and $-D \leqslant 1$, $D \geqslant -1$ contrary to assumption—or $u = \pm \frac{1}{2}$, $-Dv^2 = \frac{3}{4}$, $v = \pm \frac{1}{2}$, $D = -3$. In short, the only units are ± 1 except in the case $D = -3$, where there are six units ± 1, $\pm \frac{1}{2} \pm \frac{1}{2}\sqrt{-3}$. The solution of the problem will then be complete if it is shown that, conversely, $A \sim I$ implies that the sequence $a_0, a_1, a_2, \ldots$ must contain the

*Again, in cases where this reduction is ambiguous—that is, $r \equiv \frac{1}{2}a_0 \bmod a_0$—let the positive value $r_0 = \frac{1}{2}a_0$ be used.

the value 1. Let i be the first integer for which $a_i \leqslant a_{i+1}$. Then, as was seen above, $a_i \leqslant \sqrt{|D|/3}$. If A is principal then there exists a quadratic integer $u + v\sqrt{D}$ with divisor A_i. Then $u^2 - Dv^2 = a_i \leqslant \sqrt{|D|/3}$. Thus $\frac{1}{4}[(2u)^2 - D(2v)^2] \leqslant \sqrt{|D|/3}$ and $2u$ and $2v$ are integers. Unless $v = 0$ it follows that $\frac{1}{4}(-D) \leqslant \sqrt{|D|/3}$, $D^2 \leqslant 16|D|/3$, $|D| \leqslant 6$, $D = -3$. Thus, unless $D = -3$, it follows that $v = 0$, u is an integer with divisor A_i, and, since A_i is divisible by no integer greater than 1, $u = \pm 1$, $A_i = I$ as was to be shown. In the remaining case $D = -3$, $a_i \leqslant \sqrt{|D|/3} = 1$ and $u + v\sqrt{D}$ must be a unit, from which $A_i = I$ again follows.

Consider now the case where $D > 0$, and, for the sake of simplicity, $D \equiv 2$ or $3 \bmod 4$. The argument by which the class number was proved to be finite is still valid in this case and shows that every divisor is equivalent to one with norm at most $2\sqrt{D/3}$ in absolute value. Moreover, it does this by an explicit sequence of equivalences $A \sim A_0 \sim A_1 \sim \cdots \sim A_i$ in which $|N(A_i)| \leqslant 2\sqrt{D/3}$, so that the problem can be solved for A if it can be solved for A_i. One can assume without loss of generality, therefore, that $A = A_0$ is divisible by no integer greater than 1 and that $|N(A)| \leqslant 2\sqrt{D/3}$.

Let a_0 be the norm of A_0, let $\sqrt{D} \equiv r_0 \bmod A_0$, and let $|a_0| \leqslant 2\sqrt{D/3}$. The process above calls for reducing $r_0 \bmod a_0$ to make $|r_0| \leqslant \frac{1}{2}|a_0|$. However, the real goal is to make $a_0 a_1 = r_0^2 - D$ as small as possible in absolute value and this goal is better served by making r_0 near $\pm \sqrt{D}$ than by making it near 0. The reader may already have noticed that the steps in this process are very similar to the steps of the cyclic method of Section 1.9. In the cyclic method it was found to be effective to choose r to be the *positive* integer for which $|r^2 - D|$ is as small as possible, but in the proof that the cyclic method always produces all solutions of Pell's equation (see Exercises 9–13 of Section 1.9) it was found to be more convenient to choose r in such a way that $r^2 - D$ *is negative* but r is otherwise as large as possible. This method of choosing r will be imitated here. *Let r_0 be the largest integer $r_0 \equiv \sqrt{D} \bmod A_0$ for which $r_0^2 - D < 0$.* (There is at least one r satisfying $r \equiv \sqrt{D} \bmod A_0$ and $r^2 - D < 0$ because the distance between the roots of $x^2 - D$ is $2\sqrt{D} \geqslant |a_0|\sqrt{3} > |a_0|$.) Then, as before, let A_1 be defined by the condition that $A_0 \bar{A}_1$ is the divisor of $r_0 - \sqrt{D}$. Then the norm of A_1 is $(r_0^2 - D)/a_0$ and $\sqrt{D} \equiv -r_0 \bmod A_1$. At the next step, r_1 is to be the largest integer $r_1 \equiv \sqrt{D} \bmod A_1$ for which $r_1^2 - D < 0$, which means r_1 is the greatest integer satisfying $r_0 + r_1 \equiv 0 \bmod a_1$ for which $r_1^2 - D < 0$. Note that there always is such an r_1 because in fact $(-r_0)^2 - D < 0$ so that if necessary r_1 could be chosen to be $-r_0$.

This rule for choosing r_1, and thereafter for choosing $r_2, r_3, \ldots$ defines an infinite sequence of divisors $A_0 \sim A_1 \sim A_2 \sim \cdots$ equivalent to A_0. If the sequence of norms $a_0, a_1, a_2, \ldots$ of these divisors includes 1 then $A_i = I$ for some i and A is principal. The main theorem is that *this sufficient condition is in fact necessary*, that is, if $A_0 \sim I$ then the sequence $A_0 \sim A_1 \sim A_2 \sim \cdots$ defined above must reach $A_i = I$. This theorem will be proved in the next section.

It is also easy to show—and this will be shown in the next section as well—that the sequence $A_0 \sim A_1 \sim A_2 \sim \cdots$ must eventually begin to repeat. Therefore, in testing to see whether a given A_0 is principal it is only necessary to carry the sequence $A_0 \sim A_1 \sim A_2 \sim \cdots$ to the point where it begins repeating. If by that time $A_i = I$ has not occurred, then it never will, and by the theorem to be proved it will follow that A_0 is *not* principal.

Given a quadratic integer with divisor A_j one can, as before, find one with divisor A_{j-1} by multiplying by $r_{j-1} - \sqrt{D}$ and dividing by a_j. Thus if $A_0 \sim I$ one can find $A_0 \sim A_1 \sim \cdots \sim A_i = I$ and, starting with the quadratic integer 1 with divisor A_i, one can work backwards to find a quadratic integer with divisor A_0. Then the most general quadratic integer with divisor A_0 is a unit times this one. Thus the full solution of the problem is reduced to finding the most general quadratic integer $u + v\sqrt{D}$ which is a unit, that is, the most general solution in integers u, v of the equation $u^2 - Dv^2 = \pm 1$. This is almost the same as solving Pell's equation (Pell's equation is $u^2 - Dv^2 = +1$) and the solution can be found by essentially the same technique that was used in Section 1.9 to solve Pell's equation. The solution of this problem of finding all units—which is just the case $A = I$ of the problem under consideration—will be given in the next section for the case $D > 0$.

Finally, the modifications needed to handle the case $D > 0$, $D \equiv 1 \bmod 4$ are again minor. If A_0 is not divisible by an integer greater than 1 then there is a halfinteger r such that $r - \frac{1}{2}\sqrt{D} \equiv 0 \bmod A_0$. One can assume that the norm a_0 of A_0 has absolute value at most $\sqrt{D/3}$ and this implies that there is at least one r with the above property for which $N(r - \frac{1}{2}\sqrt{D}) = r^2 - \frac{1}{4}D$ is negative. Define r_0 to be the largest halfinteger for which $r_0 - \frac{1}{2}\sqrt{D} \equiv 0 \bmod A_0$ and $N(r_0 - \frac{1}{2}\sqrt{D}) < 0$. Define A_1 to be the divisor such that $A_0 \overline{A}_1$ is the divisor of $r_0 - \frac{1}{2}\sqrt{D}$. In the same way, define A_2 given A_1. (The existence of a halfinteger r_1 for which $r_1 - \frac{1}{2}\sqrt{D} \equiv 0 \bmod A_1$ and $N(r_1 - \frac{1}{2}\sqrt{D}) < 0$ follows from the fact that $-r_0$ is such a halfinteger.) This gives a sequence of equivalent divisors $A_0 \sim A_1 \sim A_2 \sim \cdots$. It will be shown in the next section that this sequence eventually begins to repeat. Therefore one can determine computationally whether there is an i for which $A_i = I$. If so then $A_0 \sim I$ and it is possible to use the chain of equivalences $A_0 \sim A_1 \sim \cdots \sim A_i = I$ to construct a quadratic integer with divisor A_0. To find *all* quadratic integers with divisor A_0 is then a matter of finding all *units* $u + v\sqrt{D}$, a problem which can be solved by the above method applied to $A_0 = I$, as will also be shown in the next section. Finally if the sequence $A_0 \sim A_1 \sim A_2 \sim \cdots$ begins to repeat *without* $A_i = I$ having occurred, then A_0 is not principal and this solves the problem by showing that *no* quadratic integer has divisor A_0.

Summary

Let A be a given divisor. The problem is to determine whether $A \sim I$ and, if so, to find all quadratic integers with divisor A. One can assume without loss of generality that A is divisible by no integer greater than 1. What is described above is a method of generating a sequence of divisors $A \sim A_1 \sim A_2 \sim \cdots$ equivalent to A. The method is slightly different in each of the four cases according to whether $D < 0$ or $D > 0$ and $D \equiv 2$ or $3 \bmod 4$ or $D \equiv 1 \bmod 4$. If $A_i = I$ for some i then A is of course principal. The problem of determining whether $A \sim I$ is then solved by proving that, conversely, *if A is principal then $A_i = I$ for some i*, and by proving that it is a finite computation to determine whether $A_i = I$ occurs. In cases where $D < 0$, not only was this theorem proved above, but all units were also found, so that the solution of the problem in these cases is complete. In cases where $D > 0$, the proof of the theorem and the determination of all units has been postponed to the following section.

Terminology

The above method of generating the sequence $A \sim A_1 \sim A_2 \sim \cdots$ will be called the *cyclic method*. Note that there are four different cases and that only the case $D > 0$, $D \not\equiv 1 \bmod 4$ compares to what was called the "cyclic method" in Section 1.9. Even in this case, the method corresponds more exactly to what was called the "English method" in the exercises of Section 1.9 because r is chosen so that $r^2 - D < 0$ (r as large as possible) rather than so that $|r^2 - D|$ is as small as possible ($r > 0$). However, the basic idea is the same as the one used by the ancient Indians, and it seems appropriate to remember this fact by adopting the name they gave to the process.

The computations of the cyclic method will be displayed in the form

$$r_0 \quad r_1 \quad r_2 \quad \cdots$$
$$a_0 \quad a_1 \quad a_2 \quad a_3 \quad \cdots$$

Here a_0 is the norm of A_0 and $r_0 - \sqrt{D}$ is divisible by A_0, or, in case $D \equiv 1 \bmod 4$, r_0 is a halfinteger and $r_0 - \frac{1}{2}\sqrt{D}$ is divisible by A_0. It follows from the theorem at the beginning of the section that a_0 and r_0 suffice to determine A_0 (Exercise 15). Similarly, a_j and r_j suffice to determine A_j. When $D \not\equiv 1 \bmod 4$ the sequences $a_0, a_1, a_2, \ldots$ and $r_0, r_1, r_2, \ldots$ are generated by the rules $a_{j+1} = (r_j^2 - D)/a_j$ and $r_{j+1} + r_j \equiv 0 \bmod a_{j+1}$ where r_{j+1} is chosen from among all solutions of $r_{j+1} \equiv -r_j$ $\bmod a_{j+1}$ by a set of rules which depend on the sign of D and $r^2 - D$. (If $D < 0$, simply make $|r_{j+1}|$ as small as possible. In case of a tie, choose the positive value for r_{j+1}. If $D > 0$, try to make $r_{j+1}^2 - D$ negative. If this is impossible make $|r_{j+1}|$ as small as possible. Otherwise make r_{j+1} as large as possible subject to $r_{j+1}^2 - D < 0$.) When $D \equiv 1 \bmod 4$ the rules are the same except that $r^2 - D$ is replaced by the integer $r^2 - \frac{1}{4} D$.

EXERCISES

Solve the problem of this section (determine whether the given A is principal and, if so, find all quadratic integers with divisor A) in the following cases:

1. $D = -1$, $A = (5, 2)(13, -5)$.

2. $D = -3$, $A = (31, 20)$.

3. $D = -2$, $A = (11, 3)(41, 11)(67, 20)$.

4. $D = -5$, $A = (23, 8)$.

5. $D = -5$, $A = (23, 8)^2$.

Use the format

$$r_0 \ r_1 \ \cdots$$
$$a_0 \ a_1 \ a_2 \ \cdots$$

suggested at the end of the section. Do Exercise 3 in two ways, first by finding r_0 so that A divides $r_0 - \sqrt{-2}$ and proceeding from there, and second by solving the problem for each of the three factors of A and multiplying the results. Note that Exercise 5 cannot be solved in this way.

6. Prove that if p is a prime which divides a sum of two squares in a nontrivial way (p divides $x^2 + y^2$ but divides neither x^2 nor y^2) then p is itself a sum of two squares. [Do this by showing that when $D = -1$ every divisor is principal.]

7. Prove that p divides a sum of two squares in a nontrivial way if and only if $p = 2$ or $p \equiv 1 \bmod 4$. ["Only if" is immediate from 6. For the other part use the fact that $a^{4n} - 1 \equiv 0 \bmod p$ for all a is possible only if $a^{2n} + 1 \equiv 0 \bmod p$ for some a.] Exercises 6 and 7 are essentially the theorems of Girard on representations of numbers as sums of two squares (Section 1.7). Note that the proofs are essentially the same as Euler's (Section 2.4).

8. Prove that every divisor is principal when $D = -2$ or -3 but not when $D = -5$.

In some of the exercises which follow it will be convenient to use a computational shortcut. Suppose $A_0 \sim A_1 \sim \cdots \sim A_i = I$. Define $x_i + y_i \sqrt{D} = 1$ and, working back from this, define $x_j + y_j \sqrt{D}$ by $(x_j + y_j \sqrt{D})(x_{j+1} - y_{j+1}\sqrt{D}) = r_j - \sqrt{D}$ (or $r_j - \frac{1}{2}\sqrt{D}$ when $D \equiv 1 \bmod 4$) or, what is the same, $x_j + y_j \sqrt{D} = (r_j - \sqrt{D})(x_{j+1} +$

$y_{j+1}\sqrt{D}$)/a_{j+1}. Then $x_{i-1}+y_{i-1}\sqrt{D} = r_{i-1}-\sqrt{D}$ requires no computation at all and

$$x_j + y_j\sqrt{D} = \frac{\left[(r_j + r_{j+1}) - (r_{j+1}+\sqrt{D})\right](x_{j+1}+y_{j+1}\sqrt{D})}{a_{j+1}}$$

$$= n_{j+1}(x_{j+1}+y_{j+1}\sqrt{D}) - (x_{j+2}+y_{j+2}\sqrt{D})$$

where $n_j = (r_{j-1}+r_j)/a_j$. When this equation is written in matrix form

$$\begin{bmatrix} x_j + y_j\sqrt{D} \\ x_{j+1}+y_{j+1}\sqrt{D} \end{bmatrix} = \begin{bmatrix} n_{j+1} & -1 \\ 1 & 0 \end{bmatrix}\begin{bmatrix} x_{j+1}+y_{j+1}\sqrt{D} \\ x_{j+2}+y_{j+2}\sqrt{D} \end{bmatrix}$$

it leads to the simple formula

$$\begin{bmatrix} x_0 + y_0\sqrt{D} \\ x_1 + y_1\sqrt{D} \end{bmatrix} = \begin{bmatrix} n_1 & -1 \\ 1 & 0 \end{bmatrix}\begin{bmatrix} n_2 & -1 \\ 1 & 0 \end{bmatrix} \cdots \begin{bmatrix} n_{i-1} & -1 \\ 1 & 0 \end{bmatrix}\begin{bmatrix} r_{i-1}-\sqrt{D} \\ 1 \end{bmatrix}$$

for $x_0 + y_0\sqrt{D}$. (When $D \equiv 1 \bmod 4$, replace $r - \sqrt{D}$ with $r - \frac{1}{2}\sqrt{D}$, r a halfinteger.)

In the following cases determine whether the given divisor is principal and, if so, give *one* quadratic integer with this divisor. Take for granted the theorem to be proved in the next section that if the cyclic method does not reach $A_i = I$ then the divisor is not principal.

9. $D = 67$, $A = (-1, *)$.

10. $D = 109$, $A = (-1, *)$.

11. $D = 101$, $A = (79, 38)$.

12. $D = 79$, $A = (3, 1)^6$.

13. $D = -163$, $A = (197, 25)$.

14. $D = -165$, $A = (2, *)(5, *)(151, 52)$.

15. Show that a_i and r_i determine A_i. That is, show that if A and A' are two divisors which have the same norm a and which divide the same quadratic integer $r - \sqrt{D}$ (or $r - \frac{1}{2}\sqrt{D}$ when $D \equiv 1 \bmod 4$) then $A = A'$. [The norm a determines the presence or absence of a factor $(-1, *)$.]

7.5 Validity of the cyclic method

Let A be a given divisor of quadratic integers for the determinant D (D squarefree). In the case $D < 0$ the problem of finding all quadratic integers with the divisor A was solved in the preceding section. In the case $D > 0$ a technique for solving the same problem was described, but the validity of the technique rested on some unproved assertions. This section is devoted to the proof of those assertions.

Given $D > 0$ and given a divisor A of quadratic integers for the determinant D, it was shown how to find a divisor $A_0 \sim A$ which is divisible by no integer greater than 1 and which has norm $a_0 \leqslant 2\sqrt{D/3}$ (or even, if $D \equiv 1 \bmod 4$, $a_0 \leqslant \sqrt{D/3}$). Therefore one can assume without loss of generality that the problem is to find all quadratic integers with the divisor A_0 for such an A_0. (This includes as a special case the problem of finding all quadratic integers with divisor I, that is, the problem of finding all units.)

Assume for the moment that $D \equiv 2$ or $3 \bmod 4$. Then there is an integer r such that $r \equiv \sqrt{D} \bmod A_0$ and such that $N(r - \sqrt{D}) < 0$. Let r_0 be the largest such integer. Let A_1 be defined by the condition that $A_0 \overline{A}_1$ be the divisor of $r_0 - \sqrt{D}$. Then there is an integer r such that A_1 divides $r - \sqrt{D}$ and $N(r - \sqrt{D}) < 0$, namely, $r = -r_0$. Let r_1 be the largest such integer and let A_2 be defined by the condition that $A_1 \overline{A}_2$ be the divisor of $r_1 - \sqrt{D}$. This process can then be repeated to define the sequence $A_0 \sim A_1 \sim A_2 \sim \cdots$. If $a_j = N(A_j)$ then the sequences of integers $a_1, a_2, a_3, \ldots$ and $r_1, r_2, r_3, \ldots$ can be derived from a_0, r_0 using the rules that $a_{i+1} = (r_i^2 - D)/a_i$ and that r_{i+1} is the largest integer which satisfies the conditions $r_{i+1} + r_i \equiv 0 \bmod a_{i+1}$ and $r_{i+1}^2 - D < 0$. (The fact that $\overline{A}_{i+1}$ divides $r_i - \sqrt{D}$. implies that $\sqrt{D} \equiv -r_i \bmod A_{i+1}$. Therefore $\sqrt{D} \equiv r_{i+1} \bmod A_{i+1}$ is equivalent to $r_i + r_{i+1} \equiv 0 \bmod A_{i+1}$. Since A_{i+1} is divisible by no integer, this is equivalent, by the theorem of the preceding section, to $r_i + r_{i+1} \equiv 0 \bmod a_{i+1}$.)

Theorem. *If $A_0 \sim I$ then there exists an $i > 0$ such that $N(A_i) = a_i = 1$. (In other words, if A_0 is principal then the sequence of divisors $A_0 \sim A_1 \sim A_2 \sim \cdots$ defined above reaches $A_i = I$ and the technique of the preceding section gives a method of finding a quadratic integer with the divisor A_0. The tacit assumption $D \equiv 2$ or $3 \bmod 4$ which is used in the definition of the sequence is not at all essential and, as will be shown below, when the natural modifications of the process are made in the case $D \equiv 1 \bmod 4$ the same theorem is true.)*

PROOF. The sequence of divisors $A_0 \sim A_1 \sim A_2 \sim \cdots$ must eventually begin to repeat because each divisor A_j determines all subsequent ones, and because the inequality $|a_j| \leqslant |a_j a_{j+1}| = |r_j^2 - D| < D$ shows that there are only a finite number of possibilities for A_j. Therefore there is no loss of generality (replace A_j by A_{N+j} for some large N) in assuming that A_0 itself repeats. Let this assumption be made and let k be the least positive integer such that $A_k = A_0$. What is to be shown is that under these circumstances if $A_0 \sim I$ then $A_i = I$ for some i, $0 \leqslant i < k$.

To say that $A_0 \sim I$ means that there is a quadratic integer $x_0 + y_0 \sqrt{D}$ with divisor A_0. Then multiplication by $r_0 + \sqrt{D}$ (with divisor $\overline{A}_0 A_1$)

followed by division by a_0 (with divisor $A_0\overline{A}_0$) gives a quadratic integer $x_1 + y_1\sqrt{D} = (x_0 + y_0\sqrt{D})(r_0 + \sqrt{D})/a_0$ with divisor A_1. In the same way, an entire sequence $x_0 + y_0\sqrt{D}$, $x_1 + y_1\sqrt{D}, \ldots$ of quadratic integers $x_j + y_j\sqrt{D}$ with divisors A_j can be defined. Since $A_k = A_0$, the quadratic integers $x_0 + y_0\sqrt{D}$ and $x_k + y_k\sqrt{D}$ have the same divisor and their quotient is therefore a unit, say $x_k + y_k\sqrt{D} = \varepsilon(x_0 + y_0\sqrt{D})$. Then $x_{k+1} + y_{k+1}\sqrt{D} = (x_k + y_k\sqrt{D})(r_0 + \sqrt{D})/a_0 = \varepsilon(x_1 + y_1\sqrt{D})$ and, more generally, $x_{k+j} + y_{k+j}\sqrt{D} = \varepsilon(x_j + y_j\sqrt{D})$. One can take advantage of the cyclic nature of the sequence $A_0, A_1, \ldots, A_k = A_0$ to define A_j for negative integers j as well ($A_j = A_{j+nk}$ where n is large) and when this is done one can also use the formula $x_{j+nk} + y_{j+nk}\sqrt{D} = \varepsilon^n(x_j + y_j\sqrt{D})$ to define $x_j + y_j\sqrt{D}$ with divisor A_j for negative j as well (ε is a unit and therefore is invertible). In order to prove that $A_j = I$ for some j it will suffice to prove that $x_j + y_j\sqrt{D} = \pm 1$ for some j. This can be done as follows.

The idea of the proof is that for $|j|$ large both $|x_j|$ and $|y_j|$ are large, but at one end of the sequence x_j, y_j have like signs whereas at the other end they have opposite signs. Therefore the sign of $x_j y_j$ must change somewhere. It will be shown that a change in the sign of $x_j y_j$ implies a value of j for which $x_j + y_j\sqrt{D} = \pm 1$.

The main difficulty in the proof is the proof that $x_j y_j$ does not have the same sign for all integers j, and the main difficulty in this is the proof that *the unit ε is not trivial*, that is, $\varepsilon \neq \pm 1$. This can be proved as follows.

Virtually by definition

$$\varepsilon = \frac{(r_0 + \sqrt{D})(r_1 + \sqrt{D}) \cdots (r_{k-1} + \sqrt{D})}{a_0 a_1 \ldots a_{k-1}}. \tag{1}$$

Therefore, if it can be shown that *the integers r_j are all positive*, it will follow that in $\varepsilon = u + v\sqrt{D}$, u and v are nonzero integers of like sign. If $a_j > 0$ for some j then $a_{j+1} < 0$ because $a_j a_{j+1} = r_j^2 - D < 0$. Moreover, if $a_j > 0$ then $(r_j + a_j)^2 > D$ because otherwise r_j could have been larger. Thus $r_j^2 - D + 2r_j a_j + a_j^2 > 0$, $a_j(a_{j+1} + 2r_j + a_j) > 0$, $a_{j+1} + 2r_j + a_j > 0$. Therefore $a_{j+1}(a_{j+1} + 2r_j + a_j) < 0$, $a_{j+1}^2 + 2a_{j+1}r_j + r_j^2 - D < 0$, $(-r_j - a_{j+1})^2 < D$. Since r_{j+1} is the largest integer for which $r_{j+1} \equiv -r_j \bmod a_{j+1}$ and $r_{j+1}^2 < D$, it follows that $r_{j+1} \geqslant -r_j - a_{j+1}$. Then $-r_j \leqslant r_{j+1} + a_{j+1} \leqslant r_{j+1}$. Since $r^2 - D < 0$ for $r = -r_j$ and for $r = r_{j+1}$, this implies $(r_{j+1} + a_{j+1})^2 < D$. On the other hand, $(r_{j+1} - a_{j+1})^2 > D$ because otherwise r_{j+1} could have been larger. Thus $(r_{j+1} + a_{j+1})^2 < (r_{j+1} - a_{j+1})^2$, $2r_{j+1}a_{j+1} < -2r_{j+1}a_{j+1}$, $4r_{j+1}a_{j+1} < 0$, $r_{j+1} > 0$, as was to be shown. If $a_j < 0$ then in a similar way one finds $a_{j+1} > 0$, $(r_j - a_j)^2 > D$, $a_{j+1} - 2r_j + a_j < 0$, $(-r_j + a_{j+1})^2 < D$, $r_{j+1} \geqslant -r_j + a_{j+1}$, $-r_j \leqslant r_{j+1} - a_{j+1} \leqslant r_{j+1}$, $(r_{j+1} - a_{j+1})^2 < D$, $(r_{j+1} + a_{j+1})^2 > D > (r_{j+1} - a_{j+1})^2$, $r_{j+1} > 0$. Thus $r_j > 0$ in all cases, as was to be shown.

Let $\varepsilon = u + v\sqrt{D}$. Then u and v are nonzero integers of like sign and $x_{nk} + y_{nk}\sqrt{D} = (x_0 + y_0\sqrt{D})\varepsilon^n$ for all integers n. Let $\varepsilon^n = U_n + V_n\sqrt{D}$. Then $U_{n+1} = U_n u + V_n vD$, $V_{n+1} = U_n v + V_n u$, from which it is clear inductively that U_n and V_n are nonzero integers of like sign for $n > 0$ and $|V_{n+1}| > |V_n|$ so that $|V_{n+1}|$ is arbitrarily large for n large. Now

$$x_{nk} = U_n x_0 + DV_n y_0$$
$$y_{nk} = V_n x_0 + U_n y_0.$$

If x_0 and y_0 have opposite signs, the sign of x_{nk} is the sign of the larger of the two terms $U_n x_0$ and $DV_n y_0$ in absolute value. These two absolute values can be compared by squaring and subtracting to find

$$U_n^2 x_0^2 - D^2 V_n^2 y_0^2 = (\pm 1 + DV_n^2)x_0^2 - D^2 V_n^2 y_0^2$$
$$= \pm x_0^2 + DV_n^2(x_0^2 - Dy_0^2)$$

(because $U_n^2 - DV_n^2 = N(\varepsilon^n) = \pm 1$). Since x_0^2 is fixed and V_n is arbitrarily large, this has the same sign as $x_0^2 - Dy_0^2$ for large n. Thus x_{nk} has the sign of $U_n x_0$ if $x_0^2 - Dy_0^2 > 0$ and the sign of $DV_n y_0$ if $x_0^2 - Dy_0^2 < 0$. If the same method is used to find the sign of y_{nk}, the result is that y_{nk} has the sign of $V_n x_0$ if $0 < V_n^2 x_0^2 - U_n^2 y_0^2 = V_n^2 x_0^2 - (\pm 1 + DV_n^2)y_0^2 = \pm y_0^2 + V_n^2(x_0^2 - Dy_0^2)$, which is true for large n if $x_0^2 - Dy_0^2 > 0$, and has the sign of $U_n y_0$ for large n if $x_0^2 - Dy_0^2 < 0$. Thus x_{nk} and y_{nk} have the same sign (that of $U_n x_0$ and $V_n x_0$) if $x_0^2 - Dy_0^2 > 0$ and n is large, and also have the same sign (that of $DV_n y_0$ and $U_n y_0$) if $x_0^2 - Dy_0^2 < 0$ and n is large. Thus, *if x_0 and y_0 have opposite signs then x_{nk} and y_{nk} have like signs for all sufficiently large n.* If x_0 and y_0 have like signs then $x_{-nk} + y_{-nk}\sqrt{D} = (x_0 + y_0\sqrt{D})\varepsilon^{-n}$, $x_{-nk} - y_{-nk}\sqrt{D} = \pm(x_0 - y_0\sqrt{D})\varepsilon^n$ and, because x_0 and $-y_0$ have opposite signs, the proof just given shows that x_{-nk} and $-y_{-nk}$ have like signs for all sufficiently large n. Thus in this case too the sign of $x_j y_j$ has both values. Therefore *either $x_j y_j = 0$ for some j or there is a j for which $x_j y_j$ and $x_{j+1} y_{j+1}$ have opposite signs.*

The second of these two alternatives is impossible because $x_{j+1} + y_{j+1}\sqrt{D} = (x_j + y_j\sqrt{D})(r_j + \sqrt{D})/a_j$, $(x_{j+1} + y_{j+1}\sqrt{D})(x_j - y_j\sqrt{D}) = r_j + \sqrt{D}$, from which $x_j y_{j+1} - x_{j+1} y_j = 1$. If two integers have opposite sign then their difference is at least 2; therefore this equation implies $x_j y_{j+1} x_{j+1} y_j \geqslant 0$ and it follows that $x_j y_j$ and $x_{j+1} y_{j+1}$ cannot have opposite signs. Therefore $x_j y_j = 0$ *for at least one value of* j. Then $x_{j+1} x_j - Dy_j y_{j+1}$ is equal to r_j on the one hand (by the formula $(x_{j+1} + y_{j+1}\sqrt{D}) \cdot (x_j - y_j\sqrt{D}) = r_j + \sqrt{D}$) and is equal to $x_{j+1} x_j$ or $-Dy_j y_{j+1}$ on the other (because x_j or y_j is zero). Since $|r_j| < \sqrt{D} < D$, the second of these is impossible. Therefore $y_j = 0$ and $1 = x_j y_{j+1} - x_{j+1} y_j = x_j y_{j+1}$, which shows that x_j and y_{j+1} are ± 1. Therefore $x_j + y_j\sqrt{D} = \pm 1$ and $A_j = I$, as was to be shown. This completes the proof of the theorem.

Theorem. *If $D \equiv 2$ or 3 mod 4 and if $A_0 = I$ then, by the preceding theorem, the cyclic method applied to A_0 must arrive back at I. Let k be the least positive integer such that $a_k = \pm 1$ and let ε be defined by* (1). *Then the units in the quadratic integers for the determinant D are precisely the quadratic integers $\pm \varepsilon^n$ where n is an integer.*

PROOF. To apply the cyclic method to A is the same as to apply it to $(-1, *)A$ and to multiply the result by $(-1, *)$. Therefore $(-1, *)$ is principal if and only if the cyclic method applied to $A_0 = I$ arrives at $(-1, *)$ before it arrives back at I. Therefore the ε defined in the theorem has divisor $(-1, *)$ if $(-1, *)$ is principal and otherwise has divisor I. In particular it is always a unit. If $x_0 + y_0 \sqrt{D}$ is any unit then the sequence $x_j + y_j \sqrt{D}$ constructed by the cyclic method contains ± 1. The divisor of $x_0 + y_0 \sqrt{D}$ is I or $(-1, *)$. The divisor of each $x_j + y_j \sqrt{D}$ is then A_j or $(-1, *)A_j$ where $\ldots, A_{-1}, I, A_1, A_2, \ldots$ is the cycle of divisors obtained by applying the cyclic method to I. Moreover, $x_{j+k} + y_{j+k}\sqrt{D} = \varepsilon(x_j + y_j\sqrt{D})$ and the norm of $x_j + y_j\sqrt{D}$ is $\pm a_j$. Since $a_j = \pm 1$ if and only if j is a multiple of k, the equation $x_j + y_j\sqrt{D} = \pm 1$ implies $\varepsilon^n(x_0 + y_0\sqrt{D}) = \pm 1$ and $x_0 + y_0\sqrt{D} = \pm \varepsilon^{-n}$, as was to be shown.

The units $\pm \varepsilon^n$ are all distinct because $\pm \varepsilon^n = \pm \varepsilon^m$ for $m \neq n$ would imply $\varepsilon^\mu = \pm 1$ for some positive μ, contrary to the fact that $\varepsilon^\mu = U_\mu + V_\mu \sqrt{D}$ where U_μ and V_μ are nonzero integers of like sign. Therefore the theorem gives a simple formula for the most general unit and shows how to find the most general quadratic integer with divisor A when any one quadratic integer with divisor A is known. The first theorem shows that if A_0 is the divisor of a quadratic integer then $A_j = I$ for some j in the range $0 \leqslant j < k$ and, since a quadratic integer with divisor I is known, the equivalences $A_0 \sim A_1 \sim \cdots \sim A_j = I$ make it easy to find a quadratic integer with divisor A_0.

This completes the solution, in the case $D > 0$, $D \equiv 2$ or 3 mod 4, of the problem of determining whether a given divisor is principal and, if so, of finding all quadratic integers with that divisor. Only the case $D > 0$, $D \equiv 1$ mod 4 remains. This case is substantially the same as the case just treated, and a brief sketch of the method will suffice. As was shown in the preceding section, every A is equivalent to an A_0 with norm at most $\sqrt{D/3}$. Then there is a halfinteger r such that $r - \frac{1}{2}\sqrt{D} \equiv 0 \bmod A_0$ and $N(r - \frac{1}{2}\sqrt{D}) < 0$. Let r_0 be the largest such halfinteger and let A_1 be defined by the condition that $A_0 \overline{A_1}$ be the divisor of $r_0 - \frac{1}{2}\sqrt{D}$. Then there exist halfintegers r (for example $r = -r_0$) such that $r - \frac{1}{2}\sqrt{D} \equiv 0 \bmod A_1$ and $N(r - \frac{1}{2}\sqrt{D}) < 0$. Let r_1 be the largest such r and let the process be continued in this way to give $A_0 \sim A_1 \sim A_2 \sim \cdots$. What is to be shown is that if $A_0 \sim I$ then $A_i = I$ for some $i > 0$. The proof is exactly the same as in the previous case except for the proof that $x_j y_j$ and $x_{j+1} y_{j+1}$

271

cannot have opposite signs. From $(x_{j+1} + y_{j+1}\sqrt{D})(x_j - y_j\sqrt{D}) = r_j + \frac{1}{2}\sqrt{D}$ it follows that $x_j y_{j+1} - x_{j+1} y_j = \frac{1}{2}, (2x_j)(2y_{j+1}) - (2x_{j+1})(2y_j)$ 2. Since $2x_j$, $2y_j$, $2x_{j+1}$, $2y_{j+1}$ are all integers, this implies that $x_j y_{j+1}$ and $x_{j+1} y_j$ have the same sign unless one of them is zero or unless $|(2x_j)(2y_{j+1})| = |(2x_{j+1})(2y_j)| = 1$. In the latter case $x_j, y_j, x_{j+1}, y_{j+1}$ are all $\pm \frac{1}{2}$. But then $N(r_j + \frac{1}{2}\sqrt{D}) = N(x_j + y_j\sqrt{D})N(x_{j+1} + y_{j+1}\sqrt{D}) = [(\frac{1}{2})^2 - D(\frac{1}{2})^2]^2 > 0$ contrary to assumption.

The fundamental unit ε in the case $D \equiv 1 \bmod 4$ is found by the obvious modification of the above theorem for the case $D \equiv 2$ or $3 \bmod 4$, namely:

Theorem. *If $D \equiv 1 \bmod 4$ (D squarefree and positive), if $A_0 = I$, and if*

$$\varepsilon = \frac{\left(r_0 + \frac{1}{2}\sqrt{D}\right)\left(r_1 + \frac{1}{2}\sqrt{D}\right) \cdots \left(r_{k-1} + \frac{1}{2}\sqrt{D}\right)}{a_0 a_1 \ldots a_{k-1}}$$

where the a's and r's are defined by the cyclic method and where k is the least positive integer for which $a_k = \pm 1$, then the units are the quadratic integers $\pm \varepsilon^n$ where n is an integer.

EXERCISES

1. Find all units among the quadratic integers for the determinant $D = 61$. Compare to the solution of Pell's equation $x^2 = 61y^2 + 1$.

2. Find all units among the quadratic integers for the determinant $D = 109$.

7.6 The divisor class group: examples

In Chapter 5 the notions of "class number" and "representative set" were defined, but no attempt was made to compute the class number (which is difficult even for $\lambda = 11$) or to find a representative set. Because the cyclic method gives such a simple way of determining whether a given divisor is principal, the solution of both of these problems is quite elementary in the case of divisors of quadratic integers.

As a first example, consider the case $D = 67$. Since $D \equiv 3 \bmod 4$ every divisor is equivalent to one which has norm at most $2\sqrt{67/3} < 10$ and which is divisible by no integer greater than 1. Of the primes 2, 3, 5, 7 in this range, 2 ramifies ($67 \equiv 3 \bmod 4$), 3 splits ($67 \equiv 1 = 1^2 \bmod 3$), 5 remains prime ($67 \equiv 2 \not\equiv n^2 \bmod 5$ because $n^2 \equiv 0, 1, 4 \bmod 5$) and 7 splits ($67 \equiv 4 = 2^2 \bmod 7$). In addition, because $D > 0$, the modification of Section 7.3 calls for the inclusion of the divisor $(-1, *)$ with norm -1. Thus every divisor is equivalent to one of the divisors I, $(2, *)$, $(3, \pm 1)$, $(2, *)(3, \pm 1)$, $(7, \pm 2)$, $(3, \pm 1)^2$ or to $(-1, *)$ times one of these. The cyclic method applied to $(-1, *)$ gives

$$
\begin{array}{ccccccccccc}
r = & 8 & 7 & 5 & 2 & 7 & 7 & 2 & 5 & 7 & 8 & 8 \\
a = & -1 & 3 & -6 & 7 & -9 & 2 & -9 & 7 & -6 & 3 & -1
\end{array}
$$

after which it repeats. Since $a = 1$ is not reached, $(-1, *)$ is *not* principal and the class number is at least two. On the other hand, the above computation gives the equivalences $(-1, *) \sim (3, 1) \sim (-1, *)(2, *)(3, 2) \sim (7, 2) \sim (-1, *)(3, 1)^2 \sim (2, *) \sim (-1, *)(3, 2)^2 \sim (7, -2) \sim (-1, *)(2, *) \cdot (3, 1) \sim (3, -1)$. (For example, $a_4 = -9$ so that A_4 is a divisor with norm -9. It is not divisible by 3. Therefore it is of the form $(-1, *)(3, \pm 1)^2$. Because it divides $r_4 - \sqrt{67} = 7 - \sqrt{67}$ it must be $(-1, *)(3, 1)^2$.) Multiplication of these equivalences by $(-1, *)$ gives $I \sim (-1, *)(3, 1) \sim (2, *)(3, 2) \sim \cdots \sim (-1, *)(3, -1)$. This accounts for all of the divisors above and proves that *the class number is 2 and I*, $(-1, *)$ *is a representative set.* Moreover, as the above equivalences show, $(-1, *) \sim (2, *) \sim (3, \pm 1) \sim (7, \pm 2)$, after which all the above equivalences follow from the fact that equivalence is consistent with multiplication; for example, $(-1, *)(2, *)(3, 2) \sim (-1, *)(-1, *)(-1, *) = (-1, *)$.

As a second example, consider the case $D = -165$. Here 2 ramifies ($D \equiv 3 \bmod 4$), 3 ramifies, 5 ramifies, 7 remains prime ($-165 \equiv 3 \bmod 7$ and the squares mod 7 are 0, 1, 2, 4), 11 ramifies and 13 splits ($-165 \equiv 4 = 2^2 \bmod 13$). On the other hand, every divisor is equivalent to one with norm less than $2\sqrt{165/3} = \sqrt{220} < 15$. Thus every divisor is equivalent to at least one of the divisors I, $(2, *)$, $(3, *)$, $(5, *)$, $(2, *)(3, *)$, $(2, *)(5, *)$, $(11, *)$, $(13, \pm 2)$. The reduction process applied to any one of these divisors only gives one of larger norm—for example, $(2, *)(5, *)$ divides $5 - \sqrt{-165}$, which has norm 190, from which $(2, *)(5, *) \sim (19, -5)$ —except in the case of $(13, \pm 2)$. Because $(13, 2)$ divides $2 - \sqrt{-165}$, because 13 does not divide $2 - \sqrt{-165}$, and because $N(2 - \sqrt{-165}) = 13^2$, the divisor of $2 - \sqrt{-165}$ is $(13, 2)^2$ and $(13, 2) \sim (13, -2)$. Therefore the above list of 9 divisors gives, when one of the 2 divisors $(13, \pm 2)$ is omitted, a list of 8 divisors among which only I is principal, and every divisor is equivalent to at least one of these 8.

The group structure, that is, the multiplicative structure, of the divisor classes is easy to find in this case. Let $A = (2, *)$, $B = (3, *)$, and $C = (5, *)$. Then of course $A^2 \sim B^2 \sim C^2 \sim I$, $AB = (2, *)(3, *)$, and $AC = (2, *)(5, *)$. The class of BC can be found by noting that the divisor of $\sqrt{-165}$ is $(3, *)(5, *)(11, *)$ so that $BC \sim (11, *)B^2C^2 \sim (11, *)$. Then ABC must be equivalent to the remaining divisor $ABC \sim (13, \pm 2)$ because otherwise it would be equivalent to one of the others and this would imply that one of the others was principal. (For example, if $ABC \sim B$ then $AC \sim AB^2C \sim B^2 \sim I$, contrary to the fact that $(2, *)(5, *)$ is not principal.) In the same way, no 2 of the 8 divisors I, A, B, C, AB, BC, AC, and ABC can be equivalent. Thus *the class number is 8 and the above 8 divisors are a representative set.*

(Of course it is simple to show directly that $ABC \sim (13, \pm 2)$. See Exercise 7.)

The case $D = -163$ gives a very different result. Since $D \equiv 1 \bmod 4$, every divisor is equivalent to one with norm at most $\sqrt{163/3} < 8$. Now 2 remains prime ($D \equiv 5 \bmod 8$), 3 remains prime ($D \equiv -1 \bmod 3$), 5 remains prime ($D \equiv 2 \bmod 5$), and 7 remains prime ($D \equiv 5 \bmod 7$). Therefore the *only* divisors with norm < 8 are the principal divisors I and (2), *the class number is 1 and I is a representative set*. Therefore *unique factorization holds for quadratic integers for the determinant* $D = -163$. This has as a consequence Euler's famous theorem that $x^2 + x + 41$ is prime for $x = 0, 1, 2, \ldots, 39$ (see Exercise 8).

The case $D = 79$ is one that Gauss frequently uses as an example* (*Disquisitiones Arithmeticae* Arts. 185, 186, 187, 195, 196, 198, 205, 223). When $D = 79$ every divisor is equivalent to one with norm at most $2\sqrt{79/3} < 11$. Such a divisor must be a product of $(-1, *)$, $(2, *)$, $(3, \pm 1)$, $(5, \pm 2)$, $(7, \pm 3)$. The computation to determine whether $(-1, *)$ is principal is

$$\begin{array}{rccccc} r = & 8 & 7 & 7 & 8 \\ a = & -1 & 15 & -2 & 15 & -1 \end{array}$$

and $(-1, *)$ is therefore not principal. At the same time this computation shows that $(-1, *) \sim (-1, *)(2, *)$, from which $I = (-1, *)^2 \sim (2, *)$, and $(2, *)$ is principal. In fact, $(2, *)$ is the divisor of $9 - \sqrt{79}$. The computation to determine whether $(3, 1)$ is principal is:

$$\begin{array}{rccccccc} r = & 7 & 3 & 4 & 5 & 7 & 8 & 7 \\ a = & 3 & -10 & 7 & -9 & 6 & -5 & 3 \end{array}$$

Therefore $(3, 1)$ is not principal and $(3, 1) \sim (-1, *)(2, *)(5, -2) \sim (7, -3) \sim (-1, *)(3, -1)^2 \sim (2, *)(3, 1) \sim (-1, *)(5, -2)$. Let $A = (-1, *)$ and $B = (3, 1)$. Then $B \sim A\overline{B}^2$, $B^3 \sim A(B\overline{B})^2 \sim A$. Thus $B^6 \sim I$ but $B^3 \not\sim I$. Now if $B^2 \sim I$ then $B \sim A\overline{B}^2 \sim A$, $AB \sim I$. But the computation to determine whether $AB = (-1, *)(3, 1)$ is principal is simply the above computation with the signs of the a's reversed. Therefore $AB \not\sim I$, $A \not\sim B$, $B^2 \not\sim I$. Therefore I, B, B^2, B^3, B^4, B^5 are six inequivalent[†] divisors. Moreover, $(-1, *) \sim B^3$, $(2, *) \sim I$, $(3, 1) = B$, $(3, -1) = \overline{B} \sim B^5$, $(5, -2) \sim B^4$ (because $(3, 1) \sim (-1, *)(5, -2)$ gives $(5, -2) \sim (-1, *)(3, 1) \sim B^4$), $(5, 2) \sim B^2$, $(7, -3) \sim B$ (because $A_2 = (7, -3)$ in the second com-

---

*The case $D = 79$ is of interest because it is the least positive determinant for which there is more than 1 divisor class per genus (see Section 7.9 for the definition of "genus"). This fact was latent in Lagrange's discovery that a conjecture of Euler concerning primes of the form $x^2 - Dy^2$ fails for $D = 79$ (see Exercise 8 of Section 7.9), and Gauss's interest in $D = 79$ may be related to Lagrange's counterexample.

[†]$B^4 \sim I$ would imply $B^2 \sim B^2 B^6 \sim (B^4)^2 \sim I$ and is therefore false. Similarly $B^5 \sim I$ would imply $B \sim B \cdot B^5 \sim I$ and is false. Thus no two of the divisors $B^0, B^1, \ldots, B^5$ are equivalent because otherwise it would follow that $B^j \sim I$ for some j, $0 < j < 6$.

putation above), $(7, 3) \sim B^5$. Therefore *the class number is 6 and I, B,* B^2, B^3, B^4, B^5 *is a representative set*, where $B = (3, 1)$.

Representative sets should be chosen in such a way that the multiplication table is easily described. For example, in the preceding case the formula $B^6 \sim I$ describes the whole multiplication table, and in the case $D = -165$ the relations $A^2 \sim I$, $B^2 \sim I$, $C^2 \sim I$ do the same. In the case* $D = -161$ it is a little less obvious how this can be done. In this case 2 ramifies ($-161 \equiv 3 \bmod 4$), 3 splits ($-161 \equiv 1 \bmod 3$), 5 splits ($-161 \equiv 4 \bmod 5$), 7 ramifies ($-161 \equiv 0 \bmod 7$), 11 splits ($-161 \equiv 4 \bmod 11$) and 13 remains prime ($-161 \equiv 8 \equiv -5 \bmod 13$ and the squares mod 13 are $1, 4, 9 \equiv -4, 3, -1, -3$). The divisors with norm less than $2\sqrt{|D|/3} <$ 15 divisible by no integer greater than 1 are thus I, $(2, *)$, $(3, 1)$, $(3, -1)$, $(5, 2)$, $(5, -2)$, $(2, *)(3, 1)$, $(2, *)(3, -1)$, $(7, *)$, $(3, 1)^2$, $(3, -1)^2$, $(2, *)(5, 2)$, $(2, *)(5, -2)$, $(11, 2)$, $(11, -2)$, $(2, *)(7, *)$. Every divisor is equivalent to at least one of these 16.

It is easily checked that none of these 16 other than I is principal. (For example, $(11, 2)$ divides $2 - \sqrt{-161}$ which has norm $165 = 11 \cdot 15$ and the cyclic method does not reduce $(11, 2)$.) Let $A = (3, 1)$. Then A^2 is among the 16 but A^3 is not. The reduction of A^3 is

$$r = \quad 1 \quad -1$$
$$a = 27 \quad 6$$

and $A^3 \sim (2, *)(3, -1)$. Thus $A^4 \sim (2, *)(3, -1)(3, 1) \sim (2, *)$, $A^5 \sim (2, *)(3, 1)$, $A^6 \sim (2, *)^2(3, -1)^2 \sim (3, -1)^2$, and $A^7 \sim (3, -1)^2(3, 1) \sim (3, -1)$ are all nonprincipal but $A^8 \sim I$. Not every divisor is equivalent to a power of A, however. Explicitly, $(7, *)$ is a divisor whose square is principal; therefore it can be equivalent to no powers of A other than I or A^4 and, because $A^4 \sim (2, *)$ and neither $(7, *)$ nor $(7, *)(2, *)$ is principal, $(7, *)$ is not equivalent to a power of A. Let $B = (7, *)$. Then, as in the proof of Fermat's theorem, the divisors B, BA, BA^2, ..., BA^7 are inequivalent to each other and inequivalent to I, A, A^2, ..., A^7. Thus there are at least 16 inequivalent divisors and the 16 divisors above must be inequivalent and must each be equivalent to a unique divisor of the form $A^i B^j$ ($i = 0, 1, \ldots, 7$; $j = 0, 1$). Explicitly, $AB = (3, 1)(7, *)$ reduces by the cyclic method

$$r = \quad 7 \quad 3$$
$$a = 21 \quad 10 \quad 17$$

to $AB \sim (2, *)(5, -2)$. Similarly $A^2B \sim (3, 1)(2, *)(5, -2) \sim (11, -2)$, $A^3B \sim (3, 1)(11, -2) \sim (5, 2)$, $A^4B \sim (2, *)(7, *)$, $A^5B \sim A^3B \sim (5, -2)$, $A^6B \sim (11, 2)$, $A^7B \sim (2, *)(5, 2)$.

In summary, when $D = -161$ the 16 divisors $A^i B^j$ ($i = 0, 1, \ldots, 7$; $j = 0, 1$) are a representative set where $A = (3, 1)$, $B = (7, *)$, and $A^8 \sim I$, $B^2 \sim I$. In the language of group theory, *the divisor class group is the abelian*

---

*This case is considered by Gauss in Article 307 of the *Disquisitiones Arithmeticae*.

group generated by two (unrelated) generators A and B of orders 8 *and* 2 *respectively.* The advantages of the representative set $\{A^iB^j\}$ over the previous one $(2, *), (3, \pm 1), \ldots$ should be obvious. Once a few prime divisors are expressed in terms of A and B—$(2, *) \sim A^4, (3, 1) \sim A, (5, 2) \sim A^3B, (7, *) \sim B, (11, 2) \sim A^6B$—the class of any divisor is easy to compute. For example, $(3, -1)(11, 2)^2 \sim A^{-1}(A^6B)^2 \sim A^3$. The classification of other prime divisors is also simple. For example, $(17, 3) \sim (2, *)(5, 2) \sim A^4A^3B = A^7B, (23, *) \sim (7, *) \sim B, (29, 10) \sim (3, -1)^2 \sim A^{-2} \sim A^6$, and so forth.

As a final example, consider the case $D = 985$. Here $D \equiv 1 \bmod 4$ and in fact $D \equiv 1 \bmod 8$. Therefore 2 splits. The computation to determine whether $(2, 0)$ is principal $((2, 0)$ is the prime divisor of 2 mod which $\omega = \frac{1}{2} - \frac{1}{2}\sqrt{985} \equiv 0)$ is

$$
\begin{array}{ccccccccccc}
r = & \frac{29}{2} & \frac{7}{2} & \frac{19}{2} & \frac{29}{2} & \frac{31}{2} & \frac{29}{2} & \frac{7}{2} & \frac{19}{2} & \frac{29}{2} & \frac{31}{2} & \frac{29}{2} \\
a = 2 & -18 & 13 & -12 & 3 & -2 & 18 & -13 & 12 & -3 & 2
\end{array}
$$

which not only shows that $(2, 0)$ is not principal but also shows that $(-1, *)$ is principal—because $(2, 0) \sim (-1, *)(2, 0)$—and that $(2, 0) \sim (2, 1)(3, 1)^2 \sim (13, 6) \sim (2, 0)^2(3, -1) \sim (3, 1)$. The computation to determine whether the prime divisor $(5, *)$ is principal is

$$
\begin{array}{cccccccccc}
r = & \frac{25}{2} & \frac{11}{2} & \frac{13}{2} & \frac{21}{2} & \frac{27}{2} & \frac{21}{2} & \frac{13}{2} & \frac{11}{2} & \frac{25}{2} & \frac{25}{2} \\
a = 5 & -18 & 12 & -17 & 8 & -8 & 17 & -12 & 18 & -5
\end{array}
$$

which shows that $(5, *)$ is not principal and also gives $(5, *) \sim (2, 1)(3, -1)^2 \sim (2, 0)^2(3, 1) \sim (17, 4) \sim (2, 1)^3 \sim (2, 0)^3 \sim (17, -4) \sim \cdots$. Thus $(2, 0)^6 \sim (5, *)^2 \sim I$. Let $A = (2, 0) \sim (3, 1)$. Then A^3 is not principal but A^6 is. It is easily checked that A^2 is not principal:

$$
\begin{array}{cccccccc}
r = & \frac{25}{2} & \frac{5}{2} & \frac{27}{2} & \frac{29}{2} & \frac{25}{2} & \frac{15}{2} & \frac{23}{2} & \frac{25}{2} \\
a = 6 & -15 & 16 & -4 & 9 & -10 & 19 & -6
\end{array}
$$

and it follows that I, A, A^2, A^3, A^4, A^5 are inequivalent. Moreover, $(-1, *) \sim I, (2, 0) \sim A, (2, 1) \sim A^5, (3, 1) \sim A, (3, -1) \sim A^5, (5, *) \sim A^3, (13, 6) \sim A, (13, -6) \sim A^5, (17, \pm 4) \sim A^3$. Every divisor is equivalent to one with norm $< \sqrt{D/3} < \sqrt{330} < 19$. Thus every divisor is equivalent to a product of prime divisors with norms less than 19 and the above will show that every divisor is equivalent to a power of A provided it is shown that there are no prime divisors of norm 7 or 11, that is, provided it is shown that 7 and 11 remain prime. This follows from $985 \equiv 5 \not\equiv x^2 \bmod 7, 985 \equiv -5 \equiv -4^2 \bmod 11, -1 \not\equiv x^2 \bmod 11$. Thus *the class number is 6 and every divisor is equivalent to a power of A*, where $A = (2, 0)$ and $A^6 \sim I$.

276

EXERCISES

Treat each of the following cases like those of the text—that is, find a representative set and express the multiplication table in as simple a form as possible.

1. $D = 61$.

4. $D = 305$.

2. $D = -235$.

5. $D = -129$.

3. $D = 145$.

6. $D = -105$.

7. Use the cyclic method to prove in the case $D = -165$ that $(2, *)(3, *)(5, *) \sim (13, \pm 2)$.

8. Prove that $x^2 + x + 41$ is prime for $x = 0, 1, \ldots, 39$. [Any prime divisor of $x^2 + x + 41$ must split in quadratic integers for the determinant -163. Because every divisor is principal, $\frac{1}{2}[(2x + 1) + \sqrt{-163}] = (r + s\sqrt{-163}) \cdot (u + v\sqrt{-163})$ where $r^2 + 163s^2$ is the given prime factor of $x^2 + x + 41$. This gives $(2r)(2v) + (2s)(2u) = 2$. Normally the two terms in this sum have opposite sign, which means that the terms of $ru - 163sv = \frac{1}{2}(2x + 1)$ have like sign. In this case $x \geqslant 0$ implies $x \geqslant 40\frac{1}{2}$. If $rsuv = 0$ then $v = 0$, $u = \pm 1$, and $x^2 + x + 41 = r^2 + 163s^2 = $ prime, as desired.]

7.7 The divisor class group: a general theorem

The technique that was used in the preceding section to test whether $A \sim B$ was to apply the cyclic method to $A\bar{B}$ to determine whether it is principal. There is another method for testing whether $A \sim B$ which is usually better. Let the cyclic method be applied to A and B separately to give two sequences $A \sim A_1 \sim A_2 \sim \cdots$ and $B \sim B_1 \sim B_2 \sim \cdots$ of equivalent divisors. If there exist integers j and k such that $A_j = B_k$ then of course $A \sim A_j = B_k \sim B$. The sequences $A \sim A_1 \sim A_2 \sim \cdots$, $B \sim B_1 \sim B_2 \sim \cdots$ both eventually become periodic, so that there are only finitely many A_j and B_k and the condition $A_j = B_k$ is one that can be tested computationally. The theorem to be proved in this section is that the sufficient condition $A_j = B_k$ is also necessary—that is, *if $A \sim B$ then there exist integers j and k such that $A_j = B_k$*, where $A \sim A_1 \sim A_2 \sim \cdots$ and $B \sim B_1 \sim B_2 \sim \cdots$ are the divisors obtained by applying the cyclic method to A and B.

Let a divisor A_j be called *reduced* if application of the cyclic method to A_j eventually returns to A_j. Then application of the cyclic method to A_j produces a finite cycle of equivalent reduced divisors. Such a cycle is called a *period* of reduced divisors. The cyclic method applied to any A gives a period of reduced divisors equivalent to A. Similarly, there is, for any given B, a period of reduced divisors equivalent to B. If $A \sim B$ then $A_j \sim B_k$ where A_j and B_k are reduced divisors equivalent to A and B

respectively. The cyclic method applied to A_j yields just the divisors in its period and applied to B_k yields just the divisors in its period. According to the theorem to be proved, $A_j \sim B_k$ implies that two of the divisors in these periods must coincide and therefore that the periods themselves must be *identical*, that is, consist of the same divisors in the same cyclic order. Otherwise stated, the theorem to be proved implies that *two reduced divisors are equivalent only if their periods are the same*. Conversely, if this fact is known and if $A \sim B$ then there exist reduced divisors $A_j \sim A$ and $B_k \sim B$, the period of A_j must coincide with the period of B_k, there is an i such that $A_{j+i} = B_k$, and the theorem to be proved follows.

In cases where $D < 0$ this is simple to prove. Consider first the case $D < 0$, $D \equiv 2$ or 3 mod 4. Let A and B be given equivalent divisors $A \sim B$. One can assume without loss of generality not only that A and B are reduced, but also that application of the cyclic method to either A or B increases its norm. The assumption $A \sim B$ implies that there is a quadratic integer $x + y\sqrt{D}$ with divisor $A\bar{B}$. Then $x^2 - Dy^2 = ab$ where $a = N(A)$ and $b = N(b)$. Because $a \le 2\sqrt{|D|/3}$ and $b \le 2\sqrt{|D|/3}$ (otherwise the next step of the cyclic method would reduce the norm) this gives $x^2 - Dy^2 \le 4|D|/3$. Therefore $y^2 = 0$ or 1.

Case 1: If $y^2 = 0$ then $A\bar{B}$ is the divisor of the integer x. In particular $x \equiv 0$ mod A, $x \equiv 0$ mod a, $x = ua$ for some integer u. Similarly $x = vb$ for some v. Then $ab = x^2 = uvab$, $uv = 1$, $u = v = \pm 1$. Thus a, x, and b all have the same divisor, from which $A\bar{A} = A\bar{B} = B\bar{B}$, and $A = B$ as was to be shown.

Case 2: If $y^2 = 1$ then either $x + y\sqrt{D}$ or $-x - y\sqrt{D}$ has the form $u - \sqrt{D}$ and has divisor $A\bar{B}$. Then, since application of the cyclic method to A gives $r - \sqrt{D} \equiv 0$ mod A, $|r|$ as small as possible, $a' = (r^2 - D)/a \ge a$, one has $|r| \le |u|$, $b = (x^2 - D)/a = (u^2 - D)/a \ge (r^2 - D)/a \ge a$. Similarly $a \ge b$. Thus $a = b$, $|u| = |r|$, and $|u| = |r'|$ where $r' \equiv \sqrt{D}$ mod B and $|r'|$ is as small as possible. The common value of $|r| = |u| = |r'|$ is at most $\frac{1}{2}a$. If it is $\frac{1}{2}a$ then $(r^2 - D)/a = a$ gives $\frac{1}{4}a^2 - D = a^2$, $D = -3(a/2)^2$ where a is even. Since D is squarefree, this implies $a = 2$, $D = -3$, contrary to the assumption $D \not\equiv 1$ mod 4. Therefore $|r| = |r'| < \frac{1}{2}a$. Then $r \equiv u \equiv -r'$ mod a implies $r = -r'$ and B is the successor of A in the cyclic method. This completes the proof.

The proof in the case $D < 0$, $D \equiv 1$ mod 4, is a simple modification of this proof and will be left to the reader (Exercise 2). On the other hand, the proof in the cases $D > 0$ is considerably more difficult.* Consider first the

---

*Gauss devotes Articles 188–193 of the *Disquisitiones Arithmeticae* to the proof of this theorem. His theorem is slightly more general than the one proved here, but, as will be shown in Chapter 8, the proof of the more general case is no more difficult. Dirichlet calls the theorem "the most difficult question" in the theory ([D7], §80). Dirichlet's proof, which is based on the theory of continued fractions, is very different in appearance from the one given here, but it is nearly the same in substance.

case $D > 0$, $D \equiv 2$ or $3 \bmod 4$. What is to be shown is that reduced divisors A and B in different periods cannot be equivalent. It is natural, then, to begin the proof with an analysis of the notion of the equivalence of two divisors.

The main idea of the proof which follows is to show that an equivalence between two divisors A and B gives rise to a 2×2 matrix of integers, and to show that, if A and B are reduced, then enough can be said about this 2×2 matrix to conclude that it must come from an equivalence of A with one of the A_j in its cycle given by the cyclic method. An equivalence gives rise to a 2×2 matrix as follows.

Assume that A is divisible by no integer greater than 1. Then $x + y\sqrt{D}$ is divisible by A if and only if $x + yr \equiv 0 \bmod a$ where $a = N(A)$ and $r \equiv \sqrt{D} \bmod A$. In other words, $x + y\sqrt{D}$ is divisible by A if and only if there exist integers u and v such that $x + y\sqrt{D} = au + (r - \sqrt{D})v$. Similarly, if B is divisible by no integer greater than 1 then a quadratic integer is divisible by B if and only if it is of the form $bu' + (s - \sqrt{D})v'$ where $b = N(B)$, $s \equiv \sqrt{D} \bmod B$, and u' and v' are integers. If A and B are equivalent divisors, then there is a quadratic integer $x + y\sqrt{D}$ with divisor $\bar{A}B$. The operation of multiplication by $x + y\sqrt{D}$ followed by division by a carries quadratic integers divisible by A to quadratic integers divisible by B (because it carries a quadratic integer with divisor AC to one with divisor $A\bar{C}\bar{A}B/A\bar{A} = BC$). When elements in the domain of this operation are written $au + (r - \sqrt{D})v$ and those in the range $bu' + (s - \sqrt{D})v'$ then the operation is described by a 2×2 matrix of integers

$$\begin{pmatrix} u' \\ v' \end{pmatrix} = \begin{pmatrix} \alpha & \beta \\ \gamma & \delta \end{pmatrix}\begin{pmatrix} u \\ v \end{pmatrix}$$

where α, β, γ, δ can be found by setting $u = 1$, $v = 0$ to find $b\alpha + (s - \sqrt{D})\gamma = bu' + (s - \sqrt{D})v' = [a \cdot 1 + (r - \sqrt{D}) \cdot 0](x + y\sqrt{D})/a = x + y\sqrt{D} = x + ys - y(s - \sqrt{D})$, from which $\gamma = -y$, $\alpha = (x + ys)/b$, and by setting $u = 0$, $v = 1$ to find $b\beta + (s - \sqrt{D})\delta = bu' + (s - \sqrt{D})v' = [a \cdot 0 + (r - \sqrt{D}) \cdot 1](x + y\sqrt{D})/a = [(rx - yD) + (ry - x)\sqrt{D}]/a$, from which $\delta = (x - ry)/a$ and β is the rather complicated expression $\beta = (rx - yD - sx + rsy)/ab$ which will not be needed in what follows. In summary, the operation of multiplication by $x + y\sqrt{D}$ followed by division by a, which carries quadratic integers $au + (r - \sqrt{D})v$ to quadratic integers $bu' + (s - \sqrt{D})v'$ is given by the 2×2 matrix of integers

$$\begin{bmatrix} u' \\ v' \end{bmatrix} = \begin{bmatrix} \dfrac{x + ys}{b} & \dfrac{rx - yD - sx + rsy}{ab} \\ -y & \dfrac{x - yr}{a} \end{bmatrix}\begin{bmatrix} u \\ v \end{bmatrix}.$$

The inverse of this operation is multiplication by $x - y\sqrt{D}$ followed by division by b (because $(x + y\sqrt{D})(x - y\sqrt{D})/ab = 1$). Therefore the inverse of the above 2×2 matrix is

$$
\begin{bmatrix}
\dfrac{x - yr}{a} & \dfrac{sx + yD - rx - rsy}{ab} \\[2ex]
y & \dfrac{x + ys}{b}
\end{bmatrix}
$$

because this is the same operation as before with B in place of A, A in place of B, and $x - y\sqrt{D}$ in place of $x + y\sqrt{D}$. This shows in particular that both of these matrices have determinant 1.

In the special case where $A = A_0$ and $B = A_1$ is its successor in the cyclic method, $x + y\sqrt{D} = r_0 + \sqrt{D}$ and the 2×2 matrix is

$$
\begin{pmatrix} n_1 & 1 \\ -1 & 0 \end{pmatrix}
$$

where $n_1 = (r_0 + r_1)/a_1$ and where the entry in the upper right corner need not be computed because the fact that it is 1 follows from the fact that the second row is $-1, 0$ and the determinant is 1. Therefore the operation carrying quadratic integers divisible by A_0 to quadratic integers divisible by A_j which corresponds to the string of equivalences $A_0 \sim A_1 \sim A_2 \sim \cdots \sim A_j$ is given by the 2×2 matrix

$$
\begin{pmatrix} n_j & 1 \\ -1 & 0 \end{pmatrix}\begin{pmatrix} n_{j-1} & 1 \\ -1 & 0 \end{pmatrix} \cdots \begin{pmatrix} n_2 & 1 \\ -1 & 0 \end{pmatrix}\begin{pmatrix} n_1 & 1 \\ -1 & 0 \end{pmatrix}. \tag{1}
$$

In particular, let E_A be the matrix which results when j is taken to be the least positive integer for which $A_j = A_0$. (There is such a j by virtue of the assumption that A_0 is reduced.) Given any explicit equivalence $A \sim B$ corresponding to a 2×2 matrix M, the matrix ME_A^n also corresponds to an equivalence $A \sim B$ for all positive integers n. The method of proof will be to show that for n sufficiently large the matrix ME_A^n can be written in the form (1) and to conclude from the uniqueness of such a representation that it *is* the matrix (1) for some j and, in particular, to conclude that $B = A_j$ for some j.

The first step in this program is to characterize the matrices of the form (1). It is important to notice that *the n's alternate in sign* because $n_j = (r_{j-1} + r_j)/a_j$, the r's are positive (as was shown in Section 7.5) and the a's alternate in sign (because $a_j a_{j+1} = r_j^2 - D < 0$).

Theorem. *If a 2×2 matrix of integers M has the form*

$$
\begin{pmatrix} X & Y \\ Z & W \end{pmatrix} = \pm \begin{pmatrix} n_j & 1 \\ -1 & 0 \end{pmatrix}\begin{pmatrix} n_{j-1} & 1 \\ -1 & 0 \end{pmatrix} \cdots \begin{pmatrix} n_1 & 1 \\ -1 & 0 \end{pmatrix} \tag{2}
$$

in which $n_1, n_2, \ldots, n_j$ is a sequence of integers which alternates in sign *then $XW - YZ = 1$, $|X| \geqslant |Z| \geqslant |W|$, and $|X| \geqslant |Y| \geqslant |W|$. If a matrix can be written in this form then this can be done in only one way; that is, X, Y, Z, and W determine the value of j, the alternating sequence $n_1, n_2, \ldots, n_j$, and the sign in front. Finally, the three necessary conditions $XW - YZ = 1$, $|X| \geqslant |Z| \geqslant |W|$ and $|X| \geqslant |Y| \geqslant |W|$ are also sufficient for a matrix to have a representation in the form* (2).

PROOF. $XW - YZ = 1$ because the determinant of a product is equal to the product of the determinants. In addition to the conditions $|X| \geqslant |Z| \geqslant |W|$, $|X| \geqslant |Y| \geqslant |W|$ it will be shown that the sign* of XZ is opposite to the sign of n_j. If $j = 1$ all three of these conditions obviously hold. Suppose that they hold for $j - 1$ for some $j > 1$. Then

$$\begin{pmatrix} X & Y \\ Z & W \end{pmatrix} = \begin{pmatrix} n_j & 1 \\ -1 & 0 \end{pmatrix} \begin{pmatrix} X' & Y' \\ Z' & W' \end{pmatrix} \tag{3}$$

where $|X'| \geqslant |Z'| \geqslant |W'|$, $|X'| \geqslant |Y'| \geqslant |W'|$, and the sign of $X'Z'$ is opposite to the sign of n_{j-1}, the same as the sign of n_j. Thus $X = n_j X' + Z'$ where $n_j X' Z' > 0$ and where, as a result, the two terms $n_j X'$ and Z' have the same sign. Then $|X| = |n_j||X'| + |Z'| \geqslant |X'| = |-X'| = |Z|$. The sign of $XZ = (n_j X' + Z')(-X') = -n_j(X')^2 - X'Z'$ is opposite to the sign of n_j. Moreover $Z = -X'$ and $W = -Y'$ so that $|Z| \geqslant |W|$. It remains to show that $|X| \geqslant |Y| \geqslant |W|$. From $X'W' - Y'Z' = 1$ one has $X'W'Y'Z' = (Y'Z')^2 + Y'Z'$ and $X'W'Y'Z' \geqslant 0$ because $x^2 + x \geqslant 0$ for all integers x. Thus $W'Y'$ is either zero or has the sign of $X'Z'$ and n_j. The terms of $Y = n_j Y' + W'$ do not have opposite sign and $|Y| = |n_j||Y'| + |W'| \geqslant |-Y'| = |W|$. Finally, $|Y| = |n_j||Y'| + |W'| \leqslant |n_j||X'| + |Z'| = |X|$ and the first statement of the theorem is proved.

Now suppose that a representation in the form (2) is given. Then, as was just shown, the sign of n_j is determined by the fact that it is opposite to the sign of XZ. Moreover, the equation $|X| = |n_j||X'| + |Z'| = |n_j||Z| + r$ where $0 \leqslant r \leqslant |Z|$ shows that $|n_j|$ is determined as the quotient when $|X|$ is divided by $|Z|$ unless $r = 0$ or $|Z|$, that is, unless $|Z|$ divides $|X|$ evenly. If $|Z|$ divides $|X|$ evenly then Z divides $XW - YZ = 1$ and Z must be ± 1. Therefore, unless $Z = \pm 1$, the value of n_j can be determined from X and Z (although the value of j cannot), the equation (3) can be solved for X', Y', Z', W', and the process can be repeated to find the values of n_{j-1}, n_{j-2}, and so forth, until $Z = \pm 1$ is reached. (It must be reached, by the principle of infinite descent, because $r > 0$, $|X| > |Z| = |X'|$, $|X'| > |X''|$ and so forth.) The proof of the uniqueness of the representation is therefore reduced to the case $Z = \pm 1$.

If $Z = \pm 1$ then $W = 0$ or ± 1 because $|W| \leqslant |Z|$.

Case 1: $W = 0$. In this case j must be 1 since otherwise there would be an equation (3) and then $W = 0$ would imply $-Y' = 0$, which combines

*Included in this assertion is the assertion that XZ has a sign, that is, $X \neq 0$ and $Z \neq 0$.

with $|Y'| \geqslant |W'|$ to give $Y' = W' = 0$, contrary to $X'W' - Y'Z' = 1$. Then the sign in front on the right side must be $-$ if $Z = 1$ and $+$ if $Z = -1$. Then $n_1 = n_j$ is determined by $X = \pm n_1$ (specifically $n_1 = -ZX$) and the proof of uniqueness is complete.

Case 2: $W = \pm 1$. In this case j cannot be 1 and there must be an equation of the form (3). Then $\pm 1 = Z = -X'$ and $|X'| \geqslant |Z'| \geqslant |W'|$ and $|X'| \geqslant |Y'| \geqslant |W'|$ implies $X' = \pm 1$, $Y' = \pm 1$, $Z' = \pm 1$, and $W' = \pm 1$ or 0 (because $Y' = 0$ and $Z' = 0$ are impossible). The equation $X'W' - Y'Z' = 1$ then shows that W' must be zero. Thus $Y = n_j Y' + W' = n_j(-W)$ determines n_j (and in fact, because $W' = 0$, also determines $j = 2$) and the proof of uniqueness is complete.

Now let X, Y, Z, W be given satisfying the three conditions. Consider first the exceptional cases $Z = \pm 1$. If $W = 0$ then $-YZ = 1$, $Y = -Z = \pm 1$, and the matrix itself is of the required form. If $W = \pm 1$ set $j = 2$, define n_2 by $Y = -n_2 W$, and define X', Y', Z', and W' by (3). Then $Y' = -W$, $W' = 0$, $Z' = -Y' = \pm 1$ and when the signs on both sides of (3) are changed, if necessary, the equation takes the form

$$\begin{pmatrix} X & Y \\ Z & W \end{pmatrix} = \begin{pmatrix} n_2 & 1 \\ -1 & 0 \end{pmatrix}\begin{pmatrix} n_1 & 1 \\ -1 & 0 \end{pmatrix}$$

which is of the required form (2) provided n_2 and n_1 have opposite signs. Since $Z = -n_1$ and $Z = \pm 1$, $n_1 = \pm 1$. Thus $X = \pm|n_2| - 1$ and $Y = n_2$. From $|X| \geqslant |Y|$ it follows that $X = -|n_2| - 1$, $n_1 n_2 = -|n_2|$ and n_1 and n_2 have opposite signs. Finally, consider the case $|Z| > 1$. Then $|Z|$ does not divide $|X|$ evenly. Define n_j (with the value of j undetermined) by $|X| = |n_j||Z| + r$ where $0 < r < |Z|$ and by the condition that the sign of n_j be opposite to the sign of XZ. With this definition of n_j, let (3) define X', Y', Z', and W'. Then $X = n_j X' + Z' = -n_j Z + Z'$ and $|X| = |n_j||Z| + r$ shows that Z' has the sign of X and that $|Z'| = r < |Z| = |X'|$. Therefore the sign of $X'Z' = (-Z)Z'$ is equal to the sign of $-ZX$ and opposite to the sign of XZ. Moreover, $X' = -Z$, $Y' = -W$, so that $|X'| \geqslant |Y'|$. Of course $X'W' - Y'Z' = 1$ because the product of the determinants is the determinant of the products: $1 = XW - YZ = 1 \cdot (X'W' - Y'Z')$. It will be shown next that $|Y'| > |W'|$ and $|Z'| \geqslant |W'|$. If $|Z'| < |W'|$ then $|Y'||Z'| + 1 \geqslant |Y'Z' + 1| = |X'W'| \geqslant |X'|(|Z'| + 1) = |X'||Z'| + |X'| \geqslant |Y'||Z'| + |X'|$, from which $1 \geqslant |X'| = |Z|$ contrary to assumption. Similarly, if $|Y'| < |W'|$ then $|Y'||Z'| + 1 \geqslant |X'W'| \geqslant |X'|(|Y'| + 1) \geqslant |Z'||Y'| + |X'|$, from which $1 \geqslant |X'| = |Z|$. This completes the proof that X', Y', Z', W' satisfy the three conditions.

Thus the three conditions guarantee, when $Z \neq \pm 1$, an equation of the form (3) in which X', Y', Z', W' satisfy the three conditions. Moreover, the sign of n_j is opposite to the sign of XZ and equal to the sign of $X'Z'$. The process can then be repeated, provided $Z' \neq \pm 1$, to split off another factor

$$\begin{pmatrix} X & Y \\ Z & W \end{pmatrix} = \begin{pmatrix} n_j & 1 \\ -1 & 0 \end{pmatrix}\begin{pmatrix} n_{j-1} & 1 \\ -1 & 0 \end{pmatrix}\begin{pmatrix} X'' & Y'' \\ Z'' & W'' \end{pmatrix}$$

(the value of j being as yet undetermined) in which n_j and n_{j-1} have opposite signs. Since $|Z| = |X'| > |Z'|$ the sequence of Z's decrease in absolute value $|Z| > |Z'| > |Z''| > \cdots$. By the principle of infinite descent the process must terminate and because $|Z^{(m)}| = 0$ is impossible, it must terminate with $|Z^{(m)}| = 1$. Then

$$\begin{pmatrix} X^{(m)} & Y^{(m)} \\ Z^{(m)} & W^{(m)} \end{pmatrix} = \pm \begin{pmatrix} n_1 & 1 \\ -1 & 0 \end{pmatrix} \quad \text{or} \quad \pm \begin{pmatrix} n_2 & 1 \\ -1 & 0 \end{pmatrix}\begin{pmatrix} n_1 & 1 \\ -1 & 0 \end{pmatrix}$$

where the sign of n in the leftmost factor is opposite to that of $X^{(m)}Z^{(m)}$. This then gives a representation of the given matrix in the required form and completes the proof of the theorem.

Therefore, in order to show that ME_A^n has the required form for n large it will suffice to show that it satisfies the three conditions of the theorem for n large. The matrix E_A describes the map from quadratic integers divisible by A to themselves given by multiplication by $r_0 + \sqrt{D}$ followed by division by a_0 followed by multiplication by $r_1 + \sqrt{D}$ followed by division by a_1, and so forth. Therefore it corresponds simply to multiplication by the unit

$$\varepsilon_A = \frac{(r_0 + \sqrt{D})(r_1 + \sqrt{D})\dots(r_{j-1} + \sqrt{D})}{a_0 a_1 \dots a_{j-1}}$$

(both numerator and denominator have divisor $A_0\bar{A}_1 A_1 \bar{A}_2 \dots \bar{A}_{j-1} A_{j-1} \bar{A}_0$ because $A_j = A_0$). Therefore ME_A^n corresponds to multiplication by $(x + y\sqrt{D})\varepsilon^n$ followed by division by a, where $x + y\sqrt{D}$ is the given quadratic integer with divisor $\bar{A}B$. This shows that ME_A^n is simply the new M which results when $x + y\sqrt{D}$ is replaced by the quadratic integer $(x + y\sqrt{D})\varepsilon_A^n$ with the same divisor $\bar{A}B$. This leads to the question of finding conditions on x and y under which

$$M = \begin{bmatrix} \dfrac{x + ys}{b} & \dfrac{rx - yD - sx + rsy}{ab} \\[2ex] -y & \dfrac{x - yr}{a} \end{bmatrix}$$

satisfies the conditions of the theorem. Since the determinant is 1, this is a matter of finding when the four inequalities $|X| \geqslant |Y| \geqslant |W|, |X| \geqslant |Z| \geqslant |W|$ are satisfied.

Since the coefficients of ε_A are of like signs, the proof of Section 7.5 shows that by making n large one can assume that x and y are large in absolute value and have the same sign. Then, because $x^2 - Dy^2 = ab$, $x^2 = ab + (D - r^2)y^2 + r^2y^2$ where ab is fixed and $D - r^2 > 0$, it follows that $x^2 > r^2y^2$ (when $|y|$ is sufficiently large) and $|x| > |r||y|$, $|x - ry| = |x| - r|y|$ (when x and y have like sign). Thus $|Z| > |W|$ is equivalent to $|a||y| > |x| - r|y|$, which in turn is equivalent to $(r + |a|)|y| > |x|$, $(r + |a|)^2y^2 > x^2 = ab + Dy^2$, and this holds for large $|y|$ because $(r + |a|)^2 > D$ (otherwise r could be increased). Similarly $|x + ys| = |x| + s|y|$ when x and y have like sign and $|X| > |Z|$ is equivalent to $|x| + s|y| > |b||y|$, or to $|x| > (-s + |b|)|y|$. This is of course true if $-s + |b| \leqslant 0$. Otherwise it is equivalent to $x^2 > (-s + |b|)^2y^2$ or to $ab + Dy^2 > (s - |b|)^2y^2$ and this is true for $|y|$ large because $(s - |b|)^2 < D$ (see Section 7.5, where it was shown that if $a_{j+1} < 0$ then $(r_{j+1} + a_{j+1})^2 < D$ but if $a_{j+1} > 0$ then $(r_{j+1} - a_{j+1})^2 < D$). Therefore $|X| > |Z| > |W|$ if $|y|$ is sufficiently large and if x and y have the same sign. Then $|Y||Z| = |XW - 1| \geqslant |X||W| - 1 > |Z||W| - 1$ (unless $|W| = 0$, in which case $|Y| \geqslant |W|$ is immediate) from which $|Y||Z| \geqslant |W||Z|$ and $|Y| \geqslant |W|$ (because $|Z| > |W| \geqslant 0$). Similarly $|Y||Z| \leqslant |X||W| + 1 < |X||Z| + 1$, $|Y| \leqslant |X|$, and all the inequalities are proved when $|y|$ is large and $xy > 0$.

Therefore, if A and B are reduced, and if $A \sim B$, one can assume without loss of generality that the equivalence corresponds to a matrix M of the type studied in the theorem. This argument uses nothing about ε_A other than the fact that it is a unit whose coefficients are of like sign. Therefore it shows that the equivalence $B \sim B$ corresponding to the quadratic integer $b\varepsilon_A^n$ with divisor $B\bar{B}$ gives rise to a matrix M_B which is also of the form studied in the theorem, at least if n is sufficiently large. (Actually 1 is easily seen to be sufficiently large. See Exercise 6.)

Now $M_B M = ME_A^n$ because each of these is the matrix which describes the map from quadratic integers divisible by A to quadratic integers divisible by B resulting from multiplication by $(x + y\sqrt{D})\varepsilon_A^n$ followed by division by a. (On the left side a multiplication and a division by b cancel, and on the right side n multiplications and divisions by a are cancelled.) Each of the matrices M_B, M, and E_A is of the form

$$\pm \prod \begin{pmatrix} n & 1 \\ -1 & 0 \end{pmatrix}$$

where the n's are an alternating sequence of integers. Moreover, $M_B M$ and ME_A^n are also of this form and their decompositions in this form are obtained merely by multiplying the decompositions of their factors. To prove this it is necessary to show that the sign of n in the rightmost factor of M_B is opposite to the sign of n in the leftmost factor of M, and similarly for the rightmost factor of M, and the end factors of E_A. Since E_A has an even number of factors by its definition, its two end factors have n's of opposite sign. The right end factor has $(r_0 + r_1)/a_1$ which has the sign of $-a_0 = -N(A)$. The left end factor therefore has the sign equal to that of $N(A)$. The n in the left end factor of M has the sign opposite to that of $XZ = -(x + ys)y/b$. Since x and y have like signs, this means that the sign in question is that of $N(B)$. The n in the right end factor of M has the sign equal to that of XY (Exercise 3). Since $XYZW = (YZ - 1)YZ \geqslant 0$ and $W \neq 0$, this is the same as the sign of $ZW = -y(x - yr)/a$. Therefore the rightmost factor of M has a sign opposite to the sign of the leftmost factor of E_A. The signs of the end factors of M_B can be found similarly. Then $M_B^k M = M E_A^{nk}$ is a decomposition of the type of the theorem for all k and by the uniqueness assertion of the theorem it follows that the factors in the decomposition of M are equal to the last factors of E_A^{nk} for k sufficiently large, which is to say that, for some j,

$$\pm M = \begin{pmatrix} n_j & 1 \\ -1 & 0 \end{pmatrix}\begin{pmatrix} n_{j-1} & 1 \\ -1 & 0 \end{pmatrix} \cdots \begin{pmatrix} n_1 & 1 \\ -1 & 0 \end{pmatrix}$$

where $n_1, n_2, \ldots$ is the cyclic alternating sequence obtained by applying the cyclic method to A. Therefore the matrix corresponding to the equivalence $A \sim B$ coincides with that corresponding to the equivalence $A \sim A_j$ which results from a suitable number of cycles through the period of A.

The matrix M arises on the one hand from multiplication by $x + y\sqrt{D}$ with divisor $\overline{A}B$ followed by division by a and on the other hand from multiplication by some $x' + y'\sqrt{D}$ with divisor $\overline{A}A_j$ followed by division by a. The values of x and y are determined by the matrix M and the divisor A because the last row of M is $-y, (x - yr)/a$ which, when a and r are known, easily gives x and y. Therefore $x = x'$, $y = y'$ and $\overline{A}B = \overline{A}A_j$. This shows that $B = A_j$ and completes the proof of the case $D > 0$, $D \equiv 2$ or 3 mod 4.

In the case $D > 0$, $D \equiv 1 \bmod 4$ the modifications are minor. The quadratic integers divisible by A are those of the form $au + (r - \frac{1}{2}\sqrt{D})v$ where u and v are integers. An equivalence between two divisors $A \sim B$ corresponds to a 2×2 matrix of integers, namely

$$\begin{bmatrix} \dfrac{x + 2ys}{b} & \dfrac{rx - \frac{1}{2}yD - sx + 2rsy}{ab} \\ -2y & \dfrac{x - 2yr}{a} \end{bmatrix}$$

where $x + y\sqrt{D}$ has divisor $\overline{A}B$. (Here x and y may both be integers or may both be halfintegers, but the entries of the matrix will always be integers.) The remainder of the proof is the same.

EXERCISES

1. Prove that if $D \equiv 2$ or 3 mod 4 and if A and B are divisors with equal norm which divide $r - \sqrt{D}$ then $A = B$. [This says, in essence, that a divisor is determined by the set of all things that it divides. As was seen in Section 7.3 this is not true when $D > 0$. However, it is true that if A and B divide exactly the same quadratic integers then either $A = B$ or $A = (-1, *)B$.]

2. Prove the theorem that equivalent divisors correspond to the same periods in the case $D \equiv 1$ mod 4, $D < 0$.

3. Prove that the sign of n_0 in (2) is equal to the sign of XY. [This can be done either by a simple induction or by using the fact that n_j has the sign of $-XZ$ and taking the transpose.]

4. State and prove a theorem analogous to the theorem of the text for decompositions of the form

$$\begin{pmatrix} X & Y \\ Z & W \end{pmatrix} = \begin{pmatrix} n_k & 1 \\ 1 & 0 \end{pmatrix} \begin{pmatrix} n_{k-1} & 1 \\ 1 & 0 \end{pmatrix} \cdots \begin{pmatrix} n_1 & 1 \\ 1 & 0 \end{pmatrix} \tag{4}$$

where the n's are positive. This, in essence, is the Euclidean algorithm applied to X and Z and is much easier to prove than the theorem of the text.

5. Prove that a decomposition of the form (2) can always be obtained from one of the form (4) by inserting signs appropriately and obtain in this way an alternative proof of the theorem of the text.

6. Prove that if $\varepsilon_A = u + v\sqrt{D}$ is a unit in which u and v have like sign, if B is a reduced divisor, and if M_B is the 2×2 matrix which corresponds to the map of quadratic integers divisible by B to themselves corresponding to the quadratic integer $b\varepsilon_A$ with divisor $B\bar{B}$, then the theorem applies to the matrix M_B.

7.8 Euler's theorems

In carrying out computations in the divisor theory of quadratic integers, it is necessary to determine which primes split, which remain prime, and which ramify. This amounts essentially to the problem of deciding, for a given squarefree D and for a given odd prime p prime to D, whether the congruence $D \equiv x^2$ mod p has a solution. Of course this question can be answered simply by finding all squares mod p (there are $\frac{1}{2}(p-1)$ of them) and looking to see whether D is among them. However, there are some surprising and marvelous patterns in the solution to this problem which were discovered and published by Euler over a century before Kummer's theory of ideal factorization appeared. This section is devoted to a description of the patterns which Euler discovered.

Euler did quite literally *discover* them empirically by a perspicacious study of numerical evidence and he was never able to prove* the theorems which he stated as early as the 1740s [see E7 and F6, pp. 146–151] still early in his career. As was noted by Gauss (*Disquisitiones Arithmeticae* Art. 151), the theorems imply† and are implied by the law of quadratic reciprocity. Therefore, because Gauss gave the first proof of quadratic reciprocity, credit for the first proof of them is due to Gauss.

The specific question which Euler was studying was the question of whether a given prime can divide a number of the form $x^2 + ny^2$ where n is a given integer and where x and y are relatively prime integers. The core of his discovery was the surprising fact that *the answer to this question depends only on the class of the prime* mod $4n$. That is, if p_1 and p_2 are primes that are congruent mod $4n$ then there exist relatively prime integers x_1 and y_1 such that $p_1 | x_1^2 + ny_1^2$ if and only if there exist relatively prime x_2, y_2 such that $p_2 | x_2^2 + ny_2^2$.

In order to put this statement in the terminology of this chapter, it is natural to set $n = -D$ so that $x^2 + ny^2$ is the norm of $x + y\sqrt{D}$. Only the case in which D is squarefree will be considered. (As Exercises 1 and 9 show, this involves no real loss of generality.) If p divides $4D$ then the theorem is vacuous because p is congruent to no other prime mod $4D$. Therefore p does not ramify in cases where the theorem is not vacuous and p either splits or remains prime. If p splits then $D \equiv x^2 \bmod p$ has a nonzero solution x, $p | x^2 - D \cdot 1^2$, and Euler's theorem states that if $p_1 \equiv p \bmod 4D$ then there exist relatively prime x_1, y_1 such that $p_1 | x_1^2 - Dy_1^2$. This implies that p_1 divides $(x_1 - y_1\sqrt{D})(x_1 + y_1\sqrt{D})$ without dividing either factor (since x_1 and y_1 are relatively prime, no integer, except

---

*H. J. S. Smith [S3, Art. 16] says " . . . his [Euler's] conclusions repose on induction [i.e. empirical observation] only; though in one memoir he seems to have imagined (for his language is not very precise) that he had obtained a satisfactory demonstration." In another place, however, in a paper published at the end of his life "Euler expressly observes that the theorem is undemonstrated." Students of Euler must become accustomed to such paradoxes.

†This acknowledgement by Gauss that Euler stated theorems which imply quadratic reciprocity seems often to be overlooked by historians. Gauss claims for himself only the priority for the statement of quadratic reciprocity in the simple form "if $p \equiv 1 \bmod 4$ then q is a square mod p if and only if p is a square mod q, and if $p \equiv -1 \bmod 4$ then q is a square mod p if and only if $-p$ is a square mod q." Gauss appears to have decided, however, that he was being unjust to Legendre in this claim, because only a few years later he wrote [G3] that "we must surely regard Legendre as the discoverer of this most elegant theorem." Indeed, Legendre's statement of the theorem—which is to say that p is a square mod q if and only if q is a square mod p, except in the case $p \equiv q \equiv -1 \bmod 4$, in which p is a square mod q if and only if q is *not* a square mod p—seems if anything simpler than Gauss's statement. Euler's clearest statement of the law is usually taken to be the statement in [E11], which is surely less simple than either Gauss's or Legendre's. (The English translation of the *Disquisitiones Arithmeticae* takes too many liberties with the text in Art. 151, with the result that Gauss's acknowledgement is lost. The original text or one of the other translations should be consulted.)

possibly 2, divides either factor). Therefore p_1, since it does not ramify, must split. In short, Euler's theorem implies that *if $p_0 \equiv p_1$ mod $4D$ then p_1 splits, ramifies, or remains prime if and only if p_0 splits, ramifies, or remains prime, respectively*.* As is easy to show (Exercise 1) this theorem also implies Euler's theorem. Therefore it is reasonable to think of this theorem as a *restatement* of Euler's theorem in terms of the divisor theory of quadratic integers.

This theorem, if one assumes it is true, makes it possible to determine just how primes p factor in quadratic integers $x + y\sqrt{D}$. Apart from the prime factors of $4D$, it is necessary to test only one prime from each[†] class mod $4D$ relatively prime to $4D$. For example, when $D = 2$, the fact that the primes 3 and 5 remain prime (the squares mod 3 are just $1 \not\equiv 2$ mod 3 and mod 5 are 1, $4 \not\equiv 2$ mod 5) implies by the theorem that all primes $\equiv 3$ or 5 mod 8 remain prime, and the fact that 17 and 7 split ($6^2 \equiv 2$ mod 17, $3^2 \equiv 2$ mod 7) implies that all primes $\equiv 1$ or 7 mod 8 split. The only remaining prime, 2, ramifies.

Thus, on the basis of the theorem, it is possible to determine the pattern of splitting and nonsplitting for various values of D. Some results—much less extensive than Euler's computations—are given in Table 7.8.1. One obvious conclusion from this table is that *exactly half of the classes of primes* mod $4D$ *are splitting classes,*[‡] that is, contain primes which split. A further, less apparent fact which was noticed by Euler is that *a product of splitting classes is a splitting class*, the multiplication being ordinary multiplication of integers mod $4D$. Closely related to these two observations —and in fact a consequence[§] of them—is the fact that *the square of any class relatively prime to $4D$ is a splitting class*.

A final observation which is quite obvious from the table, and which Euler of course made, is that *when $D > 0$ the negative of a splitting class is*

---

*A similar phenomenon occurred in the case of the cyclotomic integers, where the number of prime divisors of p depended only on the exponent of p mod λ and therefore only on the class of p mod λ. For more general types of algebraic integers than quadratic or cyclotomic integers this phenomenon does *not* normally occur—that is, the way in which primes p will factor is not so easy to predict. Reciprocity laws and class field theory are closely related to this special property of quadratic and cyclotomic integers.

†In specific cases it will, of course, be found that every class relatively prime to $4D$ does contain a prime because, according to a famous theorem of Dirichlet, every class relatively prime to any modulus m ($4D$ in the case at hand) contains infinitely many prime numbers. (See Section 9.7.)

‡In order to avoid any tacit appeal to Dirichlet's theorem mentioned in the note above, it would be better to state this theorem as follows: It is possible to divide the classes mod $4D$ relatively prime to $4D$ into two subsets of equal size, called *splitting classes* and *nonsplitting classes* in such a way that any prime in a splitting class splits and any prime in a nonsplitting class remains prime. The formulations of the other theorems should also be modified slightly in order to allow for the possibility of classes which contain no primes and therefore are neither splitting nor nonsplitting. This modification to allow for a situation which never arises will be left to the reader.

§See Exercise 6.

Table 7.8.1 The classes set in boldface type are splitting classes. That is, any prime congruent to a boldface number mod $4D$ splits in quadratic integers for the determinant D.

D																			
$D = -1$	**1**	3																	
$D = -2$	**1**	**3**	5	7															
$D = -3$	**1**		5	**7**		11													
$D = -5$	**1**	**3**		**7**	**9**	11	13		17	19									
$D = -6$	**1**		**5**	**7**		**11**	13		17	19		23							
$D = -7$	**1**	3	5		**9**	**11**	13	**15**	17	19		**23**	**25**	27					
$D = -10$	**1**	3		**7**	**9**	**11**	**13**		17	**19**	21	**23**		27	29	31	33	**37**	39
$D = 2$	**1**	3	5	**7**															
$D = 3$	**1**		5	7		**11**													
$D = 5$	**1**	3		7	**9**	**11**	13		17	**19**									
$D = 6$	**1**		**5**	7		11	13		17	**19**		**23**							
$D = 7$	**1**	**3**	5		**9**	11	13	15	17	**19**		23	**25**	**27**					
$D = 10$	**1**	**3**		7	**9**	11	**13**		17	19	21	23		**27**	29	**31**	33	**37**	**39**

a splitting class, whereas when $D < 0$ the negative of a splitting class is a nonsplitting class. Because taking the negative is a one-to-one correspondence and because half the classes are splitting classes, these statements are equivalent to their converses, that is, equivalent to saying that when $D > 0$ the negative of a nonsplitting class is a nonsplitting class and when $D < 0$ the negative of a nonsplitting class is a splitting class.

These theorems greatly facilitate the determination of the splitting classes. For example, when $D = 2$, the class of 1 must be a splitting class because it is a square. The class of -1 must therefore be a splitting class because $D > 0$. The remaining classes, those of ± 3, must be nonsplitting classes because only half of the classes can be splitting classes. For another example, consider the case $D = -11$. Then the class of 3 is a splitting class because $-11 \equiv 1^2$ mod 3. Therefore the classes of $1, 3, 3^2, 3^3 \equiv -17, 3^4 \equiv -7, 3^5 \equiv -21, 3^6 \equiv -19, 3^7 \equiv -13, 3^8 \equiv 5, 3^9 \equiv 15, 3^{10} \equiv 1$ mod 44 are all splitting classes. Since these comprise half of the 20 classes relatively prime to 44, the remaining classes are nonsplitting classes. Note that they are the negatives of the splitting classes. For a third example consider the case $D = -15$. The least prime to be considered is $p = 7$. Since $-15 \equiv -1$ is not a square mod 7 (the squares mod 7 are 1, 2, and 4) the class of 7 is not a splitting class. Therefore the class of -7 is a splitting class and so are those of $(-7)^2 \equiv -11$, $(-7)^3 \equiv 17$, and $(-7)^4 \equiv 1$ mod 60. This gives 4 of the 8 splitting classes. Exactly one of the classes ± 13 is a splitting class and it has not yet been determined which of them it is. The squares mod 13 are $1, 4, -4, 3, -1, -3$, which does not include the class of $-15 \equiv -2$ mod 13. Therefore 13 is a nonsplitting class and -13 a splitting class. Then $(-13)^2 \equiv -11$, $(-13)^3 \equiv 23$, $(-13)^4 \equiv (-7)^4 \equiv 1$ mod 60

289

are also splitting classes, which gives just 2 more. Moreover $(-7)(-13) \equiv -29$ and $(-7)^3(-13) \equiv 17(-13) \equiv 19$ are also splitting classes. Thus the splitting classes are the classes of $1, -7, -11, -13, 17, 19, 23, -29$ and the nonsplitting classes are their negatives.

All of these theorems of Euler are easy to deduce from the law of quadratic reciprocity (see Section 5.6 or the note below). The first theorem, which is the main one, is simply the statement that the value of the Legendre symbol* $\left(\frac{D}{p}\right)$ for odd primes $p \nmid D$ depends only on the class of $p \bmod 4D$. This can be proved by showing how quadratic reciprocity can be used to *evaluate* $\left(\frac{D}{p}\right)$ in such a way that only the class of $p \bmod 4D$ enters. For this, let D be written as a product of numbers for which quadratic reciprocity has a simple form, namely, numbers -1 or 2 or p where $p \equiv 1 \bmod 4$ or $-p$ where $p \equiv -1 \bmod 4$. That is, let $D = \varepsilon r p_1 p_2 \cdots p_\mu (-p'_1)(-p'_2) \cdots (-p'_\nu)$ where $\varepsilon = \pm 1$, $r = 1$ or 2, $p_i = $ prime that is $1 \bmod 4$ and $p'_i = $ prime that is $-1 \bmod 4$. Then by quadratic reciprocity

$$\left(\frac{D}{p}\right) = \left(\frac{\varepsilon}{p}\right)\left(\frac{r}{p}\right)\left(\frac{p_1}{p}\right) \cdots \left(\frac{p_\mu}{p}\right)\left(\frac{-p'_1}{p}\right) \cdots \left(\frac{-p'_\nu}{p}\right)$$

$$= \left(\frac{\varepsilon}{p}\right)\left(\frac{r}{p}\right)\left(\frac{p}{p_1}\right) \cdots \left(\frac{p}{p_\mu}\right)\left(\frac{p}{p'_1}\right) \cdots \left(\frac{p}{p'_\nu}\right)$$

(where $\left(\frac{1}{p}\right)$ is defined to be $+1$ for all p when ε or r is 1). The first factor on the right depends only on the class of $p \bmod 4$ and therefore only on the class of $p \bmod 4D$. The second factor is $+1$ unless $r = 2$, in which case it depends only on the class of $p \bmod 8$; since $r = 2$ implies that D is even, the class of $p \bmod 4D$ then determines its class mod 8 and therefore determines the second sign $\left(\frac{r}{p}\right)$. The other signs depend only on the class of $p \bmod p_i$ or p'_i and therefore only on the class of $p \bmod 4D$. This completes the proof that $\left(\frac{D}{p}\right)$ depends only on the class of $p \bmod 4D$.

This proof actually does more because it shows how the definition of $\left(\frac{D}{p}\right)$ can be *extended* from odd primes p that do not divide D to all integers p which are relatively prime to $2D$ (and even, if $\varepsilon = r = 1$, to those relatively prime to D). This extended definition of $\left(\frac{D}{n}\right)$ for n relatively prime to $4D$ is called *the Jacobi symbol*.[†] The class of n is a splitting class if $\left(\frac{D}{n}\right) = +1$ and a nonsplitting class if $\left(\frac{D}{n}\right) = -1$. This definition of the

---

*The Legendre symbol $\left(\frac{n}{p}\right)$ is defined in Section 5.6. It is meaningful when p is an odd prime and n is an integer not divisible by p, and it is $+1$ if n is a square mod p (the congruence $n \equiv x^2 \bmod p$ has a solution), -1 otherwise. It is simple to show that if neither n_1 nor n_2 is divisible by p then $\left(\frac{n_1 n_2}{p}\right) = \left(\frac{n_1}{p}\right)\left(\frac{n_2}{p}\right)$. The law of quadratic reciprocity states that if p and q are odd primes and $p \equiv 1 \bmod 4$ then $\left(\frac{p}{q}\right) = \left(\frac{q}{p}\right)$ whereas if $p \equiv 3 \bmod 4$ then $\left(\frac{-p}{q}\right) = \left(\frac{q}{p}\right)$. The supplementary laws state $\left(\frac{-1}{p}\right) \equiv p \bmod 4$—that is, $\left(\frac{-1}{p}\right) = +1$ if $p \equiv 1 \bmod 4$ and $= -1$ if $p \equiv -1 \bmod 4$—and that $\left(\frac{2}{p}\right)$ is $+1$ if $p \equiv \pm 1 \bmod 8$, -1 if $p \equiv \pm 3 \bmod 8$.

†See Jacobi [J1], where $\left(\frac{D}{n}\right)$ is defined, for odd numbers n relatively prime to D, to be $\left(\frac{D}{p_1}\right)\left(\frac{D}{p_2}\right) \cdots \left(\frac{D}{p_\nu}\right)$ where $n = p_1 p_2 \ldots p_\nu$ is the factorization of n into primes. This is easily seen to be equivalent to the definition given above.

splitting and nonsplitting classes does not assume that the class of n contain a prime and therefore does not rely on Dirichlet's theorem on primes in an arithmetic progression.

It is clear from the definition of the Jacobi symbol that $(\frac{D}{n_1 n_2}) = (\frac{D}{n_1})(\frac{D}{n_2})$ because this is true of each of the factors. Therefore not only is the product of two splitting classes a splitting class, but also the product of two nonsplitting classes is a splitting class and the product of a nonsplitting class and a splitting class is a nonsplitting class. In particular the square of any class relatively prime to $4D$ is a splitting class.

To prove that there are equal numbers of splitting and nonsplitting classes it will suffice to prove that there is at least one nonsplitting class because multiplication by a nonsplitting class is a one-to-one mapping of classes relatively prime to $4D$ to themselves which interchanges splitting and nonsplitting classes. In order to find a nonsplitting class one can proceed as follows. If $\varepsilon r \neq 1$ then $(\frac{\varepsilon}{p})(\frac{r}{p})$ assumes both values ± 1 and for any given n relatively prime to $4D$ there is an odd n' such that $(\frac{\varepsilon}{n'})(\frac{r}{n'})$ is opposite to $(\frac{n}{p_1}) \cdots (\frac{n}{p_\mu})(\frac{n}{p'_1}) \cdots (\frac{n}{p'_\nu})$; by the Chinese remainder theorem there is an $n'' \equiv n \bmod p_1 \ldots p_\mu p'_1 \ldots p'_\nu$ and $n'' \equiv n' \bmod 8$. Then $(\frac{D}{n''}) = -1$ as required. If $\varepsilon r = 1$ then, because $D \neq 1$, either μ or ν must be nonzero and a similar method gives an n'' with $(\frac{D}{n''}) = -1$.

It remains to show that $(\frac{D}{-n}) = (\frac{D}{n})$ when $D > 0$ and $(\frac{D}{-n}) = -(\frac{D}{n})$ when $D < 0$. In other words, it remains to show that $(\frac{D}{-1})$ is the sign of D. The factors $(\frac{r}{-1})$ and $(\frac{-1}{p_i})$ in $(\frac{D}{-1})$ are always $+1$. The factors $(\frac{-1}{p'_i})$ are always -1 and the factor $(\frac{\varepsilon}{-1})$ is $+1$ if $\varepsilon = +1$ and -1 if $\varepsilon = -1$. Therefore $(\frac{D}{-1}) = \varepsilon(-1)^\nu$, which is also the sign of D, as was to be shown.

This completes the proof that quadratic reciprocity implies all of Euler's theorems. The converse, that Euler's theorems imply quadratic reciprocity, can be proved as follows.* The supplementary laws follow immediately from Euler's theorems in the cases $D = -1$ and $D = 2$ (Exercise 4). Suppose now that $D = q$ where q is an odd prime. The square of any class relatively prime to $4q$ is a splitting class. It is simple to show that exactly a quarter of the classes relatively prime to $4q$ are squares (Exercise 3). This rule accounts, therefore, for half of the splitting classes. Because $D = q > 0$, the negative of a splitting class is also a splitting class. Therefore the negative of a square is a splitting class. There is no overlap between the squares and the negatives of the squares because $x^2 \equiv z \equiv -y^2 \bmod 4q$ for z relatively prime to $4q$ would imply that both x and y were odd and therefore would imply both $z \equiv 1$ and $z \equiv -1 \bmod 4$. Therefore the other half of the splitting classes consists of the negatives of the squares and *all of the splitting classes are accounted for.* Therefore an odd prime $p \neq q$ splits if and only if $p \equiv \pm x^2 \bmod 4q$ for some integer x. That is, *if $p \neq q$ are odd primes then q is a square* mod p *if and only if p or $-p$ is a square* mod $4q$.

---

*This proof is due to Kronecker [K1].

This is easily seen to be the law of quadratic reciprocity. If $p \equiv 1 \bmod 4$ then $-p$ cannot be a square mod 4, let alone mod $4q$, and the theorem states in this case that q is a square mod p if and only if p is a square mod $4q$. This of course implies that p is a square mod q, but, conversely, if p is a square mod q, say $p \equiv y^2 \bmod q$ then also $p \equiv (y - q)^2 \bmod q$ and, because either y or $y - q$ is odd, one can assume y is odd, $y^2 \equiv 1 \equiv p \bmod 4$, $y^2 \equiv p \bmod 4q$ (because q is odd) and p is a square mod $4q$. Thus *if $p \equiv 1 \bmod 4$ then q is a square mod p if and only if p is a square mod q.* Similarly, if $p \equiv -1 \bmod 4$ then p or $-p$ is a square mod $4q$ only if $-p$ is a square mod q, and *if $p \equiv 3 \bmod 4$ then q is a square mod p if and only if $-p$ is a square mod q.* This is quadratic reciprocity in the form that Gauss stated it.

EXERCISES

1. Given that primes congruent mod $4D$ factor in the same way (split, ramify, or remain prime) in the divisor theory of quadratic integers for the determinant D, prove that if $p_0 \equiv p_1 \bmod 4D$ and if $p_0 | x_0^2 - Dy_0^2$ for relatively prime x_0 and y_0 then there exist relatively prime x_1 and y_1 such that $p_1 | x_1^2 - Dy_1^2$. This proves Euler's theorem on divisors of $x^2 + ny^2$ in the case where $-n$ is squarefree. Prove that Euler's theorem is trivial when $-n$ is a square. Finally, deduce the case $-n = t^2 D$ for squarefree D from the case $-n = D$.

2. Extend Table 7.8.1 to cover all squarefree determinants D in the range $|D| \leqslant 15$.

3. Prove that exactly a quarter of the classes relatively prime to $4q$ (q a prime) are squares.

4. Deduce the formulas for $(\frac{-1}{p})$, $(\frac{2}{p})$, $(\frac{-2}{p})$ (the supplementary laws of quadratic reciprocity) from Euler's theorems with $D = -1, 2$, and -2 respectively. [For $D = -1$ and $D = 2$ it is not necessary to test any prime p for splitting. When $D = -2$ one test is necessary unless one uses $(\frac{-2}{p}) = (\frac{-1}{p})(\frac{2}{p})$.]

5. Prove that if A is any divisor of quadratic integers for the determinant D and if $a = N(A)$ then $(\frac{D}{a}) = +1$ or $(\frac{D}{a})$ is not defined. Thus *the image of the norm map from divisors to integers* mod $4D$ omits half of the classes relatively prime to $4D$, namely, the nonsplitting classes. [In particular, the class of -1 is a splitting class when $D > 0$.]

6. In the text it is asserted that if it is known that exactly half the classes mod $4D$ are splitting classes, and if it is known that a product of splitting classes is always a splitting class, then it follows that the square of any class is a splitting class. Prove this by proving more generally that a product of two nonsplitting classes is always a splitting class.

7. Explain the fact that $(\frac{D}{n}) = +1$ when $D = -7$ and $n = 15$ but -7 is not a square mod 15.

8. Prove that *the law of quadratic reciprocity holds for Jacobi symbols* whenever the terms are defined. That is, if p and q are odd and relatively prime, then

$$\left(\frac{q}{p}\right) = \begin{cases} \left(\dfrac{p}{q}\right) & \text{if } p \equiv 1 \bmod 4, \\[2mm] \left(\dfrac{-p}{q}\right) & \text{if } p \equiv -1 \bmod 4, \end{cases}$$

even if p and q are not prime. This shows that the computation of Exercise 4, Section 5.6, is valid.

9. Show that if $D = -n$ is a square then Euler's theorems are false for the simple reason that $p \mid x^2 + ny^2$ (x, y relatively prime) for all primes p. Show that if $D = t^2 D'$ then Euler's theorems for D' imply the same theorems for D.

7.9 Genera

The problem of finding the divisor class group for a given determinant D turns on the problem of determining whether two given divisors are equivalent. Gauss observed that there are some simple *necessary* conditions for two divisors to be equivalent and he used these necessary conditions to divide the divisor classes into what he called *genera*. (Gauss was, however, dealing with binary quadratic forms rather than divisors. See Chapter 8.) This division into genera plays an important part in his second proof of quadratic reciprocity.

Two divisors are equivalent if and only if the product of one with the conjugate of the other is principal. Gauss's necessary conditions for equivalence follow from the following simple necessary conditions for a divisor to be principal.

If A is the divisor of $x + y\sqrt{D}$ then $N(A) = x^2 - Dy^2$ where x and y are integers or, possibly, halfintegers. If p is an odd prime divisor of D then $4 \cdot N(A) = (2x)^2 - D(2y)^2 \equiv u^2 \bmod p$ where $u = 2x$ is an integer. Since p is odd, division by 2 is possible mod p and it follows that *the norm of a principal divisor must be a square* mod p for any odd prime factor p of D. Therefore, if A_1 and A_2 are equivalent divisors and if n_1 and n_2 are, respectively, their norms, then $A_1 \bar{A}_2$ is principal and $n_1 n_2$ must be a square mod p. If $n_1 n_2 \not\equiv 0 \bmod p$ it follows easily that n_1 is a square mod p if and only if n_2 is. (If $n_1 n_2 \equiv u^2$ and $u \not\equiv 0$ and $n_1 \equiv v^2$ then $v \not\equiv 0$, division by v is possible, and $n_2 \equiv (u/v)^2$ is a square mod p. By symmetry, if n_1 is not a square mod p then neither is n_2.) In terms of the Legendre symbol this can be stated simply as $\left(\frac{n_1}{p}\right) = \left(\frac{n_2}{p}\right)$. Thus *a necessary condition for* $A_1 \sim A_2$ *is* $\left(\frac{n_1}{p}\right) = \left(\frac{n_2}{p}\right)$ where $n_1 = N(A_1)$, $n_2 = N(A_2)$ and where it is assumed that A_1 and A_2 are relatively prime to p.

If $D \not\equiv 1 \bmod 4$ then there is another necessary condition for equivalence which corresponds to the prime 2. If $D \equiv 3 \bmod 4$ then, for any principal divisor A, $N(A) = x^2 - Dy^2 \equiv x^2 + y^2 \bmod 4$ and x and y are integers. If A is relatively prime to 2 then $N(A)$ is odd, x and y have opposite parity, and $N(A) \equiv 1 \bmod 4$. Thus if $A_1 \sim A_2$ and if both are relatively prime to 2 then $n_1 n_2 \equiv 1 \bmod 4$, which means either that n_1 and n_2 are both 1 mod 4 or both -1 mod 4. If $D \equiv 2 \bmod 4$ then there is another necessary condition for equivalence, but it is very different from the one in the case $D \equiv 3 \bmod 4$. If $D \equiv 2 \bmod 4$ and A is a principal divisor relatively prime to 2 then $N(A) = x^2 - Dy^2$ where x and y are integers and x is odd. Thus $x^2 \equiv 1 \bmod 8$ and $-Dy^2 \equiv 0$ or $-D \bmod 8$, according to whether y is even or odd. If $D \equiv 2 \bmod 8$ then $N(A) \equiv 1$ or $-1 \bmod 8$ and $N(A) \equiv \pm 3 \bmod 8$ is impossible. Therefore if $n_1 \equiv \pm 1 \bmod 8$ then $n_2 \equiv \pm 1 \bmod 8$ and if $n_1 \equiv \pm 3 \bmod 8$ then $n_2 \equiv \pm 3 \bmod 8$, when n_1 and n_2 are the norms of equivalent divisors relatively prime to 2. Similarly, if $D \equiv -2 \bmod 8$ then $N(A) \equiv 1$ or $3 \bmod 8$, and $n_1 \equiv 1$ or $3 \bmod 8$ if and only if $n_2 \equiv 1$ or $3 \bmod 8$, $n_1 \equiv 5$ or $7 \bmod 8$ if and only if $n_2 \equiv 5$ or $7 \bmod 8$. In summary, if $D \not\equiv 1 \bmod 4$, and if an *additional sign* is defined as in Table 7.9.1 then, for divisors A_1, A_2 relatively prime to 2, *a necessary condition for $A_1 \sim A_2$ is that n_1 and n_2 have the same additional sign*, where $n_1 = N(A_1)$ and $n_2 = N(A_2)$. (Note that the supplementary laws of quadratic reciprocity state that the additional sign is equal to $\left(\frac{-1}{n}\right)$ if $D \equiv -1 \bmod 4$, to $\left(\frac{2}{n}\right)$ if $D \equiv 2 \bmod 8$, and to $\left(\frac{-2}{n}\right)$ if $D \equiv -2 \bmod 8$.)

Let m be the number of odd prime factors of D and let ε be 0 if $D \equiv 1 \bmod 4$ and 1 otherwise. The *character* of a divisor A is defined to be a list of $m + \varepsilon$ signs ± 1 consisting of the m signs $\left(\frac{N(A)}{p_i}\right)$ (p_i an odd prime factor of D) together with the additional sign defined in Table 7.9.1 (with $n = N(A)$) in case $\varepsilon = 1$. If A is not relatively prime to D or, in the case $\varepsilon = 1$, not relatively prime to $2D$, then its character is not defined. For the sake of definiteness, let the $m + \varepsilon$ signs be written in the order $\left(\frac{N(A)}{p_1}\right)$, $\left(\frac{N(A)}{p_2}\right)$, ..., $\left(\frac{N(A)}{p_m}\right)$, where $p_1 < p_2 < \cdots < p_m$, followed by the additional sign, if any.

Because the signs, when they are defined, are the same for equivalent divisors, one can define the *character of a divisor class* provided the class contains at least one divisor whose character is defined. It is simple to show (see Section 8.3) that any divisor is equivalent to a divisor relatively prime to $2D$ and in this way to define the character of a divisor class. However, the character of a divisor class can be defined without proving this theorem, simply by defining it *one sign at a time*. To define the sign corresponding to a prime p (including $p = 2$ for the additional sign) it

Table 7.9.1

D	Additional sign is	
	$+$ if	$-$ if
$\equiv 3 \bmod 4$	$n \equiv 1 \quad \bmod 4$	$n \equiv -1 \quad \bmod 4$
$\equiv 2 \bmod 8$	$n \equiv \pm 1 \quad \bmod 8$	$n \equiv \pm 3 \quad \bmod 8$
$\equiv -2 \bmod 8$	$n \equiv 1$ or $3 \bmod 8$	$n \equiv 5$ or $7 \bmod 8$

suffices to find a divisor A in the given class which is relatively prime to p. For this it will suffice to find a divisor equivalent to $(p, *)$ which is relatively prime to p. (Note that 2 ramifies if and only if $\varepsilon = 1$.) Except in the case $p = 2$, $D \equiv 3 \bmod 4$, such a divisor is the divisor of $\sqrt{D}$ divided by $(p, *)$. In the remaining case, such a divisor is the divisor of $1 - \sqrt{D}$ divided by $(p, *) = (2, *)$.

The character of the divisor classes in the cases considered in Section 7.6 are as follows. When $D = 67$ the character contains $1 + 1$ signs. For the class of the divisor $(-1, *)$ the first sign is -1 (because -1 is not a square mod 67) and the second sign is also -1 (because $-1 \equiv -1 \bmod 4$). Thus the character is $--$. The principal class of course has character $++$.

When $D = -165$ the character consists of 4 signs, for the primes 3, 5, 11, and 2. The divisor $A = (2, *)$ has the first three signs equal to $(\frac{2}{3}) = -1$, $(\frac{2}{5}) = -1$, $(\frac{2}{11}) = -1$; the fourth sign is found by observing that the divisor of $1 - \sqrt{-165}$ is $(2, *)(83, 1)$ so that $(2, *) \sim (83, 1)$ and the required sign is $-$ because $83 \equiv -1 \bmod 4$. Thus $----$ is the character of the class of A. The divisor $B = (3, *)$ has as the last three signs of its character $(\frac{3}{5}) = -1$, $(\frac{3}{11}) = +1$, and -1 because $3 \equiv -1 \bmod 4$; its first sign is found from $(3, *) \sim (5, *)(11, *)$, $(\frac{55}{3}) = (\frac{1}{3}) = +1$. The class of B therefore has the character $+-+-$. In the same way the character of the class of $C = (5, *)$ is $--++$. The characters of the remaining classes can then be found by simple multiplication, once it is observed that *the character of the product of two classes is the product of their characters*. (Characters are multiplied by multiplying corresponding signs. For the signs corresponding to odd prime factors p of D the fact that the sign of the product is the product of the signs $(\frac{n_1 n_2}{p}) = (\frac{n_1}{p})(\frac{n_2}{p})$ was shown in Section 5.6. For the signs corresponding to the prime 2 it can be verified by inspection of Table 7.9.1. This property is closely related to the property $\chi(n_1 n_2) = \chi(n_1)\chi(n_2)$ of the "characters" in Section 6.4. In fact, the use of the word "character" in group theory derives from Gauss's use of it in this part of the *Disquisitiones Arithmeticae*.)

The characters of the various classes in the case $D = -165$ are given in Table 7.9.2. Note that each divisor class has a different character. This means that the necessary conditions for equivalence are in this case *sufficient*—two divisors with the same character are equivalent. This convenient property of $D = -165$ was already noted and exploited by Euler (see Exercise 4).

Table 7.9.2 $D = -165$, $A = (2, *)$, $B = (3, *)$, $C = (5, *)$

class of	has character	class of	has character
I	$++++$	AB	$-+-+$
A	$----$	AC	$++--$
B	$+-+-$	BC	$-++-$
C	$--++$	ABC	$+--+$

In the next case of Section 7.6, namely, $D = -163$, there is only the principal class. The character consists of only one sign (163 is prime, $-163 \equiv 1 \bmod 4$) and the character of I is $+$.

In the next case $D = 79$ the character consists of 2 signs and the character of $B = (3, 1)$ is $\left(\frac{3}{79}\right) = -\left(\frac{79}{3}\right) = -1$ and -1 (because $3 \equiv -1 \bmod 4$). The character of B^2 is $(--)^2 = ++$. Thus the classes I, B^2, B^4 have character $++$ and the remaining classes B, B^3, B^5 have character $--$. Gauss called the set of divisor classes with a given character a *genus*. Thus in the case $D = 79$ there are two genera, $\{I, B^2, B^4\}$ and $\{B, B^3, B^5\}$. In the previous cases the genera coincided with the divisor classes.

In the next example of Section 7.6, $D = -161$, the character consists of 3 signs (D has the two odd prime factors 7 and 23, and $D \equiv 3 \bmod 4$). The divisor $A = (3, 1)$ has the character $-+-$ and $B = (7, *)$ the character* $+--$. The characters of all 16 classes $A^i B^j$ can then be derived immediately. They are shown in Table 7.9.3. Here there are four genera $\{I, A^2, A^4, A^6\}$, $\{A, A^3, A^5, A^7\}$, $\{B, A^2B, A^4B, A^6B\}$, $\{AB, A^3B, A^5B, A^7B\}$. Gauss described this division of the divisor classes into genera by saying that $D = -161$ *corresponds to the classification* IV.4, meaning that there are IV genera each containing 4 classes. Similarly, $D = 79$ corresponds to the classification II.3, $D = -163$ to I.1, $D = -165$ to VIII.1, and $D = 67$ to II.1.

In the last example of Section 7.6, $D = 985 = 5 \cdot 197$, the character consists of two signs. The character of $A = (2, 0)$ is $\left(\frac{2}{5}\right)\left(\frac{2}{197}\right) = \left(\frac{2}{5}\right)\left(\frac{2}{5}\right) = --$. Thus $D = 985$ corresponds to the classification II.3 with the genera $\{I, A^2, A^4\}$ and $\{A, A^3, A^5\}$.

Closely related to Euler's observation that exactly half the classes mod $4D$ are splitting classes is Gauss's observation that *exactly half of the possible characters actually occur*. (See Table 7.9.4.) It is easy to deduce from quadratic reciprocity—or, what is essentially the same, from Euler's theorems—that *at most* half of the possible characters actually occur

Table 7.9.3 $D = -161$, $A = (3, 1)$, $B = (7, *)$

class of	character	class of	character
I	$+++$	B	$+--$
A	$-+-$	AB	$--+$
A^2	$+++$	A^2B	$+--$
A^3	$-+-$	A^3B	$--+$
A^4	$+++$	A^4B	$+--$
A^5	$-+-$	A^5B	$--+$
A^6	$+++$	A^6B	$+--$
A^7	$-+-$	A^7B	$--+$

*Properly speaking the first sign of the character of B is not defined. The character of its *class* has $+$ as its first sign because $(7, *) \sim (23, *)$ and $\left(\frac{23}{7}\right) = +1$.

Table 7.9.4

D	number of characters possible	number which occur
67	4	$2(++, --)$
-165	16	8 (those whose product is $+$)
-163	2	$1(+)$
79	4	$2(++, --)$
-161	8	$4(+++, -+-, +--, --+)$
985	4	$2(++, --)$

(Exercise 1). The method of Gauss's second proof of quadratic reciprocity is to prove that, conversely, if it is known that at most half of the possible characters occur then quadratic reciprocity can be deduced (see Section 7.11). In order to prove quadratic reciprocity it suffices, therefore, to prove that at most half of the possible characters actually occur, something which Gauss was able to do by counting *ambiguous classes* (see Section 7.10) and comparing them to the number of characters which occur.

Note that the number of characters possible is $2^{m+\varepsilon}$ and that the number of genera is equal to the number of characters which occur. Therefore Gauss's theorem on the occurrence of characters is equivalent to the statement that there are $2^{m+\varepsilon-1}$ genera. The classification IV.4, II.3, I.1, VIII.1, II.1, etc., of a determinant D can be thought of as a *factorization of the class number* $h = g \cdot n$ where $g =$ IV, II, I, VIII, II, etc. is the number of genera and $n = 4, 3, 1, 1, 1$, etc. is the number of classes per genus. (The fact that all genera contain the same number of classes is an immediate consequence of the fact that the character of a product is the product of the characters.) Gauss's theorem says that the first factor in this factorization is relatively easy to determine—$g = 2^{m+\varepsilon-1}$ where m is the number of odd prime factors of D and ε is 1 if $D \equiv 2$ or 3 mod 4, $\varepsilon = 0$ otherwise. However, the relationship between the determinant D and the class number h is a very subtle one, and the fact that the first factor g is simply related to D merely means that the second factor n is not.

EXERCISES

1. Using quadratic reciprocity, show that at most half of the possible characters actually occur. Show, in fact, that in any actually occurring character the number of minuses is even. [Use the Jacobi symbol.] Using Dirichlet's theorem on primes in an arithmetic progression, show that *exactly* half of the possible characters actually occur.

2. Prove that if $k = x^2 + ny^2$ where $n > 0$ then necessary conditions for k to be prime are (1) x is relatively prime to ny, (2) x and ny have opposite parity (except in the trivial case $x^2 + ny^2 = 2$), and (3) the only other representations $k = u^2 + nv^2$ are those in which $u = \pm x$, $v = \pm y$ or, if $n = 1$, $u = \pm y$ and $v = \pm x$.

3. Prove that if $n = 165$ then the necessary conditions of Exercise 2 are sufficient. [If k is not prime then the divisor of $x + y\sqrt{-165}$ is either of the form $A_1 A_2$ where A_1 and A_2 are relatively prime, or it is a power of a prime divisor $(p, u)^n$ where $n > 1$. In the first case $\overline{A}_1 A_2$ is principal but distinct from both $A_1 A_2$ and $\overline{A_1 A_2}$, which contradicts (3). In the second case, there is a principal divisor which has norm k and is divisible by p, and this too contradicts (3).]

4. Prove that if a and b are positive integers with the properties that ab is squarefree, $ab \not\equiv 3 \bmod 4$, and each genus for the determinant $D = -ab$ contains only one class then the following are both necessary and sufficient conditions for $k = ax^2 + by^2$ to be prime: (1) ax and by are relatively prime. (2) ax and by have opposite parity. (3) The only other representations $k = au^2 + bv^2$ are those in which $u = \pm x$, $v = \pm y$. [Assume without loss of generality that $a = p_1 p_2 \cdots p_n$ is odd. Then ak is the norm of a principal divisor of the form $(p_1, *)(p_2, *) \cdots (p_n, *)A$.] A number $n = ab$ of this sort—such as 165—is a case of what Euler called a *convenient number* (*numerus idoneus*). A full discussion of convenient numbers must deal with the cases $ab \equiv 3 \bmod 4$ and/or ab not squarefree. See Exercises 8–12 of Section 8.1.

5. Find all values of x less than 50 for which $165 + x^2$ is prime. [Apart from the obvious exclusions on the basis of conditions (1) and (2) there are 5 exclusions on the basis of condition (3), giving 7 primes in all.]

6. Prove that 5 is a convenient number. Prove that $1301 = 36^2 + 5$ is prime. [Starting with 1301 subtract successively, 5, 15, 25, 35, 45, ... and note that the progression contains only one square.]

7. Fermat's conjecture on numbers of the form $x^2 + 5y^2$ (see Section 1.7) was that if p_1 and p_2 are primes which are both 3 mod 4 and which "end in 3 or 7" then $p_1 p_2 = x^2 + 5y^2$. Prove that the conjecture is correct. [$p \equiv 3$ or 7 mod 20 implies p splits and its prime divisors are in the class (genus) not the principal class (genus).]

8. Show that a necessary condition for an odd prime p to be of the form $p = x^2 - Dy^2$ is that p or $p + D$ be a square mod $4D$. Euler guessed that this condition was also sufficient. The smallest counterexamples to this conjecture are in fact rather large. One was found by Lagrange [L2, Art. 84]. Lagrange's example is the case $D = 79$, $p = 101$ where $p + D$ is a square mod $4D$ but $p \neq x^2 - Dy^2$. He goes on to observe that one cannot answer the question of whether $p = x^2 - Dy^2$ without knowing more than the class of $p \bmod 4D$ because $101 \equiv 733 \bmod 4D$, 733 is prime, and $733 = x^2 - Dy^2$ when $D = 79$. Restate these facts in terms of the classes and genera of the prime factors of 101 and 733 and prove them. For a smaller counterexample to Euler's conjecture see Exercise 9 of Section 8.4.

7.10 Ambiguous divisors

Gauss called a divisor class (although of course in his formulation it was classes of binary quadratic forms rather than divisors that he was speaking of) an *ambiguous* class if it was its own conjugate. Other ways to phrase this definition are to say that a divisor A of the class satisfies $A \sim \overline{A}$ or, simply, that the square of the class is the principal class. Gauss discovered that the number of ambiguous classes could be found—or at least an upper bound given—directly, without using quadratic reciprocity or Euler's theorems, and that this gave enough information about the possible characters of divisor classes that one could *deduce* quadratic reciprocity (and therefore all of Euler's theorems in Section 7.8). This section is devoted to the count of ambiguous classes. The deduction of quadratic reciprocity is given in the next section.

If p is a prime which ramifies, then $(p, *)^2 \sim I$ and the class of $(p, *)$ is an ambiguous class. In addition, when $D > 0$, $(-1, *)$ lies in an ambiguous class. Therefore any product $(p_1, *)(p_2, *) \cdots (p_k, *)$ in which p_1, $p_2, \ldots, p_k$ are primes which ramify or, when $D > 0$, where p_1 may be -1, lies in an ambiguous class. In the examples of Section 7.6 it can be seen by direct examination that this accounts for all of the ambiguous classes— that is, every ambiguous class contains a divisor of the form $(p_1, *)(p_2, *) \cdots (p_k, *)$. [When $D = 67$, both classes are ambiguous; one contains I—the empty product—$(-1, *)(2, *)$, $(-1, *)(67, *)$, and $(2, *)(67, *)$, while the other contains $(-1, *)$, $(2, *)$, $(67, *)$, and $(-1, *)(2, *)(67, *)$. When $D = -165$ the divisors $A = (2, *)$, $B = (3, *)$ and $C = (5, *)$ and their products lie in all 8 possible classes. When $D = -163$ the only class is ambiguous and it contains both the empty product I and $(163, *)$. When $D = 79$ the ambiguous classes are those of I and B^3 where $B = (3, 1)$. The first contains I, $(2, *)$, $(-1, *)(79, *)$, and $(-1, *)(2, *)(79, *)$, while the other contains $(-1, *)$, $(79, *)$, $(-1, *)(2, *)$, and $(2, *)(79, *)$. When $D = -161$, the ambiguous classes are those of I, A^4, B, A^4B where $A = (3, 1)$, $B = (7, *)$. They contain I and $(7, *)(23, *)$, $(2, *)$ and $(2, *)(7, *)(23, *)$, $(7, *)$ and $(23, *)$, and $(2, *)(7, *)$ and $(2, *)(23, *)$ respectively. When $D = 985$ there are two ambiguous classes, those of I and A^3 where $A = (2, 0)$; the first contains I, $(-1, *)$, $(-1, *)(5, *)(197, *)$ and $(5, *)(197, *)$, while the second contains $(5, *)$, $(-1, *)(5, *)$, $(-1, *)(197, *)$, and $(197, *)$.]

In this section the number of ambiguous classes will be found in two steps. The first step is to prove that the above phenomenon occurs in all cases, that is, that every ambiguous class contains a divisor of the form $(p_1, *)(p_2, *) \cdots (p_k, *)$ where the p's are primes which ramify or, when $D > 0$, -1. The second is to count the number of different classes which contain divisors of this form.

Consider first the case $D < 0$ and $D \equiv 2$ or $3 \bmod 4$. Every divisor can be reduced by the cyclic method to a divisor A_0 which the cyclic method does not reduce

$$
\begin{array}{ccccccc}
& r_0 & & r_1 & & r_0 & \cdots \\
a_0 & & a_1 & & a_0 & & \cdots
\end{array}
$$

where $a_1 \geqslant a_0$, and where the test for divisibility of $x + y\sqrt{D}$ by A_0 is $x + yr_0 \equiv 0 \bmod a_0$. If $A \sim A_0$ and if A is ambiguous—that is, if $A \sim$

$\overline{A}$—then $A_0 \sim A \sim \overline{A} \sim \overline{A}_0$. If $a_1 > a_0$ then the cyclic method applied to $\overline{A}_0$ increases its norm, and it follows from the theorem of Section 7.7 that $A_0 \sim \overline{A}_0$ implies $A_0 = \overline{A}_0$. Then A_0 divides both $r_0 - \sqrt{D}$ and $r_0 + \sqrt{D}$. From this it follows that A_0 divides $2r_0$ and therefore that a_0 divides $2r_0$. Therefore, since a_0 divides $r_0^2 - D$ and, consequently, divides $(2r_0)^2 - 4D$, a_0 divides $4D$. Since all prime integer divisors of $4D$ are primes which ramify ($D \equiv 2$ or $3 \bmod 4$) it follows that A_0 has the desired form $(p_1, *)(p_2, *) \cdots (p_k, *)$. If $a_0 = a_1$ then* a different method is needed. Let B be defined by the condition that $A_0 \overline{B}$ is the divisor of $r_0 + a_0 - \sqrt{D}$. Let $N(B) = b$. Then $a_0 b = r_0^2 - D + 2a_0 r_0 + a_0^2 = a_0 a_1 + 2a_0 r_0 + a_0^2$, $b = 2(a_0 + r_0)$. Since b divides $(r_0 + a_0)^2 - D$ it must also divide $4(r_0 + a_0)^2 - 4D = b^2 - 4D$ and $b|4D$. Therefore $B \sim A_0$ is a divisor of the required form.

Consider next the case $D < 0$, $D \equiv 1 \bmod 4$. This case differs from the preceding one only in that r_0 and r_1 are halfintegers and $a_0 a_1 = [(2r_0)^2 - D]/4$. The fact that a_0 divides $2r_0$ (true if $a_1 \neq a_0$) then implies $a_0|D$, not just $a_0|4D$, and A_0 must have the desired form. In the case $a_1 = a_0$ define B by the condition that $A_0 \overline{B}$ be the divisor of $r_0 + a_0 - \frac{1}{2}\sqrt{D}$. Then $N(B)$ divides D as required.

Next consider the case $D > 0$, $D \equiv 2$ or $3 \bmod 4$. Every divisor A is equivalent to a divisor A_0 with the property that the cyclic method applied to A_0 returns to A_0. If $A \sim \overline{A}$ then of course $A_0 \sim \overline{A}_0$. It will suffice, therefore, to show that if $A_0 \sim \overline{A}_0$ then A_0 is equivalent to a divisor of the required type $(p_1, *)(p_2, *) \ldots (p_k, *)$. The main fact required for this proof is that *the cycle of $\overline{A}_0$ is the cycle of A_0 in reverse order.* That is, if the cyclic method applied to A_0 gives

$$
\begin{array}{ccccccc}
r_0 & & r_1 & \cdots & r_{m-1} & & r_0 & & r_1 & \cdots \\
& a_0 & & a_1 & \cdots & a_{m-1} & & a_0 & & a_1 & \cdots
\end{array}
$$

then applied to $\overline{A}_0$ it gives

$$
\begin{array}{ccccccc}
r_{m-1} & & r_{m-2} & \cdots & r_1 & & r_0 & & r_{m-1} & \cdots \\
& a_0 & & a_{m-1} & \cdots & a_2 & & a_1 & & a_0 & \cdots \quad .
\end{array}
$$

Clearly $\overline{A}_0$ divides $r_{m-1} - \sqrt{D}$ and $N(r_{m-1} - \sqrt{D}) < 0$. In order to show that the cyclic method applied to $\overline{A}_0$ gives $\overline{A}_{m-1}$ it is necessary and sufficient to prove that $N(r_{m-1} + |a_0| - \sqrt{D}) > 0$, that is, $(r_{m-1} + |a_0|)^2 > D$. In the proof that $r_j > 0$ in Section 7.5 it was shown that $(r_{j+1} + a_{j+1})^2 < D$ when $a_j > 0$ and $(r_{j+1} - a_{j+1})^2 < D$ when $a_j < 0$. In other words $(r_{j+1} - |a_{j+1}|)^2 < D$ for all j. With $j + 1 = m - 1$ this gives $r_{m-1}^2 - 2r_{m-1}|a_{m-1}| + a_{m-1}^2 < D$, $a_0 a_{m-1} - 2r_{m-1}|a_{m-1}| + a_{m-1}^2 < 0$, $-|a_0| - 2r_{m-1} + |a_{m-1}| < 0$, $a_0^2 + 2r_{m-1}|a_0| + a_0 a_{m-1} > 0$, $(|a_0| + r_{m-1})^2 > D$ as desired. Then, by the same token, the cyclic method applied to $\overline{A}_{m-1}$ gives $\overline{A}_{m-2}$, and so forth.

*For example, when $D = 165$ the cyclic method does not reduce $(13, 2) \sim (13, -2)$ and $(13, 2) \sim (2, *)(3, *)(5, *) \sim (13, -2) \sim (2, *)(11, *)$.

If $A_0 \sim \bar{A}_0$ then, by the theorem of Section 7.7, $\bar{A}_0$ must occur in the period of A_0, say $\bar{A}_0 = A_j$. Then $A_0 = \bar{A}_j$ and, by what was just shown, $A_1 = \bar{A}_{j-1}$, $A_2 = \bar{A}_{j-2}$, and so forth. Because $N(A_j)$ alternates in sign and $N(A_0) = N(\bar{A}_0) = N(A_j)$, j must be even, say $j = 2k$. Then $A_k = \bar{A}_{j-k} = \bar{A}_k$. Thus A_k divides both $r_k - \sqrt{D}$ and $r_k + \sqrt{D}$ and a_k divides $2r_k$. It then follows as before that a_k divides $4D$ and that A_k is of the required form.

The modifications required for the remaining case $D > 0$, $D \equiv 1 \bmod 4$ are very simple and are left to the reader.

In short, what has been shown is that a divisor is ambiguous if and only if it is equivalent to a divisor which divides $4D$ (if $D \equiv 2$ or $3 \bmod 4$) or which divides D (if $D \equiv 1 \bmod 4$). To count the inequivalent ambiguous divisors it suffices, therefore, to count the inequivalent divisors of $4D$ or D.

The number of primes which ramify is $m + \varepsilon$ because the primes which ramify are the odd prime factors of D when $D \equiv 1 \bmod 4$ and these together with 2 when $D \equiv 2$ or $3 \bmod 4$. When $D < 0$ there are therefore $2^{m+\varepsilon}$ divisors of the form $(p_1, *)(p_2, *) \cdots (p_k, *)$ (including the empty product I). Each of these is equivalent to at least one other, namely, its complement in the divisor of $\sqrt{D}$ or, if $(2, *)$ divides it but does not divide $\sqrt{D}$, its complement in the divisor of $2\sqrt{D}$. Thus *there are at most $2^{m+\varepsilon-1}$ ambiguous classes* when $D < 0$. The same is true when $D > 0$, but the proof is somewhat less elementary. There are $2^{m+\varepsilon+1}$ divisors of the form $(p_1, *)(p_2, *) \cdots (p_k, *)$ when $D > 0$ because in that case $(-1, *)$ is allowed. What is to be shown, then, is that each of these is equivalent to at least 3 others. For this it will suffice to prove that I is equivalent to at least 3 others. The cyclic method applied to I returns to I (the cyclic method applied to any principal divisor reaches I) and, because the sign of the norm alternates, the first return must occur at an even step, say $A_0 = I$, $A_{2k} = I$, $A_j \neq I$ for $0 < j < 2k$. Then A_k is of the form $(p_1, *)(p_2, *) \cdots (p_k, *)$ as was seen above. Moreover, A_k has norm with absolute value less than $|a_{k-1}a_k| = |r_k^2 - D| = D - r_k^2 < D$ and is therefore distinct from the divisor of $\sqrt{D}$ and distinct from its complement in the divisor of $\sqrt{D}$ or, if $(2, *)$ divides it but does not divide $\sqrt{D}$, in the divisor of $2\sqrt{D}$. Thus I, A_k, the divisor of $\sqrt{D}$, and the complement of A_k give 4 distinct principal divisors of the required form. Thus *there are at most $2^{m+\varepsilon-1}$ ambiguous classes* in any case.

Since $2^{m+\varepsilon}$ is equal to the number of possible characters, the theorem just proved can be reformulated as the statement that *the number of ambiguous classes is at most half the number of possible characters.*

EXERCISES

When $D < 0$ there are $2^{m+\varepsilon}$ divisors of the form $(p_1, *)(p_2, *) \cdots (p_k, *)$, and when $D > 0$ there are $2^{m+\varepsilon+1}$. In each of the following cases determine which of these divisors are principal and deduce the exact number of ambiguous *classes*.

1. $D = -30$.

4. $D = 210$.

2. $D = -31$.

5. $D = 61$.

3. $D = 31$.

6. $D = 59$.

7. Prove that if $D > 0$ is a prime other than 2, then the fundamental unit has norm -1 if and only if $D \equiv 1 \bmod 4$, norm $+1$ if and only if $D \equiv 3 \bmod 4$.

8. Prove that if $D = pq$ where p and q are primes $\equiv 3 \bmod 4$ then either $(p, *)$ or $(q, *)$ is principal but not both.

9. Show that if D is a (positive) prime that is 1 mod 4 then the cycle of I, say $I \sim A_1 \sim A_2 \sim \cdots$ contains consecutive divisors with $N(A_i) = -N(A_{i+1})$. Conclude that this gives a solution of Fermat's problem of finding a representation $D = u^2 + v^2$. Apply this method in the cases $D = 13$ and $D = 233$ (*Disquisitiones Arithmeticae*, Art. 265).

7.11 Gauss's second proof of quadratic reciprocity

Gauss claimed [G3] that he discovered the law of quadratic reciprocity entirely on his own, at a time when he was ignorant of the work of his predecessors. Although this claim strains one's credulity somewhat, there can be no doubt that Gauss's proof of the law in the *Disquisitiones Arithmeticae** was the first valid proof. Gauss in fact gave two entirely different proofs of the law in the *Disquisitiones*, one in the fourth section and one in the fifth. The proof in the fourth section, which he later said [G3] was the first one that he found, was described by Smith [S3] as being "repulsive to any but the most laborious students," at least in the form in which Gauss presents it. (Smith goes on to recommend Dirichlet's version of this proof as being done "with that luminous perspicuity by which his [Dirichlet's] mathematical writings are distinguished.") The proof in the fifth section is, in essence, the proof which is given below.

Gauss phrased this proof in terms of the theory of binary quadratic forms and he regarded the law of quadratic reciprocity as a theorem about the solution of quadratic congruences. Seen from this point of view, the proof seems very remote from the theorem and for this reason seems very unsatisfying. If, however, one phrases the proof instead in terms of the divisor theory of quadratic integers and adopts the view of Section 7.8 that the theorem is about the splitting of primes in the arithmetic of quadratic integers, then the theorem and the proof appear quite naturally related. It is not surprising, then, that it was this second proof of Gauss's which was Kummer's inspiration for his proof in 1859 of the higher reciprocity laws for regular prime exponents.

*It should be remembered that the *Disquisitiones* were published when Gauss was 24 years old.

It was noted at the end of Section 7.9 that the class number can be factored $h = gn$ where g, n is the "classification" to which D corresponds, that is, g is the number of genera and n is the number of classes per genus. At the same time, the class number can be factored $h = sa$ where s is the number of distinct classes that are squares of other classes and where a is the number of ambiguous classes. (In terms of group theory, this follows immediately from the fact that squaring is a homomorphism whose kernel has a elements and whose image has s elements.) To see this, let A_1, $A_2, \ldots, A_a$ be a list of divisors with one in each of the a ambiguous classes. Two classes have the same square if and only if $B^2 \sim C^2$ when B is a divisor from one class and C a divisor from the other, and this is true if and only if $(B\bar{C})^2 \sim I$, which is true if and only if $C \sim A_iB$ for some $i = 1, 2, \ldots, a$. Thus exactly a classes have the same square as the class of any given B, the number of different squares is $s = h/a$, and $h = sa$, as was to be shown.

Clearly $s \leqslant n$, because the square of any class is in the principal genus. Therefore $g \leqslant a$ and the estimate of a in the preceding section shows that *the number g of characters that actually occur is at most half as great as the number of possible characters*. Gauss deduced the law of quadratic reciprocity, and the supplementary laws, from this theorem. The steps of the proof that follows and their numbers are taken from Article 262 of *Disquisitiones Arithmeticae*, although the terminology has been completely changed.

I. *If p (a prime) is* -1 mod 4, *then* $(\frac{-1}{p}) = -1$. Consider quadratic integers with $D = -1$. The character consists of only one sign, namely, the character mod 4. If $p \equiv -1$ mod 4 and if $(\frac{-1}{p}) = +1$ then p would split and its prime factors would have character -1. Since the principal class has character $+1$, both characters would then occur, contrary to the above theorem. Therefore $p \equiv -1$ mod 4 implies $(\frac{-1}{p}) \neq +1$, as was to be shown.

II. *If $p \equiv 1$ mod 4, then* $(\frac{-1}{p}) = +1$. Consider $D = p$. The character consists of only one sign, namely, the quadratic character mod p. The character -1 cannot occur (by the theorem) and $(-1, *)$ must therefore have the character $+1$, as was to be shown.

III. *If $p \equiv 1$ mod 8 then* $(\frac{2}{p}) = (\frac{-2}{p}) = +1$. Consider $D = p$. Then the character consists of only one sign. Therefore $(\frac{q}{p}) = +1$ for any prime q which splits, and, since $(\frac{-1}{p}) = +1$ by II, $(\frac{-q}{p}) = +1$ as well. But 2 splits because $\frac{1}{2}(1 - \sqrt{D})$ has norm divisible by 2 and is not divisible by 2.

IV. *If $p \equiv 3$ or 5 mod 8 then* $(\frac{2}{p}) = -1$. Consider $D = 2$. Then the character consists of only one sign, namely, the sign $+1$ for divisors with norm $\equiv \pm 1$ mod 8 and the sign -1 for divisors with norm $\equiv \pm 3$ mod 8. The character -1 cannot occur and it follows that $p \equiv \pm 3$ mod 8, $(\frac{2}{p}) = +1$ is impossible.

V. *If $p \equiv 5$ or 7 mod 8 then* $(\frac{-2}{p}) = -1$. This is the same as IV with $D = -2$ instead of $D = 2$.

VI. *If $p \equiv 3$ mod 8 then* $(\frac{-2}{p}) = +1$. $(\frac{-2}{p}) = (\frac{-1}{p})(\frac{2}{p}) = (-1)^2 = +1$.

VII. If $p \equiv 7 \bmod 8$ *then* $(\frac{2}{p}) = +1$. $(\frac{2}{p}) = (\frac{-2}{p})(\frac{-1}{p}) = (-1)^2 = +1$.

VIII. *If* $p \equiv 1 \bmod 4$ *and* q *is a prime for which* $(\frac{q}{p}) = -1$ *then* $(\frac{p}{q}) = -1$. Consider $D = p$. The character consists of one sign which must always be $+$. Therefore if $(\frac{q}{p}) = -1$ then q cannot split, as was to be shown.

IX. *If* $p \equiv -1 \bmod 4$ *and* $(\frac{q}{p}) = -1$ *then* $(\frac{-p}{q}) = -1$. Consider $D = -p$. The character consists of just one sign, namely, the quadratic character mod p, and q cannot split unless this character is $+1$.

X. *If* $p \equiv 1 \bmod 4$ *and* $(\frac{q}{p}) = +1$ *then* $(\frac{p}{q}) = +1$. If $q \equiv 1 \bmod 4$ then $(\frac{p}{q}) = -1$ would imply $(\frac{q}{p}) = -1$ by VIII. If $q \equiv -1 \bmod 4$ then $(\frac{-q}{p}) = (\frac{q}{p})(\frac{-1}{p}) = 1$ by II and $(\frac{p}{q}) \neq -1$ follows from IX.

XI. *If* $p \equiv -1 \bmod 4$ *and* $(\frac{q}{p}) = +1$ *then* $(\frac{-p}{q}) = +1$. If $q \equiv 1 \bmod 4$ then $(\frac{-p}{q}) = (\frac{-1}{q})(\frac{p}{q}) = +1$ by II and VIII. In the remaining case $q \equiv -1 \bmod 4$ let $D = pq$. The character then consists of two signs, namely, the quadratic character mod p and the quadratic character mod q. The character of $(-1, *)$ is $-1, -1$ by I. Therefore only the characters $+ +$ and $- -$ occur. Since $(-1, *)(p, *)(q, *)$ is the divisor of $\sqrt{D}$, $(q, *) \sim (-1, *)(p, *)$. The character of the class of $(q, *)$ and $(-1, *)(p, *)$ is $(\frac{q}{p})$, $(\frac{-p}{q})$ and this completes the proof.

In summary, I and II imply that $(\frac{-1}{p}) \equiv p \bmod 4$, III–VIII imply that $(\frac{2}{p})$ is $+1$ if $p \equiv \pm 1 \bmod 8$ and is -1 if $p \equiv \pm 3 \bmod 8$, and VIII–XI imply that $(\frac{p}{q}) = (\frac{q}{p})$ when $p \equiv 1 \bmod 4$, $(\frac{-p}{q}) = (\frac{q}{p})$ if $p \equiv -1 \bmod 4$. Therefore the quadratic reciprocity laws and the Euler theorems of Section 7.8 are proved.

EXERCISES

1. Prove that the number of classes per genus is 1 if and only if every divisor is equivalent to a divisor of the form $(p_1, *)(p_2, *) \ldots (p_k, *)$.

2. In addition to the above proofs of VI and VII, Gauss offers the following alternative proofs. In VI let $D = 2p$. Then there are 4 characters, of which at most 2 occur. The character of $(-1, *)$ can be determined, and this shows exactly which 2 characters occur. The desired conclusion then follows from consideration of the character of the class of $(p, *) \sim (-1, *)(2, *)$. In VII, let $D = -p$ and note that 2 splits. Fill in the details of these proofs.

3. Show that the statement that exactly half of the possible characters actually occur is equivalent to the statement that every class in the principal genus is a square. (Gauss proved in Article 286 of the *Disquisitiones* that both of these statements are true by using the theory of ternary quadratic forms to *construct* a divisor with given square in the principal genus. Compare to Exercise 1, Section 7.9.)

Gauss's theory of binary quadratic forms 8

8.1 Other divisor class groups

In the preceding chapter, *divisor class groups* were defined for squarefree determinants D. The primary objective of the present chapter is to show the relationship between these groups and the *groups of classes of binary quadratic forms* which Gauss introduced in *Disquisitiones Arithmeticae*. In order to show this relationship in its simplest form, it will be necessary first to generalize the notion of divisor class group so that it includes all the cases which arise in Gauss's theory. The needed generalization can be motivated entirely on the basis of divisor theory for more general types of "quadratic integers," without bringing binary quadratic forms into the discussion at all. This is the subject of the present section. The following sections will then show how this theory is related to the theory of binary quadratic forms.

The most straightforward notion of "quadratic integers for the determinant D" is simply the set of all numbers of the form $x + y\sqrt{D}$ where x and y are integers. In the heuristic discussion which follows, this is what will be meant by a quadratic integer of the form $x + y\sqrt{D}$. In the case where $D \equiv 1 \bmod 4$ and D is squarefree, a different definition was given in the preceding chapter (denominators of 2 were allowed); in order to avoid confusion, the case $D \equiv 1 \bmod 4$ will be excluded from the following heuristic discussion. It is also natural to exclude the case where D is a square, since in that case $x + y\sqrt{D}$ is an ordinary integer. The objective is to develop a theory of divisors for quadratic integers $x + y\sqrt{D}$. Since this was already accomplished in the preceding chapter for the case where D is squarefree, the case to be considered is the case in which $D = t^2 D'$ where D' is a squarefree integer and t is a positive integer greater than 1. Then

305

the quadratic integers $x + y\sqrt{D} = x + yt\sqrt{D'}$ are *included* among the quadratic integers for the determinant D' as they were defined in the preceding chapter. Since the divisor of a quadratic integer for the determinant D' was defined in the preceding chapter, this implies a definition of the divisor of a quadratic integer $x + y\sqrt{D}$. This does not yet give a satisfactory divisor theory, however, because *the fundamental theorem fails* for the following reason.

Suppose that $x + y\sqrt{D}$ and $u + v\sqrt{D}$ have divisors A and B respectively in the theory of divisors for quadratic integers for the determinant D'. To say that $x + y\sqrt{D}$ divides $u + v\sqrt{D}$ in the arithmetic of quadratic integers for the determinant D should, of course, mean that there exist integers m and n such that $u + v\sqrt{D} = (x + y\sqrt{D})(m + n\sqrt{D})$. If this is true then $B = AC$ where C is the divisor of $m + n\sqrt{D}$, and divisibility of B by A is a *necessary* condition for divisibility of $u + v\sqrt{D}$ by $x + y\sqrt{D}$. The fundamental theorem states that this necessary condition is also sufficient, but, as examples show, this need not be the case. For example, in the case where $D = -9$, $t = 3$, $D' = -1$, $x + y\sqrt{D} = 3 + 0\sqrt{-9}$, $u + v\sqrt{D} = 3 + \sqrt{-9}$, one has $u + v\sqrt{D} = 3 + 3\sqrt{-1} = (x + y\sqrt{D})(1 + \sqrt{-1})$ and $x + y\sqrt{D}$ divides $u + v\sqrt{D}$ *when they are considered as quadratic integers for the determinant D'*—and therefore the necessary condition $A|B$ is met—but $x + y\sqrt{D}$ does *not* divide $u + v\sqrt{D}$ when they are considered as quadratic integers for the determinant D because the quotient $1 + \sqrt{-1}$ is not of the form $m + n\sqrt{D}$; therefore the condition $A|B$ is not sufficient.

The principal observation to be made is that failures of the fundamental theorem are always of the simple type of the above example—in which division removes needed factors of t—and that, in the main, the divisor theory for quadratic integers for the determinant D is valid even when D is not squarefree. To see the sense in which this is true, consider first the case in which $D' \not\equiv 1 \bmod 4$. Then the quadratic integers for the determinant D' are $\{x + y\sqrt{D'} : x, y \text{ integers}\}$ and such a quadratic integer is a quadratic integer for the determinant D if and only if t divides y. Thus, loosely speaking, one in every t quadratic integers for the determinant D' is a quadratic integer for the determinant D. The integer t is called the *index** of the quadratic integers for the determinant D in those for the determinant D'. Suppose that $x + y\sqrt{D}$ has divisor A, that $u + v\sqrt{D}$ has divisor B, and that A divides B. Then $u + v\sqrt{D} = (x + y\sqrt{D}) \cdot (m + n\sqrt{D'})$ by the fundamental theorem, and the question is whether t divides n. If $x + y\sqrt{D}$ *is relatively prime to t* then t *must divide* n. This can be proved as follows.

The given equation multiplied by $x - y\sqrt{D}$ (rationalize the denominator) is $(u + v\sqrt{D})(x - y\sqrt{D}) = (x^2 - Dy^2)(m + n\sqrt{D'})$. If $x + y\sqrt{D}$ is

*In terms of group theory, this is simply the index of a subgroup.

relatively prime to t, then so are $x - y\sqrt{D}$ and $x^2 - Dy^2$. Let k denote the integer $x^2 - Dy^2$. If $x + y\sqrt{D}$ is relatively prime to t then the integers t and k are relatively prime and there exist integers a and b such that $ak + bt = 1$. Then $m + n\sqrt{D'} = (ak + bt)(m + n\sqrt{D'}) = a[k(m + n\sqrt{D'})] + bt(m + n\sqrt{D'}) = a[(u + v\sqrt{D})(x - y\sqrt{D})] + (btm + bn\sqrt{D})$ is clearly of the form $c + d\sqrt{D}$, as was to be shown.

The condition that $x + y\sqrt{D}$ be relatively prime to t is a very natural one. If it is not fulfilled then there is a prime divisor P which divides both t and $x + y\sqrt{D}$. But then $x \equiv x + yt\sqrt{D'} = x + y\sqrt{D} \equiv 0 \bmod P$ and it follows that x is divisible by the prime integer p which P divides. Then $x + y\sqrt{D} = p(x' + yt'\sqrt{D'})$ where $x' = x/p$ and $t' = t/p$. This means that in dividing by $x + y\sqrt{D}$ it is natural to divide first by the integer p and then by the quadratic integer $x' + yt'\sqrt{D'}$ considered as a quadratic integer for the smaller determinant $D'' = (t')^2 D' < D$.

In summary, the theory of divisors for quadratic integers for the determinant D which they inherit as quadratic integers for the determinant D' is a theory in which the fundamental theorem holds, provided that one excludes division by elements $x + y\sqrt{D}$ which are not relatively prime to t. This exclusion is a natural one because an element $x + y\sqrt{D}$ which is not relatively prime to t is divisible by an *integer* and the problem can be reduced to one with smaller D.

One can go on (maintaining, for the moment, the assumptions that neither D nor D' is 1 mod 4) to define the notions of *principal* and *equivalent* for divisors of quadratic integers $x + y\sqrt{D}$ in the obvious ways. A divisor of quadratic integers for the determinant D' is said to be *principal for quadratic integers for the determinant D* if it is the divisor of some $x + y\sqrt{D}$. Two divisors A and B of quadratic integers for the determinant D' are said to be *equivalent for quadratic integers for the determinant D* if the condition "AC is principal" is equivalent to the condition "BC is principal." In other words, A is equivalent to B if, in any divisor (for quadratic integers for the determinant D') divisible by A, one can replace A by B and the new divisor is principal for quadratic integers for the determinant D if and only if the original divisor was.

As in previous cases, it follows immediately from these definitions that equivalence is reflexive, symmetric, transitive, and consistent with multiplication of divisors. Thus divisor *classes* can be multiplied, and the class of I is a multiplicative identity. However, the divisor classes do not necessarily form a *group* because there may be divisors A for which no divisor B satisfies $AB \sim I$—that is, there may be divisors A which have no multiplicative inverse. However, as with the failure of the fundamental theorem, the failure of A to have a multiplicative inverse occurs only when A is not relatively prime to t. This can be proved as follows.

If A is relatively prime to t then $A\bar{A}$ is the divisor of an integer, call it k, which is relatively prime to t. The claim is that $A\bar{A} \sim I$. Clearly if $I \cdot C$ is principal then $A\overline{A}\cdot C$ is principal—it is the divisor of $k(x+y\sqrt{D})$ when $x+y\sqrt{D}$ has divisor C. What must be shown is that if $A\overline{A}\cdot C$ is principal then $I\cdot C$ is principal. This follows immediately from the fundamental theorem because if $u+v\sqrt{D}$ has divisor $A\overline{A}C$ then division by k, which has divisor $A\overline{A}$ and is relatively prime to t, gives a quadratic integer $x+y\sqrt{D}$ with divisor C. Therefore $A\overline{A}\sim I$ and A has a multiplicative inverse, as was to be shown.

Thus, if one restricts consideration to divisors relatively prime to t the equivalence classes do form a *group*. This is the *divisor class group* for the determinant D. As in the case of squarefree determinants, this group is always finite and it can be found explicitly, for given D, using the cyclic method. These facts will be proved later in this chapter.

It remains to define the divisor class group in the case where $D'\equiv 1\bmod 4$. (If $D\equiv 1\bmod 4$ then t is odd, $t^2\equiv 1\bmod 4$, and $D'\equiv 1\bmod 4$, so that this case is included in the case $D'\equiv 1\bmod 4$.) In this case, quadratic integers for the determinant D' can be written in the form $u+v\omega$ where u and v are integers and $\omega=\frac{1}{2}(1-\sqrt{D'})$. Such a quadratic integer is of the form $x+y\sqrt{D}$ if and only if v is divisible by $2t$. The same argument as before then shows that *the fundamental theorem is true* for quadratic integers of the form $\{x+y\sqrt{D}\ :x,y\text{ integers}\}$ *in the case of division by elements relatively prime to $2t$*. When *principal* and *equivalent* are defined in the obvious ways then, exactly as before, one can multiply *equivalence classes* of divisors, this operation has a multiplicative identity, and the classes of divisors relatively prime to the index $2t$ form a *group* called the *divisor class group* for the quadratic integers $\{x+y\sqrt{D}\ :x,y$ integers$\}$.

The general case of the divisor class group can be described as follows. An *order** of quadratic integers for the determinant D' is a subset of the form $\{x+y\sqrt{D'}\ :x,y$ integers, y divisible by $s\}$ when $D'\not\equiv 1\bmod 4$ or of the form $\{x+y\omega:x,y$ integers, y divisible by $s\}$ when $D'\equiv 1\bmod 4$, $\omega=\frac{1}{2}(1-\sqrt{D'})$. The positive integer s is called the *index* of the order in the full order of all quadratic integers for the determinant D'. Chapter 7 was devoted to the divisor class groups in cases where $s=1$. For each squarefree D' and for each positive s there is one and only one order of quadratic integers for the determinant D' which has index s. When $D'\not\equiv 1\bmod 4$ this order consists of what were called quadratic integers for the determinant s^2D' above. When $D'\equiv 1\bmod 4$ and s is even—say $s=2t$ —it consists of $\{x+y\sqrt{t^2D'}\ :x,y$ integers$\}$. When $D'\equiv 1\bmod 4$ and s

308

is odd it consists of $\{x + y\sqrt{s^2 D'} \; : x, y \text{ integers or } x, y \text{ halfintegers}\}$. The cases considered above lead to the following definitions.

Let an order of quadratic integers for the determinant D' be given. Let a divisor A of quadratic integers for the determinant D' be called *principal* with respect to the order if it is the divisor of an element of the order. Let two divisors A and B be called *equivalent* with respect to the order if AC is principal with respect to the order when and only when BC is principal with respect to the order. Then multiplication of divisors is well defined for equivalence classes (the equivalence class of AB depends only on the class of A and the class of B) and the class of I is a multiplicative identity. Let s be the index of the order. Then the set of all equivalence classes of divisors relatively prime to s form a *group* called the divisor class group. More specifically, if A is relatively prime to s then $A\bar{A}$ is equivalent to I. This can be proved as follows.

Since $A\bar{A}$ is the divisor of an integer a, it is obvious that $I \cdot C$ principal implies $A\bar{A} \cdot C$ principal. Conversely, if $A\bar{A}C$ is principal then it is the divisor of an element $x + y\sqrt{D}$ of the order. Since the divisor of a divides the divisor of $x + y\sqrt{D}$, $x + y\sqrt{D} = a(u + v\sqrt{D'})$ where $u + v\sqrt{D'}$ is a quadratic integer, not necessarily in the order, whose divisor is C. To prove that $I \cdot C$ is principal it will suffice to prove that $u + v\sqrt{D'}$ is in the order. Now, by virtue of the assumption that A is relatively prime to s, a is relatively prime to s. Therefore there exist integers k and m such that $ak + ms = 1$. Then $u + v\sqrt{D'} = (ak + ms)(u + v\sqrt{D'}) = k(x + y\sqrt{D}) + ms(u + v\sqrt{D'})$ is in the order because, directly from the definition, $s(u + v\sqrt{D'})$ is in the order for any quadratic integer $u + v\sqrt{D'}$.

Exercises 1 and 6 deal with the actual computation of the divisor class group in two specific cases. More examples will be given in Section 8.5 after some efficient techniques have been developed for *classifying* divisors with respect to an order, that is, for determining whether two given divisors are equivalent with respect to a given order.

EXERCISES

1. Prove the following facts about the divisor class group for $\{x + y\sqrt{-11} \; : x, y \text{ integers}\}$: (a) $(3, 1)$ is not principal. (b) If (2) denotes the divisor of 2 (which is a prime divisor when $D' = -11$) then (2) and $(2)(3, 1)$ are principal, though $(3, 1)$ is not. (c) $(3, 1)^2$ is not principal, but $(3, 1)^3$ is principal. Therefore $(3, 1)^2 \sim (3, -1)$. (d) $(5, 2) \sim (3, 1)$. (e) $(11, *) \sim I$. (f) 7, 13, 17, 19 remain prime. (g) $(23, 9) \sim (3, 1)$. (h) 29 remains prime. (i) 31 splits into two prime divisors, one of which is equivalent to $(3, 1)$ and the other to $(3, 1)^2$. These computations strongly suggest that I, $(3, 1)$ and $(3, 1)^2$ is a representative set for the divisor class group. To prove this, one need only prove that every divisor is equivalent to one with smaller norm unless its norm is already small. (j) Any

divisor is principal for the divisor class group for $\{x + y\sqrt{-11} \; : \; x, y$ integers or halfintegers$\}$. (k) If (p, u) is not principal for the divisor class group under consideration then $(p, u) \sim (3, 1)$ or $(p, u) \sim (3, 1)^2$ can be obtained by multiplying an element with divisor (p, u) by $2(1 \pm \sqrt{-11})$ and dividing by 4.

2. In the example of Exercise 1, show that there is no divisor B such that $(2)B \sim I$. Conclude that the set of *all* equivalence classes of divisors does not form a group.

3. The cyclic method cannot be used in the most obvious way to classify divisors for the more general divisor class groups defined in this section. To see this, apply the cyclic method to the divisor $(13, 5)$ in the case $D' = -1$, $t = 6$, $D = -36$. The result is a period of two divisors *neither* of which is equivalent to $(13, 5)$. The theory of the next three sections has as its main objective the overcoming of this difficulty so that the cyclic method can be used to classify divisors in the new cases as well as the old ones.

4. Prove that if A is relatively prime to the index s of the order then A is principal if and only if $A \sim I$.

5. Show that a subset of the quadratic integers for the determinant D' is an order if and only if it has the two properties (1) sums and differences of things in the subset are in the subset and (2) the subset properly contains the ordinary integers (that is, every ordinary integer $x + 0 \cdot \sqrt{D'}$ is in the subset and at least one element of the subset is not an ordinary integer). [Consider the smallest positive integer or halfinteger that can occur as the coefficient of $\sqrt{D'}$ in an element of the subset.]

6. Find the divisor class group corresponding to the order $\{x + y\sqrt{18} \; : \; x, y$ integers$\}$. [For a given divisor, the methods of Chapter 7 make it simple to find all quadratic integers which have that divisor. The formula is of the form $\pm (x + y\sqrt{2})\varepsilon^n$ where $\varepsilon = (1 - \sqrt{2})^2$. In order to determine whether an element is in the order, it suffices to compute it mod 3, which is periodic in n. Show $(-1, *) \sim (2, *) \not\sim I$. Show that every divisor is equivalent either to I or to $(-1, *)$.]

7. If D is negative and $D \equiv -3 \bmod 8$ then the divisor class group of the order $\{x + y\sqrt{D} \; : \; x, y$ integers$\}$ contains an element of order 3 (an element that is not in the principal class, but whose cube is) except in the case $D = -3$. This fact was observed by Gauss (*Disquisitiones Arithmeticae* Art. 256, VI). Compare to Exercise 1. [The cube of the divisor of $(1 - \sqrt{D})/2$ is principal with respect to the order in question. The divisor of $(1 - \sqrt{D})/2$ is not principal unless $D = -3$.]

8. A positive integer n is said to be *convenient* if the divisor class group of the order $\{x + y\sqrt{-n} \; : \; x, y$ integers$\}$ has the property that the square of any class is the principal class. Prove that a number that is convenient in the sense of Exercise 4 of Section 7.9 is convenient in this new sense. [See Exercise 1 of Section 7.11.] Prove that if n is convenient, if $ab = n$, if $k = ax^2 + by^2$, if ax and by are relatively prime and of opposite parity, and if $k = au^2 + bv^2$ occurs only

when $u = \pm x, v = \pm y$, then k is prime. [One can assume without loss of generality that a is odd. Then the divisor of $ax + y\sqrt{-n}$ is of the form AK where $A = (p_1, *)(p_2, *) \ldots (p_n, *)$ and where K is relatively prime to the index s. If $K = K_1K_2$ then $AK_1\bar{K}_2$ is principal and gives an essentially different representation $k = au^2 + bv^2$.] Euler gave a list of 65 convenient numbers, the largest of which is 1848. No others have ever been found and it is conjectured that there are no others. Convenient numbers are also called *idoneal* or *suitable*. For Euler's list see [E10, E12]. See also Steinig [S4].

9. Euler used his method of convenient numbers to find large primes. The largest prime he found by this method was $18518809 = 197^2 + 1848 \cdot 100^2$. The amount of work needed to prove that this number is prime is an interesting measure of the difficulty—even for a computer as agile and sophisticated as Euler—of the problem of finding prime numbers of this magnitude. Prove that this number is prime. [For Euler's method see [E12]. Alternatively, one can use Gauss's method of "excludents" as follows. Let $A = 197^2 + 1848 \cdot 100^2$. The object is to show that $x^2 + 1848y^2 = A$ has only the solutions $x = \pm 197, y = \pm 100$. (Because 1848 is convenient and because 197 and $100 \cdot 1848$ are relatively prime and of opposite parity, this proves A is prime.) Mod 1848 the required equation is $x^2 \equiv 1$. Since $1848 = 8 \cdot 3 \cdot 7 \cdot 11$, this congruence is equivalent to $x^2 \equiv 1$ mod 8, $x^2 \equiv 1$ mod 3, $x^2 \equiv 1$ mod 7, and $x^2 \equiv 1$ mod 11. That is, $x \equiv 1$ mod 2 and $x \equiv \pm 1$ mod 3, 7, 11. By the Chinese remainder theorem it follows that x has one of the values $\pm 1, \pm 43, \pm 155, \pm 197$ mod $2 \cdot 3 \cdot 7 \cdot 11 = 462$. Since x can be assumed to be positive and $x < \sqrt{A}$, this eliminates all numbers other than the numbers of the 8 sequences $1 + 462k$ ($0 \leqslant k \leqslant 9$), $-1 + 462k$ ($1 \leqslant k \leqslant 9$), $43 + 462k$ ($0 \leqslant k \leqslant 9$), $-43 + 462k$ ($1 \leqslant k \leqslant 9$), $\ldots$, $-197 + 462k$ ($1 \leqslant k \leqslant 9$) which is 76 numbers, 75 in addition to the known solution $x = 197$. The method of excludents eliminates most of these possibilities as follows. Let 5 be the "excludent." The required equation mod 5 is, for the first of the 8 sequences, $(1 + 462k)^2 + 1848y^2 = A$, $(1 + 2k)^2 \equiv 4 - 3y^2$ mod 5. Since $y^2 \equiv 0, \pm 1$ it follows that $(1 + 2k)^2 \equiv 4, 1,$ or 2 mod 5. Since $z^2 \equiv 2$ is impossible, $(1 + 2k)^2 \equiv 1$ or 4 and $1 + 2k \equiv 1, 2, 3,$ or 4 mod 5. This excludes $k = 2$ and $k = 7$. Similarly, when 13 is the excludent (13 is the next prime which does not divide 1848) one finds $(1 + 7k)^2 \equiv -3 - 2y^2$ mod 13 where $y^2 \equiv 0, \pm 1, \pm 3, \pm 4$. $(1 + 7k)^2 \equiv -3, -5, -1, -9 \equiv 4, 3, -11 \equiv 2,$ or 5. Thus $(1 + 7k)^2 \equiv -3, -1, 4,$ or 3 and $1 + 7k \equiv \pm 6, \pm 5, \pm 2,$ or ± 4 mod 13. This excludes $k = 0, 4, 5,$ and 9. The excludents 3, 7, 11 can be used even though they divide 1848. For 3, for example, $x^2 + 3 \cdot 616y^2 = A$ gives $x^2 + 3y^2 \equiv 4$ mod 9. (Recall that numbers can be reduced mod 9 simply by adding digits.) Since $y^2 \equiv 0$ or 1 mod 3, $3y^2 \equiv 0$ or 3 mod 9 and $x^2 \equiv 4$ or 1 mod 9. When $x = 1 + 462k \equiv 1 + 3k$ mod 9 this gives $2 \cdot 3k \equiv 3$ or 0 mod 9, $2k \equiv 0$ or 1 mod 3. Therefore $k \not\equiv 1$ mod 3. In the same way one finds $x^2 \equiv -6, 8, 22,$ or 1 mod 49, $x \equiv 1 + 7 \cdot 3k$ mod 49, $6k \equiv -1, 1, 3, 0$ mod 7, $k \not\equiv 2, 3, 5$ mod 7. The only remaining possibilities are $k = 6$ or 8. These are eliminated by using the excludent 11 to find $x^2 \equiv 1, -10, 23, -32, -43,$ or -54 mod 121, which with $x = 1 + 462k$ gives $-4k \equiv 0, -1, 2, -3, -4, -5$ mod 11. Consider next the sequence $-1 + 462k$. From $x^2 \equiv 1$ or 4 mod 9 (see above) one finds $k \not\equiv 2$ mod 3. From $x^2 \equiv 1$ or 4 mod 5 one finds $k \not\equiv 3$ mod 5. From $x^2 \equiv 1, -6, 8,$ or 22 mod 49 one finds $-6k \equiv -1, 1, 3, 0 \bmod 7$ and $k \not\equiv 2, 4, 5 \bmod 7$. Similarly, for the excludent 11 one finds by a simple modification of the previous case $4k \equiv 0$, $-1, 2, -3, -4, -5 \bmod 11$. From among those $(1, 6, 7)$ which remain, this excludes only 1. Use of the excludent 13 gives $-1 + 7k \equiv \pm 2, \pm 4, \pm 5, \pm 6 \bmod 13$ and of the excludent 17 gives $-1 + 3k \equiv 0, \pm 1, \pm 3, \pm 6, \pm 7$, but neither of these excludes 6 or 7. The excludent 19 excludes both 6 and 7. Simple variations of these computations also eliminate all terms of the sequence $43 + 462k$, but in the sequence $-43 + 462k$ the case $k = 3$ survives all these exclusions, as well as the exclusion for the excludent 23. The reason for this is that $(-43 + 462 \cdot 3)^2 + 1848y^2 = A$ gives $y^2 = 9045$; this is impossible because $95^2 = 9025 < 9045 < 96^2$, but $y^2 \equiv 9045$ *is* possible, as the above exclusions show, mod 3, 5, 7, 11, 13, 17, 19, 23. One other of the 75 numbers to be tested survives all these exclusions—namely $x = -155 + 462 \cdot 6$—but this is a solution only if 6315 is a square, which it is not because $80^2 = 6400 > 6315 > 79^2$. The case $x = 197 + 462 \cdot 6$ survives all exclusions except the one for the excludent 23, but *only* this one does, and it is probably more efficient to exclude this one by the direct method and not to use the excludent 23.]

10. Euler stated [E12], without proof, the following criterion for determining whether a given n is convenient: The positive integer n is convenient if and only if the numbers of the set $\{x^2 + n: x$ relatively prime to n, and $x^2 + n < 4n\}$ are all either prime, or twice a prime, or the square of a prime, or a power of 2. Show that Euler's criterion correctly classifies 11 and 13. Assuming the validity of Euler's criterion, determine which numbers in the range 80–89 are convenient.

11. Prove that Euler's criterion is necessary—that is, that if n is convenient then it satisfies Euler's criterion. [Let AK be the divisor of $x + \sqrt{-n}$ where A is a product of prime divisors of 2 and K is relatively prime to 2. If $K = I$ then $x^2 + n$ is a power of 2. Otherwise K is divisible by (p, u) where p is relatively prime to the index of the order $\{x + y\sqrt{-n} : x, y$ integers$\}$. Let K_1 be K with (p, u) changed to $(p, -u)$. Then AK_1 is the divisor of $a + b\sqrt{-n}$ where $a^2 + nb^2 = x^2 + n$. Thus $b^2 = 0$ or 1. Consider first the case $b^2 = 0$. Then A and K_1 are both divisors of integers. This implies $K = (p, u)^2$ and it must be shown that $A = I$. A is the divisor of 2^k for some k. If $k > 0$ then 4 divides $x^2 + n$ and $n \equiv 3 \bmod 4$. The case $n \equiv 3 \bmod 8$ can be handled using Exercise 7. If $n \equiv 7 \bmod 8$ then $k > 1$ and 4 divides $x + \sqrt{-n}$, which is impossible. Finally, consider the case $b^2 = 1$. Then AK_1 must be the conjugate of AK, from which $A = \bar{A}$ and $K = (p, u)$. If 2 splits or remains prime, then A is the divisor of 2^k for some $k \geqslant 0$ and the case $k > 0$ can be excluded as before. If 2 ramifies then $A = (2, *)^k$ for some $k \geqslant 0$. Since 2 does not divide $x + \sqrt{-n}$, $k < 2$. Therefore $x^2 + n$ is either p or $2p$, as was to be shown.]

12. Prove that Euler's criterion is sufficient. [This appears to be an unsolved problem. Dickson [D2, vol. 1, p. 363] reports that the criterion was proved by F. Grube in 1874, but an examination of Grube's paper [G9] shows that he explicitly states he could *not* prove the sufficiency of the criterion. Gauss said in passing (*Disquisitiones Arithmeticae*, Art. 303) that Euler's criterion was "easy to prove."]

8.2   Alternative view of the cyclic method

Although the divisor class groups defined in the preceding section do coincide with those defined by Gauss, in outward appearance the theory of the preceding section bears no resemblance at all to Gauss's theory of binary quadratic forms. A bridge between the two is provided by the *cyclic method*. In fact, the computational technique of the cyclic method should perhaps be regarded as the heart of the matter, and the two theories in question—the divisor theory of quadratic integers and the theory of binary quadratic forms—should be regarded as different interpretations of the types of problems which this computational technique solves.

The fact that the cyclic method was used by the ancient Indians, long before anything as sophisticated as divisor theory had been devised, clearly indicates that the cyclic method has other interpretations than as the solution of the problem of determining whether a given divisor is principal. One alternative interpretation, which was related to the multiplication of two equations of the form $a = x^2 - Dy^2$, was given in Section 1.9. Another interpretation is the interpretation which follows. This is essentially the continued fractions approach used by Brouncker, Euler, and Lagrange, and it was therefore almost surely familiar to Gauss at the time that he was formulating his theory.

The cyclic method was used in Section 1.9 to find the solution $x = 48842$, $y = 5967$ of the equation $x^2 = 67y^2 + 1$. The computations by which this solution was found can be motivated as follows. Let the desired equation be rewritten in the form $x^2 - 67y^2 = 1$. Here x and y are to be positive integers. Clearly x must be much larger than y. In fact x is larger than $8y$ (because $(8y)^2 - 67y^2 < 0$) but less than $9y$ (because $(9y)^2 - 67y^2 > 1$). Therefore if z is defined by $x = 8y + z$ then $x > y > z > 0$ and the given problem is equivalent to the problem of finding integers y and z such that $(8y + z)^2 - 67y^2 = 1$, that is, $-3y^2 + 16yz + z^2 = 1$, and such that $y > z > 0$. Note that the trivial solution $y = 0$, $z = 1$ is excluded. Just as the previous equation $x^2 - 67y^2 = 1$ gave information on the ratio of x to y, this equation $-3y^2 + 16yz + z^2 = 1$ gives information on the ratio of y to z. With $y = z$, $2z$, $3z$, $4z$, or $5z$ the value of $-3y^2 + 16yz + z^2$ is negative and therefore too low, but with $y \geq 6z$ it is greater than 1 and therefore is too large. Therefore $y = 5z + a$, where $z > a > 0$, and the given problem is equivalent to $-3(5z + a)^2 + 16(5z + a)z + z^2 = 1$, that is, to $6z^2 - 14za - 3a^2 = 1$, where $z > a > 0$. Following this method, one is led to the sequence of equations in Table 8.2.1.

The last equation is simple to solve subject to the restriction $h > i$, namely, $h = 1$, $i = 0$. Then a solution of the next-to-last equation is $g = 5h + i = 5$, $h = 1$, a solution of the equation preceding that one is $f = 2g + h = 11$, $g = 5$, and so forth. This process of back substitution culminates in the desired solution $x = 48842$, $y = 5967$.

Table 8.2.1

$x^2 - 67y^2 = 1$	$x = 8y + z$
$-3y^2 + 16yz + z^2 = 1$	$y = 5z + a$
$6z^2 - 14za - 3a^2 = 1$	$z = 2a + b$
$-7a^2 + 10ab + 6b^2 = 1$	$a = b + c$
$9b^2 - 4bc - 7c^2 = 1$	$b = c + d$
$-2c^2 + 14cd + 9d^2 = 1$	$c = 7d + e$
$9d^2 - 14de - 2e^2 = 1$	$d = e + f$
$-7e^2 + 4ef + 9f^2 = 1$	$e = f + g$
$6f^2 - 10fg - 7g^2 = 1$	$f = 2g + h$
$-3g^2 + 14gh + 6h^2 = 1$	$g = 5h + i$
$h^2 - 16hi - 3i^2 = 1$	

The computations required to generate Table 8.2.1 are basically the same computations—though they are in a very different guise—as those that were required to generate the successive a's and r's in the format of Chapter 7:

$$D = 67$$
$$r = \quad 8 \quad 7 \quad 5 \quad 2 \quad 7 \quad 7 \quad 2 \quad 5 \quad 7 \quad 8$$
$$a = \quad 1 \quad -3 \quad 6 \quad -7 \quad 9 \quad -2 \quad 9 \quad -7 \quad 6 \quad -3 \quad 1.$$

In fact, the correspondence between these two forms of the computation can be summarized by saying that *if a_i, a_{i+1}, and a_{i+2} are three successive values of a in the cyclic method, if r_i and r_{i+1} are the intervening values of r, and if $n_{i+1} = (r_i + r_{i+1})/a_{i+1}$ then the substitution $u = |n_{i+1}|v + w$ transforms the binary quadratic form* $a_{i+1}u^2 \pm 2r_iuv + a_iv^2$ into the binary quadratic form $a_{i+2}v^2 \mp 2r_{i+1}vw + a_{i+1}w^2$, where the sign of the middle term in both quadratic forms is opposite to the sign of the first term.* This theorem, based here on a mere comparison of the two computations above, is easily verified algebraically: The coefficient of w^2 in $a_{i+1}(|n_{i+1}|v + w)^2 \pm 2r_i(|n_{i+1}|v + w)v + a_iv^2$ is obviously a_{i+1}. The coefficient of wv is $2a_{i+1}|n_{i+1}| \pm 2r_i = (\text{sign } a_{i+1})(2r_i + 2r_{i+1}) \pm 2r_i = \mp 2r_{i+1}$ when the sign in front of r_i is opposite to the sign of a_{i+1}. Finally, the coefficient of v^2 is $a_{i+1}n_{i+1}^2 \pm 2r_i|n_{i+1}| + a_i$; when this is multiplied by a_{i+1} it becomes $(r_i + r_{i+1})^2 - 2r_i(r_i + r_{i+1}) + r_i^2 - D = r_{i+1}^2 - D = a_{i+1}a_{i+2}$ and the coefficient is a_{i+2} as claimed (because $a_{i+1} \neq 0$).

One of the first problems in the theory of binary quadratic forms is the problem of finding *representations of a given integer m by a given form*[†]

---

*A *form* is a polynomial in which all terms have the same degree. A *quadratic* form is one in which this degree is two. A *binary* quadratic form is one in which there are two variables.

†Note that the coefficient of xy is assumed to be even. This is done in order to conform to Gauss's notation. There is no loss of generality in this assumption in the solution of $ax^2 + bxy + cy^2 = m$ because if b is odd then the entire equation can be doubled.

$ax^2 + 2bxy + cy^2$, that is, of finding, when integers a, b, c and m are given, integers x and y such that $ax^2 + 2bxy + cy^2 = m$. The solution of the problem of representing 1 by the binary quadratic form $x^2 - 67y^2$ that was given above can be summarized by saying that *the binary quadratic forms in Table 8.2.1 are all equivalent* in the sense that any number which can be represented by *one* of these forms can be represented by *all* of them, and the method for going from a representation by one of them to a representation by the others is to use the changes of variables in the right-hand column of the table. For example, to solve $x^2 - 67y^2 = -2$ one can note that $c = 1$ and $d = 0$ in $-2c^2 + 14cd + 9d^2$ gives a representation of -2 by this form; then $b = c + d = 1$, $a = b + c = 2$, $z = 2a + b = 5$, $y = 5z + a = 27$, and $x = 8y + z = 221$ gives the representation $(221)^2 - 67(27)^2 = -2$. This suggests yet another interpretation of the cyclic method, namely, the following one.

Two binary quadratic forms $ax^2 + 2bxy + cy^2$ and $a'u^2 + 2b'uv + c'v^2$ are said to be *equivalent* if there is an invertible change of coordinates with integer coefficients which transforms one to the other. More specifically, the forms are said to be equivalent if there exist integers α, β, γ, δ such that $\alpha\delta - \beta\gamma = \pm 1$ and such that substitution of $x = \alpha u + \beta v$ and $y = \gamma u + \delta v$ in $ax^2 + 2bxy + cy^2$ gives $a'u^2 + 2b'uv + c'v^2$. In terms of 2×2 matrices this can be stated, because

$$ax^2 + 2bxy + cy^2 = (x \quad y)\begin{pmatrix} a & b \\ b & c \end{pmatrix}\begin{pmatrix} x \\ y \end{pmatrix},$$

in the form

$$\begin{pmatrix} a' & b' \\ b' & c' \end{pmatrix} = \begin{pmatrix} \alpha & \gamma \\ \beta & \delta \end{pmatrix}\begin{pmatrix} a & b \\ b & c \end{pmatrix}\begin{pmatrix} \alpha & \beta \\ \gamma & \delta \end{pmatrix}$$

where the determinant $\Delta = \alpha\delta - \beta\gamma$ of the coordinate change is ± 1 so that its inverse

$$\begin{pmatrix} \alpha & \beta \\ \gamma & \delta \end{pmatrix}^{-1} = \begin{pmatrix} \delta/\Delta & -\beta/\Delta \\ -\gamma/\Delta & \alpha/\Delta \end{pmatrix}$$

has integer coefficients.

Then *if a_i, a_{i+1}, and a_{i+2} are three successive values of a in the cyclic method and if r_i and r_{i+1} are the intervening values of r then the binary quadratic forms $a_i x^2 + 2r_i xy + a_{i+1} y^2$ and $a_{i+1} u^2 + 2r_{i+1} uv + a_{i+2} v^2$ are equivalent.* This follows from the theorem above and the observation that $ax^2 + 2bxy + cy^2$ is equivalent both to $au^2 - 2buv + cv^2$ ($x = u, y = -v$) and to $av^2 + 2buv + cu^2$ ($x = v, y = u$). *An explicit equivalence is given by*

$$\begin{pmatrix} x \\ y \end{pmatrix} = \begin{pmatrix} 0 & -1 \\ 1 & n_{i+1} \end{pmatrix}\begin{pmatrix} u \\ v \end{pmatrix}; \qquad \begin{pmatrix} u \\ v \end{pmatrix} = \begin{pmatrix} n_{i+1} & 1 \\ -1 & 0 \end{pmatrix}\begin{pmatrix} x \\ y \end{pmatrix}$$

where $n_{i+1} = (r_i + r_{i+1})/a_{i+1}$. *In other words,*

$$\begin{pmatrix} a_i & r_i \\ r_i & a_{i+1} \end{pmatrix} = \begin{pmatrix} n_{i+1} & -1 \\ 1 & 0 \end{pmatrix}\begin{pmatrix} a_{i+1} & r_{i+1} \\ r_{i+1} & a_{i+2} \end{pmatrix}\begin{pmatrix} n_{i+1} & 1 \\ -1 & 0 \end{pmatrix}.$$

Since the determinants $a_i a_{i+1} - r_i^2 = -D$ and $1 \cdot (a_{i+1} a_{i+2} - r_{i+1}^2) \cdot 1 = -D$ of these two matrices are equal and because they are both symmetric, this equation can be checked merely by noting that the bottom row on the right is $a_{i+1} n_{i+1} - r_{i+1} = r_i$ and a_{i+1}, as desired.

Thus the cyclic method can be used to generate equivalent forms very quickly and in this way to solve problems of representations. Consider, for example, the problem of representing -3 by $13x^2 + 6xy - 4y^2$. The cyclic method then gives

$$
\begin{array}{ccccccccc}
r = & 3 & 5 & 4 & 6 & 4 & 5 & 7 \\
a = & 13 & -4 & 9 & -5 & 5 & -9 & 4 & -3.
\end{array}
$$

where $D = 61$. The last three numbers correspond to a form $4u^2 + 14uv - 3v^2$ which obviously represents -3 (when $u = 0$, $v = 1$). An explicit equivalence between this form and the given one $13x^2 + 6xy - 4y^2$ is given, according to the theorem above, by

$$
\begin{pmatrix} x \\ y \end{pmatrix} = \begin{pmatrix} 0 & -1 \\ 1 & -2 \end{pmatrix} \begin{pmatrix} 0 & -1 \\ 1 & 1 \end{pmatrix} \begin{pmatrix} 0 & -1 \\ 1 & -2 \end{pmatrix} \begin{pmatrix} 0 & -1 \\ 1 & 2 \end{pmatrix} \begin{pmatrix} 0 & -1 \\ 1 & -1 \end{pmatrix} \begin{pmatrix} 0 & -1 \\ 1 & 3 \end{pmatrix} \begin{pmatrix} u \\ v \end{pmatrix}
$$

which, with $u = 0$, $v = 1$, gives the solution $x = -37$, $y = -100$ of the given problem.

This example was somewhat contrived in that the given number -3 was one that occurred on the bottom line when the cyclic method was applied to the given form. A more general technique is to *construct* a form which obviously represents the given number and to attempt to use the cyclic method to construct an equivalence between that form and the given form.

For example, consider the representation of -217 by $7x^2 - 6xy + y^2$. The cyclic method gives the equivalent forms

$$
\begin{array}{ccccc}
-3 & 1 & 1 & \cdots \\
7 & 1 & -1 & 1 & \cdots
\end{array}
$$

where $D = 2$. A form which obviously represents -217 is one of the form $-217u^2 + 2duv + ev^2$ where d and e are integers to be found. If the above forms can be obtained by applying the cyclic method to this form then $d^2 - D = -217e$ where $D = 2$. In particular $d^2 \equiv 2 \bmod 217$. The Chinese remainder theorem can be used to solve this congruence as follows. $217 = 7 \cdot 31$. Mod 7 the congruence $d^2 \equiv 2$ implies $d \equiv \pm 3$. Mod 31 it implies $d \equiv \pm 8$. This gives four possible values $d \equiv \pm 39$, $\pm 101 \bmod 217$. For example, with $d = 39$ one finds $e = -(d^2 - D)/217 = -7$. Then

$$
\begin{array}{ccccc}
39 & 3 & 1 & 1 \\
-217 & -7 & -1 & 1 & -1
\end{array}
$$

reduces this form to the same forms obtained above. Reversing this computation and putting it together with the first one gives

$$
\begin{array}{ccccc}
-3 & 1 & 3 & 39 \\
7 & 1 & -1 & -7 & -217
\end{array}
$$

which shows that $7x^2 - 6xy + y^2$ is equivalent to $-7u^2 + 78uv - 217v^2$ and that an equivalence is given by

$$\binom{x}{y} = \begin{pmatrix} 0 & -1 \\ 1 & -2 \end{pmatrix}\begin{pmatrix} 0 & -1 \\ 1 & -4 \end{pmatrix}\begin{pmatrix} 0 & -1 \\ 1 & -6 \end{pmatrix}\binom{u}{v}.$$

With $u = 0$ and $v = 1$ this gives the representation $7(-23)^2 - 6(-23)(-40) + (-40)^2 = -217$.

The general technique for solving $ax^2 + 2bxy + cy^2 = m$ is a simple generalization of this example. Let D be defined by $b^2 - D = ac$—that is, $D = b^2 - ac$—and let the cyclic method be applied with a and c as the first two values of a and with b as the first value of r. This will lead eventually* to a *period*, a repetition of the numbers generated by the cyclic method. For the given m, consider the congruence $d^2 \equiv D$ mod m. For any solution d of this congruence (the problem of finding all solutions is a finite computation), define e by $D = d^2 - em$ and consider the binary quadratic form $mu^2 + 2duv + ev^2$. The cyclic method applied to this form also leads to a period. If this period coincides with the period found above then a solution $ax^2 + 2bxy + cy^2 = m$ can be found because the forms $ax^2 + 2bxy + cy^2$ and $mu^2 + 2duv + ev^2$, being equivalent to the same thing, are equivalent to each other and, in fact, an explicit equivalence is easily given so that the representation can be found merely by setting $u = 1$, $v = 0$.

It can be shown that this technique will always produce a representation provided one is possible.† That is, if the equation $ax^2 + 2bxy + cy^2 = m$ has a solution then there must be at least one solution d of the congruence $d^2 \equiv D$ mod m with the property that the cyclic method applied to $mu^2 + 2duv + ev^2$ (where $e = (d^2 - D)/m$) leads to the same period as the cyclic method applied to $ax^2 + 2bxy + cy^2$. The main objective here, however, is not the complete solution of $ax^2 + 2bxy + cy^2 = m$ (for this see Exercises 7 and 8 of Section 8.4) but rather a clear picture of the connection between the cyclic method and binary quadratic forms. The cyclic method thus becomes the link between the theory of divisors and the theory of binary quadratic forms. The resulting connection between divisors and binary quadratic forms is the subject of the next section.

*This assumes that $D = b^2 - ac$ is *not a square* so that $r^2 - D = 0$ never occurs. Gauss, with his usual thoroughness, dealt also with the case where D is a square (*Disquisitiones Arithmeticae*, Art. 215) but the solution involves neither the cyclic method nor any class group, and it is not of interest here.

†More precisely, the method produces a *proper* representation $ax^2 + 2bxy + cy^2 = m$, that is, a representation in which x and y are relatively prime, provided one is possible. An improper representation, one in which $x = dx'$, $y = dy'$ where $d > 1$ and x', y' are relatively prime, is possible only if m has a square factor $m = d^2m'$ and m' has a proper representation $ax'^2 + 2bx'y' + cy'^2 = m'$ by the given form. Thus if the above method fails to provide a representation one should then try to find a square factor d^2 of m and a representation of m/d^2 by the given form. If this too fails then there is no representation. The proof of these statements is considered in Exercise 7 of Section 8.4.

EXERCISES

1. Prove that a change of variables $x' = ax + by, y' = cx + dy$ with integer coefficients a, b, c, d is invertible, in the sense that the inverse also has integer coefficients, if and only if the determinant $ad - bc$ is ± 1.

2. Use the method of the text to find the two essentially distinct representations $x^2 + y^2 = 65$.

3. Use the method of the text to find a representation of 121 by $4x^2 + 2xy + 5y^2$.

4. Find a representation of 23 by $15x^2 + 40xy + 27y^2$.

5. Find a representation of 129 by $42x^2 + 118xy + 81y^2$.

6. Find a representation of 91 by $x^2 + xy + y^2$.

8.3 The correspondence between divisors and binary quadratic forms

In order to establish a connection between the divisor class groups defined in Section 8.1 and the groups which Gauss defined in connection with his theory of binary quadratic forms, it is necessary to establish a connection between divisors and binary quadratic forms. In the simplest case—the one where D is squarefree and congruent to 2 or 3 mod 4—the discussion of the preceding section makes it clear how this should be done.

Let $D \not\equiv 1$ mod 4 be a squarefree integer. If A is a divisor of quadratic integers for the determinant D which is divisible by no integer greater than 1, then to apply the cyclic method to A is the same as to apply the cyclic method to the binary quadratic form $ax^2 + 2bxy + cy^2$ where $a = N(A)$, $b \equiv \sqrt{D}$ mod A, and $c = (b^2 - D)/a$, except that the condition $b \equiv \sqrt{D}$ mod A does not determine* b but only determines it mod a. Thus, a mapping from divisors divisible by no integer greater than 1 to binary quadratic forms with $b^2 - ac = D$ can be defined by the condition that b be chosen according to the rules of the cyclic method (choose b from among the class mod a which satisfies $b \equiv \sqrt{D}$ mod A by the condition that $|b|$ be as small as possible—choosing the positive value in case of a tie —unless $b^2 < D$ is possible, in which case choose b to be as large as possible subject to $b^2 < D$).

This mapping from (some) divisors of quadratic integers for the determinant D to (some) binary quadratic forms with[†] $b^2 - ac = D$ is a one-to-one correspondence and is easy to invert. Given a binary quadratic form $ax^2 + 2bxy + cy^2$ with $b^2 - ac = D$, it can come from the divisor A only if $a = N(A)$ and $b \equiv \sqrt{D}$ mod A. If $D < 0$ this implies $a > 0$ because

---

*There is at least one integer b which satisfies this condition because, as was shown in Section 7.4, $\sqrt{D}$ is congruent to an integer mod A whenever A is divisible by no integer greater than 1.

†Gauss called $b^2 - ac$ the *determinant* of the binary quadratic form $ax^2 + 2bxy + cy^2$ and this terminology will be followed here.

norms are always positive when $D < 0$. With this restriction, that a be positive if D is negative (which obviously implies that c is also positive), it is easy to see that the conditions $a = N(A)$, $b \equiv \sqrt{D} \bmod A$ determine a unique divisor A as follows. Because $b - \sqrt{D}$ is divisible by no integer, its divisor is a product of prime divisors in which primes which remain prime do not occur and in which only one of the two prime factors of a prime which splits can occur. Thus a divisor which divides $b - \sqrt{D}$ is completely determined by its norm (including the presence or absence of $(-1, *)$ when $D > 0$) and every factor of $N(b - \sqrt{D}) = b^2 - D = ac$ is the norm of a divisor which divides $b - \sqrt{D}$ provided, in the case $D < 0$, that the factor is positive.

The theorem of Section 7.7 implies that *equivalent divisors correspond to equivalent binary quadratic forms* under the above correspondence, because it shows that if two divisors are equivalent then they can be transformed by the steps of the cyclic method to the same divisor; the steps of the cyclic method transform binary quadratic forms to equivalent binary quadratic forms, so that this shows that forms corresponding to equivalent divisors are equivalent to the same form (one obtained by reducing to the periods of each divisor) and are therefore equivalent to each other.

However, *the converse is not necessarily true.* In fact, if A is a divisor not divisible by an integer greater than 1 and if $ax^2 + 2bxy + cy^2$ is the corresponding binary quadratic form then $\bar{A}$ corresponds to $ax^2 - 2bxy + cy^2$ or to a form equivalent to this one (if the rules of the cyclic method choose a $b' \equiv -b \bmod a$ other than $b' = -b$). Thus A and $\bar{A}$ correspond to equivalent binary quadratic forms (because $x \mapsto x, y \mapsto -y$ transforms $ax^2 + 2bxy + cy^2$ to $ax^2 - 2bxy + cy^2$) even though A and $\bar{A}$ are usually not equivalent divisors. One of the keys to Gauss's development of his theory was his realization that *the definition of equivalence of binary quadratic forms is not sufficiently restrictive.* The correct definition of equivalence of binary quadratic forms can be found by translating the definition of equivalence of divisors. If two binary quadratic forms correspond to equivalent divisors then each can be transformed into a third by the cyclic method, which not only gives an equivalence between the two but also gives an equivalence which is a composition of equivalences of the form

$$\begin{pmatrix} x \\ y \end{pmatrix} = \begin{pmatrix} 0 & -1 \\ 1 & n_{i+1} \end{pmatrix} \begin{pmatrix} u \\ v \end{pmatrix}$$

and their inverses. Since each of these equivalences has determinant $+1$, this shows that the two given forms must be equivalent under an equivalence which has determinant $+1$.

Definition. Two binary quadratic forms $ax^2 + 2bxy + cy^2$ and $a'u^2 + 2b'uv + c'v^2$ are said to be *properly equivalent* if there is a change of coordinates $x = \alpha u + \beta v, y = \gamma u + \delta v$ for which $\alpha\delta - \beta\gamma = 1$ which transforms the one into the other. Otherwise stated, they are properly equivalent if there exist integers α, β, γ, δ satisfying

$$\begin{pmatrix} a' & b' \\ b' & c' \end{pmatrix} = \begin{pmatrix} \alpha & \gamma \\ \beta & \delta \end{pmatrix}\begin{pmatrix} a & b \\ b & c \end{pmatrix}\begin{pmatrix} \alpha & \beta \\ \gamma & \delta \end{pmatrix} \quad \text{and} \quad \det\begin{pmatrix} \alpha & \beta \\ \gamma & \delta \end{pmatrix} = 1. \quad (1)$$

Then *equivalent divisors correspond to properly equivalent forms*. As will be shown below, the converse of this theorem is also true; that is, *if the forms corresponding to two divisors are properly equivalent then the divisors are equivalent.*

It is precisely this distinction between equivalence and proper equivalence of forms which Kummer felt was artificial and which he believed that his theory of ideal complex numbers explained by showing that $ax^2 + 2bxy + cy^2$ and $ax^2 - 2bxy + cy^2$ represent "nothing other than two ideal factors of one and the same number" [K7, p. 324]. He does not explain in what way $ax^2 + 2bxy + cy^2$ and $ax^2 - 2bxy + cy^2$ represent divisors (ideal numbers) but he undoubtedly had in mind something akin to the above correspondence in which conjugate divisors correspond to forms $ax^2 \pm 2bxy + cy^2$, or at least to forms properly equivalent to these.

In summary, there is a natural one-to-one correspondence between equivalence classes of divisors of quadratic integers for the determinant D and *proper* equivalence classes of binary quadratic forms with determinant D, provided $D \not\equiv 1 \bmod 4$ is squarefree and provided that, when $D < 0$, consideration is restricted to forms in which a and c are positive. What is to be shown in the remainder of this section is that when D is not squarefree or when $D \equiv 1 \bmod 4$ the proper equivalence classes of forms still correspond one-to-one in a natural way to equivalence classes of divisors if ordinary equivalence is replaced by equivalence with respect to a suitable order as it was defined in Section 8.1.

In the correspondence above, the binary quadratic form $ax^2 + 2bxy + cy^2$ corresponds to the divisor which divides $b - \sqrt{b^2 - ac}$ and has norm a. Here it is assumed that $D = b^2 - ac$ is squarefree and $\not\equiv 1 \bmod 4$, and it is assumed that $a > 0$ if $D < 0$, but the same rule can be used in *all* cases provided that a, b, c are such that the two conditions "A divides $b - \sqrt{b^2 - ac}$" and "$N(A) = a$" determine a unique divisor A. A simple way to find natural conditions on a, b, c which assure this is to draw on the experience of Section 8.1, which showed that in dealing with quadratic integers $x + y\sqrt{D}$ (x, y integers) it is natural to exclude divisors which are not relatively prime to the index s of the order $\{x + y\sqrt{D} : x, y \text{ integers}\}$; this index is $s = t$ if $D = t^2 D'$ where D' is squarefree and $\not\equiv 1 \bmod 4$, and it is $s = 2t$ if $D = t^2 D'$ where D' is squarefree and $\equiv 1 \bmod 4$. This prompts one to require that $ax^2 + 2bxy + cy^2$ satisfy the condition that a is relatively prime to the largest square factor in $b^2 - ac$ and, in addition, that a is odd (relatively prime to 2) when $b^2 - ac \equiv 1 \bmod 4$. (If $b^2 - ac = t^2 D'$ then $b^2 - ac \equiv 1 \bmod 4$ implies t is odd and $D' \equiv 1 \bmod 4$. Conversely, if $D' \equiv 1 \bmod 4$, then either t is even, in which case the first condition

320

already requires that a be odd, or t is odd, in which case $b^2 - ac \equiv$ 1 mod 4.) In addition, if $b^2 - ac < 0$ then $a > 0$ is a necessary condition for $a = N(A)$.

Theorem. *Let a, b, c be integers, and let $b^2 - ac = t^2 D'$, where D' is squarefree. If a is relatively prime to t, if a is odd in the case $b^2 - ac \equiv$ 1 mod 4, and if $a > 0$ ih the case $b^2 - ac < 0$, then the conditions $N(A) = a$ and $b \equiv \sqrt{D}$ mod A determine a unique divisor A of quadratic integers for the determinant D'.*

   1) *If a', b', c' is another triple of integers, if $(b')^2 - a'c' = b^2 - ac$, and if a' satisfies the same conditions as a, then the binary quadratic forms $ax^2 + 2bxy + cy^2$ and $a'x^2 + 2b'xy + c'y^2$ are properly equivalent if and only if the corresponding divisors are equivalent with respect to the order $\{x + y\sqrt{D} : x, y \text{ integers}, D = b^2 - ac\}$.*

   2) *If $ax^2 + 2bxy + cy^2$ is a binary quadratic form for which the above conditions on a are not satisfied, then necessary and sufficient conditions that $ax^2 + 2bxy + cy^2$ be properly equivalent to a binary quadratic form $a'x^2 + 2b'xy + c'y^2$ in which a' does satisfy these conditions are that the integers $a, 2b,$ and c have no common divisor greater than 1 and that, if $b^2 - ac < 0$, a be positive.*

Gauss called a binary quadratic form $ax^2 + 2bxy + cy^2$ *properly primitive* if $a, 2b,$ and c have no common divisor greater than 1. (It is *primitive* if a, b, c have no common divisor greater than 1.) It follows from 2) that any form properly equivalent to a properly primitive one is itself properly primitive. It is therefore meaningful to speak of a properly primitive proper equivalence class of forms. Similarly, if $b^2 - ac < 0$ and $a > 0$ then every form properly equivalent to $ax^2 + 2bxy + cy^2$ has these same properties. Gauss called such a form *positive.* (In modern terminology it is called *positive definite.*) It is meaningful, therefore, to speak of a positive proper equivalence class of forms. Part 2) of the theorem states that a proper equivalence class of forms which is properly primitive and, in case $D < 0$, positive, contains forms which correspond to divisors. Part 1) states that different forms in the same class correspond to equivalent divisors. Therefore the theorem shows that there is a mapping from proper equivalence classes of forms (properly primitive and, if $D < 0$, positive) to equivalence classes of divisors (for quadratic integers in the order $\{x + y\sqrt{D} : x, y$ integers$\}$). This mapping is one-to-one because, by 1), forms which correspond to equivalent divisors must be properly equivalent. It is onto the set of divisor classes in the divisor class group because every such class contains a divisor A which is relatively prime to the index of the order and which is divisible by no integer greater than 1, and such an A is the image of $ax^2 + 2bxy + cy^2$ where $a = N(A)$, $b \equiv \sqrt{D}$ mod A, $c = (b^2 - D)/a$. Thus the theorem establishes a one-to-one correspondence between elements of the divisor class group for the order $\{x + y\sqrt{D} : x, y$ integers$\}$

and proper equivalence classes of binary quadratic forms $ax^2 + 2bxy + cy^2$ which have determinant D, which are properly primitive, and which, in the case $D < 0$, are positive. *This is the desired correspondence between Kummer's theory and Gauss's.* The remainder of this section is devoted to the proof of the theorem.

PROOF. Let a, b, c satisfy the conditions stated at the beginning of the theorem. Then $b - \sqrt{D} = b - t\sqrt{D'}$ is a quadratic integer for the determinant D' and its norm is $b^2 - D = ac$. For each prime factor p of a, p does not divide $b - t\sqrt{D'}$ (because the condition that a is relatively prime to t implies p does not divide t, and if $p = 2$ then t must be odd and $b^2 - ac \not\equiv 1 \bmod 4$, in which case 2 does not divide $b - t\sqrt{D'}$). Therefore p does not remain prime; if p ramifies then its prime divisor divides $b - \sqrt{D}$ with multiplicity exactly 1, and if p splits then only one of its two prime divisors divides $b - \sqrt{D}$. Therefore $b - \sqrt{D}$ is divisible by one and only one divisor with norm a. This divisor is the divisor A of the first part of the theorem.

Statement 2) will be proved before statement 1). If any odd prime divides a, $2b$, and c then it divides a, b, and c. It then follows from the definition of proper equivalence that it divides a', b', and c' for any form $a'x^2 + 2b'xy + cy^2$ equivalent to the given form. Since its square divides $D = b^2 - ac = t^2 D'$, it divides t as well as a', and a' does not have the required properties. If 2 divides a, $2b$, and c then $a' = a\alpha^2 + 2b\alpha\gamma + c\gamma^2$ is even and either $b^2 - ac \equiv 0 \bmod 4$ (if b is even) or $b^2 - ac \equiv 1 \bmod 4$ (if b is odd), in either of which cases a' does not have the required properties. Finally, if $a < 0$ and $b^2 - ac < 0$ then $a' = a^{-1}[(a\alpha + b\gamma)^2 - (b^2 - ac)\gamma^2] < 0$ and a' does not have the required properties. This shows that the given conditions are necessary in order for there to be an $a'x^2 + 2b'xy + c'y^2$ with the required properties.

To prove that they are sufficient, note first that the above argument shows that if $a > 0$ and $b^2 - ac < 0$ then $a' > 0$ always. What remains to be shown, then, is that if a, $2b$, c are relatively prime then there is a properly equivalent form in which a' is relatively prime to t and, if $b^2 - ac \equiv 1 \bmod 4$, in which a' is odd. Gauss (*Disquisitiones Arithmeticae*, Art. 228) proved not only that this is true but that if a, $2b$, c are relatively prime and if k is any given integer then there is a properly equivalent form in which a' is relatively prime to k. For this, let α be the product of those primes p which divide k but do not divide c, and let γ be the product of those primes p which divide k and c but not a. (The empty product is 1. That is, if no prime satisfies the required properties the "product" is to be taken to be 1.) Then α and γ are relatively prime and there is an integer δ such that $\delta\alpha \equiv 1 \bmod \gamma$. Define β by $\beta = (\alpha\delta - 1)/\gamma$. Then $\alpha\delta - \beta\gamma = 1$ and (1) defines a form properly equivalent to $ax^2 + 2bxy + cy^2$ in which $a' = a\alpha^2 + 2b\alpha\gamma + c\gamma^2$. It is to be shown that a' is relatively prime to k. If p is a prime divisor of k which does not divide c then, because it divides α but not γ, it does not divide a'. If p is a prime divisor of k which divides c but not a then, because it divides γ but not α, it does not divide a'. Finally if p is a prime divisor of k which divides both a and c then, because it divides neither α nor γ nor (because a, $2b$, and c are relatively prime) $2b$, it does not divide a'. Thus a' is relatively prime to k, as was to be shown.

For the proof of statement 1) assume that $ax^2 + 2bxy + cy^2$ and $a'x^2 + 2b'xy + c'y^2$ are two forms with determinant $D = t^2 D'$ in which a and a' both satisfy the required conditions. Then they correspond to divisors A and A' respectively. It will be shown first that if A and A' are equivalent as divisors of the order $\{x + y\sqrt{D} : x, y \text{ integers}\}$ then the forms are properly equivalent. It follows from the definition of equivalence and the fact that $A'\overline{A'}$ is principal that there is an element $u + v\sqrt{D}$ of the order with divisor $A\overline{A'}$. Then multiplication by $u + v\sqrt{D}$ followed by division by a' carries elements of the order divisible by A' to elements of the order divisible by A. Now an element $z + w\sqrt{D}$ of the order is divisible by A' if and only if $z + wb' - w(b' - \sqrt{D})$ is divisible by A', and this is true if and only if $z + wb' \equiv 0 \bmod a'$. In other words, $z + w\sqrt{D}$ is divisible by A' if and only if it is of the form $a'x + (b' - \sqrt{D})y$ $(x = (z + wb')/a', y = -w)$. Similarly, an element $z + w\sqrt{D}$ of the order is divisible by A if and only if it is of the form $ax + (b - \sqrt{D})y$ where x and y are integers. Then, as in Section 7.7, multiplication by $u + v\sqrt{D}$ followed by division by a' carries $a'x + (b' - \sqrt{D})y$ to $ax' + (b - \sqrt{D})y'$ where $x' = \alpha x + \beta y$, $y = \gamma x + \delta y$, and

$$
\begin{bmatrix} \alpha & \beta \\ \gamma & \delta \end{bmatrix} = \begin{bmatrix} \dfrac{u + vb}{a} & \dfrac{b'u - Dv - bu + bb'v}{aa'} \\ -v & \dfrac{u - vb'}{a'} \end{bmatrix}. \tag{2}
$$

Because multiplication by $u + v\sqrt{D}$ followed by division by a' followed by multiplication by $u - v\sqrt{D}$ followed by division by a is the identity $a'x + (b' - \sqrt{D})y \mapsto a'x + (b' - \sqrt{D})y$ (the norm of $u + v\sqrt{D}$ is aa'), this matrix can be inverted by changing $u + v\sqrt{D}$ to $u - v\sqrt{D}$ and interchanging A and A'. Since this interchanges α and δ and reverses the signs of β and γ, the determinant of the matrix (2) is 1. In order to prove that the forms are properly equivalent, as desired, it suffices, then, to show that the first of the equations (1) is satisfied. This can be done by direct computation. $a\alpha^2 + 2b\alpha\gamma + c\gamma^2 = a^{-1}[(u + vb)^2 + 2b(u + vb)(-v) + acv^2]$ $= a^{-1}[u^2 + v^2(b^2 - 2b^2 + ac)] = a^{-1}[aa'] = a'$. $a\alpha\beta + b\alpha\delta + b\beta\gamma + c\gamma\delta = a\alpha\beta + b\alpha\delta + b(\alpha\delta - 1) + c\gamma\delta = -b + (aa')^{-1}[(u + vb)(b'u - Dv - bu + bb'v) + 2b(u + vb)(u - vb') + ac(-v)(u - vb')] = -b + (aa')^{-1}[u^2(b' - b + 2b) + uv(-D + bb' + bb' - b^2 + 2b^2 - 2bb' - ac) + v^2(-bD + b^2b' - 2b^2b' + acb')] = -b + (aa')^{-1}[(b + b')u^2 - (b + b')Dv^2] = -b + (b + b') = b'$. Then $c'' = a\beta^2 + 2b\beta\delta + c\delta^2$ is equal to c', as desired, because $b^2 - ac = D = b'^2 - a'c'$ by assumption and $b^2 - ac = b'^2 - a'c''$ by (1) and the facts already proved; since $a' \neq 0$, $c'' = c'$ follows.

Suppose, finally, that the forms are properly equivalent. It is to be shown that the divisors are then equivalent or, what is the same, that there is an element $u + v\sqrt{D}$ whose divisor is $A\overline{A'}$. The argument above suggests that one try $v = -\gamma$ and $u = \alpha a + \gamma b$ (from $\alpha = (u + vb)/a$), or $u = \delta a' - \gamma b'$ (from $\delta = (u - vb')/a'$). The two definitions of u coincide because, by equation (1),

$$\begin{pmatrix} \delta & -\gamma \\ -\beta & \alpha \end{pmatrix} \begin{pmatrix} a' & b' \\ b' & c' \end{pmatrix} = \begin{pmatrix} a & b \\ b & c \end{pmatrix} \begin{pmatrix} \alpha & \beta \\ \gamma & \delta \end{pmatrix} \tag{3}$$

from which $\delta a' - \gamma b' = a\alpha + b\gamma$. With these definitions of u and v, one has $u + v\sqrt{D} = \alpha a + \gamma(b - \sqrt{D}) = \delta a' - \gamma(b' + \sqrt{D})$, which is obviously divisible by both A and $\overline{A'}$. It has norm $(a\alpha + b\gamma)^2 - D\gamma^2 = a[a\alpha^2 + 2b\alpha\gamma + c\gamma^2] = aa'$. If a and a' are relatively prime, this suffices to prove that the divisor of $u + v\sqrt{D}$ is $\overline{A'}A$ and therefore that $A \sim A'$, as was to be shown. If a and a' are not relatively prime then the technique used to prove part 2) of the theorem shows that there is a third form $a''x^2 + 2b''xy + c''y^2$ properly equivalent to both of the given forms in which a'' is relatively prime to both a and a' as well as t. Then, if A'' is the corresponding divisor, it has been shown that $A \sim A''$ and $A' \sim A''$. Thus $A \sim A'$ and the proof of the theorem is complete.

EXERCISES

For each of the following determinants, give one binary quadratic form from each properly primitive proper equivalence class:

1. $D = 67$.

2. $D = -165$. (Include negative proper equivalence classes.)

3. $D = 79$.

4. $D = -161$.

5. $D = -11$. (See Exercise 1, Section 8.1.)

6. $D = 18$. (See Exercise 6, Section 8.1.)

8.4 The classification of forms

In order to be able to compute the divisor class groups that were defined in Section 8.1 it is necessary to be able to determine, given two divisors A and A', whether they are equivalent—assuming, of course, that they are relatively prime to the index s of the order being considered. The theorem of the preceding section shows that this problem can be reduced to the problem of determining whether two given binary quadratic forms are properly equivalent. In case $s = 1$ this problem was solved—using the terminology of divisors rather than of binary quadratic forms—in Chapter

7 using the cyclic method. In cases where $s > 1$, it is still possible to solve the problem using the cyclic method, but, because the cyclic method applied to a divisor relatively prime to s need not yield a divisor relatively prime to s, it is more natural to describe the solution in terms of binary quadratic forms than in terms of divisors. With this modification of the terminology, the theorem and proof are virtually identical to those of Chapter 7.

Theorem. *Let $a_0x^2 + 2r_0xy + a_1y^2$ be a given binary quadratic form with determinant $r_0^2 - a_0a_1 = D$ not a square. Define two sequences of integers $a_0, a_1, a_2, \ldots$ and $r_0, r_1, r_2, \ldots$ as follows. Given r_i and a_{i+1}, r_{i+1} satisfies $r_i + r_{i+1} \equiv 0 \bmod a_{i+1}$; it is the solution r_{i+1} of this congruence for which $|r_{i+1}|$ is smallest—with the positive value chosen in case of a tie—unless there are solutions for which $r_{i+1}^2 < D$, in which case r_{i+1} is the largest solution of the congruence for which $r_{i+1}^2 < D$. Given a_i and r_i, a_{i+1} is defined by $a_{i+1} = (r_i^2 - D)/a_i$. Then the binary quadratic forms $a_ix^2 + 2r_ixy + a_{i+1}y^2$ are all properly equivalent to the original form $a_0x^2 + 2r_0xy + a_1y^2$. Moreover, these forms begin to repeat once i is sufficiently large. The cycle of forms which repeat is called the period of the given form. Two binary quadratic forms are properly equivalent if and only if they have the same period.*

This theorem is the linchpin of the entire subject. The remainder of this section is devoted to its proof. Briefly put, the proof is exactly the same as in Chapter 7 except that it must be translated from the language of divisors to the language of binary quadratic forms. The assumption that D is squarefree is simply not used in the proof of Chapter 7 except to define the correspondence between forms and divisors.

The theorem as it is stated above applies only to binary quadratic forms in which the middle term is even. However, in the proof which follows *the case in which r_0 is a halfinteger will also be allowed*. Then, as in Chapter 7, all of the r's generated by the cyclic method are halfintegers. The notation of chapter 7 will be altered somewhat, however, in that D will be taken to be $b^2 - ac$ in all cases, so that when b is a half-integer D is $\frac{1}{4}$ of an integer that is congruent to 1 mod 4. In all cases, then, $b - \sqrt{D}$ is a quadratic integer.

If $a_0, r_0,$ and a_1 have a common factor then that factor divides all successive r's and a's. Another form $a_0'x^2 + 2r_0'xy + a_1'y^2$ can be equivalent to this one only if a_0', r_0' and a_1' are all divisible by the common factor of $a_0, r_0,$ and a_1. The theorem therefore reduces to the case in which all such common factors have been removed. Moreover, if the integers $a_0, 2r_0,$ and a_1 are all even then r_0 is an integer and all terms in the cyclic method can be divided by 2 (using the fact that r_0 may be a halfinteger). Therefore *the general case reduces to the case* in which $a_0x^2 + 2r_0xy + a_1y^2$ *is properly primitive*. This justifies the assumption in the proof which follows that each of the forms $a_i x^2 + 2r_i xy + a_{i+1} y^2$ is properly primitive.

PROOF. By infinite descent, $|a_i| \leqslant |a_{i+1}|$ must occur infinitely often. When it does occur, $|a_i|^2 \leqslant |a_i a_{i+1}| \leqslant r_i^2 + |D|$. Either $r_i^2 < D$ or $|r_i| \leqslant \frac{1}{2}|a_i|$. This shows that when it does occur a_i is restricted to a finite number of values $(a_i^2 \leqslant 2|D|$ or $a_i^2 \leqslant \frac{1}{4} a_i^2 + |D|$, $|a_i| \leqslant 2\sqrt{|D|/3}$) and that r_i is restricted to a finite number of values. Consequently the triple of integers (a_i, r_i, a_{i+1}) can have only a finite number of values when $|a_i| \leqslant |a_{i+1}|$. Therefore some triple must occur twice. Then, since each triple uniquely determines its successor, the entire list of forms following one which repeats must repeat in cyclic fashion *ad infinitum*.

It was shown in Section 8.2 that each form is properly equivalent to its successor in the sequence. Therefore all the forms in the sequence are properly equivalent to the original one and it is obvious that two forms with the same period are properly equivalent. The real substance of the theorem is the statement that *if two forms are properly equivalent then they have the same period*.

Let a form be called *reduced* if it belongs to its own period—that is, if application of the cyclic method to this form brings one back to this form itself. Every form is properly equivalent to a reduced form with the same period. Therefore the statement to be proved is equivalent to the statement that *if two reduced forms are properly equivalent, then each is in the period of the other*.

If $D < 0$ then this can be proved as follows.* One can assume without loss of generality that the given forms $ax^2 + 2bxy + cy^2$ and $a'x^2 + 2b'xy + c'y^2$ are not only reduced but also satisfy $|a| \leqslant |c|$, $|a'| \leqslant |c'|$, because each period contains such a form. What is to be shown is that

$$\begin{pmatrix} a' & b' \\ b' & c' \end{pmatrix} = \begin{pmatrix} \alpha & \gamma \\ \beta & \delta \end{pmatrix}\begin{pmatrix} a & b \\ b & c \end{pmatrix}\begin{pmatrix} \alpha & \beta \\ \gamma & \delta \end{pmatrix}; \quad \det\begin{pmatrix} \alpha & \beta \\ \gamma & \delta \end{pmatrix} = 1 \quad (1)$$

together with $D = b^2 - ac < 0$, $|a| \leqslant |c|$, and $|a'| \leqslant |c'|$, implies that the two forms lie in the same period. Since $aa' = a(a\alpha^2 + 2b\alpha\gamma + c\gamma^2) = (a\alpha + b\gamma)^2 - D\gamma^2 \geqslant |D|\gamma^2 \geqslant 0$ and a and a' are nonzero (otherwise $D = b^2 - ac$ would be a square), $aa' > 0$. Therefore a and a' have the same sign. Since both signs can be reversed if necessary, one can assume that both a and a' are positive. By $b^2 - ac = D < 0$ it follows that $c > 0$. Similarly $c' > 0$. Since $|b| \leqslant \frac{1}{2}a$, $a \leqslant c$ one has $b^2 - ac = D$, $a^2 \leqslant ac = b^2 + |D| \leqslant \frac{1}{4}a^2 + |D|$, $\frac{3}{4}a^2 \leqslant |D|$, $a \leqslant 2\sqrt{|D|/3}$ and, similarly, $a' \leqslant 2\sqrt{|D|/3}$. Therefore the above inequality $aa' \geqslant |D|\gamma^2$ implies $\frac{4}{3}|D| \geqslant |D|\gamma^2$ and $\gamma^2 = 0$ or 1. *Case 1*. If $\gamma^2 = 0$ then $\alpha\delta = 1$, $\alpha = \delta = \pm 1$, $a' = a\alpha^2 + 2b\alpha\gamma + c\gamma^2 = a$, and $b' = a\alpha\beta + b\alpha\delta + b\beta\gamma + c\gamma\delta = a\alpha\beta + b$, from which $b \equiv b'$ mod a. Since

---

*Compare this proof to the corresponding proof in Section 7.7. See also §65 of Dirichlet's *Vorlesungen über Zahlentheorie* [D7].

$|b| < \frac{1}{2} a$ or $b = \frac{1}{2} a$ and $|b'| < \frac{1}{2} a' = \frac{1}{2} a$ or $b' = \frac{1}{2} a$ the condition $b \equiv b'$ mod a implies $b = b'$. Then $c = c'$ and the two forms are identical. *Case 2.* If $\gamma^2 = 1$, then, by interchanging the two forms if necessary, one can assume $\gamma = 1$. (Inversion reverses the sign of γ.) Then $aa' = (a\alpha + b\gamma)^2 - D\gamma^2 = r^2 - D$ where $r = a\alpha + b \equiv b$ mod a. By the choice of b, then, $|r| \geqslant |b|$, $a \leqslant c = (b^2 - D)/a \leqslant (r^2 - D)/a = aa'/a = a'$. Similarly, $aa' = (a'\delta - b'\gamma)^2 - D\gamma^2 = (r')^2 - D$ where $r' = a'\delta - b' \equiv -b'$ mod a'. Then $|r'| \geqslant |b'|$ and $a' \leqslant c' = [(b')^2 - D]/a' \leqslant [(r')^2 - D]/a' = a$. Therefore $a = c = a' = c'$. Because $r = a\alpha + b\gamma = a'\delta - b'\gamma = r'$ [see (3) of Section 8.3] is $\equiv b$ mod a and $\equiv -b'$ mod a', it follows that $b + b' \equiv 0$ mod a. Since b and b' are both less than $\frac{1}{2} a$ in absolute value, unless they are equal to $\frac{1}{2} a$, this implies either that $b = b' = \frac{1}{2} a$ and the forms coincide, or that $b' = -b$ and their common absolute value is less than $\frac{1}{2} a$. It then follows that the cyclic method applied to either of these forms gives the other and they are in the same period, as was to be shown. This completes the proof in the case $D < 0$.

When $D > 0$ the proof is more difficult. The main tool is the theorem of Section 7.7 concerning decompositions of the form

$$\begin{pmatrix} X & Y \\ Z & W \end{pmatrix} = \pm \begin{pmatrix} n_j & 1 \\ -1 & 0 \end{pmatrix} \begin{pmatrix} n_{j-1} & 1 \\ -1 & 0 \end{pmatrix} \cdots \begin{pmatrix} n_1 & 1 \\ -1 & 0 \end{pmatrix} \quad (2)$$

in which the integers $n_1, n_2, \ldots, n_j$ alternate in sign. In fact, the entire proof is a straightforward modification of the proof of the case $s = 1$ in Section 7.7.

Suppose that M is a 2×2 matrix of integers such that

$$\begin{pmatrix} a' & b' \\ b' & c' \end{pmatrix} = M^t \begin{pmatrix} a & b \\ b & c \end{pmatrix} M; \qquad \det M = 1$$

where $a'x^2 + 2b'xy + c'y^2$ and $ax^2 + 2bxy + cy^2$ are two reduced forms and where M^t denotes the transpose of M. The first step is to show that M can be replaced, without loss of generality, by a matrix which has a decomposition of the form (2). For this, note first that the cyclic method applied to the reduced form $ax^2 + 2bxy + cy^2$ gives a proper equivalence of this form with itself

$$\begin{pmatrix} a & b \\ b & c \end{pmatrix} = E_1^t \begin{pmatrix} a & b \\ b & c \end{pmatrix} E_1$$

$$E_1 = \begin{pmatrix} m_k & 1 \\ -1 & 0 \end{pmatrix} \begin{pmatrix} m_{k-1} & 1 \\ -1 & 0 \end{pmatrix} \cdots \begin{pmatrix} m_1 & 1 \\ -1 & 0 \end{pmatrix} \quad (3)$$

where k is the number of forms in the period of $ax^2 + 2bxy + cy^2$. Therefore $E_1^n M$ has the same properties as M for every positive integer n. It will be shown that when n is sufficiently large $E_1^n M$ has a decomposition (2).

As in the statement of the theorem, let $a = a_0$, $c = a_1$, and $b = r_0$. Then the formulas of the preceding section show that each factor of E_1 can

327

be regarded as the mapping from quadratic integers of the form $a_i x + (r_i - \sqrt{D})y$ to those of the form $a_{i+1}x + (r_{i+1} - \sqrt{D})y$ given by multiplication by $r_i + \sqrt{D}$ followed by division by a_i. [This operation carries $a_i \cdot 1 + (r_i - \sqrt{D}) \cdot 0$ to $r_i + \sqrt{D} = [(r_{i+1} + r_i)/a_{i+1}]\, a_{i+1}] - (r_{i+1} - \sqrt{D})$ where $(r_{i+1} + r_i)/a_{i+1} = m_{i+1}$ and it carries $a_i \cdot 0 + (r_i - \sqrt{D}) \cdot 1$ to $(r_i^2 - D)/a_i = a_{i+1}$.] As in Chapter 7, the r_i are all positive and the a_i alternate in sign. Therefore the m_i alternate in sign, and E_1 is given as a decomposition in the form (2).

Moreover, the formulas of the preceding section show that M corresponds to the mapping from quadratic integers of the form $ax + (b - \sqrt{D})y$ to those of the form $a'x + (b' - \sqrt{D})y$ given by multiplication by $u + v\sqrt{D}$ followed by division by a where $u + v\sqrt{D} = \alpha a + \gamma(b - \sqrt{D}) = \delta a' - \gamma(b' + \sqrt{D})$, α, β, γ, and δ being the coefficients of M. In addition, $u^2 - Dv^2 = (\alpha a + \gamma b)^2 - D\gamma^2 = a^2\alpha^2 + 2ab\alpha\gamma + b^2\gamma^2 - b^2\gamma^2 + ac\gamma^2 = a[a\alpha^2 + 2b\alpha\gamma + c\gamma^2] = aa'$. Similarly, E_1 corresponds to the mapping from quadratic integers of the form $ax + (b - \sqrt{D})y$ to themselves given by multiplication by $U + V\sqrt{D}$ followed by division by a where U and V are integers. Here $U^2 - DV^2 = a^2$. Since E_1 also corresponds to multiplication by $r_0 + \sqrt{D}$ followed by division by a_0 followed by multiplication by $r_1 + \sqrt{D}$ followed by division by a_1, and so forth, it follows that

$$\frac{U + V\sqrt{D}}{a} = \frac{(r_0 + \sqrt{D})(r_1 + \sqrt{D}) \cdots (r_{k-1} + \sqrt{D})}{a_0 a_1 \ldots a_{k-1}}$$

and, in particular, since the r_i are positive, U and V have the same sign.

If one assumes, as one may, that $ax^2 + 2bxy + cy^2$ is *properly primitive*, then $(U + V\sqrt{D})/a$ is a unit of the form $\varepsilon = g + h\sqrt{D}$ where g and h are integers. This follows from the equation

$$\begin{pmatrix} \delta' & -\gamma' \\ -\beta' & \alpha' \end{pmatrix}\begin{pmatrix} a & b \\ b & c \end{pmatrix} = \begin{pmatrix} a & b \\ b & c \end{pmatrix}\begin{pmatrix} \alpha' & \beta' \\ \gamma' & \delta' \end{pmatrix}$$

where α', β', γ', δ' are the coefficients of E_1. These give $\delta' a - \gamma' b = a\alpha' + b\gamma'$ and $\delta' b - \gamma' c = a\beta' + b\delta'$, from which $a(\delta' - \alpha') = 2b\gamma'$ and $a\beta' = -c\gamma'$. Therefore a divides $a\gamma'$, $2b\gamma'$, and $c\gamma'$. Because there are integers k, l, m such that $ka + l(2b) + mc = 1$, it follows that a divides γ'. Since $U + V\sqrt{D} = \alpha' a + \gamma'(b - \sqrt{D})$, this shows that a divides both U and V. Thus $(U + V\sqrt{D})/a = g + h\sqrt{D}$ and $g + h\sqrt{D}$ is a unit because its norm is $(U^2 - DV^2)/a^2 = 1$.

Thus the matrix E_1 corresponds to the mapping of quadratic integers of the form $ax + (b - \sqrt{D})y$ to themselves given by multiplication by $g + h\sqrt{D}$. The matrix $E_1^n M$ therefore corresponds to the mapping from quadratic integers of the form $ax + (b - \sqrt{D})y$ to those of the form $a'x + (b' - \sqrt{D})y$ given by multiplication by $(u + v\sqrt{D})(g + h\sqrt{D})^n$ followed by division by a. With $u' + v'\sqrt{D} = (u + v\sqrt{D})(g + h\sqrt{D})^n$

one can then show, as in Chapter 7, that for large n the signs of u' and v' are the same and the absolute value of v' is large. Then, again as in Chapter 7, the corresponding 2×2 matrix $E_1^n M$ is one in which $|X| \geqslant |Z| \geqslant |W|$ and $|X| \geqslant |Y| \geqslant |W|$ and has a decomposition (2). Therefore one can assume without loss of generality that M itself has this form.

Multiplication by $g + h\sqrt{D}$ is a mapping from quadratic integers of the form $a'x + (b' - \sqrt{D})y$ to themselves (a quadratic integer $w + z\sqrt{D}$ has this form if and only if $w + zb' \equiv 0$ mod a', something which is true for $(g + h\sqrt{D})(w + z\sqrt{D})$ if it is true for $w + z\sqrt{D}$ because $D \equiv b'^2$ mod a'). Therefore it corresponds to a 2×2 matrix E_2 for which $E_1 M = ME_2$. The matrix E_2 has a decomposition (2) and the signs of the decompositions match in such a way that both $E_1 M$ and ME_2 are decompositions (2). The equation $E_1^n M = ME_2^n$ for all $n > 0$ and the uniqueness of decompositions (2) then imply that the factors of M are a truncation of the decomposition of E_1^n for large n. It then follows from (2) that $\pm M$ is the same transformation as results from applying the cyclic method to $ax^2 + 2bxy + cy^2$. Therefore $a'x^2 + 2b'xy + c'y^2$ is in the period of $ax^2 + 2bxy + cy^2$, as was to be shown.

EXERCISES

1. Prove that the number of properly primitive proper equivalence classes of forms of a given determinant is finite. Conclude that in all cases *the divisor class group is finite.*

2. Show that if $ax^2 + 2bxy + cy^2$ is a reduced form, if $D = b^2 - ac > 0$, and if the unit $\varepsilon = g + h\sqrt{D}$ is found as in the proof of the theorem, then ε has norm 1 and the only units with norm 1 in the order $\{x + y\sqrt{D}\}$ are $\pm \varepsilon^n$. Show that there is a unit with norm -1 only if every form $ax^2 + 2bxy + cy^2$ is improperly equivalent to its negative $-ax^2 - 2bxy - cy^2$ and therefore properly equivalent to $-ax^2 + 2bxy - cy^2$. When this is the case, give an algorithm for finding a unit with norm -1 and show that its square is the unit ε defined above. Summarize these facts by giving a procedure for finding, for any order $\{x + y\sqrt{D}\}$, a *fundamental unit*—that is, a unit ε with the property that the formula $\pm \varepsilon^n$ comprehends every unit once and only once.

3. Find a reduced form properly equivalent to $x^2 - 67y^2$ and use Exercise 2 to find the most general solution of Pell's equation $u^2 = 67v^2 + 1$.

4. Find a reduced form properly equivalent to $x^2 + xy - 15y^2$ and deduce, using Exercise 2, a fundamental unit for the order $\{x + y\sqrt{61} : x, y$ integers or halfintegers$\}$.

5. Find the divisor class group corresponding to the order $\{x + y\sqrt{99} : x, y$ integers$\}$. [-1 and 2 ramify, 5 splits, $s = 3$. The divisors I, $(-1, *)$, $(2, *)$, and $(-1, *)(2, *)$ are inequivalent. The test to determine whether $(5, 1)$ is principal not only shows that it is not but also shows it is not equivalent to any of the 4 divisors above, but that $(5, 1) \sim (-1, *)(2, *)(5, 1)$. This implies that the formula $(-1, *)^i (5, 1)^j$ for $i = 0, 1$; $j = 0, 1, 2, 3$ gives 8 inequivalent divisors. These

correspond to 8 periods of reduced forms. Show that every reduced form lies in one of these periods.]

6. Find the divisor class group corresponding to the order $\{x + y\sqrt{-99} : x, y$ integers$\}$.

7. Prove that if α and γ are given relatively prime integers then there exist integers β and δ such that
$$\det\begin{pmatrix} \alpha & \beta \\ \gamma & \delta \end{pmatrix} = 1.$$

Deduce from this the fact that if $ax^2 + 2bxy + cy^2$ represents m *properly* (with relatively prime x and y) then it is properly equivalent to a form whose first coefficient is m. Conclude that the method of Section 8.2 produces a representation $ax^2 + 2bxy + cy^2$ whenever such a representation is possible.

8. Describe in detail (write a computer program if possible) a procedure for finding all solutions of an equation of the form $ax^2 + 2bxy + cy^2 = m$, when a, b, c, and m are integers in some stated range.

9. Prove that $3 \neq x^2 - 37y^2$, $151 = x^2 - 37y^2$ even though $3 \equiv 151 \bmod 4 \cdot 37$. This gives a counterexample to Euler's conjecture of Exercise 8, Section 7.9, which is smaller than Lagrange's counterexample.

8.5 Examples

As was seen in Section 7.6, the class number in the case* $D = -163$ is 1. If one considers the order of index 2, which is $\{x + y\sqrt{-163} : x, y$ integers$\}$, then the class number is greater than 1. In fact, the divisor $(41, 1)$, which is the divisor of $\frac{1}{2}(1 - \sqrt{-163})$ when the denominators of 2 are allowed, is not principal when they are not allowed because it corresponds to the binary quadratic form $41x^2 + 2xy + 4y^2$ and the cyclic method

$$
\begin{array}{ccc}
1 & -1 & 1 \cdots \\
41 & 4 & 41 \cdots
\end{array}
$$

shows that this form is not properly equivalent to the form $x^2 + 163y^2$ which corresponds to the divisor I.

[This is perhaps an appropriate point to pause to discuss the curious etymology of the adjective "principal" in the phrase "principal divisor" or "principal ideal." The simplest binary quadratic form with determinant D is $x^2 - Dy^2$. Gauss called this form, naturally enough, the *principal* form with determinant D. Its proper equivalence class, the class of the principal form, he called the *principal class*. This class corresponds to the class of the divisor I under the correspondence of Section 8.3. This divisor class, which is the identity element of the divisor class group, is therefore naturally called the principal class. A divisor is then in the principal class if and only if it is the divisor of an actual element and an ideal is in the principal class

*In the notation used in the preceding section, this case would be the case $D = -163/4$.

if and only if it is the set of all elements divisible by some actual element. Thus an ideal in a ring is said to be principal if it is the set of all elements divisible by some given element of the ring.]

Let $A = (41, 1)$. Then A^2 is a divisor with norm $41^2 = 1681$ and, since A^2 divides $(1 - \sqrt{-163})^2 = -162 - 2\sqrt{-163} = -2(81 + \sqrt{-163})$, $\sqrt{-163} \equiv -81 \bmod A^2$. Therefore the cyclic method applied to A^2 gives

$$
\begin{array}{ccccc}
-81 & 1 & -1 & 1 & \cdots \\
1681 & 4 & 41 & 4 & \cdots
\end{array}
$$

and shows that $A^2 \sim \overline{A}$, from which $A^3 \sim I$. That the three classes, those of I, A, and A^2, account for all the classes in the divisor class group, can be proved as follows. Any divisor relatively prime to 2 corresponds to a form $ax^2 + 2bxy + cy^2$ in which a, b, c are integers, $a > 0$, $c > 0$, and $b^2 - ac = -163$. The cyclic method can be used to reduce any such form to another in which $a \leqslant c$, $|b| \leqslant \frac{1}{2}a$. Then $a^2 \leqslant ac = b^2 + 163 \leqslant \frac{1}{4}a^2 + 163$, $a^2 \leqslant (4 \cdot 163)/3$, $a \leqslant 14$, $|b| \leqslant 7$. The problem is then to find small factors of $b^2 + 163$ for $b = 0, 1, \ldots, 7$. The fact is that, apart from $a = 1$, $b = 0$, the only small factors are $a = 2$ or 4. One way to see this is to note that for $p = 3, 5, 7$, and 11 the determinant $D = -163$ is not a square mod p, so that $p | a$ implies $a \nmid b^2 + 163$. Therefore a can be divisible only by the prime 2 and $8 \nmid a$ because $b^2 + 163 \equiv b^2 + 3 \not\equiv 0 \bmod 8$. If $a = 2$ then $b = \pm 1$, $c = 82$ and $ax^2 + 2bxy + cy^2$ is not properly primitive. This leaves just the three cases $a = 1$, $b = 0$, and $a = 4$, $b = \pm 1$. Therefore, there are just three classes, as was to be shown. Thus the class number is* 3 for the order $\{x + y\sqrt{-163} : x, y = \text{integers}\}$ and the divisors A, A^2, and $A^3 \sim I$ are a representative set.

A case considered by Gauss (*Disquisitiones Arithmeticae* Arts. 223, 224, and 226) is the case of the order $\{x + y\sqrt{-235} : x, y = \text{integers}\}$. Every form is equivalent to one $ax^2 + 2bxy + cy^2$ in which $c \geqslant a > 0$, $|b| \leqslant \frac{1}{2}a$. These inequalities imply $a^2 \leqslant 4 \cdot 235/3$, $a \leqslant 17$. Any prime divisor of a must be one modulo which -235 is a square. This trivially includes the primes 2 and 5; it excludes the primes 3, 7, 11, 17 and includes 13 because $-235 \equiv -1 \equiv 5^2 \bmod 13$. Let $A = (13, 5)$. That this divisor is not principal follows from

$$
\begin{array}{ccc}
5 & -5 & 5 \cdots \\
13 & 20 & 13 \cdots
\end{array}
$$

Its square has norm $N(A^2) = 169$ and divides $(5 - \sqrt{-235})^2 = -210 - 10\sqrt{-235}$, from which $\sqrt{-235} \equiv -21 \bmod A^2$. The test to see whether A^2 is principal is therefore

$$
\begin{array}{ccc}
-21 & 1 & -1 \cdots \\
169 & 4 & 59 \cdots
\end{array}
$$

and the conclusion is that A^2 is not principal but it is equivalent to the

---

*The German edition [G2] of *Disquisitiones Arithmeticae* erroneously gives (p. 351) the classification II.3 rather than I.3 in this case.

divisor $(59, -1)$. Thus $A^3 \sim A(59, -1)$. The latter divisor has norm $13 \cdot 59$ $= 767$ and divides $(5 - \sqrt{-235})(1 + \sqrt{-235}) = 240 + 4\sqrt{-235}$, from which the cyclic method shows

$$
\begin{array}{ccccc}
-60 & & 0 & & 0 \\
767 & 5 & & 47 & 5 \cdots
\end{array}
$$

that $A^3 \sim (5, *)$. Since $(5, *)^2$ is the divisor of 5, it follows that $A^6 \sim I$ but, as was seen above, $A^2 \nsim I$ and $A^3 \nsim I$. Therefore the six divisors A, A^2, A^3, A^4, A^5, $A^6 \sim I$ are all in distinct classes. That these six classes include *all* the classes follows from the fact that they include the reduced forms in which $a = 1$, $b = 0$; $a = 4$, $b = \pm 1$; $a = 5$, $b = 0$; and $a = 13$, $b = \pm 5$. The only other possible values of $a \leqslant 17$ are $a = 2$ or $a = 10$, both of which lead to forms that are not properly primitive.

Consider next the case of the order $\{x + y\sqrt{60} : x, y = \text{integers}\} = \{x + 2y\sqrt{15} : x, y = \text{integers}\}$. Here, since $D > 0$, the divisor $(-1, *)$ is to be considered. The computation

$$
\begin{array}{ccccc}
7 & 4 & 4 & 7 & 7 \cdots \\
-1 & 11 & -4 & 11 & -1 \cdots
\end{array}
$$

shows that $(-1, *)$ is not principal. When $D = 15$ the prime 3 ramifies. Since it is relatively prime to 2, $(3, *)$ lies in a class of the divisor class group being sought. This class is not the principal class by virtue of

$$
\begin{array}{ccccccc}
6 & 2 & 5 & 5 & 2 & 6 & 6 \cdots \\
3 & -8 & 7 & -5 & 7 & -8 & 3 \ldots
\end{array}
$$

and the same computation shows not only that the divisor $(-1, *)(3, *)$ with norm -3 is not principal, but also that the four divisors I, $(-1, *)$, $(3, *)$, and $(-1, *)(3, *)$, since they all lead to different periods of forms, are in different classes. The square of any of these classes is the class of I, and the product of any two of them is therefore easy to find. In order to show that this is a complete description of the divisor class group in question, it will suffice to prove that these four classes include all the possibilities. Every form is properly equivalent to one $ax^2 + 2bxy + cy^2$ in which $a \leqslant 2\sqrt{60/3} < 9$. Moreover, $b^2 \equiv 60 \bmod a$, $b^2 < 60$, and $(b + |a|)^2 > 60$. In the above classes are included cases in which $a = \pm 1$, ± 3, ± 4, ± 5, ± 7, ± 8. There is only one possible value of b when $a = \pm 1$, $a = \pm 3$, $a = \pm 5$. If $a = 4$ then $b^2 \equiv 0 \bmod 4$, $b \equiv 0$ or $2 \bmod 4$ and the case $b \equiv 2 \bmod 4$, $b = 6$ is not included above. In this case $c = (b^2 - D)/a = -6$ and the form is not properly primitive. Similarly, $a = -4$, $b = 6$ leads to a form that is not properly primitive. If $a = \pm 7$ then there are just 2 possible values of b, $b \equiv 2$ or $5 \bmod a$, and all these possibilities are included above. If $a = \pm 8$ then $b^2 \equiv 4 \bmod 8$ has 2 solutions $b \equiv 2$ or $6 \bmod 8$, and all possibilities are included above. If $a = \pm 2$ then $b = 6$ and $c = \mp 12$ and the form is not properly primitive. Finally, if $a = \pm 6$ then

$b^2 \equiv 0 \bmod 6$, $b \equiv 0 \bmod 6$, $b = 6$, $c = \mp 4$ and again the form is not properly primitive. Therefore the divisor class group is the four element group described above.

Finally, consider another case covered by Gauss (*Disquisitiones Arithmeticae* Art. 226), the case of the order $\{x + y\sqrt{45} : x, y \text{ integers}\}$. In this case the index s is 6 because $45 = 3^2 \cdot 5$ and $5 \equiv 1 \bmod 4$. Again $(-1, *)$ is not principal. The cyclic method shows, however, that $(5, *) \sim (-1, *)$ and therefore that $(-1, *)(5, *) \sim I$. In fact, a process of elimination like the one used above shows that *the divisor class group contains just two elements, the classes of I and* $(-1, *)$. ($a = \pm 2, \pm 3$ lead to forms that are not properly primitive. $a = \pm 7$ is impossible because 45 is not a square mod 7. $a = \pm 5, \pm 6$ lead to forms that are already included. $a = \pm 4$ implies $b = 3$ or $b = 5$, all of which lead to periods already found above.)

EXERCISES

Find the divisor class groups (i.e. a representative set and a multiplication table) corresponding to the following orders:

1. $\{x + y\sqrt{-99} : x, y \text{ integers}\}$. (*Disquisitiones Arithmeticae* Art. 226. Cyclic of order 6.)

2. $\{x + y\sqrt{-117} : x, y \text{ integers}\}$. (The class number is 8.)

3. $\{x + y\sqrt{-531} : x, y \text{ integers}\}$. (*Disquisitiones Arithmeticae* Art. 255. The class number is 18.)

4. $\{x + y\sqrt{305} : x, y \text{ integers}\}$. (The class number is 4.)

5. $\{x + y\sqrt{305} : x, y \text{ integers or halfintegers}\}$.

6. $\{x + y\sqrt{-59} : x, y \text{ integers}\}$. (Compare to Exercise 3.)

7. $\{x + y\sqrt{-59} : x, y \text{ integers or halfintegers}\}$.

8. $\{x + y\sqrt{-531} : x, y \text{ integers or halfintegers}\}$.

9. Look up either of the following tables of binary quadratic forms and verify them for various D's:

   Cayley, A., Tables des formes quadratiques binaires pour les déterminants négatifs depuis $D = -1$ jusqu'à $D = -100$, pour les déterminants positifs non carrés depuis $D = 2$ jusqu'à $D = 99$ et pour les treize déterminants négatifs irréguliers qui se trouvent dans le premier millier, *Jour. für Math.* (Crelle), vol. LX (1862), pp. 357–372. [Also in *Mathematical Papers* of Cayley, vol. 5, pp. 141–156.]

   Ince, E. L., *Cycles of Reduced Ideals in Quadratic Fields*, British Assn. for the Advancement of Science, Mathematical Tables, Vol. IV, Cambridge University Press, 1966.

10. Verify various entries in Table III of Cohn, H., *A Second Course in Number Theory*, Wiley, New York, 1962.

11. Prove the following lemma which was used without proof in Section 3.3: If a and b are relatively prime, then every odd factor of $a^2 - 5b^2$ can be written in the form $p^2 - 5q^2$.

12. The divisor class groups in Gauss's theory can be divided into *genera* as follows. The group corresponds to an order of the form $\{x + y\sqrt{D} : x, y$ integers$\}$. For each odd prime factor of D define a *character* on the group exactly as in Section 7.9. If $D \equiv 1 \bmod 4$ these characters comprise the total character. If $D \equiv 2$ or $3 \bmod 4$ the total character contains an additional sign as defined in Table 7.9.1. If $D \equiv 0 \bmod 4$ there are two cases. When $D \equiv 4 \bmod 8$ the odd numbers of the form $x^2 - Dy^2$ are all 1 or 5 mod 8, that is, are all 1 mod 4, and the additional sign is $+$ if $n \equiv 1 \bmod 4$, $-$ if $n \equiv 3 \bmod 4$ (as when $D \equiv 3 \bmod 4$). When $D \equiv 0 \bmod 8$ the odd numbers of the form $x^2 - Dy^2$ are all 1 mod 8 and there are *two* additional signs in the character, which can be taken to be the signs $\left(\frac{-1}{n}\right)$ and $\left(\frac{2}{n}\right)$. In each of the above examples in the text and the exercises (except Exercises 5, 7, and 8), and in the cases of the orders $\{x + y\sqrt{D} : x, y$ integers$\}$ where $D = 24, 8, -8, -16, -24$, find the partition of the divisor class group into genera, and find all characters which occur. (For example, when $D = 60$ the divisor class group consists of 4 genera each containing 1 class—the classification is IV.1 in the notation of Section 7.9—and the characters which actually occur are $+ + +$, $+ - -$, $- + -$, and $- - +$.) Note that in all cases, half the number of possible characters actually occur, just as in Section 7.11. Gauss proved that this is always the case (*Disquisitiones Arithmeticae*, Arts. 257–261).

8.6 Gauss's composition of forms

As the computations of the preceding section show, in the process of using the cyclic method to find the elements of the divisor class group, it is simplest to think of the classes as being not equivalence classes of divisors but proper equivalence classes of binary quadratic forms. It might seem natural, then, to discard the notion of divisor classes and, following Gauss, to formulate the entire theory in terms of binary quadratic forms. True, there is Kummer's objection that the notion of proper equivalence is more natural when it is seen from the point of view of the theory of divisors, but this simplicity is hardly enough to justify the whole machinery of divisor theory.

The advantage of the theory of divisors is that it gives an explanation of the operation of *multiplication* of classes. From the point of view of divisor classes, the multiplication could not be more natural: the class of A times the class of B is the class of AB. From the point of view of proper equivalence classes of forms, on the other hand, the multiplication is extremely difficult to describe. This section is devoted to a mere sketch of the many steps required to define this multiplication. It is the operation which Gauss called *composition of forms*.

The fact that Kummer does not point out this advantage of his approach but restricts himself to saying that his theory "has the greatest

analogy to the very difficult section *De compositione formarum* of Gauss"
[K7, p. 325] surely indicates that he had not worked out in any detail the
"theory of complex numbers of the form $x + y\sqrt{D}$ " which he cites in the
previous paragraph. Other testimony of the close relationship between
divisor theory for quadratic integers and the composition of forms is
contained in a letter which Kummer wrote to Kronecker on 14 June, 1846
[K5, p. 68]. In this letter he said, "[Dirichlet] recounted and showed to me,
specifically from oral and written remarks of Gauss, that Gauss had
already used in his own private work something like ideal factors at the
time that he was completing the section on composition of forms in the
Disquisitiones Arithmeticae, but that he was never able to put it on a firm
foundation; he says in particular, in a note to his article on the decomposition of polynomials into linear factors, that: 'If I wanted to proceed with
the use of imaginaries in the way that earlier mathematicians have done,
then one of my earlier researches which is very difficult could have been
done in a very simple way.' Gauss later told Dirichlet that he was referring
here to the composition of forms."

The invention of the notion of *composition of forms* was Gauss's great
contribution to the theory of binary quadratic forms. (The use of the cyclic
method to solve equations $ax^2 + 2bxy + cy^2 = m$ was well known to
Lagrange and Euler, and it was included in the *Disquisitiones Arithmeticae*
primarily as a *resumé* of what had gone before.) A particular case of
composition of forms which was known long before Gauss is the formula

$$(x^2 + ny^2)(u^2 + nv^2) = (xu - nyv)^2 + n(xv + yu)^2 \qquad (1)$$

showing that the product of two numbers of the form $x^2 + ny^2$ itself has
this form. Another example is the formula

$$(2x^2 + 2xy + 3y^2)(2u^2 + 2uv + 3v^2) = X^2 + 5Y^2 \qquad (2)$$

where

$$X = 2xu + xv + yu - 2yv$$
$$Y = xv + yu + yv.$$

This formula can be used to prove Fermat's conjecture about numbers of
the form $x^2 + 5y^2$ (see Exercise 1).

In general, a binary quadratic form $\alpha x^2 + 2\beta xy + \gamma y^2$ is said to be a
composition of two other forms $ax^2 + 2bxy + cy^2$ and $a'x^2 + 2b'xy + c'y^2$ if
there is an equation of the form

$$(ax^2 + 2bxy + cy^2)(a'u^2 + 2b'uv + c'v^2) = \alpha X^2 + 2\beta XY + \gamma Y^2$$

in which X and Y are bilinear functions of (x, y) and (u, v), that is,

$$X = pxu + p'xv + p''yu + p'''yv$$
$$Y = qxu + q'xv + q''yu + q'''yv,$$

where $p, p', p'', p''', q, q', q''$, and q''' are given integers, and in which the

335

six determinants

$$P = \begin{vmatrix} p & p' \\ q & q' \end{vmatrix}, \qquad Q = \begin{vmatrix} p & p'' \\ q & q'' \end{vmatrix}, \qquad R = \begin{vmatrix} p & p''' \\ q & q''' \end{vmatrix},$$

$$S = \begin{vmatrix} p' & p'' \\ q' & q'' \end{vmatrix}, \qquad T = \begin{vmatrix} p' & p''' \\ q' & q''' \end{vmatrix}, \qquad U = \begin{vmatrix} p'' & p''' \\ q'' & q''' \end{vmatrix}$$

have no common factor greater than 1. This last condition on P, Q, R, S, T, U is a type of nondegeneracy condition which, like all the other aspects of the theory, Gauss does not motivate in any way.

In a *tour de force* of algebraic manipulation, Gauss shows that, in any composition of forms as defined above, the integers $(P, R - S, U)$ are necessarily proportional to the integers $(a, 2b, c)$ and the integers $(Q, R + S, T)$ are proportional to $(a', 2b', c')$. For example, in the composition (1) the p's and q's are

$$\begin{matrix} 1 & 0 & 0 & -n \\ 0 & 1 & 1 & 0 \end{matrix}$$

from which $P = 1, Q = 1, R = 0, S = 0, T = n, U = n$ and $(P, R - S, U) = (1, 0, n)$ and $(Q, R + S, T) = (1, 0, n)$. Similarly in the composition (2) the p's and q's are

$$\begin{matrix} 2 & 1 & 1 & -2 \\ 0 & 1 & 1 & 1 \end{matrix}$$

from which $P = 2, Q = 2, R = 2, S = 0, T = 3, U = 3, (P, R - S, U) = (2, 2, 3)$ and $(Q, R + S, T) = (2, 2, 3)$. Gauss says if the first of these constants of proportionality is positive that the form $a'x^2 + 2b'xy + c'y^2$ enters the composition *directly* and if the constant of proportionality is negative then the form enters *inversely*. Similarly, the form $ax^2 + 2bxy^2 + cy^2$ enters the composition *directly* or *inversely* according to whether the constant of proportionality between $(Q, R + S, T)$ and $(a', 2b', c')$ is positive or negative. Thus, in the compositions above both forms enter directly. On the other hand, in the composition

$$(x^2 + ny^2)(u^2 + nv^2) = (xu + nyv)^2 + n(xv - yu)^2$$

one has for the p's and q's

$$\begin{matrix} 1 & 0 & 0 & n \\ 0 & 1 & -1 & 0 \end{matrix}$$

from which $P = 1, Q = -1, R = 0, S = 0, T = -n, U = n, (P, R - S, U) = (1, 0, n), (Q, R + S, T) = (-1, 0, -n)$ and the first form enters the composition *inversely*, the second one *directly*.

In what follows, only compositions in which both factors enter *directly* will be considered. Briefly put, what Gauss proved was that this composition operation is well defined on the set of *proper* equivalence classes of binary quadratic forms for a given determinant (and it was for this specific reason that the distinction between proper and improper equivalence was indispensible to his theory) and that it gives to the set of classes of properly primitive forms the structure of a commutative *group*. Spelled out in more detail, he showed, among other things, the following.

(1) If F can be written as a composition of f and f', if F' can be written as a composition of f and f'', and if f' and f'' are properly equivalent, then F and F' are properly equivalent.

(2) If F can be written as a composition of f and f' then F can be written as a composition of f' and f. (This is simple.)

(3) Up to proper equivalence, composition is *associative*. That is, if F is a composition of f and f' and $\mathcal{F}$ is a composition of F and f'', and F' is a composition of f' and f'' and $\mathcal{F}'$ a composition of f and F', then $\mathcal{F}$ and $\mathcal{F}'$ are properly equivalent. (This is extraordinarily difficult. See the *Disquisitiones Arithmeticae* Art. 240.)

(4) Given two forms f and f' with the same determinant, it is possible to find a third form F with the same determinant which is a composition of f and f'. (This is by no means obvious.)

(5) Given two properly primitive forms f and f' there is a third properly primitive form f'' such that the composition of f and f'' is properly equivalent to f'. (See Art. 251.)

Note that Gauss, with magisterial generality, defines the composition of *any* two forms, but that, by virtue of (1), in an actual computation of the composed *class* one can replace the given forms by properly equivalent ones if that is convenient. One can make use of this in the computation of the composition of two properly primitive forms $ax^2 + 2bxy + cy^2$ and $a'x^2 + 2b'xy + c'xy$ with the same determinant $b^2 - ac = b'^2 - a'c'$ as follows.

As was seen in Section 8.3, one can assume without loss of generality that a is odd and relatively prime to $D = b^2 - ac$. Similarly one can assume that a' is odd and relatively prime to D and relatively prime to a. The change of variable $x' = x + ny$, $y' = y$ gives a new form properly equivalent to the original one in which a is the same and b is changed to $b + na$. Therefore b can be changed to any number congruent to it mod a. Similarly, b' can be changed to any number congruent to it mod a'. Since by the Chinese remainder theorem there is an integer B which satisfies $B \equiv b \bmod a$ and $B \equiv b' \bmod a'$, one can assume without loss of generality that the forms to be composed are of the form $ax^2 + 2Bxy + cy^2$ and $a'x^2 + 2Bxy + c'y^2$ where a and a' are odd integers relatively prime to each other and to D. Since $B^2 - D = ac = a'c'$ and since a and a' are relatively prime, $B^2 - D = aa'e$ for some integer e and the forms to be composed are simply $ax^2 + 2Bxy + a'ey^2$ and $a'x^2 + 2Bxy + aey^2$.

In short, one can assume without loss of generality that a and a' are odd integers relatively prime to each other and to D, and that $b = b'$, in which case $c = a'e$ and $c' = ae$ for some integer e. The problem, therefore, is to compose these two particular forms. Gauss announces after much difficult computation (*Disquisitiones Arithmeticae* Art. 243) that in this case the form $aa'x^2 + 2bxy + ey^2$ is the composition of $ax^2 + 2bxy + a'ey^2$ and $a'x^2 + 2bxy + aey^2$. This can be verified directly by writing

$$(ax^2 + 2bxy + a'ey^2)(a'u^2 + 2buv + aev^2) = aa'X^2 + 2bXY + eY^2 \quad (3)$$

where X and Y are to be found. Since $aa'e = b^2 - D$, the first factor on the left can be written as

$$\frac{1}{a}\left[(ax + by)^2 - (b^2 - b^2 + D)y^2\right] = \frac{1}{a}N\left[(ax + by) - y\sqrt{D}\,\right].$$

The other two forms can be rewritten in similar ways, and multiplication of equation (3) by aa' puts it in the form

$$N\left[ax + by - y\sqrt{D}\,\right]N\left[a'u + bv - v\sqrt{D}\,\right] = N\left[aa'X + bY - Y\sqrt{D}\,\right].$$

Since the norm of a product is the product of the norms, this suggests

$$aa'X + bY - Y\sqrt{D} = (ax + by - y\sqrt{D})(a'u + bv - v\sqrt{D})$$

$$= -\sqrt{D}\left[axv + byv + a'yu + byv\right]$$

$$+ aa'xu + abxv + a'byu + b^2yv + Dyv,$$

$$Y = axv + byv + a'yu + byv,$$

$$aa'X = aa'xu + abxv + a'byu + b^2yv + Dyv - bY$$

$$= aa'xu + 0 \cdot xv + 0 \cdot yu - b^2yv + Dyv,$$

$$X = xu - eyv.$$

(Gauss would surely disapprove of the use of $\sqrt{D}$ in this computation.) The steps of this computation are reversible and with this *definition* of X and Y one has the formula (3) and the p's and q's are

$$\begin{matrix} 1 & 0 & 0 & -e \\ 0 & a & a' & 2b \end{matrix}$$

from which $P = a$, $Q = a'$, $R = 2b$, $S = 0$, $T = ea$, $U = ea'$; thus (3) describes a composition (a and a' are relatively prime) in which both forms enter directly.

That Gauss's composition operation coincides with the natural composition of divisors can be proved as follows. Again it suffices to consider the case in which a, a' are odd and relatively prime to D and to each other, and $b = b'$. Then in the theorem of Section 8.3 the form $ax^2 + 2bxy + cy^2$ corresponds to the divisor A which satisfies $N(A) = a$, $b \equiv \sqrt{D} \mod A$. Similarly, $a'x^2 + 2bxy + c'y^2$ corresponds to A' where $N(A') = a'$ and $b \equiv \sqrt{D} \mod A'$. Then $N(AA') = aa'$ and $b \equiv \sqrt{D} \mod AA'$ (because the divisor of $b - \sqrt{D}$ is divisible by one and only one divisor of norm a and by one and only one divisor of norm a' and because a and a' are relatively prime). This divisor corresponds to the form $aa'x^2 + 2bxy + ey^2$ where $e = (b^2 - D)/aa'$, as was to be shown.

1. Use formula (2) to prove Fermat's conjecture that the product of two primes p_1, $p_2 \equiv 3$ or $7 \bmod 20$ is of the form $p_1 p_2 = x^2 + 5y^2$. [Prove that if $p \equiv 3$ or $7 \bmod 20$ then there is a binary quadratic form $px^2 + 2dxy + ey^2$ with determinant -5. This form is equivalent either to $x^2 + 5y^2$ or $2x^2 + 2xy + 3y^2$. Since p has no representation $p = x^2 + 5y^2$ it must have a representation $p = 2x^2 + 2xy + 3y^2$.] Note that the conjecture says, in essence, that the divisor class group when $D = -5$ is the group with 2 elements.

2. Find a square (under composition) of the form $4x^2 + 9y^2$. Find the class of the square by finding a properly equivalent form which corresponds to a divisor and by squaring the corresponding divisor. [See Exercise 3 of Section 8.1.]

8.7 Equations of degree 2 in 2 variables

The most general first degree equation in two variables $ax + by + c = 0$, in which the coefficients a, b, c and the desired solution x, y are all integers, is solved by the Euclidean algorithm (see Exercise 1). The most general second degree equation in two variables $ax^2 + bxy + cy^2 + dx + ey + f = 0$ can be solved by completing the square in order to reduce the problem to one of the form $\alpha x^2 + 2\beta xy + \gamma y^2 = m$ and by then using the cyclic method as in Section 8.2.

Lagrange [L1] carries out the completion of the square as follows. First write the equation as an equation in x with coefficients which are polynomials in y, that is, $ax^2 + (by + d)x + (cy^2 + ey + f) = 0$. Multiply by $4a$ so that the equation becomes

$$(2ax + by + d)^2 - (by + d)^2 + 4a(cy^2 + ey + f) = 0.$$

Let $t = 2ax + by + d$ and write this equation as $\alpha y^2 + 2\beta y + \gamma = t^2$ where $\alpha = b^2 - 4ac$, $\beta = bd - 2ae$, and $\gamma = d^2 - 4af$. Multiply by α to find $(\alpha y + \beta)^2 - \beta^2 + \alpha\gamma = \alpha t^2$ or, more simply, $u^2 - \alpha t^2 = m$ where

$$\alpha = b^2 - 4ac,$$
$$m = \beta^2 - \alpha\gamma = (bd - 2ae)^2 - (b^2 - 4ac)(d^2 - 4af)$$

and where t and u are the integers

$$t = 2ax + by + d,$$
$$u = \alpha y + \beta = (b^2 - 4ac)y + bd - 2ae.$$

Every solution x, y of the original equation gives a unique solution t, u of the new equation $u^2 - \alpha t^2 = m$. Conversely, a solution of the new equation comes from a solution of the original one if and only if the rational numbers

$$y = \frac{u - \beta}{\alpha}, \qquad x = \frac{1}{2a}\left[t - b\frac{u - \beta}{\alpha} - d\right] = \frac{\alpha t - bu + b\beta - d\alpha}{2a\alpha} \qquad (1)$$

are integers, that is, if and only if

$$u \equiv \beta \bmod \alpha, \qquad \alpha t - bu + b\beta - d\alpha \equiv 0 \bmod 2a\alpha. \tag{2}$$

Thus one can find the most general solution of the given equation by using the cyclic method to solve $u^2 - \alpha t^2 = m$, by discarding all solutions which fail to satisfy the congruences (2), and by using the equations (1) to define x and y for those solutions which do satisfy the congruences.

(If $a = 0$ then, unless b and c are also both zero, an invertible change of coordinates can be used to convert the problem to one in which $a \neq 0$. If $\alpha = 0$ the problem has a simple solution—see Exercise 2.)

This cannot yet be accepted as a solution of the problem because the operation of discarding those (t, u) which do not satisfy the congruences involves an infinite number of steps. If this were acceptable as a "solution" then one would be on the verge of accepting as a "solution" the set of all pairs of integers (x, y) which remain after those which fail to satisfy $ax^2 + bxy + cy^2 + dx + ey + f = 0$ have been discarded. However, the discarding process for the above congruences can be reduced to a finite number of steps as follows.

If $\alpha < 0$, or if α is a square and $m \neq 0$, then the equation $u^2 - \alpha t^2 = m$ has a finite number of solutions. They can be enumerated explicitly and those which fail to satisfy the congruences can be discarded. If α is a square and $m = 0$ then there are infinitely many solutions but they have the simple form $u = \pm kt$ ($k^2 = \alpha$, $t = $ integer) and the question of whether this solution satisfies the congruences (2) depends only on the sign $\pm k$ and on the class of $t \bmod 2a\alpha$. Therefore one can give two (finite) lists of classes mod $2a\alpha$ such that the most general solution is $u = kt$ where t is in the first list of classes or $u = -kt$ where t is in the second list.

Finally, if α is a positive integer not a square then $u^2 - \alpha t^2 = m$ either has no solution at all or it has a finite number of infinite sequences of solutions of the form

$$u + t\sqrt{\alpha} = (X + Y\sqrt{\alpha})(U + T\sqrt{\alpha})^n \tag{3}$$

where $U + T\sqrt{\alpha}$ is a fundamental unit, where n ranges over all integers, and where $X + Y\sqrt{\alpha}$ ranges over a finite number of quadratic integers. It will suffice to show that in any one sequence of the form (3) it is possible to enumerate those values of n for which t and u satisfy the congruences (2). Since there are only finitely many possible values of $(U + T\sqrt{\alpha})^n \bmod 2a\alpha$, there must be distinct integers i and j such that $(U + T\sqrt{\alpha})^i \equiv (U + T\sqrt{\alpha})^j \bmod 2a\alpha$. Because $U + T\sqrt{\alpha}$ has an inverse, namely, $\pm(U - T\sqrt{\alpha})$, it follows that $(U + T\sqrt{\alpha})^k \equiv 1 \bmod 2a\alpha$ where $k = i - j$. Then (3) gives values of t and u which satisfy the congruences if and only if it does when n is changed to $n + k$. Thus, in order to test *all* solutions of the form (3), it suffices to test just those solutions (3) in which $n = 0, 1, 2, \ldots, k - 1$.

This completes the solution of $ax^2 + bxy + cy^2 + dx + ey + f = 0$. The solution given here is essentially the one given by Gauss in *Disquisitiones Arithmeticae* Art. 216. However, Gauss uses a more symmetrical change of variable to complete the square, and this results in neater formulas.

EXERCISES

1. Solve the equation $ax + by + c = 0$ as follows. Show first that there is no solution unless c is a multiple of the greatest common divisor d of a and b. The Euclidean algorithm gives $d = au + bv$. Use this to construct *one* solution of $ax + by + c = 0$ in any case where the equation has a solution. Show that the difference between two solutions is a solution of $ar + bs = 0$. Complete the solution by finding the most general solution of $ar + bs = 0$.

2. Reduce the solution of the equation $ax^2 + bxy + cy^2 + dx + ey + f = 0$ to the case $a \neq 0$, $\alpha = b^2 - 4ac \neq 0$ as follows. If $a = 0$ but $c \neq 0$ then interchange of x and y gives $a \neq 0$. If $a = c = 0$ but $b \neq 0$ then $x \mapsto x$, $y \mapsto x + y$ is an invertible change of coordinates which makes $a \neq 0$. If $a = b = c = 0$ then of course the equation has degree 1. This shows that the given equation can be reduced to one of the form $\alpha y^2 + 2\beta y + \gamma = t^2$ and the further reduction of the text can be used unless $\alpha = 0$, that is, unless the equation takes the form $2\beta y + \gamma = t^2$ where $t = 2ax + by + d$. Use these two equations to express x and y in terms of t and t^2. (The case $\beta = 0$ must be excluded and treated separately.) Show that t gives rise to a solution of the original equation if and only if $t + 4\alpha\beta$ does; conclude that all solutions of the original equation can be enumerated in a finite number of steps.

3. Write a computer program which solves $ax^2 + bxy + cy^2 + dx + ey + f = 0$ in integers for given integers a, b, c, d, e, f. [You may prefer to use Gauss's method for completing the square.]

4. Find all solutions of $x^2 + 8xy + y^2 + 2x - 4y + 1 = 0$. [*Disquisitiones Arithmeticae* Art. 221.]

5. Find all solutions of $3x^2 + 4xy - 7y^2 = 12$. [*Disquisitiones Arithmeticae* Art. 212.]

9 Dirichlet's class number formula

9.1 The Euler product formula

This chapter is devoted to Dirichlet's formula for the number of elements in the divisor class groups of Chapters 7 and 8. In Gauss's formulation of the theory, these groups are, of course, groups of proper equivalence classes of binary quadratic forms, the group operation being composition. Although Dirichlet followed Gauss's formulation, it will be convenient here to refer to them instead as divisor class groups.

As was stated in Section 6.2, the main idea behind Dirichlet's formula is the Euler product formula

$$\sum_A \frac{1}{N(A)^s} = \prod_P \frac{1}{1 - \dfrac{1}{N(P)^s}} \tag{1}$$

where P ranges over all prime divisors and A over all divisors. (For the divisor class groups of Chapter 8 the ranges of P and A are restricted to prime divisors and divisors relatively prime to the index of the order of quadratic integers under consideration. This case will be considered in Section 9.6. Until then, the divisor class groups that are considered will be those of the full order of all quadratic integers for the squarefree determinant D—that is, the divisor class groups of Chapter 7.)

The Euler product formula is, in essence, simply a novel way of stating that a divisor A can be written in one and only one way as a product of powers of distinct prime divisors P, because formally

$$\frac{1}{1 - \dfrac{1}{N(P)^s}} = 1 + \frac{1}{N(P)^s} + \frac{1}{N(P^2)^s} + \cdots \tag{2}$$

and the infinite product in (1) can be expanded by choosing the term 1 in all but a finite number of factors and choosing terms $N(P^j)^{-s}$ in the others. The actual proof of formula (1) for real numbers $s > 1$ is easily accomplished in the same way as in section 6.3: The product on the right converges because $\Sigma N(P)^{-s} \leqslant \Sigma 2p^{-s} \leqslant 2\Sigma n^{-s} < \infty$. Any finite sum of terms $\Sigma N(A)^{-s}$ on the left is less than some partial product on the right while, on the other hand, any partial product on the right is equal to an infinite sum of terms, all of which occur on the left (multiplication of several absolutely convergent series of the form (2)). That is, $\Sigma N(A)^{-s} \leqslant \Pi(1 - N(P)^{-s})^{-1} \leqslant \Sigma N(A)^{-s}$.

As in Chapter 6, the method for deriving the class number formula will be to multiply the Euler product formula by $s - 1$ and to let $s \downarrow 1$. On the right the limit will be a number $L(1, \chi)$ that can be written as an infinite sum depending on D, a sum which can also be evaluated quite explicitly. On the left the sum can be broken into h sums, one for each divisor class, and, in the limit as $s \downarrow 1$, each of these h sums is the same. Therefore the limit on the left is equal to $h \lim(s - 1)\Sigma N(A)^{-s}$ where the sum is over *principal* divisors A. This limit can be evaluated by writing it as a sum over quadratic integers and by replacing it with an integral which can be evaluated by integral calculus.

Of course Dirichlet did not base his original proof of the class number formula on the Euler product formula in the form (1) because his work preceded even Kummer's development of the theory of ideal cyclotomic integers, not to mention the much later development of ideal quadratic integers. However, it is fair to say that formula (1) is *essentially* to be found in Dirichlet's work ([D7], §90) and that all that was lacking was the vocabulary to state (and prove) it in the simple form (1).

As was the case with the definition and basic properties of the class number, there are a great many special cases—$D > 0$ or $D < 0$, $D \equiv 1$ or $D \not\equiv 1 \bmod 4$, D squarefree or $D = t^2 D'$—to be considered. It is neither enlightening nor useful to try to give a unified treatment of them. Instead, in the sections that follow, the cases will be considered one at a time. What is important is the single idea of the product formula (1), plus the handful of techniques that are used to evaluate the limits of the two sides times $s - 1$ as $s \downarrow 1$.

9.2 First case

The simplest case of the class number formula is the case $D < 0$, $D \not\equiv 1 \bmod 4$, and D squarefree. In this case the right side of the Euler product formula is

$$\prod_{P}\left(1 - \frac{1}{N(P)^s}\right)^{-1}$$

$$= \prod_{p \text{ ramifies}}\left(1 - \frac{1}{p^s}\right)^{-1} \prod_{p \text{ splits}}\left(1 - \frac{1}{p^s}\right)^{-2} \prod_{\substack{p \text{ remains} \\ \text{prime}}}\left(1 - \frac{1}{p^{2s}}\right)^{-1}$$

$$= \prod_{p}\left(1 - \frac{1}{p^s}\right)^{-1} \prod_{p \text{ splits}}\left(1 - \frac{1}{p^s}\right)^{-1} \prod_{\substack{p \text{ remains} \\ \text{prime}}}\left(1 + \frac{1}{p^s}\right)^{-1}$$

$$= \zeta(s) \prod_{p}\left(1 - \left(\frac{D}{p}\right)\frac{1}{p^s}\right)^{-1}$$

where $\zeta(s)$ is the ordinary Riemann zeta function (see Section 6.3) and where $\left(\frac{D}{p}\right)$ is 0, 1, or -1 according to whether p ramifies, splits, or remains prime in quadratic integers for the determinant D. (If $p = 2$ or if $p | D$ then $\left(\frac{D}{p}\right) = 0$. Otherwise $\left(\frac{D}{p}\right)$ is equal to the Legendre symbol.)

Since $\lim (s - 1)\zeta(s)$ as $s \downarrow 1$ is 1 (see Section 6.3) the limit of the right side times $s - 1$ as $s \downarrow 1$ is simply the limit as $s \downarrow 1$ of $\prod(1 - \left(\frac{D}{p}\right)p^{-s})^{-1}$. As was noted in Section 7.8, the definition of $\left(\frac{D}{p}\right)$ can be extended from prime integers p to all integers relatively prime to $4D$ in such a way that $\left(\frac{D}{n_1 n_2}\right) = \left(\frac{D}{n_1}\right)\left(\frac{D}{n_2}\right)$. Therefore one can rewrite the product in question formally as a sum

$$\prod_{p}\left(1 - \left(\frac{D}{p}\right)\frac{1}{p^s}\right)^{-1}$$

$$= \prod_{p}\left(1 + \left(\frac{D}{p}\right)\frac{1}{p^s} + \left(\frac{D}{p^2}\right)\frac{1}{p^{2s}} + \cdots + \left(\frac{D}{p^j}\right)\frac{1}{p^{js}} + \cdots\right)$$

$$= \sum_{n}\left(\frac{D}{n}\right)\frac{1}{n^s}$$

where the product is over all primes p which do not divide $4D$ and where the sum is over all integers n relatively prime to $4D$. For $s > 1$ this can easily be proved by rearrangement of absolutely convergent series.

It was also noted in Section 7.8 that $\left(\frac{D}{n}\right)$ is a *character* mod $4D$, that is, not only is $\left(\frac{D}{n_1 n_2}\right) = \left(\frac{D}{n_1}\right)\left(\frac{D}{n_2}\right)$ but also $\left(\frac{D}{n+4D}\right) = \left(\frac{D}{n}\right)$. Since the sum of its values is 0 (half the classes are splitting classes and half are not) summation by parts can be used as in Section 6.5 to show that the series $\Sigma\left(\frac{D}{n}\right)n^{-s}$ converges conditionally for $s > 0$ and defines a continuous function of s for $s > 0$. Therefore

$$\lim_{s \downarrow 1} (s - 1) \prod_{P}\left(1 - \frac{1}{N(P)^s}\right)^{-1} = \sum_{n=1}^{\infty}\left(\frac{D}{n}\right)\frac{1}{n}$$

where $\binom{D}{n}$ is defined to be 0 if n is not relatively prime to $4D$, and where the conditionally convergent series on the right is summed in the natural order. It is traditional to denote this number by $L(1, \chi)$ where χ denotes the character $\chi(n) = \binom{D}{n}$ mod $4D$ and, as in Chapter 6, $L(s, \chi) = \Sigma \chi(n) n^{-s}$.

This completes the evaluation of the limit as $s{\downarrow}1$ of $s - 1$ times the right side of the Euler product formula

$$\sum \frac{1}{N(A)^s} = \prod \left(1 - \frac{1}{N(P)^s}\right)^{-1}.$$

Consider now the left side of the formula. Exactly as in Section 6.8, the limit of the sum of $N(A)^{-s}$ over any two equivalence classes of divisors A is the same, so that

$$\lim_{s\downarrow 1}(s - 1)\sum N(A)^{-s} = h\lim_{s\downarrow 1}(s - 1)\sum_{A \text{ principal}} N(A)^{-s}$$

where h is the class number. Now except in the case $D = -1$, there are precisely two units ± 1, and every principal divisor is the divisor of two and only two quadratic integers. Thus

$$\sum_{A \text{ principal}} N(A)^{-s} = \frac{1}{2}\sum_{(x, y) \neq (0, 0)} (x^2 - Dy^2)^{-s}$$

except when $D = -1$, in which case the factor $\frac{1}{2}$ becomes $\frac{1}{4}$. As $s{\downarrow}1$ the sum on the right approaches infinity in the same way that the integral

$$\iint_{x^2 - Dy^2 > 1} (x^2 - Dy^2)^{-s} \, dx \, dy$$

does. More specifically, the difference between these two quantities remains bounded as $s{\downarrow}1$. This is simple to prove (Exercise 3).

The integral can be evaluated by change of variables

$$\iint_{x^2 - Dy^2 > 1} (x^2 - Dy^2)^{-s} \, dx \, dy$$

$$= \iint_{-Dz^2 - Dy^2 > 1} (-Dz^2 - Dy^2)^{-s} d(\sqrt{-D}\, z) \, dy$$

$$= |D|^{-s}|D|^{1/2} \iint_{z^2 + y^2 > |D|^{-1}} (z^2 + y^2)^{-s} \, dz \, dy$$

$$= |D|^{(1-2s)/2} \iint_{r^2 > |D|^{-1}} r^{-2s} r \, dr \, d\theta$$

$$= |D|^{(1-2s)/2} \cdot 2\pi \left. \frac{r^{2-2s}}{2 - 2s} \right|_{|D|^{-1/2}}^{\infty}$$

$$= \frac{\pi}{s - 1} \cdot |D|^{(1-2s)/2} |D|^{s-1}.$$

Therefore

$$\lim_{s\downarrow 1}(s-1)\sum_A N(A)^{-s} = h\cdot\frac{1}{2}\cdot\lim_{s\downarrow 1}\pi\cdot|D|^{(1-2s)/2}|D|^{s-1} = \frac{h\pi}{2\sqrt{|D|}}$$

except in the case $D = -1$ when the 2 in the denominator becomes a 4.

In conclusion, then, when $D < 0$, $D \not\equiv 1 \bmod 4$, and D is squarefree, the formula is

$$\frac{h\pi}{2\sqrt{|D|}} = \sum_{n=1}^{\infty}\left(\frac{D}{n}\right)\frac{1}{n}$$

except in the case $D = -1$, when the 2 in the denominator on the left becomes a 4.

Examples. When $D = -1$ the value of $\left(\frac{D}{n}\right)$ is 1 when $n \equiv 1 \bmod 4$, -1 when $n \equiv 3 \bmod 4$, and 0 when n is even. The class number is 1 because unique factorization holds for the Gaussian integers. The formula therefore is

$$\frac{\pi}{4} = 1 - \frac{1}{3} + \frac{1}{5} - \frac{1}{7} + \cdots$$

a well-known formula of calculus discovered by Leibniz. The proof here, based on properties of factorization of Gaussian integers, was in essence given by Gauss [G2, pp. 655–677]. It is of course entirely different from the usual proof, which amounts to justifying the use of the formula arctan $x = x - \frac{1}{3}x^3 + \frac{1}{5}x^5 - \cdots$ when $x = 1$. If Leibniz's formula is taken as known, then Dirichlet's formula proves that $h = 1$.

When $D = -2$ the class number is 1 and the formula gives

$$\frac{\pi}{2\sqrt{2}} = 1 + \frac{1}{3} - \frac{1}{5} - \frac{1}{7} + \frac{1}{9} + \frac{1}{11} - \frac{1}{13} - \cdots .$$

This remarkable formula was known to Newton (see his famous letter of 24 October 1676 to Oldenberg [N2]). Again, if Newton's formula is proved by another method it can be used to prove that $h = 1$ when $D = -2$.

When $D = -5$ the class number is 2 and the formula becomes

$$\frac{\pi}{\sqrt{5}} = 1 + \frac{1}{3} + \frac{1}{7} + \frac{1}{9} - \frac{1}{11} - \frac{1}{13} - \frac{1}{17} - \frac{1}{19} + \frac{1}{21} + \cdots$$

where the sum is over all integers relatively prime to 20 and the sign is $+$ if $n \equiv 1, 3, 7$, or 9 mod 20, $-$ if $n \equiv 11, 13, 17$, or 19. An alternative proof of this formula, which then allows the deduction that $h = 2$, will be given in Section 9.5.

In the general case, the class number formula relates h to $\sum\left(\frac{D}{n}\right)\frac{1}{n} = L(1, \chi)$. If h is known this gives the value of $L(1, \chi)$. Conversely, if $L(1, \chi)$ can be found, the value of h can be deduced. In Section 9.5 a method for evaluating $L(1, \chi)$ will be given. In some cases this will be a very simple way of evaluating h, while in others it will be less simple than the straightforward evaluation by the techniques of Chapter 7.

EXERCISES

1. For each D in the range $-5 > D > -20$ to which the argument of this section applies, describe the character $\left(\frac{D}{n}\right)$ explicitly. In each case find the value of $L(1, \chi)$ by computing h.

2. Is $L(1, \chi)$ independent of the order in which the series is summed?

3. Prove that the sum of $(x^2 - Dy^2)^{-s}$ over all points of the xy-plane with integer coefficients (other than $x = 0, y = 0$) differs from the integral of $(x^2 - Dy^2)^{-s}\, dx\, dy$ over the exterior of the ellipse $x^2 - Dy^2 = 1$ by an amount which remains bounded as $s\downarrow 1$. [See Section 6.12.]

4. Prove that in the limit as $s\downarrow 1$ the sum of $N(A)^{-s}$ over a divisor class is the same for all classes. [See Section 6.13.]

9.3   Another case

In this section the class number formula will be derived for cases in which D is *positive*, squarefree, and $\not\equiv 1 \bmod 4$. Recall that in cases where $D > 0$ the definition of "divisor" was extended to include $(-1, *)$. Therefore, in the Euler product formula

$$\sum \frac{1}{N(A)^s} = \prod \left(1 - \frac{1}{N(P)^s}\right)^{-1}, \tag{1}$$

when P ranges over all prime divisors, A does not range over all divisors but only over divisors A which do not include $(-1, *)$, that is to say, only divisors A for which $N(A) > 0$.

The formula

$$\lim_{s\downarrow 1}(s-1)\prod\left(1 - N(P)^{-s}\right)^{-1} = \sum_{n=1}^{\infty}\left(\frac{D}{n}\right)\frac{1}{n}$$

follows in exactly the same way as in the case $D < 0$ of the preceding section. Here the function $\chi(n) = \left(\frac{D}{n}\right)$ is a character mod $4D$, that is, $\chi(n_1 n_2) = \chi(n_1)\chi(n_2)$ and $\chi(n + 4D) = \chi(n)$; the number on the right side of this equation is also written $L(1, \chi)$.

The formula

$$\lim_{s\downarrow 1}(s-1)\sum_{N(A)>0} N(A)^{-s} = h\lim_{s\downarrow 1}(s-1)\sum_{\substack{A\text{ principal}\\ N(A)>0}} N(A)^{-s} \tag{2}$$

also follows as before, once it is shown that each of the h classes contains divisors with positive norm; this follows immediately from the observation that the divisor of $\sqrt{D}$ is $(-1, *)A$ where A has positive norm, so that $(-1, *) \sim A$ where $N(A) > 0$.

To evaluate the limit on the right side of (2) the procedure is, as before, to rewrite the sum as a sum over quadratic integers and then to replace this sum by an integral. The first step, therefore, is to find a way of choosing, for each principal divisor A with $N(A) > 0$, a quadratic integer whose divisor is A. This can be done as follows.

Let ε be the fundamental unit of the quadratic integers for the determinant D, so that $\varepsilon = u + v\sqrt{D}$ where $u > 0$ and $v > 0$ and the most general unit is $\pm \varepsilon^n$ (n an integer), and let $E = \varepsilon$ if $N(\varepsilon) = 1$, $E = \varepsilon^2$ if $N(\varepsilon) = -1$. Then if $x + y\sqrt{D}$ is any quadratic integer with divisor A, the most general quadratic integer with divisor A is $\pm E^n(x + y\sqrt{D})$ where n is an integer. Since $E = U + V\sqrt{D}$ where U and V are positive integers, it is simple to show, using the technique of Section 7.5, that for any given $x + y\sqrt{D}$ the coefficients x' and y' of $x' + y'\sqrt{D} = E^n(x + y\sqrt{D})$ have like signs for n sufficiently large (Exercise 2). Therefore every quadratic integer $x + y\sqrt{D}$ has the same divisor as at least one quadratic integer in the "first quadrant" and one can choose, for any principal divisor A, a quadratic integer $x + y\sqrt{D}$ with divisor A such that $x \geqslant 0$ and $y \geqslant 0$. Because, as above, $E^n(x - y\sqrt{D})$ is in the first or the third quadrant for n sufficiently large (that is, the coefficients of $E^n(x - y\sqrt{D})$ have like sign for large n), its conjugate $E^{-n}(x + y\sqrt{D})$ is not in the first quadrant for large n. Let n be the least positive integer such that $E^{-n}(x + y\sqrt{D})$ is not in the first quadrant. Then $x' + y'\sqrt{D} = E^{-n+1}(x + y\sqrt{D})$ has the same divisor A as $x + y\sqrt{D}$, is in the first quadrant $x'y' \geqslant 0$ (because either $n = 1$ and $x' + y'\sqrt{D} = x + y\sqrt{D}$ or $n > 1$ and $x' + y'\sqrt{D} = E^{-(n-1)}(x + y\sqrt{D})$ is in the first quadrant by the choice of n) but $E^{-1}(x' + y'\sqrt{D}) = E^{-n} \cdot (x + y\sqrt{D})$ is not in the first quadrant. For each principal divisor A there is *only* one such $x' + y'\sqrt{D}$ because if $x' + y'\sqrt{D}$ and $x'' + y''\sqrt{D}$ have the same divisor, if both are in the first quadrant, and if E^{-1} carries both out of the first quadrant, then, in the first place $x' + y'\sqrt{D} = \pm E^n(x'' + y''\sqrt{D})$ for some integer n. By interchanging $x' + y'\sqrt{D}$ and $x'' + y''\sqrt{D}$ if necessary, one can assume n is nonnegative. Therefore, because all coefficients are nonnegative, $x' + y'\sqrt{D} = E^n(x'' + y''\sqrt{D})$ where $n \geqslant 0$. If $n \geqslant 1$ then $E^{-1}(x' + y'\sqrt{D}) = E^{n-1}(x'' + y''\sqrt{D})$ is in the first quadrant, contrary to assumption. Therefore $n = 0$, as was to be shown. In conclusion, *for each principal divisor A there is one and only one quadratic integer $x + y\sqrt{D}$ with divisor A which lies in the first quadrant ($xy \geqslant 0$) but which is carried out of the first quadrant by multiplication by E^{-1}.*

Now $E^{-1} = U - V\sqrt{D}$ where U and V are positive, so that $x + y\sqrt{D}$ must have the property that $(U - V\sqrt{D})(x + y\sqrt{D})$ is not in the first quadrant, that is, $Ux - VyD$ or $Uy - Vx$ or both must be negative. If $x + y\sqrt{D}$ has positive norm and if $Uy - Vx \geqslant 0$ then $(Uy - Vx)x \geqslant 0$, $Uxy - Vx^2 \geqslant 0$, $Uxy - Vx^2 + V(x^2 - Dy^2) > 0$, $y(Ux - VyD) > 0$, and $Ux - VyD > 0$. Therefore $Uy - Vx < 0$ and $0 \leqslant y < (V/U)x$. Conversely, if $x + y\sqrt{D}$ satisfies $0 \leqslant x$, $0 \leqslant y < (V/U)x$ then $x + y\sqrt{D}$ lies in the first quadrant but $E^{-1}(x + y\sqrt{D})$ does not. Therefore *a principal divisor A for which $N(A) > 0$ is the divisor of one and only one quadratic integer $x + y\sqrt{D}$ for which $0 \leqslant x$ and $0 \leqslant Uy < Vx$.* Therefore

$$\sum_{\substack{A \text{ principal} \\ N(A)>0}} N(A)^{-s} = \sum_{x=1}^{\infty} \sum_{0 < Uy < Vx} (x^2 - Dy^2)^{-s}. \tag{3}$$

It is simple to show that the sum on the right differs by a bounded amount as $s \downarrow 1$ from the integral

$$\iint_{\substack{0 < Uy < Vx \\ x^2 - Dy^2 > 1}} (x^2 - Dy^2)^{-s} \, dx \, dy = \iint_{\substack{0 < Uy < Vz\sqrt{D} \\ Dz^2 - Dy^2 > 1}} (Dz^2 - Dy^2)^{-s} d(z\sqrt{D}) \, dy$$

$$= D^{(-2s+1)/2} \iint_{\substack{0 < Uy < Vz\sqrt{D} \\ z^2 - y^2 > 1/D}} (z^2 - y^2)^{-s} \, dz \, dy.$$

This integral can be computed by using the change of variable $z = r \cosh\theta$, $y = r \sinh\theta$, $z^2 - y^2 = r^2$, $dz\, dy = r\, dr\, d\theta$. For each fixed r, θ goes from 0 (where $z = r$, $y = 0$) to the point where $Uy = Vz\sqrt{D}$, $U \sinh\theta = V\sqrt{D} \cosh\theta$, $U(e^{2\theta} - 1) = V\sqrt{D}(e^{2\theta} + 1)$, $e^{2\theta}(U - V\sqrt{D}) = U + V\sqrt{D}$, $e^{2\theta}E^{-1} = E$, $e^{2\theta} = E^2$, $\theta = \log E$. Therefore the integral is

$$= D^{(-2s+1)/2} \log E \left.\frac{r^{2-2s}}{2 - 2s}\right|_{D^{-1/2}}^{\infty}$$

$$= D^{(-2s+1)/2} \log E \frac{D^{s-1}}{2(s-1)}.$$

Therefore the quantity (3) times $s - 1$ approaches the limit $(\log E)/2\sqrt{D}$ as $s \downarrow 1$ and multiplication of (1) by $s - 1$ followed by passage to the limit as $s \downarrow 1$ gives the desired formula

$$\frac{h \log E}{2\sqrt{D}} = \sum_{n=1}^{\infty} \left(\frac{D}{n}\right)\frac{1}{n}.$$

Here E is the fundamental unit with norm 1; that is, $E = \varepsilon$ if $N(\varepsilon) = 1$ and $E = \varepsilon^2$ if $N(\varepsilon) = -1$.

  Examples. If $D = 2$ then $\varepsilon = 1 + \sqrt{2}$, $N(\varepsilon) = -1$, $h = 1$, and $\left(\frac{D}{n}\right)$ is 1 for $n \equiv \pm 1 \bmod 8$, -1 for $n \equiv \pm 3 \bmod 8$, and 0 for n even. Therefore the formula is

$$\frac{\log(1 + \sqrt{2})}{\sqrt{2}} = 1 - \frac{1}{3} - \frac{1}{5} + \frac{1}{7} + \frac{1}{9} - \frac{1}{11} - \cdots.$$

If $D = 3$ then $\varepsilon = 2 + \sqrt{3}$, $N(\varepsilon) = 1$, $h = 2$, and $\left(\frac{D}{n}\right)$ is 1 for $n \equiv 1$ or 11 mod 12, -1 for $n \equiv 5$ or 7 mod 12, and 0 otherwise. Therefore

$$\frac{\log{(2 + \sqrt{3}\,)}}{\sqrt{3}} = 1 - \frac{1}{5} - \frac{1}{7} + \frac{1}{11} + \frac{1}{13} - \frac{1}{17} - \cdots .$$

If $D = 7$ then $\varepsilon = 8 + 3\sqrt{7} = E$, $h = 2$ and the formula is

$$\frac{\log{(8 + 3\sqrt{7}\,)}}{\sqrt{7}} = \sum_{n=1}^{\infty} \chi(n) \cdot \frac{1}{n}$$

where $\chi(n) = 1$ if $n \equiv \pm 1$, ± 3, or ± 9 mod 28, $\chi(n) = -1$ if $n \equiv \pm 5$, ± 11, or ± 13 mod 28, $\chi(n) = 0$ in all other cases, and where, of course, the conditionally convergent series $\Sigma \chi(x) n^{-1}$ is summed in the natural order.

EXERCISES

1. For each D in the range $7 < D < 20$ to which the argument of this section applies, find all 3 of the items h, E, and χ which enter into the class number formula.

2. Prove that if $X = x + y\sqrt{D}$ is any quadratic integer and if E is as in the text then the two coefficients of $E^n X$ have the same sign for all sufficiently large integers n.

3. Prove that the sum in (3) can be replaced by an integral in the way that it was done in the text.

9.4 $D \equiv 1 \bmod 4$

If D is squarefree and $D \equiv 1 \bmod 4$ then quadratic integers $x + y\sqrt{D}$ may have denominators of 2—more precisely, 2 divides $1 - \sqrt{D}$ and every quadratic integer can be written in the form $u + v \cdot \frac{1}{2}(1 - \sqrt{D}\,)$ where u and v are integers—and the prime 2 does not ramify. In fact (see Section 7.1) 2 splits if $D \equiv 1 \bmod 8$ and 2 remains prime if $D \equiv 5 \bmod 8$. Thus the character $\chi(n) = \left(\frac{D}{n}\right)$ is no longer zero for all even numbers, and, as is easily shown, $\chi(n)$ is zero if and only if n is not relatively prime to D. Moreover, χ is a character mod D, that is, $\chi(mn) = \chi(m)\chi(n)$ and $\chi(n + D) = \chi(n)$. With this χ, the formula

$$\lim_{s \downarrow 1}(s - 1) \prod_{P} \left(1 - N(P)^{-s}\right)^{-1} = \prod_{p \nmid D} \left(1 - \left(\frac{D}{p}\right)\frac{1}{p}\right)^{-1} = L(1, \chi)$$

still holds.

On the other side of the equation, $\lim\,(s - 1)\Sigma N(A)^{-s}$ can be found in exactly the same way except that the quadratic integers $x + y\sqrt{D}$ considered as points in the xy-plane are twice as dense (in each square $\{(x, y) : x_0 \leqslant x < x_0 + 1, y_0 \leqslant y < y_0 + 1\}$ there are exactly 2 quadratic integers) so that the integral is doubled. The final class number formula in the case $D \equiv 1 \bmod 4$ is therefore

$$h \frac{\pi}{\sqrt{|D|}} = L(1, \chi)$$

when $D < 0$ (except when $D = -3$, when, because there are 6 units instead of 2, the left side becomes $h\pi/3\sqrt{3}$) and

$$h \frac{\log E}{\sqrt{D}} = L(1, \chi)$$

when $D > 0$ (where E is the fundamental unit ε if $N(\varepsilon) = 1$ and where $E = \varepsilon^2$ if $N(\varepsilon) = -1$).

For example, when $D = -3$ the formula is

$$1 \cdot \frac{\pi}{3\sqrt{3}} = 1 - \frac{1}{2} + \frac{1}{4} - \frac{1}{5} + \frac{1}{7} - \frac{1}{8} + \cdots .$$

This formula is given by Euler in his *Introductio in Analysin Infinitorum* [E6], §176. (I do not know whether anyone knew it before Euler.) When $D = -7$ the formula is

$$1 \cdot \frac{\pi}{\sqrt{7}} = 1 + \frac{1}{2} - \frac{1}{3} + \frac{1}{4} - \frac{1}{5} - \frac{1}{6} + \frac{1}{8} + \cdots$$

because in this case $\chi(n) = 0$ if $n \equiv 0 \bmod 7$, 1 if $n \equiv 1, 2,$ or 4 mod 7, and -1 if $n \equiv 3, 5,$ or 6, and because in this case the class number is 1 (2 has the actual factors $\frac{1}{2}(1 \pm \sqrt{-7})$ while 3 and 5 remain prime).

If $D = 5$ then $\varepsilon = \frac{1}{2}(1 + \sqrt{5})$, $E = \varepsilon^2$, $h = 1$ (both 2 and 3 remain prime) and the formula is

$$\frac{2 \log \frac{1}{2}(1 + \sqrt{5})}{\sqrt{5}} = 1 - \frac{1}{2} - \frac{1}{3} + \frac{1}{4} + \frac{1}{6} - \frac{1}{7} - \cdots$$

If $D = 13$ (of course $D = 9$ is not to be considered) then $\varepsilon = \frac{1}{2}(3 + \sqrt{13})$, $E = \varepsilon^2$, $h = 1$ (2, 5, 7, and 11 remain prime and 3 has the actual factorization $3 = (4 - \sqrt{13})(4 + \sqrt{13})$) and the formula is

$$\frac{2 \log \frac{1}{2}(3 + \sqrt{13})}{\sqrt{13}} = \sum_{13 \nmid n} \pm \frac{1}{n}.$$

where the sign is $+$ if $n \equiv \pm 1, \pm 3, \pm 4 \bmod 13$, $-$ if $n \equiv \pm 2, \pm 5, \pm 6$. Of course it almost goes without saying that the conditionally convergent series $\sum \pm (1/n)$ is to be summed in the natural order.

EXERCISES

1. Write out the class number formula in the case $D = -23$.

2. Write out the class number formula in the case $D = 65$.

9.5 Evaluation of $\sum \left(\dfrac{D}{n}\right)\dfrac{1}{n}$

If the class number formula is to be used to compute the class number h then an independent method of evaluating the sum

$$L(1, \chi) = \sum_{n=1}^{\infty} \left(\frac{D}{n}\right)\frac{1}{n}$$

must be found. Here χ is the character mod $4D$ (or mod D if $D \equiv 1 \bmod 4$) defined by $\chi(n_1 n_2) = \chi(n_1)\chi(n_2)$ and by

$$\chi(p) = \begin{cases} 0 & \text{if } p \text{ ramifies} \\ 1 & \text{if } p \text{ splits} \\ -1 & \text{if } p \text{ remains prime} \end{cases}$$

in quadratic integers for the squarefree determinant D. Table 9.5.1 lists the values of $\chi(n)$ for $|D| \leqslant 7$ and $1 \leqslant n \leqslant 28$.

Let $m = |4D|$ if $D \not\equiv 1 \bmod 4$ and let $m = |D|$ if $D \equiv 1 \bmod 4$. Then χ is a character mod m and the arguments of Section 6.5 show that the desired number can be written in the form

$$L(1, \chi) = c_1 \log \frac{1}{1-\alpha} + c_2 \log \frac{1}{1-\alpha^2} + \cdots + c_m \log \frac{1}{1-\alpha^m} \quad (1)$$

where α is a primitive mth root of unity, where $\log (1/(1 - z))$ denotes the function defined by the series $\sum(z^n/n)$ (summed in the natural order) for all complex numbers $|z| \leqslant 1$ except $z = 1$, and where

$$c_j = \frac{1}{m} \left[\chi(1)\alpha^{-j} + \chi(2)\alpha^{-2j} + \cdots + \chi(m)\alpha^{-mj} \right]. \quad (2)$$

Here $c_m = 0$ (the sum of the values of χ is zero because half the classes mod m are splitting classes and half are nonsplitting) and the divergent series $\log(1/(1 - \alpha^m))$ is not in fact present in (1). Because $\log(1/(1 - z))$ can be computed using formula (7) of Section 6.5, these two formulas sum the infinite series $L(1, \chi)$ and reduce its evaluation to a finite calculation.

A further simplification of the evaluation of $L(1, \chi)$ can be made using the following remarkable theorem: *The Fourier transform $(c_1, c_2, \ldots, c_m)$ of the character $(\chi(1), \chi(2), \ldots, \chi(m))$ is a multiple of $(\chi(1), \chi(2), \ldots, \chi(m))$ itself.* That is, if the c's are defined by (2) then there is a number μ such that $c_j = \mu\chi(j)$ for $j = 1, 2, \ldots, m$. This can be proved as follows. If j is relatively prime to m then there is an integer k such that

Table 9.5.1 (Compare to Table 7.8.1.)

D	1	2	3	4	5	6	7	8	9	10	11	12	13	14	15	16	17	18	19	20	21	22	23	24	25	26	27	28
−1	+	0	−	0	+	0	−	0	+	0	−	0	+	0	−	0	+	0	−	0	+	0	−	0	+	0	−	0
−2	+	0	+	0	−	0	−	0	+	0	+	0	−	0	−	0	+	0	+	0	−	0	−	0	+	0	+	0
−3	+	−	0	+	−	0	+	−	0	+	−	0	+	−	0	+	−	0	+	−	0	+	−	0	+	−	0	+
−5	+	0	+	0	0	0	+	0	+	0	−	0	−	0	0	0	−	0	−	0	+	0	+	0	0	0	+	0
−6	+	0	0	0	+	0	+	0	0	0	+	0	−	0	0	0	−	0	−	0	0	0	−	0	+	0	0	0
−7	+	+	−	+	−	−	0	+	+	−	+	−	−	0	+	+	−	+	−	−	0	+	+	−	+	−	−	0
2	+	0	−	0	−	0	+	0	+	0	−	0	−	0	+	0	+	0	−	0	−	0	+	0	+	0	−	0
3	+	0	0	0	−	0	−	0	0	0	+	0	+	0	0	0	−	0	−	0	0	0	+	0	+	0	0	0
5	+	−	−	+	0	+	−	−	+	0	+	−	−	+	0	+	−	−	+	0	+	−	−	+	0	+	−	−
6	+	0	0	0	+	0	−	0	0	0	−	0	−	0	0	0	−	0	+	0	0	0	+	0	+	0	0	0
7	+	0	+	0	−	0	0	0	+	0	−	0	−	0	−	0	−	0	+	0	0	0	−	0	+	0	+	0

$jk \equiv 1 \mod m$ and

$$c_1 = \frac{1}{m}\left[\chi(1)\alpha^{-1} + \chi(2)\alpha^{-2} + \cdots + \chi(m)\alpha^{-m}\right]$$

$$= \frac{1}{m}\left[\chi(jk)\alpha^{-jk} + \chi(2jk)\alpha^{-2jk} + \cdots + \chi(mjk)\alpha^{-mjk}\right]$$

$$= \chi(j)\cdot\frac{1}{m}\left[\chi(k)\alpha^{-jk} + \chi(2k)\alpha^{-2jk} + \cdots + \chi(mk)\alpha^{-jmk}\right]$$

$$= \chi(j)\cdot\frac{1}{m}\left[\chi(1)\alpha^{-j} + \chi(2)\alpha^{-2j} + \cdots + \chi(m)\alpha^{-jm}\right]$$

$$= \chi(j)c_j$$

(because $k, 2k, \ldots, mk$ is a representative set mod m). Since $\chi(j) = \pm 1$ it follows that $c_j = \chi(j)c_1$ and with $\mu = c_1$ it follows that $c_j = \mu\chi(j)$ for j relatively prime to m. Since $\chi(j) = 0$ when j is not relatively prime to m, in order to prove the theorem it is both necessary and sufficient to prove that $c_j = 0$ when j is not relatively prime to m. Let $j = pv$ where p is a prime divisor of m, say $m = pq$. Then

$$c_j = \frac{1}{m}\left[\chi(1)\alpha^{-pv} + \chi(2)\alpha^{-2pv} + \cdots + \chi(m)\alpha^{-mpv}\right].$$

If $rv \equiv sv \mod q$ then $\alpha^{-rpv} = \alpha^{-spv}$ and c_j is a sum of q terms

$$c_j = \frac{1}{m}\left[\left(\sum_{t \equiv 1 \mod q} \chi(t)\right)\alpha^{-pv}\right.$$

$$\left. + \left(\sum_{t \equiv 2} \chi(t)\right)\alpha^{-2pv} + \cdots + \left(\sum_{t \equiv 0} \chi(t)\right)\alpha^{-mpv}\right]$$

and the theorem will follow if it is shown that these sums over classes mod q are all zero; that is, if it is shown that the sum of the values of χ over the p integers t in a congruence class mod q, $0 \leqslant t < m$, is zero. This property of the characters $\chi(j) = (\frac{D}{j})$ is easy to prove from a direct examination of the formulas for $(\frac{D}{j})$ given in Section 7.8 (Exercise 1). This completes the proof of the theorem.

The evaluation of the constant μ is very easy, at least up to sign. One need only observe that

$$\chi(j) = c_1\alpha^j + c_2\alpha^{2j} + \cdots + c_m\alpha^{mj}$$

$$= \mu\left[\chi(1)\alpha^j + \chi(2)\alpha^{2j} + \cdots + \chi(m)\alpha^{mj}\right]$$

$$= \mu\sum_{v=1}^{m} \chi(v)\alpha^{vj}.$$

Substitution of this formula in itself gives

$$\chi(j) = \mu \sum_{\nu=1}^{m} \left[\mu \sum_{\lambda=1}^{m} \chi(\lambda) \alpha^{\lambda\nu} \right] \alpha^{\nu j}$$

$$= \mu^2 \sum_{\lambda=1}^{m} \left[\sum_{\nu=1}^{m} \alpha^{\lambda\nu} \alpha^{j\nu} \right] \chi(\lambda).$$

The inner sum is zero unless $\lambda + j \equiv 0 \bmod m$, in which case it is m. [Every power of α other than 1 is a root of $1 + x + x^2 + \cdots + x^{m-1} = (1 - x^m)/(1 - x)$.] Therefore $\chi(j) = \mu^2 m \chi(-j) = \chi(j) \mu^2 m \chi(-1)$, and

$$\mu = \pm \frac{1}{\sqrt{\chi(-1)m}}.$$

Using these facts about the coefficients c_j in (1) gives

$$L(1, \chi) = \pm \frac{1}{\sqrt{\chi(-1)m}} \sum_{j=1}^{m} \chi(j) \log \frac{1}{1 - \alpha^j}. \tag{3}$$

The ambiguity of the sign presents no problem because the formula

$$L(1, \chi) = \lim_{s \downarrow 1} (s - 1) \prod_{P} \left(1 - \frac{1}{N(P)^s} \right)^{-1}$$

shows immediately that $L(1, \chi) \geqslant 0$. In further reducing the formula for $L(1, \chi)$ it is natural to separate the cases $D < 0$ and $D > 0$.

Case 1, $D < 0$. In this case $\chi(-1) = -1$ and the real parts of the logs in (3) cancel. The imaginary parts can be found using the formula

$$\log \frac{1}{1 - e^{i\theta}} = -\log \left(2 \sin \frac{\theta}{2} \right) + \frac{i}{2} (\pi - \theta)$$

$(0 < \theta < 2\pi)$ of Section 6.5. With $\alpha = e^{2\pi i/m}$ one finds then

$$L(1, \chi) = \pm \frac{1}{i\sqrt{m}} \cdot \frac{i}{2} \cdot \sum_{j=1}^{m} \chi(j) \left(-\frac{2\pi j}{m} \right) + \text{const.} \sum_{j=1}^{m} \chi(j)$$

$$= \pm \frac{\pi}{m\sqrt{m}} \sum \chi(j) j.$$

Since $L(1, \chi) \geqslant 0$ the final formula is therefore

$$L(1, \chi) = \frac{\pi}{m\sqrt{m}} |1 + \chi(2)2 + \chi(3)3 + \cdots + \chi(m)m|. \tag{4}$$

This number is relatively simple to compute for any given $D < 0$.

Examples: When $D = -1$, $m = 4$ and

$$L(1, \chi) = \frac{\pi}{4\sqrt{4}} |1 - 3| = \frac{\pi}{4}$$

which is Leibniz's formula. When $D = -2$, $m = 8$ and

$$L(1, \chi) = \frac{\pi}{8\sqrt{8}} |1 + 3 - 5 - 7| = \frac{\pi}{2\sqrt{2}}$$

which is Newton's formula given in Section 9.2. When $D = -3$, $m = 3$ and

$$L(1, \chi) = \frac{\pi}{3\sqrt{3}} |1 - 2| = \frac{\pi}{3\sqrt{3}}$$

as was noted in the preceding section. When $D = -5$

$$L(1, \chi) = \frac{\pi}{20\sqrt{20}} |1 + 3 + 7 + 9 - 11 - 13 - 17 - 19| = \frac{\pi}{\sqrt{5}},$$

when $D = -7$, $L(1, \chi) = \pi (7\sqrt{7})^{-1} |1 + 2 - 3 + 4 - 5 - 6| = \pi/\sqrt{7}$, and so forth. Some interesting simplifications of this formula are given in the exercises. (See the summary at the end of the exercises.)

Case 2, $D > 0$. In this case $\chi(-1) = 1$ and the imaginary parts of the logs in (3) cancel. The real parts are $-\log 2 - \log (\sin (\theta/2))$ and the log 2 makes no contribution because $\Sigma \chi(j) = 0$. Therefore

$$L(1, \chi) = \pm \frac{1}{\sqrt{m}} \log \left| \frac{\prod \sin \frac{\pi j}{m}}{\prod \sin \frac{\pi j}{m}} \right|$$

where in the numerator j ranges over integers between 0 and m for which $\chi(j) = 1$ and in the denominator j ranges over such integers for which $\chi(j) = -1$. The explicit value of $L(1, \chi)$ in this case is much more difficult to find than in the case $D < 0$.

Examples: When $D = 2$, $m = 8$ and

$$L(1, \chi) = \pm \frac{1}{2\sqrt{2}} \log \frac{\sin \frac{\pi}{8} \sin \frac{7\pi}{8}}{\sin \frac{3\pi}{8} \sin \frac{5\pi}{8}}.$$

This can be evaluated explicitly by setting $x = \sin(\pi/8)$ and $y = \sin(3\pi/8) = \cos(\pi/8)$ to find $x^2 + y^2 = 1$, $-x^2 + y^2 = \cos 2(\pi/8) = \sqrt{2}/2$, $2y^2 = 1 + (\sqrt{2}/2) = (2 + \sqrt{2})/2$, $2x^2 = (2 - \sqrt{2})/2$, $L(1, \chi) = \pm(2\sqrt{2})^{-1} \cdot \log (x^2/y^2) = \pm(2\sqrt{2})^{-1} \log [(2 - \sqrt{2})/(2 + \sqrt{2})] = \log [(2 + \sqrt{2})^2/2]/ 2\sqrt{2} = \log (1 + \sqrt{2})/\sqrt{2}$ as was found in Section 9.3.

Similarly, when $D = 3$ the formula gives

$$L(1, \chi) = \pm \frac{1}{2\sqrt{3}} \log \frac{\sin \frac{\pi}{12} \sin \frac{11\pi}{12}}{\sin \frac{5\pi}{12} \sin \frac{7\pi}{12}}$$

$$= \pm \frac{1}{2\sqrt{3}} \log \frac{2x^2}{2y^2} = \pm \frac{1}{2\sqrt{3}} \log \frac{1 - \frac{\sqrt{3}}{2}}{1 + \frac{\sqrt{3}}{2}}$$

$$= \frac{1}{2\sqrt{3}} \log \frac{2 + \sqrt{3}}{2 - \sqrt{3}} = \frac{1}{2\sqrt{3}} \log \frac{(2 + \sqrt{3})^2}{1} = \frac{\log (2 + \sqrt{3})}{\sqrt{3}}$$

where $x = \sin (\pi/12)$ and $y = \sin (5\pi/12)$.

In the case $D = 5$ it is more convenient to leave the formula in the form (3), after noting that the imaginary part is zero. Then, using a computation from Section 6.5, one finds

$$L(1, \chi) = \pm \frac{1}{\sqrt{5}} \log \frac{(1 - \alpha^2)(1 - \alpha^3)}{(1 - \alpha)(1 - \alpha^4)}$$

$$= \frac{1}{\sqrt{5}} \log \frac{3 + \sqrt{5}}{2} = \frac{2}{\sqrt{5}} \log \frac{1 + \sqrt{5}}{2}.$$

This formula was also derived in Section 6.5 and in Section 9.4.

In the case $D = 7$ neither of these methods leads to a very convenient derivation of the value $L(1, \chi) = \log (8 + 3\sqrt{7})/\sqrt{7}$ that was found in Section 9.3 by evaluating h directly and using the class number formula to find $L(1, \chi)$.

EXERCISES

1. Prove that if D is squarefree, if $m = |D|$ or $|4D|$ according to whether $D \equiv 1$ or $D \not\equiv 1$ mod 4, and if $m = pq$ where p is prime, then the sum of $\left(\frac{D}{t}\right)$ over the p integers $0 \leqslant t < m$ in a congruence class mod q is zero. [Write $\left(\frac{D}{n}\right) = \left(\frac{\sigma}{n}\right)\left(\frac{n}{p_1}\right)\left(\frac{n}{p_2}\right) \cdots \left(\frac{n}{p_\mu}\right)\left(\frac{n}{p_1'}\right)\left(\frac{n}{p_2'}\right) \cdots \left(\frac{n}{p_\nu'}\right)$ where $D = (-1)^\delta 2^\varepsilon p_1 p_2 \cdots p_\mu p_1' p_2' \cdots p_\nu'$, where δ and ε are 0 or 1, where the p's are 1 mod 4, where the p''s are 3 mod 4, where $\sigma = (-1)^{\delta + \nu} 2^\varepsilon \equiv D$ mod 4, where $\left(\frac{\sigma}{n}\right) = 1$ when $\sigma = 1$, and where $\left(\frac{\delta}{n}\right) = 0$ when $\sigma \neq 1$, $n =$ even. If $p|m$ and p is odd then all but one of the factors of $\left(\frac{D}{n}\right)$ have the same value for $n + q$ as for n. (If $2|\sigma$ then $8|q$. If $\sigma = -1$ then $4|q$.) The required sum is therefore a constant times the sum of $\left(\frac{t}{p}\right)$ which is zero. If $p|m$ and $p = 2$, consider two cases. If $\varepsilon = 1$ then q is 4 times an odd number and $\left(\frac{\delta}{n}\right) = -\left(\frac{\sigma}{n+q}\right)$ ($\sigma = \pm 2$) while the other factors of $\left(\frac{D}{n}\right)$ are the same for $n + q$ as for n. If $\varepsilon = 0$ then $D \not\equiv 1$ mod 4, $q \equiv 2$ mod 4 and again the first factor changes sign while the others remain unchanged.]

2. In the case $D = -1$ find the c's and show that the evaluation of $L(1, \chi)$ given in this section amounts to an evaluation of $-i \log (\frac{1}{2}\sqrt{2} + \frac{1}{2}\sqrt{2}\, i)$ and therefore to an evaluation of arctan 1.

3. When $D < 0$ and $D \equiv 1 \bmod 4$ the computation of $L(1, \chi)$ using formula (4) can be simplified as follows. The terms of the sum can be paired so that it becomes a sum over $1, 2, \ldots, \frac{1}{2}(|D| - 1)$ or they can be paired so that it becomes a sum over $2, 4, \ldots, |D| - 1$. This expresses the sum in two ways in terms of $\Sigma = (\frac{D}{1}) + (\frac{D}{2}) + \cdots + (\frac{D}{v})$ and $\Sigma' = (\frac{D}{1}) + (\frac{D}{2})2 + \cdots + (\frac{D}{v})v$ where $v = \frac{1}{2}(|D| - 1)$. These can be combined to eliminate Σ' and *it follows that if* $(\frac{D}{2}) = +1$ *or if* $D = -3$ *then* $h = |\Sigma|$ *whereas if* $(\frac{D}{2}) = -1$ *and* $D \neq -3$ *then* $h = \frac{1}{3}|\Sigma|$. Fill in the proof of this theorem.

4. Use the method of Exercise 3 to evaluate h for $D = -7, -11, -15, -19, -23, -31, -35, -39, -43, -47$. This is a good exercise in developing techniques for evaluating $(\frac{D}{n})$. There are many shortcuts. Rarely, if ever, is it necessary to evaluate Legendre symbols other than the trivial ones $(\frac{D}{2})$, $(\frac{D}{3})$, and $(\frac{D}{5})$.

5. If D is negative, squarefree, and not 1 mod 4, and if $D \neq -1$, show that $h = |\Sigma|$ where Σ is the sum of $(\frac{D}{n})$ over all n in the range $1 \leqslant n < |D|$. [Use $\chi(n + 2D) = -\chi(n)$ and $\chi(4D - n) = -\chi(n)$.] In these cases the evaluations of $(\frac{D}{n})$ are more tedious. Further simplifications are given in the exercises which follow. Evaluate h for $D = -5, -6, -10, -13, -14, -17, -21$ using this formula.

6. Let D be negative, squarefree, and -1 mod 4. The computation of the class number by the method of the preceding exercise requires more—and more difficult—evaluations of Legendre symbols than the following ingenious method of Dirichlet. (See [D7], Art. 106.) Let $D = -P$ where P is positive (not necessarily prime) and 1 mod 4. Assume, moreover, that $P \neq 1$. Then $(\frac{D}{n}) = (\frac{-1}{n})(\frac{n}{P})$ where $(\frac{n}{P})$ is the product of $(\frac{n}{p_i})$ where p_i ranges over the (necessarily distinct) prime factors of P. Define $\Psi(r)$ for integers r by

$$\Psi(r) = \sum_{s=1}^{P-1} \left(\frac{s}{P}\right) \log \frac{1}{1 - i^r\gamma^s}$$

where $\gamma = e^{2\pi i/P}$ and where the log is defined by $\log (1/(1 - z)) = \Sigma z^n/n$ and therefore has imaginary part between $-\frac{1}{2}\pi$ and $\frac{1}{2}\pi$. Formula (3) shows that $L(1, \chi) = \pm im^{-1/2}[0\Psi(0) + 1 \cdot \Psi(1) + 0 \cdot \Psi(2) + (-1)\Psi(3)]$ because every mth root of unity can be written in just one way as $i^r\gamma^s$; $r = 0, 1, 2, 3$; $s = 0, 1, 2, \ldots, P - 1$. Thus $h = \pm i\pi^{-1}[\Psi(1) - \Psi(3)]$. Define $K(r) = \Psi(r) - \Psi(-r)$. Then $h = \pm i\pi^{-1}K(1)$. The identity $K(r) = \Sigma(\frac{s}{P}) \log (1 - i^{-r}\gamma^{-s}/1 - i^r\gamma^s)$ shows not only that $K(r)$ is purely imaginary but also that it is equal to $-\Sigma(\frac{s}{P})(2\pi ik/4P)$ where k is the integer determined by the conditions $0 < k < 4P$, $k \equiv Pr + 4s$. Therefore for $r = 0, 1, 2, 3$ the difference $K(r + 1) - K(r)$ is a sum of terms equal to $-(\frac{s}{P})(2\pi iP/4P)$ unless $Pr + 4s < 4P \leqslant P(r + 1) + 4s$, in which case the term is $-(\frac{s}{P})(2\pi i(P - 4P)/4P)$. (Note the resemblance of this argument to Kummer's argument in Section 6.16.) Therefore $K(r + 1) - K(r) = 2\pi i\Sigma(\frac{s}{P})$ where the sum is over all values of s for which $Pr + 4s < 4P \leqslant P(r + 1) + 4s$. This is a sum over precisely one quarter of the integers $1, 2, \ldots, P - 1$. Specifically, if Q_j for $j = 1, 2, 3, 4$ denotes the sum of $(\frac{s}{P})$ over the jth quarter

$(j-1)(P-1)/4 < s \leqslant j(P-1)/4$ then $K(r+1) - K(r) = 2\pi i Q_{4-r}$. Since $K(3)$ $= -K(1)$ it follows that $K(1) = \frac{1}{2}[K(1) - K(3)] = -\frac{1}{2}[K(3) - K(2) + K(2) - K(1)] = -\frac{1}{2}[2\pi i Q_2 + 2\pi i Q_3]$. Since $Q_1 + Q_2 + Q_3 + Q_4 = 0$ and $Q_1 = Q_4$ it follows that $K(1) = 2\pi i Q_1$. Thus $h = \pm 2Q_1$. Use this formula to find h for $D = -5, -13, -17, -21, -29, -33, -37, -41, -53$.

7. Dirichlet also gave reductions like that of Exercise 6 in the remaining cases where D is squarefree and negative. This exercise is devoted to the case $D \equiv 2 \bmod 8$ and the next exercise is devoted to the sole remaining case $D \equiv -2 \bmod 8$. Let $D \equiv 2 \bmod 8$ and let $D = -2P$ where $P \equiv -1 \bmod 4$ is a product of distinct odd primes. Then $\binom{D}{n} = \binom{2}{n}\binom{n}{P}$ where $\binom{n}{P}$ is defined as before. Let $\beta = e^{2\pi i/8}$, $\gamma = e^{2\pi i/P}$, and

$$\Psi(r) = \sum_{s=1}^{P-1} \left(\frac{s}{P}\right) \log \frac{1}{1 - \beta^r \gamma^s} .$$

Then $h = \pm i\pi^{-1}[\Psi(1) - \Psi(3) - \Psi(5) + \Psi(7)]$. Let $K(r) = \Psi(r) + \Psi(-r)$. Then $K(r) = -\Sigma(\frac{s}{P})(2\pi i k/8P)$ where $k \equiv Pr + 8s \bmod 8P$ and $0 \leqslant k < 8P$. Therefore for $r = 0, 1, \ldots, 7$ one has $K(r+1) - K(r) = 2\pi i C$ where C is the sum of $(\frac{s}{P})$ over all values of s for which $Pr + 8s < 8P \leqslant P(r+1) + 8s$. Let C_j be the sum of $(\frac{s}{P})$ over all s in the jth "octant" $(j-1)/8 < s/P < j/8$. Then $K(r+1) - K(r) = 2\pi i C_{8-r}$. This gives $h = \pm i\pi^{-1}[K(1) - K(3)] = \pm i\pi^{-1}[K(3) - K(2) + K(2) - K(1)] = \pm 2[C_6 + C_7] = \pm 2[C_2 + C_3]$. Use this formula to find h for $D = -6, -14, -22, -30, -38$.

8. Finally, consider $D \equiv -2 \bmod 8$. Let $D = -2P$, $P \equiv 1 \bmod 4$, $P \neq 1$. Define $\Psi(r)$ as before and show $h = \pm i\pi^{-1}[\Psi(1) + \Psi(3) - \Psi(5) - \Psi(7)]$. With $K(r) = \Psi(r) - \Psi(-r)$ one has $K(r+1) - K(r) = 2\pi i C_{8-r}$, where C_j is the sum of $(\frac{s}{P})$ over the jth "octant," and $h = \pm i\pi^{-1}[K(1) + K(3)]$. The calculation of $K(1) + K(3)$ seems to require a further identity, namely, $K(r) + K(r+4) = (\frac{2}{P})K(2r)$. This follows from the analogous identity for Ψ, which in turn follows from the identity $\log(1/(1-\alpha)) + \log(1/(1+\alpha)) = \log(1/(1-\alpha^2))$. (This last identity is a little less obvious than it appears, because of the ambiguity of the imaginary part of the log. Nonetheless, it can be proved on general principles, without calculation.) In the case where $(\frac{2}{P}) = 1$ one has, then, $K(1) + K(5) = K(2)$, $K(3) + K(7) = K(6)$, $K(1) + K(3) = K(2) - K(5) + K(6) - K(7)$, from which $h = \pm 2[C_1 - C_4]$. If $(\frac{2}{P}) = -1$ then $3K(1) + 3K(3) = K(1) - K(2) + K(1) - K(5) + K(3) - K(6) + K(3) - K(7)$, from which again $h = 2|C_1 - C_4|$. Use this formula to compute h in the cases $D = -10, -26, -34, -42$.

9. Anyone who does the calculations of the preceding exercises must be struck by the fact that although the formula that was proved had the form $h = \pm n$ where n was an integer that could be computed—thereby determining h because h is by its definition positive—the computations in fact always yield *positive* integers n so that the formula is, in all cases tested, simply $h = n$. This fact that the ambiguous sign is always $+$ is one aspect of a question to which Gauss devoted great effort. In the first section of his article [G4] on this subject he said, " . . . the problem will be solved when this one sign is determined. However, this study, which at first glance seems a very simple one, leads directly to totally unexpected difficulties, and its pursuit, which has come this far without any

obstacle, absolutely requires further methods." This is not the place to develop Gauss's "further methods" which, with their extensions by Dirichlet and many others, form a very important chapter in number theory.* This exercise is devoted, rather, to the connection between the problem of the sign in the class number formula and the sign whose determination Gauss was studying. Specifically, Gauss proved that if n is a positive integer and $r = e^{2\pi i/n}$ then

$$1 + r + r^4 + r^9 + \cdots + r^{(n-1)^2} = \begin{cases} (1+i)\sqrt{n} & \text{for } n \equiv 0 \bmod 4 \\ \sqrt{n} & \text{for } n \equiv 1 \bmod 4 \\ 0 & \text{for } n \equiv 2 \bmod 4 \\ i\sqrt{n} & \text{for } n \equiv 3 \bmod 4 \end{cases} \tag{5}$$

([G4], Art. 19). It is relatively simple in each case to prove the *square* of this formula—for example, to prove that when $n \equiv 0 \bmod 4$ the square of the sum on the left is $2in$—and the whole problem is to show that the sign of the square root is always the one that is indicated.

(a) Deduce from formula (5) that the formula $\theta_0 - \theta_1 = \pm\sqrt{\pm\lambda}$ of Section 4.5, Exercise 4, (where $D = \pm\lambda \equiv 1 \bmod 4$, λ = prime) can be strengthened to $\theta_0 - \theta_1 = \sqrt{\lambda}$ if λ = prime, $\lambda \equiv 1 \bmod 4$ and $\theta_0 - \theta_1 = i\sqrt{\lambda}$ if λ = prime and $\lambda \equiv 3 \bmod 4$, provided α is taken to be the *particular* λth root of unity $e^{2\pi i/\lambda}$.

(b) The main formula to be proved is $m\bar{\mu} = \sqrt{m}$ if $D > 0$ and $m\bar{\mu} = i\sqrt{D}$ if $D < 0$. Otherwise stated,

$$\alpha + \left(\frac{D}{2}\right)\alpha^2 + \left(\frac{D}{3}\right)\alpha^3 + \cdots + \left(\frac{D}{m}\right)\alpha^m = \begin{cases} \sqrt{m} & \text{if } D > 0 \\ i\sqrt{m} & \text{if } D < 0 \end{cases} \tag{6}$$

where D is squarefree, $m = |D|$ if $D \equiv 1 \bmod 4$, $m = 4|D|$ if $D \not\equiv 1 \bmod 4$, and where $\alpha = e^{2\pi i/m}$. Deduce this from (a) in the case where m is prime.

(c) Prove (6) when $D = -1, -2, 2$.

(d) For the case $|D|$ = prime, $D \equiv -1 \bmod 4$, deduce (6) from (b) and (c). [A sum over all $s \bmod 4p$ can be written as a sum over all $s = 4a + pb$ where a ranges over integers mod p and b ranges over integers mod 4.]

(e) For cases $D = 2p$ (p = prime) deduce (6) from (b) and (c). [When $p \equiv -1 \bmod 4$ this uses $i^2 = -1$.] Deduce (6) in a similar way when $D = -2p$.

(f) Finally, by an argument like that of (e), show that if (6) holds for a particular D and if it holds for m = prime then it holds for Dp where p is an odd prime which does not divide D.

(g) Show that if $D < 0$ and if Gauss's formula (5) is given (in fact it suffices to know (5) for *prime* n) then the class number is equal to $-[\binom{D}{1} + \cdots + \binom{D}{m}m]/m$.

(h) Show that if (5) is given then the class number for $D < 0$ is given by the formulas in the summary below.

(i) If p is prime and $p \equiv -1 \bmod 4$ then the first half of the classes mod p, namely, $1, 2, \ldots, \frac{1}{2}(p-1)$, contains more squares than nonsquares mod p.

*See Gauss's original article on the subject and Smith's summary of it ([S3], Art. 20). See also Dirichlet's entirely different proof ([D6] or [D7] Arts. 111–114). A very interesting and simple proof was given by Mordell [M1] using contour integration. Almost any book on advanced number theory will contain some proof of the facts that are used here. For a recent source with rather complete references see [W4].

If $p \equiv 1$ mod 4 then the first half contains an equal number of squares and nonsquares (namely, $\frac{1}{4}(p-1)$ of each) but the first quarter $1, 2, \ldots, \frac{1}{4}(p-1)$ of the classes contains more squares than nonsquares. Deduce these statements.

(j) Conversely, show that (i) implies the validity of (5) for prime n. Thus a proof of (i) implies immediately a proof of (5) for prime n and conversely.

SUMMARY

If P is a positive squarefree number which is odd and greater than 1 and if C_j denotes the sum of $(\frac{s}{P}) = \amalg(\frac{s}{p_i})$ (where $P = p_1 p_2 \ldots p_k$ is the prime factorization of P) over all s in the range $(j-1)/8 < s/P < j/8$ (the jth "octant") then the class numbers h corresponding to the following D's are:

If $P \equiv 3$ mod 4, $D = -P$ then $h = (C_1 + C_2 + C_3 + C_4)/(2 - (\frac{D}{2}))$, except in the case $D = -3$, where $h = 1$.
If $P \equiv 1$ mod 4, $D = -P$ then $h = 2(C_1 + C_2)$.
If $P \equiv 3$ mod 4, $D = -2P$ then $h = 2(C_2 + C_3)$.
If $P \equiv 1$ mod 4, $D = -2P$ then $h = 2(C_1 - C_4)$.

If $D = -1$ or -2 then $h = 1$. As will be shown in the next section (Exercise 3), the denominator $2 - (\frac{D}{2})$ in the case $D \equiv 1$ mod 4 disappears if one considers the order $\{x + y\sqrt{D} : x, y \text{ integers}\}$, as Dirichlet did, rather than the full order of quadratic integers for the determinant D.

9.6   Suborders

The formulas of Sections 9.2–9.4 cover all the divisor class groups of Chapter 7—that is, the divisor class groups corresponding to the full order of *all* quadratic integers for various squarefree determinants D. It will be shown in this section that for the more general divisor class groups of Chapter 8 the class number h is a multiple $h = mh_0$ of the class number h_0 of the full order and that the multiplier m is simple to compute.

Recall that for any given squarefree determinant D and for every positive integer* t there is a unique suborder of index t in the order of all quadratic integers for the determinant D. (Specifically, this suborder is $\{x + yt\omega : x, y \text{ integers}\}$ where ω is $\sqrt{D}$ if $D \not\equiv 1$ mod 4 and ω is $\frac{1}{2}(1 - \sqrt{D})$ if $D \equiv 1$ mod 4.) The corresponding class group is formed by considering all divisors relatively prime to t and considering A to be equivalent to B if $A\bar{B}$ is the divisor of an element of the suborder. Let h be the number of equivalence classes and let h_0 be the number of equivalence classes in the case $t = 1$. The objective is to show that $h = mh_0$ where m is an integer which can easily be computed.

*This integer was denoted by s in Chapter 8. In this section s will be used to denote the variable which approaches 1.

As is to be expected from experience with the cases where $t = 1$, the formula for h comes out of the observation that

$$h \lim_{s \downarrow 1} (s-1) \sum_{\substack{A \approx I \\ (N(A),\, t)=1 \\ N(A)>0}} N(A)^{-s} = \lim_{s \downarrow 1} (s-1) \sum_{\substack{(N(A),\, t)=1 \\ N(A)>0}} N(A)^{-s}$$

$$= \lim_{s \downarrow 1} (s-1) \prod_{(N(P),\, t)=1} \left(1 - N(P)^{-s}\right)^{-1}$$

where $A \approx B$ means that $A\bar{B}$ is the divisor of a quadratic integer of the suborder ($A \sim B$ will mean that $A\bar{B}$ is the divisor of a quadratic integer in the full order, so that $A \approx B$ implies $A \sim B$) and where $(N(A), t) = 1$ means that the integers $N(A)$ and t are relatively prime. The last of these three numbers can also be written in the form

$$\lim_{s \downarrow 1} (s-1) \prod_{(N(P),\, t) \neq 1} \left(1 - N(P)^{-s}\right) \prod_{P} \left(1 - N(P)^{-s}\right)^{-1}$$

$$= \prod_{(N(P),\, t) \neq 1} \left(1 - N(P)^{-1}\right) \lim_{s \downarrow 1} (s-1) \sum_{N(A)>0} N(A)^{-s}$$

$$= \prod_{(N(P),\, t) \neq 1} \left(1 - N(P)^{-1}\right) h_0 \lim_{s \downarrow 1} (s-1) \sum_{\substack{A \sim I \\ N(A)>0}} N(A)^{-s}$$

because the limit of the finite product over prime divisors P not relatively prime to t can be taken separately. Note that since $A \approx I$ implies $A \sim I$ the sum on the left side of the equation

$$h \lim_{s \downarrow 1} (s-1) \sum_{\substack{A \approx I \\ (N(A),\, t)=1 \\ N(A)>0}} N(A)^{-s}$$

$$= h_0 \prod_{(N(P),\, t) \neq 1} \left(1 - N(P)^{-1}\right) \lim_{s \downarrow 1} (s-1) \sum_{\substack{A \sim I \\ N(A)>0}} N(A)^{-s} \qquad (1)$$

is the sum that one obtains by deleting all terms in the sum on the right for which $A \not\approx I$ or $(N(A), t) \neq 1$. The problem is, in essence, to compute the ratio of these two sums.

Case 1, $D < 0$. Assume first that D is not -1 or -3 so that the only units are ± 1. Then every $A \sim I$ is the divisor of just two quadratic integers and every $A \approx I$ is the divisor of just two quadratic integers in the suborder. Thus the problem is to find the ratio, in the limit as $s \downarrow 1$, of the sum $\sum N(x + y\sqrt{D})^{-s}$ over all quadratic integers to the sum $\sum N(x + y\sqrt{D})^{-s}$ over just those quadratic integers which lie in the suborder and are relatively prime to t. Let points of the xy-plane with integer coordinates correspond to quadratic integers by $(x, y) \leftrightarrow x + y\sqrt{D}$ when $D \not\equiv 1 \bmod 4$ and by $(x, y) \leftrightarrow x + y\frac{1}{2}(1 - \sqrt{D})$ when $D \equiv 1 \bmod 4$. Let the points of the xy-plane be subdivided into squares t on a side. Then each square corresponds to t^2 terms of the larger sum, and these t^2 terms are all of the same order of magnitude. Since the answer to the questions of whether $x + y\omega$ ($\omega = \sqrt{D}$ or $\frac{1}{2}(1 - \sqrt{D})$) is in the suborder of index t and whether it is relatively prime to t depends only on the classes of x and $y \bmod t$, it is clear that the number ν of these t^2 terms which correspond to terms of the smaller sum is the same for all squares t on a side. In fact, each square t on a side contains exactly t quadratic integers of the suborder (y is fixed and x ranges over t values) of which $\phi(t)$ are relatively prime to t, where $\phi(t)$ denotes the number of positive integers less than t relatively prime to t (an element $x + y\omega$ which is in the suborder has the form $x + \nu t\omega$ and is relatively prime to t if and only if x is relatively prime to t); thus $\nu = \phi(t)$. It is plausible intuitively and not difficult to prove rigorously (Exercise 5) that, since each block of t^2 terms of the larger sum contains $\phi(t)$ terms of the smaller sum and since these terms are all of the same order of magnitude, the ratio of the larger sum to the smaller is, in the limit as $s \downarrow 1$, the ratio of t^2 to $\phi(t)$. That is,

$$h = h_0 \prod_{(N(P),\, t) \neq 1} (1 - N(P)^{-1}) \frac{t^2}{\phi(t)} .$$

For example, for the quadratic integers $\{x + y\sqrt{-163} : x, y \text{ integers}\}$ one had $h_0 = 1$ (see Section 7.6), $t = 2$, there is just one prime divisor not relatively prime to t, namely the prime 2 with norm 4, and $\phi(t) = \phi(2) = 1$. Thus

$$h = 1 \cdot \left(1 - \frac{1}{4}\right) \frac{2^2}{1} = 3$$

as was seen in Section 8.5. For the quadratic integers $\{x + y\sqrt{-63} : x, y \text{ integers}\}$ one had $D = -7$, $t = 6$, and

$$h = 1 \cdot \left(1 - \frac{1}{2}\right)\left(1 - \frac{1}{2}\right)\left(1 - \frac{1}{9}\right) \cdot \frac{6^2}{\phi(6)}$$

$$= \frac{1}{4} \cdot \frac{8}{9} \cdot \frac{36}{2} = 4.$$

In fact, the techniques of Chapter 8 easily show that in this case the 4 divisors I, $(11, 2)$, $(11, 2)^2$, $(11, 2)^3$ are a representative set for the divisor class group.

This formula can be further simplified using the well-known and easily proved formula

$$\frac{\phi(t)}{t} = \prod_{p \mid t} \left(1 - \frac{1}{p}\right)$$

(Exercise 4). Since

$$\left(1 - N(P)^{-1}\right) = \left(1 - \frac{1}{p}\right)\left(1 - \left(\frac{D}{p}\right)\frac{1}{p}\right)$$

this shows that, in the case $D < 0$, $D \neq -1$, $D \neq -3$ considered above,

$$h = h_0 t \prod_{p \mid t}\left(1 - \left(\frac{D}{p}\right)\frac{1}{p}\right). \tag{2}$$

If $D = -1$ then h is half the number given by this formula because then the sum of $N(A)^{-s}$ over all principal divisors is $\frac{1}{4}$ of the sum of $N(x + y\sqrt{D})^{-s}$ over all nonzero quadratic integers rather than equal to $\frac{1}{2}$ this sum, whereas the sum of $N(A)^{-s}$ over all divisors $A \approx I$, $(N(A), t) = 1$, is still $\frac{1}{2}$ the sum over all $x + y\sqrt{D}$ in the suborder relatively prime to t because the suborder has just the two units ± 1 (the units $\pm\sqrt{-1}$ are in no proper suborder). Similarly, if $D = -3$ then h is one-third of the number given by the formula (2)

If $D > 0$ then easy modifications of the above arguments (Exercise 6) show that

$$h = h_0 \cdot \frac{1}{k} \cdot t \prod_{p \mid t}\left(1 - \left(\frac{D}{p}\right)\frac{1}{p}\right) \tag{3}$$

where h_0 is the class number of the full order of all quadratic integers for the determinant D, where h is the class number of the suborder of index t, and where k is the integer defined by $E = E_0^k$; here $E_0 = \varepsilon$ or ε^2 is the fundamental unit with norm 1 in the full order and E is the fundamental unit of norm 1 in the suborder of index t.

For example, when $D = 5$ and $t = 2$ one has $\varepsilon = \frac{1}{2}(1 + \sqrt{5})$, $E_0 = \varepsilon^2 = \frac{1}{2}(3 + \sqrt{5})$, $E_0^2 = \frac{1}{2}(7 + 3\sqrt{5})$, $E_0^3 = 9 + 4\sqrt{5} = E$. Therefore

$$h = 1 \cdot \frac{1}{3} \cdot 2\left(1 + \frac{1}{2}\right) = 1.$$

When $D = 11$ and $t = 3$ one has $\varepsilon = 10 + 3\sqrt{11} = E_0 = E$ and

$$h = h_0 \cdot \frac{1}{1} \cdot 3\left(1 + \frac{1}{3}\right) = 4h_0.$$

Since $h_0 = 2$ it follows, then, that $h = 8$. This can be verified by direct computation (Exercise 1).

EXERCISES

1. Find a representative set for the divisor class group of $\{x + y\sqrt{99} : x, y$ integers$\}$ and show, in particular, that the class number is 8.

2. For the order $\{x + y\sqrt{117} : x, y$ integers$\}$ find h both by using (3) and by finding the divisor class group explicitly.

3. Show that the denominator in the first case of the formula in the exercises of the preceding section (see Summary) disappears when one considers the order $\{x + y\sqrt{D} : x, y$ integers$\}$ rather than the full order. In other words, if $D < 0$, D squarefree, and $D \equiv 1 \bmod 4$ then the class number of this order is the number of integers n in the set $1, 2, \ldots, \frac{1}{2}(|D| - 1)$ for which $(\frac{D}{n}) = +1$ minus the number for which $(\frac{D}{n}) = -1$. This was first proved by Dirichlet. In the case where in addition $|D| =$ prime it had been conjectured by Jacobi. (See [D7], §104, for references.)

4. Prove the formula

$$\frac{\phi(t)}{t} = \prod_{p|t}\left(1 - \frac{1}{p}\right).$$

   [If t is prime it is obvious. If t is a power of a prime it is almost as obvious. If $t = uv$ where u and v are relatively prime then its truth for t follows from its truth for u and v by the Chinese remainder theorem. These observations suffice to prove the formula.]

5. Prove that in the limit as $s \downarrow 1$ the ratio of the large sum to the small sum is, as stated in the text, $t^2/\phi(t)$. [Let $\Sigma_0(s)$ denote the large sum and $\Sigma_1(s)$ the small sum. Let $\Sigma_2(s)$ denote the sum of $N(x + y\sqrt{D})^{-s}$ over the midpoints of all the squares t on a side into which the quadratic integers were divided in the argument of the text. Show that $\Sigma_0(s) - t^2\Sigma_2(s)$ and $\Sigma_1(s) - \phi(t)\Sigma_2(s)$ remain bounded as $s \downarrow 1$ but that $\Sigma_2(s)$ is unbounded as $s \downarrow 1$. The desired conclusion then follows.]

6. Prove formula (3). [Use the technique of Section 9.3 for writing a sum over all principal divisors as a sum over a subset of the quadratic integers.]

9.7 Primes in arithmetic progressions

It would seem foolish to come this far in the study of Dirichlet's work without giving at least a sketch of his famous theorem on primes in arithmetic progressions, despite the fact that this theorem has no connection at all with Fermat's Last Theorem and practically none with the divisor theory of quadratic integers.

The theorem states that *if a is any integer and if b is any integer relatively prime to a then there are infinitely many primes p which satisfy the congruence $p \equiv b \bmod a$,* that is, which lie in the arithmetic progression $ax + b$ ($x =$ integer). (Of course if b is not relatively prime to a then $p \equiv b \bmod a$ has at most one solution.)

The proof of this theorem begins with the generalization of the Euler product formula ($\Sigma n^{-s} = \prod(1 - p^{-s})^{-1}$) that was already used in Section 9.2, namely, the formula

$$\sum_{n=1}^{\infty} \frac{\chi(n)}{n^s} = \prod_{p \text{ prime}} \frac{1}{1 - \dfrac{\chi(p)}{p^s}}. \tag{1}$$

365

This formula is a formal identity for any function $\chi(n)$ which satisfies $\chi(mn) = \chi(m)\chi(n)$. (Expand the factors $(1 - \chi(p)p^{-s})^{-1}$ on the right as infinite series $\Sigma\chi(p^k)p^{-ks}$ and multiply them formally.) If χ is such a function from positive integers to real numbers—or even to complex numbers—which is bounded, then both sides of (1) are convergent for $s > 1$ and the equation is literally true.

The functions $\chi(n) = \left(\frac{D}{n}\right)$ considered above are only a particular type of function χ satisfying the two conditions $\chi(mn) = \chi(m)\chi(n)$ and $\chi =$ bounded. In studying primes in arithmetic progressions mod a it is natural to consider such functions χ which satisfy two further conditions, $\chi(n + a) = \chi(n)$ for all n, and $\chi(n) = 0$ for all n not relatively prime to a. A function χ which satisfies all these conditions and which is not identically zero is called a *character* mod a. It is a simple exercise to construct all such characters χ for a given a. For example, if $a = 20$ then the characters χ can be found as follows. Since $\chi(n) = \chi(1 \cdot n) = \chi(1)\chi(n)$ for all n, $\chi(1) \neq 1$ would imply $\chi(n) = 0$ for all n. Therefore $\chi(1) = 1$. All values of χ will be known once $\chi(3)$, $\chi(7)$, $\chi(9)$, $\chi(11)$, $\chi(13)$, $\chi(17)$, and $\chi(19)$ are known. Since $19^2 \equiv (-1)^2 \equiv 1 \bmod 20$, $\chi(19)^2 = \chi(19^2) = \chi(1) = 1$. Therefore $\chi(19) = \pm 1$. Once $\chi(19)$ is known it will suffice to determine $\chi(3)$, $\chi(7)$, and $\chi(9)$ because $\chi(11) = \chi(-9) = \chi(19)\chi(9)$, $\chi(13) = \chi(19)\chi(7)$, and $\chi(17) = \chi(19)\chi(3)$. Clearly $\chi(9) = \pm 1$. Moreover, $\chi(9) = \chi(3)^2$. Therefore $\chi(3) = 1$, -1, i, or $-i$. Once $\chi(3)$ is known then not only is $\chi(9) = \chi(3)^2$ determined but so is $\chi(7) = \chi(3)^3$. This shows that $\chi(3)$ has one of four values, $\chi(19)$ has one of two values, and these two values $\chi(3)$ and $\chi(19)$ suffice to determine all the others. It is easy to check (see Exercise 2) that all $4 \times 2 = 8$ possible choices of $\chi(3)$ and $\chi(19)$ lead to characters. Therefore the eight characters χ given in Table 9.7.1 are the only characters mod 20.

The particular character defined by '$\chi(n) = 1$ if n is relatively prime to a' is called the *principal character* mod a and is denoted χ_0. For this character the function (1) takes the form

$$\prod_{p \nmid a} \frac{1}{1 - p^{-s}} = \prod_{p \mid a} (1 - p^{-s})\zeta(s).$$

In particular, as $s \downarrow 1$ this function approaches $+\infty$ because $\zeta(s)$ approaches $+\infty$ and the finite product in front of $\zeta(s)$ approaches a positive constant. (As was seen in the preceding section, this constant is $\phi(a)/a$.)

Let $L(s, \chi)$ denote the function defined by (1) for $s > 1$ and for χ as above. Then it was just seen that $\lim L(s, \chi_0) = +\infty$ as $s \downarrow 1$. By contrast, if χ is any character mod a *other* than the principal character χ_0 then not only is $\lim L(s, \chi)$ not infinite but in fact the series on the left side of (1) converges conditionally for $s > 0$ and defines a continuous function of s for $s > 0$. This fact follows as before from summation by parts (see Section

Table 9.7.1

	$\chi(1)$	$\chi(3)$	$\chi(7)$	$\chi(9)$	$\chi(11)$	$\chi(13)$	$\chi(17)$	$\chi(19)$
χ_0	1	1	1	1	1	1	1	1
χ_1	1	i	$-i$	-1	-1	$-i$	i	1
χ_2	1	-1	-1	1	1	-1	-1	1
χ_3	1	$-i$	i	-1	-1	i	$-i$	1
χ_4	1	1	1	1	-1	-1	-1	-1
χ_5	1	i	$-i$	-1	1	i	$-i$	-1
χ_6	1	-1	-1	1	-1	1	1	-1
χ_7	1	$-i$	i	-1	1	$-i$	i	-1

6.5) once it is shown that $\sum_{j=1}^{a}\chi(j) = 0$. This follows from the identity

$$\chi(k)\sum_{j=1}^{a}\chi(j) = \sum_{j=1}^{a}\chi(kj) = \sum_{j'=1}^{a}\chi(j')$$

(k relatively prime to a) because this shows that if $\sum\chi(j) \neq 0$ then $\chi(k) = 1$ for all k relatively prime to a.

Thus there is a finite collection of these functions $L(s, \chi)$, one of which approaches ∞ as $s\downarrow 1$ and the remainder of which approach finite limits.

Now consider the logarithms of these functions. By the expansions on the right side of (1) one has, for $s > 1$,

$$\log L(s, \chi) = \sum_{p} \log \frac{1}{1 - \chi(p)p^{-s}}$$

$$= \sum_{p}\left[\sum_{m}\frac{1}{m}\left(\chi(p)p^{-s}\right)^{m}\right]$$

$$= \sum_{p}\sum_{m}\frac{1}{m}\chi(p^{m})p^{-ms}$$

where the ambiguous imaginary part of $\log L(s, \chi)$ is defined by this equation. The terms of this series in which $m \geqslant 2$ converge for $s > \frac{1}{2}$ and approach a finite limit as $s\downarrow 1$. Therefore when one ignores them one finds

$$\log L(s, \chi) \sim \sum_{p}\frac{\chi(p)}{p^{s}} \tag{2}$$

where the sign $\sim$ means that the difference between the two sides remains bounded as $s\downarrow 1$.

In order to prove Dirichlet's theorem it will suffice to prove that

$$\lim_{s\downarrow 1}\sum_{\substack{p \equiv b \\ \bmod a}}\frac{1}{p^{s}} \tag{3}$$

is not finite, because if the series contained only a finite number of terms then its limit would be a finite sum $\Sigma(1/p)$. Now the fact is that the sum (3) is, for each b relatively prime to a, a combination of the sums on the right side of (2). For example, if $a = 20$ and $b = 7$ then $\chi_0 + i\chi_1 - \chi_2 - i\chi_3 + \chi_4 + i\chi_5 - \chi_6 - i\chi_7$ is a function f which satisfies $f(n + 20) = f(n)$, $f(n) = 0$ when n is not relatively prime to 20, $f(7) = 8$, and, as can be checked directly, $f(1) = f(3) = f(9) = f(11) = f(13) = f(17) = f(19) = 0$. More generally, a simple trick shows that for any k relatively prime to 20, $\Sigma \chi_i(k)^{-1}\chi_i = f$ is the function $f(k) = 8$, $f(j) = 0$ for $j \not\equiv k$ mod 20 (Exercise 7). Therefore

$$8 \sum_{\substack{p \equiv 7 \\ \text{mod } 20}} \frac{1}{p^s} = \sum_p \left[\sum_{\nu=1}^{8} \chi_\nu(7)^{-1}\chi_\nu(p) \right] \frac{1}{p^s}$$

$$= \sum_{\nu=1}^{8} \chi_\nu(7)^{-1} \left[\sum_p \frac{\chi_\nu(p)}{p^s} \right]$$

$$\sim \sum_{\nu=1}^{8} \chi_\nu(7)^{-1} \log L(s, \chi_\nu)$$

where again $\sim$ means that the difference between the two sides is bounded as $s \downarrow 1$.

Since $\log L(s, \chi_0)$ is real and approaches ∞ as $s \downarrow 1$, in order to prove that the limit (3) (when $a = 20$, $b = 7$) approaches ∞ as $s \downarrow 1$ it will suffice to prove that for $\nu \neq 0$, Re $\log L(s, \chi_\nu)$ remains bounded as $s \downarrow 1$, that is, that Re $\log L(s, \chi_\nu) \sim 0$ for $\nu \neq 0$. Since $L(s, \chi_\nu)$ approaches $L(1, \chi_\nu)$ as $s \downarrow 1$, it will suffice to prove that $L(1, \chi_\nu) \neq 0$ because this implies, by the fact that Re $\log$ is a continuous function defined for all nonzero complex numbers, that Re $\log L(s, \chi_\nu)$ approaches the finite limit Re $\log L(1, \chi_\nu)$ and in particular that it is bounded.

In exactly the same way, one can show that

$$\phi(a) \sum_{\substack{p \equiv b \\ \text{mod } a}} \frac{1}{p^s} \sim \sum_\chi \chi(b)^{-1} \log L(s, \chi)$$

where $\phi(a)$ is the number of characters (which, as can be shown, is also the number of integers less than a relatively prime to a) and where the sum on the right is over all $\phi(a)$ characters. If one can show that $L(1, \chi) \neq 0$ for $\chi \neq \chi_0$ it will follow that

$$\text{Re} \log L(s, \chi) \sim 0$$

for $\chi \neq \chi_0$ and therefore, because $\chi_0(b) = 1$, that

$$\sum_{\substack{p \equiv b \\ \text{mod } a}} \frac{1}{p^s} \sim \frac{1}{\phi(a)} \log L(s, \chi_0).$$

368

This will prove not only that there are infinitely many primes in the arithmetic progression $ax + b$ but also that they are rather evenly distributed among the $\phi(a)$ possible progressions mod a inasmuch as the difference between the sums (3) for two different values of b remains bounded as $s\downarrow 1$ (because both differ by bounded amounts from $\phi(a)^{-1}$ $\log L(s, \chi_0) \sim \phi(a)^{-1} \log \zeta(s) \sim -\phi(a)^{-1} \log(s-1)$).

In short, *in order to prove Dirichlet's theorem it will suffice to prove that* $L(1, \chi) \neq 0$ *for any nonprincipal character* χ mod a. In the case $a = 20$ this can be done as follows. Summation of (2) over all $\phi(a)$ characters gives

$$\sum \log L(s, \chi) \sim \phi(a) \sum_{\substack{p \equiv 1 \\ \bmod a}} p^{-s}.$$

The sum on the right, which is real-valued, is bounded below—in fact it is nonnegative—as $s\downarrow 1$. Therefore $\prod L(s, \chi)$ is bounded away from zero as $s\downarrow 1$. Exactly as in Section 6.7, it follows from this that *at most one* of the functions $L(s, \chi)$ can be zero at $s = 1$, ($L(s, \chi_0)(s-1)$ remains bounded as $s\downarrow 1$) and therefore that $L(1, \bar{\chi})$ cannot be zero unless χ coincides with $\bar{\chi}$. Therefore it will suffice to prove $L(1, \chi) \neq 0$ for *real* nonprincipal characters χ—that is, for the characters χ_2, χ_4, and χ_6 in the example at hand.

The fact that $L(1, \chi_4) \neq 0$ follows immediately from the class number formula of Section 9.2 in the case $D = -5$ because χ_4 is precisely the character which occurs in that formula. The class number formula in the case $D = 5$ shows that the number

$$1 - \frac{1}{2} - \frac{1}{3} + \frac{1}{4} + \frac{1}{6} - \frac{1}{7} - \frac{1}{8} + \frac{1}{9} + \cdots$$

$$= \prod_p \frac{1}{1 - \left(\dfrac{D}{p}\right)\dfrac{1}{p}} = \frac{1}{1 - \left(\dfrac{D}{2}\right)\dfrac{1}{2}} \prod_{p \neq 2} \frac{1}{1 - \left(\dfrac{D}{p}\right)\dfrac{1}{p}} = \frac{2}{3} L(1, \chi_2)$$

is nonzero. ($\chi_2(n)$ is zero for even n and coincides with $\left(\frac{D}{n}\right)$ for odd n, where $D = 5$.) Therefore $L(1, \chi_2) \neq 0$. Finally $\chi_6(p) = \left(\frac{-1}{p}\right)$ except that $\chi_6(5) = 0$ and $\left(\frac{-1}{5}\right) = 1$. Therefore

$$L(1, \chi_6) = \prod_p \frac{1}{1 - \chi_6(p)p^{-1}} = \left(1 - \left(\frac{-1}{5}\right)5^{-1}\right)\prod_p \frac{1}{1 - \left(\dfrac{-1}{p}\right)\dfrac{1}{p}}$$

$$= \frac{4}{5}\left(1 - \frac{1}{3} + \frac{1}{5} - \frac{1}{7} + \cdots\right)$$

$$= \frac{\pi}{5} \neq 0.$$

This completes the proof that

$$\sum_{\substack{p \equiv b \\ \bmod 20}} \frac{1}{p^s} \sim \frac{1}{8} \log(s-1)$$

for all 8 possible values of b mod 20.

In the general case it is a relatively simple algebra problem (Exercise 6) to prove that a real character χ mod a is essentially the same (except for a few primes) as a character $\left(\frac{D}{n}\right)$ which occurs in a class number formula; since the class number is not zero, it then follows that $L(1, \chi) \neq 0$ for any such χ and therefore, as was seen above, that

$$\sum_{\substack{p \equiv b \\ \bmod a}} \frac{1}{p^s} \sim \frac{1}{\phi(a)} \log \frac{1}{s-1}$$

for all $\phi(a)$ possible values of b mod a, where $\sim$ means that the difference between the two sides remains bounded as $s{\downarrow}1$. In particular, the number of terms on the left is infinite and Dirichlet's theorem follows.

EXERCISES

1. Prove that if n is relatively prime to a then there is a positive integer j such that $n^j \equiv 1$ mod a. Prove that the nonzero values of characters mod a must all be $\phi(a)$th roots of unity.

2. Show that if $a = \alpha\beta$ where α and β are relatively prime integers then any character mod a is the product of a uniquely determined character mod α and a uniquely determined character mod β. Moreover, the product is real only if the factors are real. This shows that in order to find all characters, or all real characters mod a it suffices to be able to solve the problem in the particular cases where a is a prime power. [Define χ_α by $\chi_\alpha(k) = \chi(s)$ when $s \equiv k$ mod α and $s \equiv 1$ mod β.]

3. Show that if $a = p^m$ where p is a prime other than 2 then there are $\phi(a) = a(p - 1)/p$ characters mod a. Of these, exactly two—the principal character and one other—are real-valued. The one other is the Legendre symbol $\chi(n) = \left(\frac{n}{p}\right)$. Use the class number formula to prove for this χ that $L(1, \chi) \neq 0$. [If $m = 1$ let γ be a primitive root mod p. Then χ is completely determined by $\chi(\gamma)$, and χ is real only if $\chi(\gamma) = \pm 1$. Therefore this case is easy. The case $m > 1$ is similarly easy if one can show that there is an integer whose powers mod p^m include all $(p - 1)p^{m-1}$ classes mod p^m relatively prime to p^m. Next consider the case $m = 2$. Let γ be a primitive root mod p. Then γ is a primitive root mod p^2 if and only if $\gamma^{p-1} \not\equiv 1$ mod p^2 because the least power of γ which is 1 mod p^2 must divide $\phi(p^2) = (p - 1)p$. If γ is any primitive root mod p and if $\gamma^{p-1} \equiv 1$ mod p^2 then $(\gamma + p)^{p-1} \not\equiv 1$ mod p^2. Therefore there is always a primitive root mod p^2, even if $p = 2$. For such a primitive root $\gamma^{p-1} = 1 + hp$ where $h \not\equiv 0$ mod p. Then $\gamma^{(p-1)p} \equiv 1 + hp^2$ mod p^3 provided $p \neq 2$. It follows that γ is then a primitive root mod p^3 and $\gamma^{(p-1)pp} \equiv 1 + hp^3$ mod 4. The continuation of this process shows that γ is a primitive root mod p^m for all $m \geqslant 2$ if and only if it is a primitive root mod p^2. Therefore there exist primitive roots mod p^m for all m.]

4. If $a = 2$ then there is only the principal character. If $a = 4$ then there are two characters, the principal character and $\chi(n) = \left(\frac{-1}{n}\right)$. If $a = 2^m$ for $m > 2$ then there are 2^{m-1} characters of which 4 are real, namely, the principal character and the characters $\chi(n) = \left(\frac{-1}{n}\right)$, $\chi(n) = \left(\frac{2}{n}\right)$, $\chi(n) = \left(\frac{-2}{n}\right)$. Prove that $L(1, \chi) \neq 0$

for each of the 3 nonprincipal real characters χ. [From $5 \equiv 1 + 4$ mod 8, $5^2 \equiv 1 + 8$ mod 16, $5^4 \equiv 1 + 16$ mod 32, $5^8 \equiv 1 + 32$ mod 64, $5^{16} \equiv 1 + 64$ mod 128, ... it is clear that mod 2^m the order of 5—that is, the least power of 5 that is 1 mod 2^m—is not divisible by 2^{m-3}. Thus the order of 5 is 2^{m-2} or 2^{m-1}. If its order were 2^{m-1} then every odd integer would be a power of 5 mod 2^m. This is impossible because 3 is not a power of 5 mod 8. Show that every integer is congruent to just one of the integers $3^\varepsilon 5^j$ for $\varepsilon = 0$ or 1 and $j = 0, 1, 2, \ldots$, $2^{m-2} - 1$. This gives 2^{m-1} characters of which 4 are real.]

5. Show that the number of real characters mod a can be found as follows. Let $a = 2^\mu a'$ where a' is odd and divisible by ν distinct primes. If $\mu = 0$ or 1 then the number of real characters is 2^ν, if $\mu = 2$ then it is $2 \cdot 2^\nu$, and if $\mu \geqslant 3$ then it is $4 \cdot 2^\nu$.

6. Show that for each of the real characters found in Exercise 5 there is an integer D such that the character coincides with $\chi_D(n) = \left(\frac{D}{n}\right)$ except that there may be a finite number of primes p such that the character is zero for multiples of p whereas χ_D is not. Conclude that $L(1, \chi)$ is equal to $L(1, \chi_D)$ times a finite number of nonzero factors. Since $L(1, \chi_D) \neq 0$ by the class number formula, this proves that $L(1, \chi) \neq 0$ for all real nonprincipal characters χ mod a.

7. Show that if k is relatively prime to a then the sum of $\chi(k)^{-1}\chi(j)$ over all characters χ is 0 if $j \not\equiv k$ mod a. (Of course it is $\phi(a)$ if $j \equiv k$ mod a.) [It will suffice to prove that the sum of $\chi(j)$ over all characters χ is zero whenever $j \not\equiv 1$ mod a. Since the product of two characters is a character, the sum of $\chi(j)$ over all χ is the same as the sum $\chi'(j)\chi(j)$ over all χ when χ' is any particular character. If this sum is not zero then $\chi'(j)$ must be 1 for all characters χ'. It suffices to prove, then, that if $j \not\equiv 1$ mod a then there is a character χ' mod a for which $\chi'(j) \neq 1$. By Exercise 2 it suffices to prove this in the case where a is a prime power. If $a = p^m$ and $p \neq 2$ define χ' by $\chi'(\gamma) = e^{2\pi i/\phi(a)}$ where γ is a primitive root mod a. Then $\chi'(j) = 1$ only if $j \equiv 1$ mod a. If $a = 2^m$ then one of the *real* characters χ' satisfies $\chi'(j) \neq 1$ unless $m \geqslant 3$ and $j \equiv 1$ mod 8. In this case the character χ' defined by $\chi'(3) = 1$, $\chi'(5) = e^{2\pi i/\sigma}$, $\sigma = 2^{m-2}$ satisfies $\chi'(j) \neq 1$.]

Appendix: The natural numbers

A.1 Basic properties

Number theory is first and foremost the study of the properties of the *natural numbers*, which are the numbers 1, 2, 3, . . . that occur in the process of *counting*. It is the process of counting—for example, counting the number of times that a certain repetitive operation is performed —which gives the natural numbers their meaning and which underlies the relations among them. One says that $k = m + n$ if performing an operation m times and then performing it n times amounts to performing it k times in all. One says that $k = mn$ if m repetitions of the act of performing the operation n times amounts to performing it k times in all. One says that $k < m$ if performing an operation k times amounts to doing it fewer than m times.

The fundamental facts about the natural numbers include the following. To say that $k < m$ is the same as to say that there is a natural number n for which $m = k + n$. If $k < m$ then $k + n < m + n$ for all n and, conversely, if $k + n < m + n$ for some n then $k < m$. Similarly, $k < m$ if and only if $kn < mn$. If $k < m$ and $m < n$ then $k < n$. For any k and m exactly one of the three relations $k < m$, $k = m$, $k > m$ holds. Thus, to say that $k \leqslant m$ means that k is not greater than m. The operations of addition and multiplication are associative and commutative, and multiplication is distributive over addition:

$$(k + m) + n = k + (m + n), \qquad (km)n = k(mn),$$
$$k + m = m + k, \qquad km = mk,$$
$$k(m + n) = km + kn.$$

One can *cancel* from sums and products; that is, if $k + n = m + n$ then

$k = m$, and if $kn = mn$ then $k = m$. (The fact that one can cancel from products makes the natural numbers simpler than the integers. As one must perpetually remind oneself, the cancellation law for products holds for products of integers only with the additional stipulation that $n \neq 0$.) The number 1 is the least number. Thus $n + 1$ is the least number greater than n. It is the successor of n. If $n > 1$ then $n = 1 + m$ for some m. Then m is less than n and is the greatest number less than n. It is the predecessor of n. The number 1 is an identity for multiplication—that is, $1 \cdot n = n$.

The principle of infinite descent states that a decreasing sequence of natural numbers must terminate. This principle can be made the basis of most proofs relating to the natural numbers. Consider, for example, the fundamental facts about *division*: *If k and m are natural numbers with $k \leqslant m$ then either k divides m—that is, there is a natural number q for which $m = qk$—or there are uniquely determined natural numbers q and r for which $m = qk + r$ and $r < k$.* To prove this, note first that 1 divides any natural number, so that one can assume $k > 1$. Then $mk > m$. Since $m \geqslant k > 1$ there is a predecessor, call it $m - 1$, of m. It may or may not satisfy $(m - 1)k > m$. If it does then $m - 1 > 1$ and the predecessor has a predecessor, call it $m - 2$. Then $(m - 2)k > m$ may or may not hold. Repetition of this process would lead to an infinite descending sequence of numbers unless a stage were reached at which one had a number n (the predecessor of the predecessor of ... of m) for which nk was not greater than m but $(n + 1)k > m$. Therefore by the principle of infinite descent there is such an integer n. Then either $nk = m$ or $nk < m$. If $nk = m$ then k divides m and the first alternative holds. If $nk < m$ then $m = nk + r$ for some r. Since $(n + 1)k > m$, $nk + k > nk + r$, it follows that $k > r$. Finally, if $m = qk + s$ where $s < k$ then $q = n$ because otherwise $q < n$ or $q > n$; if $q > n$ then $q = n + a$ for some a, $nk + r = m = qk + s = nk + ak + s$, $r = ak + s > k$ contrary to assumption and, in the same way, $n > q$ is contrary to assumption. Thus $q = n$ and $qk + r = m = qk + s$, from which it follows that $r = s$ and the proposition is proved.

Perhaps the most important aspect of the natural numbers which goes beyond these elementary facts of arithmetic is the *Euclidean algorithm** for finding the greatest common divisor of two natural numbers. This algorithm, which can be summarized by the phrase "measure the larger by the smaller," consists of the following procedure. Let k, m be given natural numbers, and assume without loss of generality that $k \leqslant m$. Then either k divides ("measures") m or $m = qk + r$ where $r < k$. In the latter case set $m' = k$, $k' = r$, and repeat the procedure (measure the larger by the smaller) to find that either k' divides m' or $m' = q'k' + r'$ where $r' < k'$. In the latter case set $m'' = k'$, $k'' = r'$ and repeat again. Continue until a stage is reached in which the division produces no remainder. Otherwise stated, set $a_0 = m$, $a_1 = k$, and define a sequence $a_0 > a_1 > a_2 > \cdots > a_n$ by the

---

*Elements, Book VII, Propositions 1 and 2.

conditions

$$a_0 = q_1 a_1 + a_2 \qquad a_1 > a_2$$
$$a_1 = q_2 a_2 + a_3 \qquad a_2 > a_3$$
$$\vdots \qquad\qquad \vdots$$
$$a_{n-2} = q_{n-1} a_{n-1} + a_n \qquad a_{n-1} > a_n$$
$$a_{n-1} = q_n a_n.$$

It follows from the principle of infinite descent that a stage must be reached at which division produces no remainder, because otherwise the sequence $a_0 > a_1 > a_2 > \cdots$ would decrease indefinitely. The number a_n is then the greatest common divisor (or "greatest common measure") of the two given numbers a_0, a_1. This means that a_n divides both a_0 and a_1 (the last equation shows that a_n divides a_{n-1}, then the next-to-last equation shows that a_n divides a_{n-2}, and continuing up the list of equations shows that a_n divides all the a's) and that any number which divides both a_0 and a_1 divides a_n (the first equation shows that such a number must divide a_2, the second equation then shows that it must divide a_3, and so forth until the next-to-last equation shows that it must divide a_n). Thus the Euclidean algorithm not only proves the existence of a greatest common divisor of two numbers but also gives an explicit method for *finding* the greatest common divisor by successive divisions.

Two natural numbers a and b are said to be *congruent* modulo a third natural number c, written $a \equiv b \bmod c$, if there are natural numbers m and n such that $a + mc = b + nc$. Other ways of saying the same thing are to say that a and b lie in the same arithmetic progression with modulus c or to say that division of a by c leaves the same remainder as division of b by c. It is easy to see that the congruence relation has the following properties. Congruence modulo c is an *equivalence relation*, that is, it is reflexive ($a \equiv a \bmod c$ for all a), symmetric ($a \equiv b \bmod c$ if and only if $b \equiv a \bmod c$) and transitive (if $a \equiv b \bmod c$ and $b \equiv d \bmod c$ then $a \equiv d \bmod c$). It is consistent with both addition and multiplication of numbers, which is to say that if $a \equiv b \bmod c$ then $a + d \equiv b + d \bmod c$ for all d and $ad \equiv bd \bmod c$ for all d. This means that congruence classes mod c can be added and multiplied in an obvious way: To add (multiply) two congruence classes, choose a number in each and add (multiply) them. The congruence class of the result is independent of the choices (if $a \equiv a'$ and $b \equiv b'$ then $a + b \equiv a + b' \equiv a' + b'$ and $ab \equiv ab' \equiv a'b'$) and it is therefore valid to call this congruence class the sum (product) of the two given classes. The operation of addition of congruence classes has a much stronger property than the cancellation property, namely, the property that for any given a, b, c there is a number x such that $a + x \equiv b \bmod c$ and any two solutions x of this problem are congruent mod c. This follows from the observation that $b + nc > a$ for large enough n (say $n = a$) so that

$b + nc = a + x$, from which $a + x \equiv b \bmod c$ follows. If $a + x \equiv a + x'$ then $x \equiv x'$ by the cancellation law of addition. Another way to say the same thing is to say that with respect to addition the congruence classes mod c form a *group*. The identity element of this group is the congruence class of all numbers divisible by c because if k is divisible by c then $a + k \equiv a \bmod c$ for all a. Since the group operation is denoted $+$ it is natural to let 0 denote this identity element and to let $k \equiv 0 \bmod c$ mean that c divides k. When it comes to multiplication, the situation is much less simple. The congruence $ax \equiv b \bmod c$ may or may not have a solution, and, when it does, the solution may or may not be unique.

The solution of this congruence $ax \equiv b \bmod c$ can be handled using the very important *Chinese remainder theorem*.* This theorem, which is the next step beyond the Euclidean algorithm, can be formulated as follows. Two natural numbers are said to be *relatively prime* if their greatest common divisor is 1. *Let c and d be relatively prime. If a and b are given natural numbers, then there is a solution x of the problem* "$x \equiv a \bmod c$ and $x \equiv b \bmod d$". *Moreover, if x, x' are both solutions then $x \equiv x' \bmod cd$.* This is the Chinese remainder theorem. It can be proved as follows.

If c or d is 1 then one of the congruences $x \equiv a \bmod c$, $x \equiv b \bmod d$ is vacuous and the statements to be proved are obvious. The basic assumption that c and d are relatively prime is the assumption that the Euclidean algorithm must eventually arrive at 1, say $c_0 = c$, $c_1 = d$, $c_0 = q_1 c_1 + c_2$, $c_1 = q_2 c_2 + c_3, \ldots$, $c_0 > c_1 > \cdots > c_n = 1$. Since the theorem is true for $c = c_{n-1}$ and $d = c_n = 1$, it will suffice to show that the case $c = c_m$, $d = c_{m+1}$ implies the case $c = c_{m-1}$, $d = c_m$. To this end, let a, b be given. The task is to find a natural number x for which $x \equiv a \bmod c_m$ and $x \equiv b \bmod c_{m-1}$. Since $x + c_m c_{m-1}$ has the same properties as x, one can assume without loss of generality that $x > b$. Then $x = b + kc_{m-1}$ for some k. Hence $x = b + k(q_m c_m + c_{m+1}) \equiv b + kc_{m+1} \bmod c_m$ and $x \equiv a \bmod c_m$. Therefore k must satisfy $b + kc_{m+1} \equiv a \bmod c_m$. Let y be a solution of $y \equiv b \bmod c_{m+1}$ and $y \equiv a \bmod c_m$. (Such a y exists by the induction hypothesis.) Since $y + c_m c_{m+1}$ has the same two properties as y, one can assume without loss of generality that $y > b$. Then $y = b + j$ for some j and $j \equiv 0 \bmod c_{m+1}$; that is, $y = b + kc_{m+1}$ for some k. Now *define* x by $x = b + kc_{m-1}$. Then mod c_m one has $x = b + k(q_m c_m + c_{m+1}) \equiv b + kc_{m+1} = y \equiv a$ and mod c_{m-1} one has $x = b + kc_{m-1} \equiv b$, as desired. This proves that there is always an x with the desired properties. If x and x' both solve $x \equiv a \bmod c$, $x \equiv b \bmod d$, and if one assumes that $x > x'$, then $x = x' + k$ where $k \equiv 0 \bmod c$ and $k \equiv 0 \bmod d$, that is, k is divisible both by c and by d. What is to be shown, then, is that if k is divisible by both c and d and if c and d are relatively prime then k is divisible by cd. By the portion of the theorem already proved there is a number x such that $x \equiv 0 \bmod c$ and $x \equiv 1 \bmod d$. In other words, $x = n_1 c$ and $x = n_2 d + 1$.

*I do not know how this name originated. It may come from the fact that the theorem was known in very early times in China. See Dickson [D2, vol. 2, p. 57].

Since k is divisible by c, $k = qc$ for some q. Then $n_1 k = n_1 qc = qx = q(n_2 d + 1) = n_2 qd + q$. Since d divides k it follows that d divides q, that is, $q = q'd$ and $k = q'cd$, as was to be shown.

It is very simple to show that if c and d are not relatively prime then $x \equiv a \bmod c$ and $x \equiv b \bmod d$ *never* has a unique solution $x \bmod cd$—that is, either there is no solution or there is more than one. For this it suffices to note that if the greatest common divisor of c and d is D and say $c = Dc'$, $d = Dd'$, then the integer $m = Dc'd'$ is divisible both by c and by d but is not divisible by cd because $cd = D^2 c'd' > m$. Therefore if $x \equiv a \bmod c$, $x \equiv b \bmod d$ then $x + m$ also satisfies these congruences but $x + m \not\equiv x \bmod cd$. Therefore if there is a solution there is more than one, as was to be shown.

Return now to the solution of the congruence $ax \equiv b \bmod c$. *If a and c are relatively prime then this problem can be solved* by using the Chinese remainder theorem to solve $y \equiv 0 \bmod a$, $y \equiv b \bmod c$ and setting $y = ax$. In particular, the congruence $az \equiv 1 \bmod c$ has a solution. Therefore if $ax \equiv b$ and $ax' \equiv b \bmod c$ it follows that $x \equiv azx \equiv zb \equiv zax' \equiv x' \bmod c$ and *the solution of $ax \equiv b \bmod c$ is unique* mod c. On the other hand, *if a and c are not relatively prime then the congruence $ax \equiv b \bmod c$ never has a unique solution* mod c—either it has no solution or it has more than one mod c. To see this it suffices to write $a = Da'$, $c = Dc'$ where $D > 1$ and to note that $ac' = a'c \equiv 0 \bmod c$ whereas $c' \not\equiv 0 \bmod c$ because $c' < c$. Then $ax \equiv b \bmod c$ implies $a(x + c') \equiv b \bmod c$ but $x + c' \not\equiv x \bmod c$. Therefore if there is a solution there is more than one.

A natural number is said to be *prime* if it is relatively prime to all numbers less than itself. (In the case of 1, where this condition is vacuous, the number 1 is by special definition *not* prime.) By what was just shown, this is the same as saying that p is prime if and only if $p > 1$ and the congruence $ax \equiv b \bmod p$ has a unique solution x for all b and for all $a \not\equiv 0 \bmod p$. Therefore, in arithmetic modulo a prime one can always *divide* by nonzero elements. In particular, prime natural numbers p have the basic property that if p divides uv then either p divides u or p divides v. (If $uv \equiv 0 \bmod p$ and $u \not\equiv 0 \bmod p$ then by division $v \equiv 0 \bmod p$.)

The fundamental theorem of arithmetic states that every natural number greater* than 1 can be written as a product of prime numbers in exactly one way. More precisely, given $n > 1$ there is a finite sequence $p_1, p_2, \ldots, p_m$ of prime numbers ($m \geqslant 1$) such that $n = p_1 p_2 \cdots p_m$ and if $p_1 p_2 \cdots p_m = p'_1 p'_2 \cdots p'_\mu$ then $m = \mu$ and the sequence $p'_1, p'_2, \ldots, p'_\mu$ is merely a rearrangement of the sequence $p_1, p_2, \ldots, p_m$. This theorem can be proved as follows.

Given $n > 1$, try dividing n by $2, 3, 4, \ldots, n$. Let $p \leqslant n$ be the first one of these numbers for which the division leaves no remainder, say $n = pq$.

---

*1 itself should be regarded as a product of *no* primes. It is in order to make this theorem true that 1 is not regarded as a prime.

Then $n > q$. Moreover, p must be prime because otherwise either $p = 1$, which is excluded at the outset, or there would be a number $a < p$ that was not relatively prime to p. Let d be the greatest common divisor of a and p. Then $1 < d$ and $d \leqslant a < p$; since d divides p it divides $n = pq$ and this would contradict the definition of p. Therefore p is prime. If $q > 1$ the same argument gives $q = p'q'$ where p' is prime and $q > q'$. Thus $n = pp'q'$. If $q' > 1$ then the process can be repeated to find $n = pp'p''q''$ where $q'' < q'$. By the principle of infinite descent this process must terminate with a representation of n as a product of primes. If $p_1 p_2 \cdots p_m = p_1' p_2' \cdots p_\mu'$ then $p_1' p_2' \cdots p_\mu' \equiv 0 \bmod p_1$. If $p_1' \neq p_1$ then p_1 and p_1' are relatively prime and the congruence can be divided by p_1' to give $p_2' p_3' \cdots p_\mu' \equiv 0 \bmod p_1$. Similarly, if $p_2' \neq p_1$ then $p_3' p_4' \cdots p_\mu' \equiv 0 \bmod p_1$. In this way the assumption $p_1' \neq p_1$, $p_2' \neq p_1, \ldots, p_\mu' \neq p_1$ would imply $1 \equiv 0 \bmod p_1$. Since $1 \not\equiv 0 \bmod p_1$ (because 1 is not a prime) it follows that $p_j' = p_1$ for some j. By rearranging the sequence $p_1', \ldots, p_\mu'$ if necessary, one can assume that $p_1' = p_1$. Then by the cancellation law $p_2 p_3 \cdots p_m = p_2' p_3' \cdots p_\mu'$. If $m = 1$ this gives $1 = p_2' p_3' \cdots p_\mu'$, from which $\mu = 1$ and the sequences $p_1, p_2, \ldots, p_m$ and $p_1', p_2', \ldots, p_\mu'$ are identical. Otherwise the same argument as before shows that p_2 must occur among $p_2', p_3', \ldots, p_\mu'$. By rearrangement one can assume $p_2' = p_2$ and conclude that $p_3 p_4 \cdots p_m = p_3' p_4' \cdots p_\mu'$. If $m = 2$ then $\mu = 2$ and the sequences are identical. Otherwise $p_4 p_5 \cdots p_m = p_4' p_5' \cdots p_\mu'$ after rearrangement, and so forth. By the principle of infinite descent one finally arrives at the conclusion that by rearrangement of the sequence $p_1', p_2', \ldots, p_\mu'$ the two sequences can be made to be identical.

The natural numbers are certainly the objects most entitled to the name "number" and indeed the ancient Greeks used the term "number" only for natural numbers. Even in Fermat's time, fractions, irrational quantities, zero, and negative quantities were not generally accepted as "numbers." However, there are many situations in which it is convenient to have a more inclusive notion of numbers. For example, it was convenient after the first few sections of this book to deal with the *integers* rather than the natural numbers. The integers are the natural numbers $1, 2, 3, \ldots$ together with their negatives $-1, -2, -3, \ldots$ and 0. In terms of the integers, Fermat's Last Theorem can be stated more symmetrically in the form "if p is an odd prime then the equation $x^p + y^p + z^p = 0$ is impossible in integers x, y, z except in the trivial case in which one of the three is zero." This form of the theorem would have facilitated Euler's proof of the case $p = 3$ because it would have eliminated the need to separate the cases of $x^3 + y^3 = z^3$ into the one in which z is even and the one in which z is odd.

The basic properties of the integers are simple extensions of those of the natural numbers, and because integers are today introduced in elementary school it hardly seems necessary to discuss the integers any further here.

EXERCISES

1. Prove that if d divides both a and b then it divides $a + b$. Moreover, it divides $qa + rb$ for all q and r.

2. Prove that if a divides b and b divides a then $a = b$.

3. Prove from the Euclidean algorithm that if d is the greatest common divisor of a and b then either there exist natural numbers u and v such that $ua = d + vb$ or there exist natural numbers u and v such that $d + ua = vb$. [Work backwards through the algorithm as in the proof of the Chinese remainder theorem in the text.] Conclude that if a and b are relatively prime then either $ax \equiv 1 \bmod b$ or $ax + 1 \equiv 0 \bmod b$ has a solution x. In the second case show that $ax \equiv 1 \bmod b$ has a solution x. Therefore if a and b are relatively prime the congruences $y \equiv 0 \bmod a$, $y \equiv 1 \bmod b$ always have a solution. Deduce the Chinese remainder theorem.

4. Find all solutions of the following pairs of congruences:
   (a) $x \equiv 3 \bmod 46$, $\quad x \equiv 4 \bmod 52$
   (b) $x \equiv 2 \bmod 5$, $\quad x \equiv 3 \bmod 7$
   (c) $x \equiv 7 \bmod 56$, $\quad x \equiv 2 \bmod 295$.

5. Show that there exist natural numbers y and z such that the solution to the problem of the Chinese remainder theorem is given by the formula $x = ay + bz$. Thus y and z are the arithmetic equivalent of a "partition of unity," showing that an integer x can be found which takes on arbitrarily assigned "values at c and d."

6. Prove that if $c_1, c_2, \ldots, c_n$ are pairwise relatively prime and if $a_1, a_2, \ldots, a_n$ are arbitrary integers then there is an integer x such that $x \equiv a_i \bmod c_i$ for $i = 1, 2, \ldots, n$.

7. Let a, b, and c be given *integers*. Describe a procedure, using the Euclidean algorithm, for finding all (integer) solutions x and y of the equation $ax + by + c = 0$. [If a or b is zero the problem is trivial. One can assume without loss of generality that a and b are positive. If d is their greatest common divisor then the problem has a solution if and only if d divides c. The problem is also very simple when $c = 0$ and a and b are relatively prime.]

A.2 Primitive roots mod p

A natural number g is called a *primitive root* modulo a prime p if the powers $g, g^2, g^3, \ldots$ exhaust all possibilities mod p for integers relatively prime to p. That is, g is a primitive root mod p if for every natural number k either $k \equiv 0 \bmod p$ or there is a natural number j such that $k \equiv g^j \bmod p$. In Chapters 4, 5, and 6, frequent use is made of the fact that *for every prime p there is at least one primitive root* mod p. The proof of this fact which follows is one of the two that Gauss gives in the *Disquisitiones Arithmeticae*. Gauss contends (Art. 56) that his predecessors did not rigorously state and prove the theorem although they knew (or believed they knew) it was true. (See, however, the Foreword to volume 3 of Euler's *Opera* (1), p. xxix.)

It was shown in Section 1.8 that for every integer $a \not\equiv 0 \bmod p$ there is a natural number d with the properties that $a^d \equiv 1 \bmod p$ and that $a^j \equiv 1 \bmod p$ implies $d|j$. This natural number d is called the *order* of a mod p. By Fermat's theorem the order of a mod p divides $p - 1$. What is to be shown is that there is at least one a whose order mod p is $p - 1$ itself. Gauss proves this by actually constructing such an integer.

Let $p - 1 = p_1 p_2 \cdots p_m$ be the decomposition of $p - 1$ as a product of primes. Let q be one of the primes which occurs in the list $p_1, p_2, \ldots, p_m$ and let ν be the multiplicity with which it occurs. The differencing argument of Section 2.4 shows that the congruence $x^{(p-1)/q} - 1 \equiv 0 \bmod p$ cannot be satisfied by all of the integers $x = 1, 2, 3, \ldots, p - 1$. (If it were, then, by taking differences, one could conclude that p divided the factorial of $(p - 1)/q$ which is impossible because p is prime and is greater than all the factors of $(p - 1)/q$ factorial.) Therefore, there is an integer b, $0 < b < p$, such that $b^{(p-1)/q} \not\equiv 1 \bmod p$. Let c be the integer mod p obtained by raising b to the power $(p - 1)/q^\nu$. The claim is that the order of c is q^ν. Because c to the power q^ν is $b^{p-1} \equiv 1 \bmod p$ by Fermat's theorem, the order of c must divide q^ν. This implies that the order of c is divisible by no prime other than q; that is, the order of c is a power of q. To prove that it is q^ν it suffices to prove that c to the power $q^{\nu-1}$ is *not* 1 mod p, and this follows immediately from the way that c and b were chosen.

Let such an integer c be found for each prime divisor q of $p - 1$, say c_1, $c_2, \ldots, c_n$ and let g be their product. (Here $n \leqslant m$ is the number of *distinct* prime factors of $p - 1$.) The claim is that g has order $p - 1$. If g did not have order $p - 1$ then its order would be a proper divisor of $p - 1$. Thus there would be a prime divisor q of $p - 1$ which divided the order of g fewer times than it divided $p - 1$. Then the order of g would divide $(p - 1)/q$ and it would follow that $g^\mu \equiv 1 \bmod p$ where $\mu = (p - 1)/q$. Without loss of generality one can assume that the c corresponding to this q is c_1. Then $c_2^\mu \equiv c_3^\mu \equiv \cdots \equiv c_n^\mu \equiv 1 \bmod p$ because the orders of all these integers divide $\mu = (p - 1)/q$. This would then give $1 \equiv g^\mu \equiv c_1^\mu$ which would be a contradiction because q^ν does not divide μ. Therefore g has order $p - 1$, as was to be shown.

EXERCISES

1. By trial-and-error find primitive roots for several primes. Extensive tables of primitive roots can be found in Jacobi's famous *Canon Arithmeticus* [J3] and in various other books.

2. Show that if γ is a primitive root mod p then γ^k is a primitive root mod p if and only if k is relatively prime to $p - 1$. Find all primitive roots of all primes less than 20.

3. Read and reformulate in your own words Gauss's other proof of the existence of primitive roots. [*Disquisitiones Arithmeticae* Art. 54.]

Answers to exercises

1.3 **1.** $2p^2 = 4961 + 8161$ gives $p = 81$, then $q = 40$.

1.6 **1.** See Ex. 3, Sec. 1.3. **2.** If $x^4 - y^4 = z^2$ then the Pythagorean triangle $(x^4 - y^4, 2x^2y^2, x^4 + y^4)$ would have area a square.

1.7 **1.** x, y have opposite parities. If x is even then x^2 is of the form $4k$, if odd then of the form $4k + 1$. **2.** If $p = x^2 + 3y^2$ then $x = 3k \pm 1$ and $x^2 = 3n + 1$. If $3m + 1$ is prime then m is even. **3.** $(2n + 1)^2$ is of the form $8k + 1$; $2y^2$ is either $8k$ or $8k + 2$. **4.** n satisfies Girard's conditions if and only if any prime p of the form $4n + 3$ which divides n divides it with even multiplicity.

1.8 **1.** $p = 74k + 1$. When $p = 149$ the remainders of $2^5, 2^8, 2^{16}, 2^{32}, 2^{37}$ are 32, 107, $-24, -20$, and $-44 \neq 1$ respectively. When $p = 223$ the remainder of 2^{37} is 1. **2.** $a^{2n+1} + b^{2n+1} = (a + b)(a^{2n} - a^{2n-1}b + a^{2n-2}b^2 - \cdots + b^{2n})$. If $p \nmid x$ then $x^{p-1} = kp + 1$. **3.** $n = 3, 4, 5, 6, 8, 10, 12, 15, 16, 17, 20, 24, 30$.

1.9 **3.** $pQ - Pq = (p^2 + pqr - pqr - q^2A)/|k| = \text{sign}(k)$. $Q(QA + PR) = P^2 - K + PQR$ is divisible by K. Therefore so is $QA + PR$. **4.** $P - rQ = -qs/|k| = \pm qK$. Since $P + RQ = nK$, $K|(r + R)Q$ and $K|(r + R)$. **5.** For the cyclic method the second $r = 13$ is determined by $5|(12 + r)$ and $r^2 - 149 = \text{small}$. The next k is $(13^2 - 149)/(-5) = -4$. The sequence of k's is $1, -5, -4, 7, -7, 4, 5, -1, 5, 4, -7, 7, -4, -5, 1$. For the English method the second r is 8 and the sequence of k's is $1, -5, 17, -4, 7, -7, 4, \ldots, 17, -5, 1$. **6.** $|K|\mathcal{P} = PR + QA = P(R + r) - rP + QA = Pn|K| - rP + QA$. Thus $\mathcal{P} = nP + x$ where $|k||K|x = -r(pr + qA) + A(p + qr) = p(A - r^2) = -ps$. Since $|k||K| = |s|$, $x = -\text{sign}(s) \cdot p$. The computation of $\mathcal{Q}$ is similar. **7.** $P - rQ$ is divisible by both k and K (Ex. 4). Thus $pQ + qP$ is divisible by k and K if and only if $pQ + qrQ$ is. Both $p + qr$ and $pr + Aq$ are divisible by k (Ex. 3). **9.** K and k have opposite signs and $kK = r^2 - A$. When $k < 0$, $(-r + K)^2 - A = K(k - 2r + K)$ has the same sign as $k - 2r + K = [(r + |k|)^2 - A]/k < 0$; $(-r + 2K)^2 - A = K(k - 4r + 4K)$ is positive because $k - 2r + 2K = k^{-1}[(r + |k|)^2 - A - (A - r^2)]$. The case $k > 0$ is similar, $(R^2 - A)/K = [(r + |k|)^2 - A]/k$ because both are $k + 2r \, \text{sign}(k) + K$. **10.** The next r, call it R, is determined by $R = -r + nK$, $R^2 < A$, R as large as possible. If $K<0$ then $kK+2rK+K^2<0$, $r^2-A+2rK+K^2<0$, $(-r+|K|)^2<A$. This also holds when $K>0$. Thus $R \geqslant -r+|K|$, $-r \leqslant R-|K| \leqslant R$ and $(R-|K|)^2<A$. On the other hand, $(R+|K|)^2>A$. With $\mathcal{K}=(R^2-A)/K$ it follows that $R^2-2R|K|+K^2<A$, $K(\mathcal{K}-2R\,\text{sign}(K)+K)<0$. If $K<0$ this gives $\mathcal{K}+2R+K>0$ and if $K>0$ it gives $\mathcal{K}-2R+K<0$. The opposite inequalities follow in the same way from $(R+|K|)^2>A$. **11.** r is of the form $-R+nK$. This together with the fact that $r^2<A$, $(r+|K|)^2>A$ determines r when R and K are known. The inequalities $-A<kK<0$ show that there are only finitely many possibilities for k. The first k is 1 so the first repeat is 1. **13.** The equation $x_{i+1}y_i-y_{i+1}x_i=\pm 1$ is easily proved, first for positive i and then for all i. The equation $x_{i+1}x_i-Ay_{i+1}y_i=\text{sign}(k_i)r_i$ (where r_i is the r which is used in the step from x_i, y_i to x_{i+1}, y_{i+1}) is easily proved for positive i. It then follows for all i $(n_{i+1}|k_{i+1}|=r_i+r_{i+1})$. Then $(x_i^2-Ay_i^2)(x_{i+1}^2-Ay_{i+1}^2)=(x_ix_{i+1}-Ay_iy_{i+1})^2-A(x_{i+1}y_i-y_{i+1}x_i)^2=r_i^2-A=k_ik_{i+1}$ and $x_i^2-Ay_i^2=k_i$ follows for all i.

1.10 **1.** The equation is $(1+x)(1+x^2)=y^2$. Then $x \geqslant -1$. Since $1+x^2$ is not a square unless $x=0$, $1+x$ and $1+x^2$ are not relatively prime. In fact, $1+x=2u^2$, $1+x^2=2v^2$ when $x>0$. Then (u^2,u^2-1,v) is a primitive Pythagorean triple except when $u=1$, $x=1$. If $x>1$ and u is odd then $u^2=p^2-q^2$, $u^2-1=2pq$, $v=p^2+q^2$ where p, q are relatively prime and of opposite parity and $p>q$. $1=(p-q)^2-2q^2$ and $p-q$ is a square. Thus $(q^2+1)^2=q^4+t^4$, which is impossible. If u is even then $u^2=2pq$, $u^2-1=p^2-q^2$ and $v=p^2+q^2$. Then p is even and in fact $p=2t^2$. Then $1=(p+q)^2-2p^2$, $8t^4=(p+q-1)(p+q+1)$, $2t^4=a(a+1)$. One of the numbers a, $a+1$ is a 4th power and the other is twice a 4th power, from which $b^4-2c^4=\pm 1$. If the sign is $+$ then, with $d=c^2$, $(d^2+1)^2=d^4+b^4$ which is impossible. Therefore $(d^2-1)^2=d^4-b^4$. Clearly $b \neq 0$. Therefore $b^4=d^4$. Since b and c are relatively prime, $b=\pm 1$, $c=\pm 1$, $a=b^4=1$, $a+1=2c^4=2$, $p+q=3$, $p=2$, $u^2=4$, $x=7$.

2.3 **1.** Every unit is of the form $\pm(1+\sqrt{2})^n$ for some integer n. $((1+\sqrt{2})^{-1}=-(1-\sqrt{2}).)$ The only units of the form $x+y\sqrt{-41}$ or $x+y\sqrt{-7}$ are ± 1. The equation $8^2=1+7\cdot 3^2$ gives the units $\pm(8+3\sqrt{-7})^n$ where $(8+3\sqrt{-7})^{-1}=8-3\sqrt{-7}$. These are the only units in this case because $x^2=-1+7y^2$ is impossible mod 8. The only units of the form $x+y\sqrt{.-1}$ are ± 1 and $\pm\sqrt{-1}$. **2.** $4\sqrt{2}+\sqrt{5}=(19+6\sqrt{10})(2\sqrt{2}-\sqrt{5})^3$.

2.4 **1.** Let $P=p^2+q^2=a^2+b^2$. Change sign of q, if necessary, so that P divides $pb+aq$. Either $ap-bq=0$ and $aq+bp=\pm P$ or $ap-bq=\pm P$ and $aq+bp=0$. Thus either $a=\pm q$ or $a=\pm p$. Similar arguments work in the other cases. **2.** By the binomial theorem $(x+1)^n-x^n=nx^{n-1}+$ terms of lower degree. (Also $x^n-(x-1)^n=nx^n+\cdots$.) Thus the first differences of $ax^n+bx^{n-1}+\cdots+c$ are $anx^{n-1}+\cdots$. **8.** The indicated division is possible if and only if P divides $(ap \mp bq)+(aq \pm bp)i$. **10.** By Ex. 1, $p=u^2+v^2$. If r is the inverse of v mod p then $a^2 \equiv -1$ mod p is solved by $a=ru$. Similarly, $b^2 \equiv -2$ mod p has a solution by Ex. 7. Then $2^{(p-1)/2} \equiv (a^2b^2)^{(p-1)/2} \equiv 1$ mod p by Fermat's theorem. Since 64 is the least power of 2 that is 1 mod p, $(p-1)/2=64k$, $p=1+128k$. **11.** $2^{32}-1=(2^{16}+1)(2^8+1)(2^4+1)(2^2+1)(2+1)(2-1)$. Therefore $257=2^8+1$ does not divide $2^{32}-1+2$.

2.5 **1.** From $x+\sqrt{-2}=(a+b\sqrt{-2})^3$, $b(3a^2-2b^2)=1$, $b=\pm 1$, $a=\pm 1$. Then $x=\pm 7$ or ± 5. **4.** $(1 \pm \sqrt{-3})^2$ is divisible by 2. **5.** **(a)**

$(2+\sqrt{-3})(1\pm 2\sqrt{-3})$ gives the only two representations $4^2+3\cdot 5^2$, $8^2+3\cdot 3^2$. **(b)** $(2+\sqrt{-3})(2\pm\sqrt{-3})$ gives $49=7^2=1^2+3\cdot 4^2$. **(c)** Look for representations of $112=336/3$. They come from $4(2+\sqrt{-3})$ and $2(2+\sqrt{-3})\cdot(1\pm\sqrt{-3})$. The representations are $336=3\cdot 8^2+3^2\cdot 4^2=3\cdot 2^2+3^2\cdot 6^2=3\cdot 10^2+3^2\cdot 2^2$.

3.2 **2.** Use the fact that an integer $\not\equiv 0 \bmod p$ is invertible $\bmod p$. **3.** Mod $4\cdot 13+1$ the 13th powers are 0, ± 1, ± 30. Mod $6\cdot 17+1$ the 17th powers are 0, ± 1, ± 46, ± 47 and the conditions of Sophie Germain's theorem are *not* satisfied. However, mod $8\cdot 17+1$ the 17th powers are 0, ± 41, ± 37, ± 10, ± 1. For 19 the first prime to try is 191, mod which the 19th powers are 0, ± 7, ± 49, ± 39, ± 1. **4.** $(j+5)^5\equiv j^5 \bmod 25$. Thus the nonzero 5th powers mod 25 are 1, $2^5\equiv 7$, $3^5\equiv(-2)^5\equiv -7$, $4^5\equiv -1$.

4.2 **3.** Take $a_2=0$. Then $A=2a_0-a_1$, $B=a_1$. **4.** $A=b+c$, $B=b-c$. The powers of the unit θ_0 are distinct because $\theta_0^n=\pm a_n\mp a_{n+1}\theta_0\neq 1$ where $a_{n+1}=a_n+a_{n-1}$. **6.** $f(\alpha)=g(\alpha)$ means $f(\alpha)$ is identical to $g(\alpha)+c(1+\alpha+\cdots+\alpha^{\lambda-1})$. Thus $f(1)=g(1)+c\lambda$. $Nf(\alpha)\equiv f(1)f(1^2)\cdots f(1^{\lambda-1})=f(1)^{\lambda-1}\equiv 0$ or 1 mod λ. **8.** Composition of $\alpha\mapsto\alpha^j$ and $\alpha\mapsto\alpha^k$ gives $\alpha\mapsto\alpha^n$ where $n=jk$. If neither j or k is 0 mod λ, $n\not\equiv 0$ mod λ. The inverse of $\alpha\mapsto\alpha^j$ is $\alpha\mapsto\alpha^m$ where $mj\equiv 1$ mod λ. **9.** If $k\equiv -1$ mod p then $\lambda\equiv 0$ mod p, $\lambda=p$. If $k^\lambda\equiv -1$ and $k\not\equiv -1$ then $(-k)^\lambda\equiv 1$, $\lambda|(p-1)$ by Fermat's theorem. **14.** $Nf(\alpha)\cdot Ng(\alpha)=1$ implies $Nf(\alpha)=\pm 1$. Thus $Nf(\alpha)=1=g(\alpha)f(\alpha)$. Cancel $f(\alpha)$. **15.** $r(\alpha)=0$ because $h(\alpha)=f(\alpha)=0$. If $f(X)$ is divisible by p then f/p has the same properties as f. $(j-1)h(j)=j^\lambda-1$. See Ex. 1 of Sec. 3.2. **16.** If $f(\alpha)g(\alpha)=0$ and $ah(X)=q(X)f(X)+r(X)$ then $r(\alpha)g(\alpha)=0$ and $r(\alpha)$ can replace $f(\alpha)$ if $r(X)\neq 0$. **18.** If $p(\alpha)$ is prime and $p(\alpha)=f(\alpha)g(\alpha)$ then $p(\alpha)$ divides $f(\alpha)$ or $g(\alpha)$ and the other is a unit.

4.3 **3.** $h(\alpha)$ irreducible but not prime implies $q(\alpha)h(\alpha)=f(\alpha)g(\alpha)$ where $f(\alpha)$, $g(\alpha)$ are not divisible by $h(\alpha)$. Write $q(\alpha)$, $f(\alpha)$, $g(\alpha)$ as products of irreducibles and observe that $h(\alpha)$ does not occur on the right. Although it appears obvious that every cyclotomic integer is a product of irreducibles—just keep factoring until further factoring is impossible—to prove this in a constructive way is not simple. **5.** $x^\lambda+y^\lambda\equiv 0$ mod p. If $y\not\equiv 0$ mod p this gives $(-x/y)^\lambda\equiv 1$ mod p and $\lambda|(p-1)$. Otherwise $x\not\equiv 0$ mod p. **6.** **(a)** $1,11$, 31, 61, 11^2, $5\cdot 11$, 211, $5\cdot 41$, $11\cdot 41$, $11\cdot 191$, $11\cdot 191$. **(b)** 43, 127, 547, 1093, 463, $29\cdot 71$, $29\cdot 113$, $43\cdot 127$, $7\cdot 379$, 14197, $29\cdot 449$, $7\cdot 1597$, 10039, 10501. **8.** $p|N(\alpha-k)$ for some k. Thus $h(\alpha)$ divides a binomial and the result follows from the theorems of the text. **10.** **(a)** $\alpha\equiv k$ mod $h(\alpha)$. Thus $k^\lambda\equiv\alpha^\lambda=1$ mod $h(\alpha)$, $k^\lambda\equiv 1$ mod p. $k\not\equiv 1$ mod p because $\alpha\equiv 1$ mod $h(\alpha)$ would imply $p|N(\alpha-1)=\lambda$. For the same reason $k^2,k^3,\ldots,k^{\lambda-1}\not\equiv 1$ mod p. Thus $k^i\not\equiv k^{i+j}$ for $0<j<\lambda$. If $m^\lambda\equiv 1$ and $m\not\equiv 1$ mod p then $p|N(m-\alpha)$, $h(\alpha)$ divides one of the factors $m-\alpha^j$ of $N(m-\alpha)$, $m\equiv\alpha^j\equiv k^j$ mod $h(\alpha)$, $m\equiv k^j$ mod p. **(b)** Let $p-1=\lambda\mu$. Then $k^\lambda\equiv 1$ mod p has at most λ solutions by Ex. 1, Sec. 3.2. Also $k^\mu\equiv 1$ mod p has at most μ solutions. Let $a^\mu\not\equiv 1$ mod p and set $b=a^\mu$. Then all powers of b are solutions. If d is the least integer for which $b^d\equiv 1$ mod p then $d|\lambda$, $d\neq 1$, $d=\lambda$. **11.** Mod those for which $1+7+7^2+7^3+7^4=(7^5-1)/(7-1)=2801\equiv 0$ mod p, that is, only for $p=2801$. **12.** In any case, $f(X)-g(X)=q(X)(X^{\lambda-1}+\cdots+1)+r(X)$ where $\deg r<\lambda-1$. Then $r(\alpha)=0$ implies $r(X)=0$.

4.4 **8.** A good control on the computation of $f(k), f(k^2), f(k^4), \ldots, f(k^7)$ mod 599 is the fact that their sum is 10 mod 599. **10.** Since $(\alpha-1)|\lambda| Ng(\alpha)$, $\alpha - 1$ prime, $\alpha - 1$ divides some conjugate of $g(\alpha)$. Then some conjugate of $\alpha - 1$ divides $g(\alpha)$. $\alpha - 1$ divides all its conjugates. **11.** $\alpha - 1$ times either side of the equation is λ, so the two sides are equal. Alternatively, see pp. 240–241.

4.5 **1.** See Table 4.7.1. **2.** $\theta_0 = \alpha + \alpha^2 + \alpha^4$. $\theta_0\theta_1 = 2$. $f(\alpha)f(\alpha^2)f(\alpha^4) = a + b\theta_0$ for some integers a, b. $Nf(\alpha) = (a + b\theta_0)(a + b\theta_1) = a^2 - ab + 2b^2 = \frac{1}{4}[(2a - b)^2 + 7b^2]$. **3.** $(\theta_0 - \theta_1)^2 = \lambda$ if $\lambda \equiv 1$ mod 4, $= -\lambda$ if $\lambda \equiv 3$ mod 4. **4.** $(\theta_0 + \theta_1)^2 - (\theta_0 - \theta_1)^2 = 4\theta_0\theta_1$. Therefore it suffices to find $\theta_0\theta_1$. By direct computation, $\theta_0\theta_1 = a + b\theta_0 + b\theta_1$ where a is 0 or $(\lambda - 1)/2$. The parity of a determines which. See pp. 176–177. **6.** Define periods using $\gamma = 3$. Then $\eta_0^2 = 4 + 2\eta_1 + \eta_2$, $\eta_0\eta_1 = 2\eta_0 + \eta_2 + \eta_3$, $\eta_0\eta_2 = \eta_0 + \eta_1 + \eta_2 + \eta_3 = -1$, $\eta_0\eta_3 = \eta_1 + \eta_2 + 2\eta_3$. **7.** If and only if, when p is written $p \equiv g^k$, where g is a primitive root mod λ, $e|k$. In other words, $p^f \equiv 1$ mod λ. **8.** η_0 is the period containing α.

4.6 **2.** $X(X - 1)\cdots(X - 10) = X^{11} - 55X^{10} + 1320X^9 - 18150X^8 + 157773X^7 - 902055X^6 + 3416930X^5 - 8409500X^4 + 12753576X^3 - 10628640X^2 + 3628800X$. Corollary: If p is prime then $(p - 1)! \equiv -1 \bmod p$. This is Wilson's theorem. **3.** When $\lambda = 13$ and $p = 3$, $\eta_0^3 = 6 + \eta_0 + 3\eta_1 + 3\eta_3$. **4.** If $k \equiv 1 \bmod p$ then $\lambda \equiv 0 \bmod p$. Otherwise $k^\lambda \equiv 1 \bmod p$ and $\lambda|(p - 1)$. **5.** When $\lambda = 5$, $\theta_0^2 = 2 + \theta_1, \theta_0^2 + \theta_0 = 1$. $u = 4$ satisfies $u^2 + u \equiv 1 \bmod 19$. $(\theta_0 - 4)(\theta_1 - 4) = 19$. **6.** See Sec. 4.7. $\eta_0\eta_1\eta_2\eta_3 = 3$ when $\lambda = 13$. **7.** If $f(\alpha)|g(\alpha)$ then $Nf(\alpha)|p^f$ and $Nf(\alpha)$ is a power of p. Since $Nf(\alpha) \equiv 0$ or 1 mod λ, $Nf(\alpha) = 1$ or p^f.

4.7 **1.** $(\eta_0 + 3)/(\eta_2 - \eta_1 + 1) = -\eta_1$. $1/(-\eta_1) = 1 + \eta_2$. **3.** One factorization is $(3 - \eta_0 - 2\eta_2)(3 - \eta_2 - 2\eta_0) = 17 - 12\theta_0, (17 - 12\theta_0)(17 - 12\theta_1) = 61$. **4.** $109 = (10 - \theta_0)(10 - \theta_1)$. **5.** $53 = N(1 + \alpha + \alpha^3)$. The factorization of 103 requires the solution $u = 12$ of $u^6 + u^5 - 5u^4 - 4u^3 + 6u^2 + 3u - 1 \equiv 0$ mod 103. It is then relatively easy to find that the product of the 6 distinct conjugates of $2(\alpha + \alpha^{12}) + (\alpha^6 + \alpha^7) - 1$ is 103. 3 was factored in the text. 5 is the product of the 3 conjugates of $-1 - (\alpha + \alpha^8 + \alpha^{12} + \alpha^5)$. $17 = (4 - \theta_0)(4 - \theta_1)$. 7 is irreducible.

4.8 **1.** (a) and (b) are both $\alpha^2 - \eta_0\alpha + 1 = 0$ where $\eta_0 = \alpha + \alpha^4$ for (a) and $\eta_0 = \alpha + \alpha^6$ for (b). (c) is $\alpha^5 - \theta_0\alpha^4 - \alpha^3 + \alpha^2 + \theta_1\alpha - 1 = 0$. (d) is $\alpha^3 - \eta_0\alpha^2 + \eta_2\alpha - 1 = 0$. (e) is $\alpha^5 - \eta_0\alpha^4 + (\eta_1 + \eta_2)\alpha^3 - (\eta_4 + \eta_5)\alpha^2 + \eta_3\alpha - 1 = 0$. **2.** The product of the first two is $\alpha^6 - 11\alpha^5 - 25\alpha^4 - 155\alpha^3 + 71\alpha^2 - 17\alpha + 1 \equiv \alpha^6 - 11\alpha^5 + 4\alpha^4 - 10\alpha^3 + 13\alpha^2 + 12\alpha + 1$ and the product of the second two has the same coefficients in the opposite order. The final product is $\alpha^{12} + \alpha^{11} - 115\alpha^{10} - 115\alpha^9 + 59\alpha^8 - 57\alpha^7 + 552\alpha^6 - 57\alpha^5 + \cdots \equiv \alpha^{12} + \alpha^{11} + \alpha^{10} + \cdots + \alpha + 1$. **3.** $(X^2 - 4X + 1)(X^2 + 5X + 1)$ mod 19. $(X^2 + 26X + 1)(X^2 - 25X + 1)$ mod 59. $(X + 4)(X - 16)(X - 18)(X - 10)$ mod 41 are all $X^4 + X^3 + X^2 + X + 1$. $(X^2 + 3X + 1)(X^2 + 5X + 1)(X^2 + 6X + 1)$ mod 13 and $(X + 13)(X + 5)(X - 7)(X + 4)(X + 6)(X + 9)$ mod 29 are $X^6 + X^5 + \cdots + 1$. Mod 2, $X^{30} + X^{29} + \cdots + 1$ is the product of the 6 factors $X^5 + X^3 + 1$, $X^5 + X^3 + X^2 + X + 1$, $X^5 + X^4 + X^3 + X + 1$, $X^5 + X^2 + 1$, $X^5 + X^4 + X^3 + X^2 + 1$, $X^5 + X^4 + X^2 + X + 1$. **4.** A vector space over a field with p elements has p^k elements, where k is its dimension. **5.** $7\alpha^3 + 45\alpha^2 + 83\alpha + 90 = (38\theta_0 + 83)\alpha + (45 - 7\theta_0) = (\theta_1 - 5)(3\theta_0 - 8)[(2 + \theta_0) \cdot \alpha + 1]$. **6.** In the polynomial $P(\alpha)$ replace each coefficient by an integer congruent to it mod $h(\alpha)$ and call the result $Q_h(\alpha)$. Then $Q_h(\alpha) \equiv P(\alpha) = 0 \bmod h(\alpha)$ and it is easy to show that $Q_h(\alpha)$ has the required properties.

4.9 **1.** $\Psi(\eta) = (1 - \eta_0)(1 - \eta_1)\eta_2(1 - \eta_3)\eta_4\eta_5$. Mod 2. $(1 - \eta_0)(1 - \eta_1)\eta_2 \equiv \eta_0 + \eta_3 + \eta_4 \equiv (1 - \eta_3)\eta_4\eta_5$. Thus $\Psi(\eta) \equiv (\eta_0 + \eta_3 + \eta_4)^2 \equiv \eta_0 + \eta_3 + \eta_4$. **2.** $\Psi(\eta) = (1 - \eta_0) \cdot$

$(-1-\eta_0)(1-\eta_1)\eta_1(1-\eta_2)\eta_2(-1-\eta_3)\eta_3 \equiv \eta_0 - \eta_1 + \eta_3 \bmod 3$. **5.** The coefficients are invariant under σ and are therefore integers. $\lambda = 5$: $X^2 + X - 1$. $\lambda = 7$: $X^2 + X + 2$ and $X^3 + X^2 - 2X - 1$. $\lambda = 11$: $X^2 + X + 3$ and $X^5 + X^4 - 4X^3 - 3X^2 + 3X + 1$. $\lambda = 13$: $X^2 + X - 3, X^3 + X^2 - 4X + 1, X^4 + X^3 + 2X^2 - 4X + 3, X^6 + X^5 - 5X^4 - 4X^3 + 6X^2 + 3X - 1$. Perhaps the easiest way to find these is to compute $\eta_0^2, \eta_0^3, \ldots, \eta_0^e$ in terms of the η's and to eliminate the other η's. **6.** $(X - \eta_i - 1)(X - \eta_i - 2) \cdots (X - \eta_i - p) \equiv (X - \eta_i)^p - (X - \eta_i) \equiv (X^p - X) - (\eta_i^p - \eta_i) \equiv X^p - X \equiv (X-1)(X-2)\cdots(X-p) \bmod p$. Multiplication of these polynomial congruences for $i = 1, 2, \ldots, e$ gives $\phi(X-1)\phi(X-2)\cdots\phi(X-p) \equiv (X-1)^e(X-2)^e \cdots (X-p)^e \bmod p$. To say that j is a root of $\phi(X) \equiv 0 \bmod p$ with multiplicity exactly k means that $\phi(X) \equiv q(X)(X-j)^k \bmod p$ and $q(j) \not\equiv 0 \bmod p$, i.e. $X - j$ does not divide $q(X) \bmod p$. Then $\phi(X-1)\phi(X-2)\cdots\phi(X-p) \equiv Q(X-1)Q(X-2)\cdots Q(X-p)(X-1)^t(X-2)^t \cdots (X-p)^t$ where t is the total number of roots of $\phi(X) \equiv 0$ counted with multiplicities and $Q(X) \equiv 0$ has no roots. If $a(X)b(X) \equiv c(X)b(X) \bmod p$ and $b(X) \not\equiv 0 \bmod p$ then $a(X) \equiv c(X) \bmod p$. Thus $Q(X-1)Q(X-2)\cdots Q(X-p) \equiv [(X-1)(X-2)\cdots(X-p)]^{e-t}$. Since $Q(X) \equiv 0 \bmod p$ has no roots, $e = t$ as was to be shown. **7.** $\phi(X) = X^2 - 1$. **8.** Compute $\eta_0^2, \eta_0^3, \ldots, \eta_0^e$ in terms of η's, write these relations as congruences on the u's, and use the known value of u_0. This gives $e - 1$ inhomogeneous linear "equations" in $e - 1$ unknowns. If the determinant of the coefficients is nonzero mod p this uniquely determines $u_1, u_2, \ldots, u_{e-1} \bmod p$. However, it is not obvious that if $u_0, u_1, \ldots, u_{e-1}$ satisfy these congruences they satisfy *all* congruences derived from relations among the η's. **9.** $p - \eta_i$ is never divisible by p (because its coefficients are not congruent mod p) unless $e = 1, f = \lambda - 1, \eta = -1$, in which case $0 - \eta$ is not divisible by p. **11.** $\alpha^{-1} = \sigma^k \alpha$ where $k = (\lambda - 1)/2$. Thus α^{-1} lies in η_k and the problem is to evaluate $ef/2 \bmod e$. If f is even $ef/2 \equiv 0 \bmod e$. Otherwise $ef/2 \equiv e/2 \bmod e$.

4.11 **1.** See the proof of the second theorem of Sec. 4.12. **2.** Use the method of Ex.1.

4.12 **1.** Let $\phi(\eta) = u_j - \eta_0$. If u_j occurs only once (mod p) in the list $u_1, u_2, \ldots, u_e$ then only one prime divisor of p divides $\phi(\eta)$. Therefore the same is true of $\phi(\eta) + kp$ for all k. Note that $(\phi(\eta) + pk)(\sigma\phi(\eta) + pk)\cdots(\sigma^{e-1}\phi(\eta) + pk) \equiv K + pkM \bmod p^2$ where $K = \prod \sigma^i\phi(\eta)$ and $M = \sum(K/\sigma^i\phi(\eta))$. All but one of the terms of the sum M are divisible by the given prime divisor of p. Thus $M \not\equiv 0 \bmod p$ and there is exactly one value of $k \bmod p$ for which $(K/p) + kM \equiv 0 \bmod p$. Thus there is exactly one value of k for which $N(\phi(\eta) + pk)$ is divisible by a higher power of p than p^f. **2.** From Ex. 1, Sec. 4.9, $\Psi(\eta) = \eta_0 + \eta_3 + \eta_4$. Then $\phi(\eta) \equiv \eta_1 + \eta_2 + \eta_5$. Either $\psi(\eta) = \phi(\eta)$ or $\psi(\eta) = \phi(\eta) + 2$. By computation $\phi(\eta) \cdot \sigma^2\phi(\eta) \cdot \sigma^4\phi(\eta) = 9 - 5\theta_1$. Thus $N\phi(\eta) = 2^5 \cdot 163^5$ and the prime divisors of 2 are $(2, \eta_1 + \eta_2 + \eta_5), (2, \eta_2 + \eta_3 + \eta_0), \ldots, (2, \eta_0 + \eta_1 + \eta_4)$. **3.** Because $N(1 - \alpha + \alpha^{-2})$, they can be written as $(47, 1 - \alpha + \alpha^{-2})$ and its conjugates. Alternatively, because $N(\alpha - 4) = (4^{23} - 1)/(4 - 1) = 23456248059221$ is not divisible by 47^2, they can be written as $(47, \alpha - 4)$ and its conjugates. **4.** The factorization can be put in this form if and only if at least one of the integers $u_1, u_2, \ldots, u_e$ occurs with multiplicity 1. Thus it is impossible in Ex. 2, possible in Ex. 3. **5.** Let $A = (p_1, \psi_1)^{\mu_1} \ldots$ as in the text, and let $n = p_1^{\mu_1} p_2^{\mu_2} \cdots p_m^{\mu_m}$. Then congruence mod n implies congruence mod A and there are at most $n^{\lambda - 1} < \infty$ congruence classes mod n. **6.** See Ex. 2, Sec. 4.11.

4.13 **1.** (iii) If a and b are cyclotomic integers for which $ab \in \mathcal{I}$ then $a \in \mathcal{I}$ or $b \in \mathcal{I}$. **2.** $\Sigma b_i g_i$ always lies in $\mathcal{I}$. Conversely, if $x \in \mathcal{I}$ then $x \equiv g_i \bmod g_1$ for some g_i and $x = g_i + b_1 g_1$ for some b_1. **3.** $\mathcal{I}$ corresponds to the greatest common divisor of $g_1, g_2, \ldots, g_n$ unless $\mathcal{I} = \{0\}$.

4.14 **1.** If A is a power of a prime divisor, say $A = P^\nu$, and if σA divides $g(\alpha)$ then p^ν divides $g(\alpha) \cdot \sigma \Psi(\eta)^\nu$, which implies that p^ν divides $\sigma^{-1} g(\alpha) \cdot \Psi(\eta)^\nu$ and therefore that A divides $\sigma^{-1} g(\alpha)$. If A, B are relatively prime and $\sigma(AB) = \sigma(A)\sigma(B)$ divides $g(\alpha)$ then both $\sigma(A)$ and $\sigma(B)$ divide $g(\alpha)$, both A and B divide $\sigma^{-1} g(\alpha)$, and, therefore, AB divides $\sigma^{-1} g(\alpha)$. This shows that, for any A, $g(\alpha) \equiv 0 \bmod \sigma A$ implies $\sigma^{-1} g(\alpha) \equiv 0 \bmod A$. Then $g(\alpha) \equiv 0 \bmod \sigma^i A$ implies $\sigma^{-1} g(\alpha) \equiv 0 \bmod \sigma^{i-1} A, \ldots, \sigma^{-i} g(\alpha) \equiv 0 \bmod A$. Conversely, $\sigma^{-i} g(\alpha) \equiv 0 \bmod \sigma^{(\lambda-1)-i}(\sigma^i A)$ implies $g(\alpha) = \sigma^{-(\lambda-1)+i} \sigma^{-i} g(\alpha) \equiv 0 \bmod \sigma^i A$. Thus $(g_1 - g_2) \equiv 0 \bmod \sigma^i A$ is equivalent to $\sigma^{-i}(g_1 - g_2) \equiv 0 \bmod A$, as was to be shown. **2.** The norm of a product is the product of the norms. The proof is easy for prime divisors. **3.** Let AC be the divisor of g_n. Then B and C are relatively prime. Write $A = A' A_B A_C$ where A_B contains all the prime factors of A that occur in B, and A_C all those that occur in C. Find y so that $y \equiv 1 \bmod A_B B$ and $y \equiv 0 \bmod A_C C$. Then $\phi y \equiv \phi \bmod A_B B$ and $\phi y \equiv 0 \equiv \phi \bmod A' A_C$, so that $\phi y \equiv \phi \bmod AB$. Similarly $\phi y \equiv 0 \bmod AC$, $\phi y = b_n g_n$.

4.15 **2.** Suppose $P = (p, \psi)$ is not the divisor of any cyclotomic integer. If $\psi = \psi_1 \psi_2$ then either $P = (p, \psi_1)$ or $P = (p, \psi_2)$. Assume, therefore, that ψ is irreducible. Let $N\psi = p^j p_1^{\nu_1} \cdots p_n^{\nu_n}$. Write $p, p_1, \ldots, p_n$ as products of irreducibles. Then $N\psi$ is a product of irreducibles in two essentially different ways.

5.2 **1.** $\lambda = 31$: $(a + b\theta_0)(a + b\theta_1) = a^2 - ab + 8b^2$. Even a very brief table gives $(1 + \theta_0) = (2, 1)^3, (2 + \theta_0) = (2, 0)(5, -2), (3 + \theta_0) = (2, 1)(7, -3)$. Thus $(2, 1) \sim (5, -2) \sim (7, 2)$ and $(2, 1)^3 \sim I$. As before, $(2, 1) \nsim I$ and therefore $(2, 1)^2 \nsim I$. Since $\frac{1}{4}p^2 + \frac{1}{2}p + 8 < p^2$ when $p \geqslant 5$, every divisor is a product of prime divisors of 2 or 3. The exponent of 3 mod 31 is 30, so 3 is prime. Therefore $I, (2, 1), (2, 1)^2 \sim (2, 0)$ is a representative set. $\lambda = 39$: $(a + b\theta_0)(a + b\theta_1) = a^2 - ab + 10b^2$. Again only $p = 2, 3$ enter. $(3 + \theta_1) = (2, 1)^4, (2 + \theta_1) = (2, 0)^2(3, 1)$. Thus $(2, 1)^2 \sim (3, 1)$. Since neither $(2, 1)$ nor $(3, 1) \sim I$, a representative set is $I, (2, 1), (2, 1)^2 \sim (3, 0)$, and $(2, 1)^3 \sim (2, 0)$. $\lambda = 43$: $a^2 - ab + 11b^2$. Again only $p = 2, 3$ enter. A short table of values gives prime values and no factors of 2 or 3. In fact, 2 has exponent 14 and 3 has exponent 42, so that e is odd in both cases and there is no $(2, u)$ or $(3, u)$. Thus all divisors are principal and I is a representative set.

5.3 **1.** (1) If $f(\alpha)$ has divisor A and $g(\alpha)$ has divisor B then $f(\alpha) g(\alpha)$ has divisor AB. (2) If $f(\alpha)$ has divisor A and $h(\alpha)$ has divisor AB then $f(\alpha)$ divides $h(\alpha)$ and the quotient has divisor B. (3) If $A \sim I$ then A is principal because I is (replace A by I in A). If A is principal then, by (1) and (2), C is principal if and only if AC is; thus $A \sim I$. (4) If $AC \sim BC \sim I$ and $AD \sim I$ then $(BD)(AC) = (AD)(BC) \sim I$ by (1), therefore $BD \sim I$ by definition of $AC \sim I$. By symmetry, $BD \sim I$ implies $AD \sim I$. Therefore $A \sim B$. Now suppose $A \sim B$. By (7), there is a C such that $AC \sim I$. Then $BC \sim I$ by definition. (5) and (6) are immediate. (7) $N(A)$ is the divisor of an integer; therefore the complement B of A in $N(A)$ has the required property. (8) Let C be such that $AC \sim I$ and set $M = BC, N = AC$. **2.** See Ex. 2, Sec. 4.15. **3.** When $g(\alpha) = a_1 \alpha + a_2 \alpha^2 + \cdots + a_{\lambda-1} \alpha^{\lambda-1}$ one has $g(\alpha) g(\alpha^{-1}) = a_1^2 + a_1 a_2 \alpha^{-1} + a_1 a_3 \alpha^{-2} + \cdots + a_2 a_1 \alpha + a_2^2$

$+ a_2a_3\alpha^{-1} + \cdots$. The sum of the $\lambda - 1$ conjugates is then $(\lambda - 1)(a_1^2 + a_2^2 + \cdots + a_{\lambda-1}^2) - (a_1a_2 + a_1a_3 + \cdots + a_2a_1 + a_2a_3 + \cdots) = \lambda\Sigma a_i^2 - (\Sigma a_i)^2$. If $|a_i| \leqslant c$ in all cases then $Ng(\alpha)$ is a product of $(\lambda - 1)/2$ positive real numbers whose arithmetic mean is at most $[\lambda(\lambda - 1)c^2 - 0]/(\lambda - 1) = \lambda c^2$. Since their geometric mean is $(Ng(\alpha))^{2/(\lambda-1)} \leqslant \lambda c^2$ the result follows.

5.5 **1.** From $\alpha \equiv 1$ follows $g(\alpha) \equiv g(1) = \text{int. mod } \alpha - 1$. If $g(\alpha) \equiv a_0 \bmod (\alpha - 1)$, let $h(\alpha) = [g(\alpha) - a_0]/(\alpha - 1)$. Then $h(\alpha) \equiv a_1 \bmod (\alpha - 1)$ and $g(\alpha) \equiv a_0 + a_1 \cdot (\alpha - 1)\bmod(\alpha - 1)^2$. If $[g(\alpha) - a_0 - a_1(\alpha - 1)]/(\alpha - 1)^2 \equiv a_2 \bmod (\alpha - 1)$ then $g(\alpha) \equiv a_0 + a_1(\alpha - 1) + a_2(\alpha - 1)^2 \bmod (\alpha - 1)^3$, and so forth. If $a_0 + a_1(\alpha - 1) + \cdots + a_{\lambda-2}(\alpha - 1)^{\lambda-2} \equiv b_0 + b_1(\alpha - 1) + \cdots + b_{\lambda-2}(\alpha - 1)^{\lambda-2}\bmod(\alpha - 1)^{\lambda-1}$ then $c_0 + c_1(\alpha - 1) + \cdots \equiv 0 \bmod (\alpha - 1)^{\lambda-1}$ where $c_i = a_i - b_i$. Then $c_0 \equiv 0 \bmod (\alpha - 1)$, from which $c_0 \equiv 0 \bmod \lambda$, $c_0 \equiv 0 \bmod(\alpha - 1)^{\lambda-1}$. Then $c_1(\alpha - 1) \equiv 0 \bmod (\alpha - 1)^2, (\alpha - 1)|c_1, c_1 \equiv 0 \bmod(\alpha - 1)^{\lambda-1}$. Then $c_2 \equiv 0 \bmod (\alpha - 1)^{\lambda-1}$, and so forth. **2.** Eliminate x from $(\alpha^k - \alpha^{-k})x \equiv (\alpha^{1-k} - \alpha^{k-1})y, (\alpha^{2k} - \alpha^{-2k})x \equiv (\alpha^{2-2k} - \alpha^{2k-2})y \bmod \lambda$ by multiplying the first congruence by $\alpha^k + \alpha^{-k}$ and subtracting it from the second. Use the fact that $(\alpha - 1)^2|(\alpha^\mu - \alpha^\nu)$ only if $\alpha^\mu = \alpha^\nu$. If $(\alpha - \alpha^{-1})x \equiv 0 \bmod \lambda$ then $a \equiv 0 \bmod \lambda$. If $(\alpha - \alpha^{-1})x \equiv (\alpha - \alpha^{-1})y \bmod \lambda$ then $a \equiv b \bmod \lambda$.

5.6 **1.** The computation in abbreviated form is: $1 + 3 = 4 + 5 = 9 + 7 = 16 + 9 = 25 + 11 = 36 + 13 = 49 + 15 = 64 + 17 = 81 \equiv 2 + 19 = 21 + 21 = 42 + 23 = 65 + 25 = 90 \equiv 11 + 27 = 38 + 29 = 67 + 31 = 98 \equiv 19 + 33 = 52 + 35 = 87 \equiv 8 + 37 = 45 + 39 = 84 \equiv 5 + 41 = 46 + 43 = 89 \equiv 10 + 45 = 55 + 47 = 102 \equiv 23 + 49 = 72 \equiv -7 + 51 = 44 + 53 = 97 \equiv 18 + 55 = 73 \equiv -6 + 57 = 51 + 59 = 110 \equiv 31$. Since $59 \equiv 29 + 30$ this gives $31 \equiv 30^2 \bmod 79$. **2.** The inequality involving y comes from $0 < x^2 < (\tfrac{1}{2}m)^2$. When $A = 31, m = 79$, the range of y is $0 \leqslant y \leqslant 19$. From $31 + 79y \equiv 0$ or $1 \bmod 3$, $1 + y \not\equiv 2 \bmod 3$ follows $y \not\equiv 1 \bmod 3$. This excludes 1, 4, 7, 10, 13, 16, 19. Similarly, $31 + 79y \equiv 0$ or $1 \bmod 4$ excludes $y \equiv 0$ or $1 \bmod 4$; $31 + 79y \equiv 0, 1,$ or $4 \bmod 5$ excludes $y \equiv -1$ or $-2 \bmod 5$. Only $y = 2, 6, 11, 15$ remain. $31 + 79y \equiv 0, 1, 2,$ or $4 \bmod 8$ excludes 2. $31 + 79y \equiv 0, 1, 2,$ or $4 \bmod 7$ excludes 15. Since $6 \cdot 79 + 31 = 505$ is not a square, the answer must be $y = 11$. Indeed, $31 + 79 \cdot 11 = 900 = 30^2$. **3.** $(\tfrac{22}{97}) = (\tfrac{2}{97})(\tfrac{11}{97}) = (\tfrac{1}{7})(\tfrac{97}{11}) = (\tfrac{9}{11}) = (\tfrac{3}{11})^2 = +1$. $y = 0, 1, \ldots,$ or 24. $E = 8$ gives $y \equiv 2, 3,$ or $6 \bmod 8$. $E = 9$ gives $y \equiv 0, 2, 3, 6 \bmod 9$. $E = 5$ gives $y \equiv 1, 2, 4 \bmod 5$. This leaves $y = 2, 6,$ or 11. **4.** $(\tfrac{79}{101}) = (\tfrac{101}{79}) = (\tfrac{22}{79}) = (\tfrac{2}{79})(\tfrac{11}{79}) = (\tfrac{7}{1})[-(\tfrac{79}{11})] = -(\tfrac{2}{11}) = -(\tfrac{2}{3}) = +1$. $(\tfrac{97}{139}) = (\tfrac{139}{97}) = (\tfrac{42}{97}) = (\tfrac{2}{97})(\tfrac{3}{97})(\tfrac{7}{97}) = (\tfrac{1}{7})(\tfrac{97}{3})(\tfrac{97}{7}) = (\tfrac{1}{3})(\tfrac{6}{7}) = (\tfrac{6}{7}) = -1$. $(\tfrac{91}{139}) = (\tfrac{7}{139})(\tfrac{13}{139}) = -(\tfrac{139}{7})(\tfrac{139}{13}) = -(\tfrac{6}{7})(\tfrac{9}{13}) = -(\tfrac{6}{7}) = +1$. **6.** $(\tfrac{2}{p}) = (\tfrac{-1}{p})(\tfrac{-2}{p})$ and $(\tfrac{-2}{p})$ is $+1$ for $p \equiv 1$ or $3 \bmod 8$, -1 for $p \equiv 5$ or $7 \bmod 8$. **8.** Prove Euler's criterion separately, as in Ex. 7. Deduce rule (1). Then if $p \equiv 1 \bmod 4$ one has $(\tfrac{p}{q}) = (\tfrac{q}{p})$. If $p \equiv 3$ and $q \equiv 1 \bmod 4$ then $(\tfrac{q}{p}) = (\tfrac{p}{q}) = (\tfrac{-1}{q})(\tfrac{p}{q})$. If $p \equiv q \equiv 3 \bmod 4$ then $(\tfrac{q}{p}) = -(\tfrac{p}{q}) = (\tfrac{-1}{p})(\tfrac{p}{q})$. **9.** Both $p - 1$ and $q - 1$ are even. If $p \equiv 1$ or $q \equiv 1 \bmod 4$ then $\tfrac{1}{2}(p - 1) \cdot \tfrac{1}{2}(q - 1)$ is even. If both $p \equiv 3$ and $q \equiv 3 \bmod 4$ then $\tfrac{1}{2}(p - 1) \cdot \tfrac{1}{2}(q - 1)$ is odd. **10.** The first congruence implies $24u^2 - 16u \equiv 56, u^2 - 16u + 64 - 64 \equiv 10, (u - 8)^2 \equiv 5 \bmod 23$. Since $(\tfrac{5}{23}) = (\tfrac{23}{5}) = (\tfrac{3}{5}) = -1$, this is impossible. The second implies $35u^2 + 30u \equiv 10, -2u^2 + 30u \equiv 10, u^2 - 15u \equiv -5, u^2 + 22u + 121 \equiv 116, (u + 11)^2 \equiv 5 \bmod 37$. Since $(\tfrac{5}{37}) = (\tfrac{37}{5}) = (\tfrac{2}{5}) = -1$ this too is impossible. **11.** If $x^2 \equiv B$ has a solution then $B^{(p-1)/2} \equiv x^{p-1} \equiv 1 \bmod p$. Therefore $B^{(p+1)/2} \equiv B \bmod p$ and the square of $x \equiv B^{(p+1)/4}$ is $B \bmod p$. The computation of $31^{20} \bmod 79$ is simple: $31^2 \equiv 13, 31^4 \equiv 13^2 \equiv 11, 31^8 \equiv 121 \equiv 42, 31^{16} \equiv 42^2 \equiv 26, 31^{20} \equiv 11 \cdot 26 \equiv 49 \equiv -30$.

6.3 **1.** When the product is written as a product over $p \equiv 1 \bmod \lambda$ times one over $p \not\equiv 1 \bmod \lambda$, the second factor is less than $\prod(1 - p^{-2s})^{-(\lambda-1)} < \prod(1 - p^{-2})^{1-\lambda} = \zeta(2)^{\lambda-1} < \infty$ and greater than 1. The first factor is the product over all $p \equiv 1 \bmod \lambda$ of $(1 - p^{-s})^{-(\lambda-1)}$. Its log is $(\lambda - 1)\Sigma(p^{-s} + \frac{1}{2}p^{-2s} + \frac{1}{3}p^{-3s} + \cdots) = (\lambda - 1)\Sigma p^{-s} + \text{finite}$, where the sum is over $p \equiv 1 \bmod \lambda$, because $\log(1-x)^{-1} - x \leqslant \frac{1}{2}x^2$ for $0 \leqslant x \leqslant \frac{1}{2}$. **2.** $1000 < \zeta(s) < 1 + 1000$. **3.** As in Ex. 1, $\Sigma n^{-s} = \Sigma p^{-s} + \text{finite}$. Since $p^{-s} < p^{-1}, \Sigma p^{-1} < \infty$ would imply Σp^{-s} was bounded as $s \downarrow \infty$ and thus that Σn^{-s} was bounded as $s \downarrow \infty$.

6.4 **1.** The powers of β are distinct roots of $X^n - 1$ so $X^n - 1 = (X - \beta)(X - \beta^2)\cdots(X - \beta^n)$. Set $X = x/y$ and multiply by y^n. If β is an μth root for some μ, $n = \mu\nu$, and the product is $(x^\mu - y^\mu)^\nu$. **2.** Suppose $p \equiv \gamma^j \bmod \lambda$. Then β^j is a primitive fth root of 1 and, by Ex. 1, $\prod_k(1 - \beta^{jk}p^{-s}) = (1^f - (p^{-s})^f)^e = \prod(1 - N(P))^{-s}$ where P ranges over prime divisors of p. Rearrangement of the product is valid by absolute convergence. **3.** χ_0 is 1 except 0 for multiples of 13. $\chi_6(n) = 0$ for $n \equiv 0 \bmod 13$, $= 1$ for $n \equiv \pm 1, \pm 3, \pm 4 \bmod 13$, $= -1$ for all other n. The first 10 nonzero terms correspond to $n = 1, 2, 3, 4, 6, 8, 9, 12, 16, 18$. **4.** $\chi(2)^3 = 0$ implies $\chi(n) = 0$ for n even. $\chi(3)^2 = 1$ and $\chi(5)^2 = 1$ imply $\chi(3) = \pm 1, \chi(5) = \pm 1$. $\chi(7) = \chi(3)\chi(5)$ implies the values of $\chi(3)$ and $\chi(5)$ determine χ.

6.5 **1.** $(1 - \theta_0)(1 - \theta_1) = 1$ implies $\log(1 - \theta_0) = -\log(1 - \theta_1)$. **2.** By (8), $L(1, \chi_3) = (i\pi m_3/\lambda)(1 - 3 + 2 - 6 + 4 - 5)$ where $m_3 = (\alpha - \alpha^3 + \alpha^2 - \alpha^6 + \alpha^4 - \alpha^5)/7 = (\theta_0 - \theta_1)/7$. By geometry, θ_0 is in the upper halfplane, θ_1 in the lower. From $(\theta_0 - \theta_1)^2 = -7$, $\theta_0 - \theta_1 = i\sqrt{7}$ and $L(1, \chi_3) = \pi/\sqrt{7} = 1.1874\ldots$. **3.** $\sigma^\mu\alpha$ is α to the power γ^μ. Since $(\gamma^\mu)^2 = \gamma^{\lambda-1} \equiv 1 \bmod \lambda$ and $\gamma^\mu \not\equiv 1 \bmod \lambda$, $\gamma^\mu \equiv -1 \bmod \lambda$. **4.** $\lambda = 3$: $L(1, \chi_1) = i\pi(\omega - \omega^2)(1 - 2)/3^2 = \pi/3\sqrt{3}$. $\lambda = 5$: $L(1, \chi_1) = i\pi(\alpha + i\alpha^2 - \alpha^4 - i\alpha^3)(1 - 2i - 4 + 3i)/5^2 = (2\pi\sin(2\pi/5) + 2\pi i\sin(4\pi/5))(3 - i)/25$. $L(1, \chi_2) = (\sqrt{5}/5)\log((3 + \sqrt{5})/2)$. $L(1, \chi_3) = \overline{L(1, \chi_1)}$. $\lambda = 7$: $L(1, \chi_3) = \pi/\sqrt{7}$. $L(1, \chi_1) = i\pi(\alpha + \beta\alpha^3 + \beta^2\alpha^2 + \beta^3\alpha^6 + \beta^4\alpha^4 + \beta^5\alpha^5)(\beta^{-1} + 3\beta^{-2} + 2\beta^{-3} + 6\beta^{-4} + 4\beta^{-5} + 5\beta^{-6})/49$ where $\beta = \exp(2\pi i/6)$, $\alpha = \exp(2\pi i/7)$. $L(1, \chi_5) = \overline{L(1, \chi_1)}$. $L(1, \chi_2) = -2(\alpha + \omega\alpha^3 + \omega^2\alpha^2 + \alpha^6 + \omega\alpha^4 + \omega^2\alpha^5)(\log|1 - \alpha| + \omega^{-1}\log|1 - \alpha^3| + \omega^{-2}\log|1 - \alpha^2|)/7$ where $\omega = \exp(2\pi i/3)$. $L(1, \chi_4) = \overline{L(1, \chi_2)}$.

6.9 **1.** $\theta^j = (-1)^j[a_{j-1} - a_j\theta]$ where the sequence $\ldots, a_{-2}, a_{-1}, a_0, a_1, a_2, \ldots$ is defined by $a_0 = 0, a_1 = 1, a_{i+1} = a_i + a_{i-1}$. Clearly, then, $\theta^j \neq \pm \theta^i$ unless $i = j$ and the sign is $+$. Let $a + b\theta$ be a unit. If $a + b\theta$ has norm -1 then $(a + b\theta)\theta$ has norm 1. If b is odd then the coefficient of θ in either $(a + b\theta)\theta$ or $(a + b\theta)\theta^2$ is even. Therefore it suffices to show that if $N(a + b\theta) = 1$ and if b is even then $a + b\theta = \pm \theta^j$. Since $N(a + 2c\theta) = N((a - c) + c(\theta_0 - \theta_1)) = (a - c)^2 - 5c^2$ gives a solution of Pell's equation, it is of the form $(a - c) + c\sqrt{5} = \pm(9 + 4\sqrt{5})^j$ where $\theta_0 - \theta_1 = \sqrt{5}$. Since $9 + 4\sqrt{5} = 13 + 8\theta_0 = \theta^6$, this shows that $a + b\theta = \pm \theta^{6j}$. **4.** $1 + 2\eta_0 = -\eta_0^2\eta_1^{-1}$. **6.** $1 + \rho^{i-j} + \rho^{2i-2j} + \cdots + \rho^{-i+j}$ is 0 if $i \not\equiv j \bmod n$, 1 if $i \equiv j \bmod n$. **8.** Let $e_0 = (1 - \sigma^2\alpha) \cdot (1 - \sigma\alpha)^{-1}$ for some fixed σ and let $e_{i+1} = \sigma e_i$. The argument of the text shows that $\log|e_{j+\mu}| = \log|e_j|$ (here $\mu = 8$) and that the system of equations $\log|\sigma^i E(\alpha)| = \Sigma_{j=0}^{\mu-1} r_j \log|e_{j+i}|$ defines the r's implicitly as functions of the unit $E(\alpha)$. (In the case $\lambda = 7$ of the text, $\eta_0 = \alpha^{-1}e_0, \log|\eta_i| = \log|e_i|$.) When $\lambda = 17$, choose $\sigma: \alpha \to \alpha^3$. Then $\eta_0 = \alpha^{-1}(1 - \alpha^4)(1 - \alpha^2)^{-1} = \alpha^{-1}(1 - \sigma^{12}\alpha) \cdot (1 - \sigma^{14}\alpha)^{-1} = \alpha^{-1}e_{11}^{-1}e_{12}^{-1}, \log|\eta_0| = -\log|e_3| - \log|e_4|$; i.e., for $E(\alpha) = \eta_0, r_3 =$

$r_4 = -1$ and the other r's are 0. If $E(\alpha)$ is a product of powers of η's then the sum of the corresponding r's is even. Therefore e_0 is not a product of powers of η's. Since every unit of the form (4) is a product of e's, it suffices to express the e's—in fact e_0—in terms of η's. If 2 is a primitive root mod λ then $e_0 = (1 - \alpha^4)(1 - \alpha^2)^{-1} = 1 + \alpha^2 = \alpha^{-1}\eta_0$ and if -2 is a primitive root mod λ (as when $\lambda = 7$) then $e_0 = (1 - \alpha^4)(1 - \alpha^{-2})^{-1} = -\alpha^4(1 - \alpha^{-4})(1 - \alpha^{-2})^{-1} = -\alpha^3\eta_0$.

6.10 **2.** If $\Phi(e_1) - \Phi(e_2)$ has integer entries then e_1 is equivalent to $e_3 e_2$ where e_3 is a real unit for which $\Phi(e_3)$ has zero entries. Since $Le_3 = 0$, $e_3 = \pm 1$.

6.14 **1.** By direct computation, $P = -1, 2 \cdot 5, -2^2 \cdot 7^2, -2^4 \cdot 11^4, 2^5 \cdot 13^5$ for $\lambda = 3, 5, 7, 11, 13$ respectively.

6.15 **1.** Prove the existence, for each k, of the kth *cyclotomic polynomial*, a polynomial Φ_k with integer coefficients and with leading coefficient 1 whose roots are precisely the primitive kth roots of unity. This is done by dividing $X^k - 1$ by Φ_j for all proper divisors j of k. In other words, $\Phi_k(X) = (X - \beta_1)(X - \beta_2) \cdots (X - \beta_{\phi(k)})$. For each fixed j, $\sum \beta_i^j$ is an integer. This can be proved either by observing that β_i^j ranges over all primitive (k/d)th roots, where $d = $ g.c.d.(j, k), giving each of them the same number of times, or by using the general theorem that every symmetric polynomial is a polynomial in the elementary symmetric polynomials. Thus $g(\beta_1) + g(\beta_2) + \cdots + g(\beta_{\phi(n)})$ is on the one hand an integer and on the other hand $\phi(n) g(\beta_1)$. Thus $g(\beta_1) = r$ where r is rational (with a denominator which divides $\phi(n)$). Let the polynomial $\psi(X) = g(X) - r$ be divided by $\Phi_n(X)$ to give $\psi(X) = q(X)\Phi_n(X) + r(X)$. Then $r(X)$ has degree less than that of $\Phi_n(X)$ and $r(\beta_i) = 0$ for all β_i. Then $r(X) = 0$ and $g(X) - r = q(X)\Phi_n(X)$, from which it follows that r is an integer. **3.** Set $I_k(x) = \int_x^{x+1} t^k \, dt$. Then $I_k(x)$ is a polynomial of degree k with leading coefficient one. Two polynomials are equal for all x only if they are identical. The desired equation is equivalent to a system of equations $a_N = $ known, $a_{N-1} + q a_N = $ known, $a_{N-2} + r a_{N-1} + s a_N = $ known,... where the known quantities are all 0 except that the known quantity which begins with a_n is 1. These equations then imply $N \geqslant n$. Conversely, if $N \geqslant n$, there is one and only one set of a's satisfying these equations (because the system is triangular). **4.** It suffices to show that the two sides have equal derivatives. The derivative in both cases is $\sum_{k=0}^{n-1} n! B_k x^{n-1-k} / k!(n-k-1)!$, by direct computation in one case and by induction in the other. **5.** $(-x)^n = \int_{-x}^{-x+1} B_n(t) \, dt = \int_{1+x}^{x} B_n(1-u) \, d \cdot (1-u) = \int_x^{x+1} B_n(1-u) \, du$, $x^n = \int_x^{x+1}(-1)^n B_n(1-t) \, dt$. Thus $B_{2n+1}(\tfrac{1}{2}) = (-1) B_{2n+1}(\tfrac{1}{2})$, $B_{2n+1}(\tfrac{1}{2}) = 0$. Then $B_{2n+1} = 0$ for $n > 0$ follows from $B_{2n+1}(\tfrac{1}{2}) - B_{2n+1} = 2^{-2n}(1 - 2^{2n+1}) B_{2n+1}$. **6.** $B_3(t) = t(t - \tfrac{1}{2})(t-1)$ is 0 at $0, \tfrac{1}{2}, 1$, positive between 0 and $\tfrac{1}{2}$, negative between $\tfrac{1}{2}$ and 1. $B_5(t)$ is 0 at $0, \tfrac{1}{2}, 1$. It cannot have other zeros in $0 \leqslant t \leqslant 1$ because if it did its derivative $5 B_4(t)$ would have 3 zeros and $20 B_3(t)$ would have 2 zeros inside the interval. Similarly, the only zeros of $B_{2n+1}(t)$ on $0 \leqslant t \leqslant 1$ are at $0, \tfrac{1}{2}, 1$. Thus $B_{2n}(t)$ is monotone on $0 \leqslant t \leqslant \tfrac{1}{2}$ and has just one 0 there. If $B_{2n}(0) > 0$ then $B_{2n}(t)$ is decreasing and $B_{2n-1}(t) < 0$ for $0 < t < \tfrac{1}{2}$; it decreases from 0 to a minimum value and then increases to 0 at $\tfrac{1}{2}$, so its derivative is negative at 0 and $B_{2n-2} < 0$. Similarly, if $B_{2n} < 0$ then $B_{2n-2} > 0$. **7.** $B_n(x) - B_n = n[1 + 2^{n-1} + 3^{n-1} + \cdots + (x-1)^{n-1}] = $ integer. **8.** $\psi(x) = x^6 + x^7 + x^9 + x^{10} + x^{11} + x^{13} + $

$x^{15} + x^{16} + x^{19} + x^{20} + x^{21} + x^{22} + x^{23} + x^{26} + x^{30} + x^{35} + x^{36}$. Using a table of powers of 2, namely, $2, 4, 8, 16, -5, -10, \ldots, 14, -9, -18, 1$ one finds $\psi(2) \equiv -10 + 17 - 6 - 12 + 13 + 15 - 14 + 9 - 2 - 4 - 8 - 16 + 5 + 3 + 11 + 7 - 18 + 1 = -9$. For $\psi(2^n)$ construct a table of powers of 2^n. When $n = 31, 2^n \equiv -15$ one finds $\psi(2^{31}) \equiv -10 + 2 + 6 - 16 + 18 + 17 + 14 + 12 + 15 - 3 + 8 - 9 - 13 - 7 + 11 - 4 - 5 + 1 \equiv 0 \mod 37$. **9.** A good check on the construction of Pascal's triangle $\mod 13$ is the fact that the row for the exponent 12 is $1, -1, 1, -1, \ldots, -1, 1$. The congruences $B_0 = 1, B_0 + 2B_1 \equiv 0, B_0 + 3B_1 + 3B_2 \equiv 0, B_0 + 5B_1 - 3B_2 - 3B_3 + 5B_4 \equiv 0, \ldots$ are easily solved to find $B_0 \equiv 1, B_1 \equiv 6, B_2 \equiv -2, B_4 \equiv 3, B_6 \equiv -4, B_8 \equiv 3, B_{10} \equiv 5$. (Of course $B_{2n+1} \equiv 0$ for $n > 0$.) The direct evaluation of $B_{10}(7)$ shows it is $\equiv -4 \mod 13$. Also $B_{10}(\frac{1}{2}) = B_{10} + 2^{-9}(1 - 2^{10})B_{10} \equiv -4$.

6.16   **1.** $\psi(\gamma^n) \equiv (\gamma_{n+1} - 1)B_{n+1}(n+1)^{-1} \mod \lambda$ for $0 < n \leqslant \lambda - 3$. **2.** When $\lambda = 13$, $\psi(2^n) \equiv -3, 0, -5, 0, -3, 0, 3, 0, -2, 0, 3, 6$ for $n = 1, 2, \ldots, 12$.

6.17   **1.** If h is the order of G then a^h is the identity for all a in G. If $\lambda \nmid h$ then $m\lambda = nh + 1$ and $a = a \cdot a^{hn} = a^{m\lambda}$. **2.** Set $e_2(\alpha) = \alpha^m E(\alpha)$ where $E(\alpha)$ is real. Then $\pm \alpha^k = e_2(\alpha)^\lambda = \alpha^{m\lambda} E(\alpha)^\lambda$ gives $|E(\alpha)| = 1, E(\alpha) = \pm 1$. Another method is to note that $\alpha^k \equiv$ integer $\mod \lambda$ implies $1 + k(\alpha - 1) \equiv$ integer $\mod(\alpha - 1)^2$, $k \equiv 0 \mod \lambda$, $(\pm e_2(\alpha))^\lambda = 1$, and $\pm e_2(\alpha)$ is a root of $X^\lambda = 1$. **4.** By infinite descent it will suffice to show: *If $p \mid o(G)$, G Abelian, and G has no element of order p then there is an Abelian G' with $p \mid o(G')$, G' Abelian, G' has no element of order p, and $o(G') < o(G)$.* Let a be any element of G other than e. Let v be the order of a. Then $p \nmid v$ because otherwise $a^{v/p}$ would have order p. Let $G' = G/(a)$. Then $p \mid o(G')$, $o(G') < o(G)$, and G' is Abelian. It is to be shown that if G' had an element of order p—say the coset of b—then G would have one. If there is such b then $b^p = a^k$. Let $n = \text{g.c.d. } (v, k), k = nK$, $v = nN$. Then $e = a^{vK} = a^{Nk} = b^{pN}$. Then b^N has order p unless $b^N = e$. But $b^N = e$ would imply $p \mid N \mid v$.

7.1   **1.** $x^2 - Dy^2 = (x + y\sqrt{D})(x - y\sqrt{D})$. If p is prime and $p \mid (x^2 - Dy^2)$ it divides one of the factors and therefore both factors. Conversely, if $p \mid (x + y\sqrt{D}) \cdot (u + v\sqrt{D})$ then $p \mid (x^2 - Dy^2)(u^2 - Dv^2)$, and $p \mid (x^2 - Dy^2)$ or $p \mid (u^2 - Dv^2)$; the given implication then shows p is prime. **2.** A quadratic integer $u + v\sqrt{D}$ in which v is 0 is an ordinary integer. **5.** $-10, -7, -6, -5, -3, -2, -1, 2, 3, 5, 6, 7, 10$. **6.** Show that $u^2 \equiv -1 \mod p$ for odd p has a solution if and only if $p \mid (a^2 + b^2)$ for some relatively prime a, b. Thus 2 ramifies, primes $\equiv 1 \mod 4$ split, primes $\equiv 3 \mod 4$ remain prime. **7.** 2 ramifies, primes $p \equiv 1$ or 3 $\mod 8$ split, others remain prime. **8.** Primes $\equiv \pm 1 \mod 8$ split. **9.** If $p \equiv 1 \mod 3$ then $f = 1, e = 2$, and p has 2 distinct prime factors, i.e. p splits. If $p \equiv 2 \mod 3$ then $e = 1$, i.e. p remains prime. $(3) = (\alpha - 1)^2$, i.e. 3 ramifies. **10.** 5 ramifies. If $p \equiv 2$ or 3 $\mod 5$ then p is prime as a cyclotomic integer and *a fortiori* as a quadratic integer. If $p \equiv 1$ or 4 $\mod 5$ and if A is a prime divisor of p then $\theta_0, \theta_1 \equiv$ integers $\mod A$, $\theta_0 - \theta_1 \equiv u \mod A$, $u^2 \equiv (\theta_0 - \theta_1)^2 = 5 \mod A$, $u^2 \equiv 5 \mod p$ and p splits. **11.** 2 and 5 ramify. $(\frac{-5}{p}) = (\frac{-1}{p})(\frac{p}{5})$ depends on the class of $p \mod 4 \cdot 5$. In fact, it is $+1$ if $p \equiv 1, 3, 7, 9 \mod 20$, -1 if $p \equiv 11$, 13, 17, 19 $\mod 20$. **12.** In each case the problem is to factor 2 and 3. For example, when $D = -5$, 2 ramifies, 3 splits, and there are 3 divisors $(2, *)^2(3, \pm 1)^2$ with norm 36. **13.** It suffices to prove $\omega \equiv$ integer $\mod (p, u)$, but $\omega \equiv (1 - p)\omega = [(1 - p)/2](1 - \sqrt{D}) = \frac{1}{2}(1 - p)(1 - u) =$ integer $\mod (p, u)$.

7.2 1. If $x+y\sqrt{D}$ has divisor A and $u+v\sqrt{D}$ divisor AB then $x+y\sqrt{D}$ divides $u+v\sqrt{D}$ and the quotient has divisor B. **2. (a)** $(4+7\sqrt{3})=(131,u)$ where $4+7u\equiv 0$ mod 131, i.e. $u\equiv -38$. **(b)** $(11,3)(17,-7)$. **(c)** $(3,1)(3,-1)\cdot$ $(5,*)$. **(d)** $(3,*)(7,*)$. **(e)** I **(f)** $(5,2)(5,-2)(2,*)$. **3.** Immediate from Proposition 2.

7.3 1. $1-\sqrt{2}$, $(1-\sqrt{2}^{\,2})=3-2\sqrt{2}$, $(1-\sqrt{2})^3=7-5\sqrt{2}$ have divisors $(-1,*)$, I, $(-1,*)$ respectively. **2.** $u^2-3v^2=-1$ would give $u^2+v^2\equiv 3$ mod 4. **3.** $2+\sqrt{5}$ has norm -1.

7.4 1. When $A_0=(5,2)(13,-5)$, $\sqrt{-1}\equiv r$ mod A_0 if and only if $r\equiv 2$ mod 5, $r\equiv -5$ mod 13, i.e. $r\equiv -18$ mod 65. From

$$
\begin{array}{ccc}
 & -18 & -2 \\
65 & 5 & 1
\end{array}
$$

A_2 is the divisor of 1, A_1 the divisor of $-2-\sqrt{-1}$, A_0 the divisor of $(-18-\sqrt{-1})(-2-\sqrt{-1})/5$. Thus the 4 solutions are $\pm(7+4\sqrt{-1})$, $\pm(-4+7\sqrt{-1})$. **2.** $r_0=-11/2$, $a_1=1$. The 6 solutions are $\pm\frac{1}{2}\cdot$ $(-11-\sqrt{-3})$, $\pm(2+3\sqrt{-3})$, $\pm\frac{1}{2}(7-5\sqrt{-3})$. **3.** The solution of $r\equiv 3(11)$, $r\equiv 11(41)$, $r\equiv 20(67)$ is $r\equiv 12147$ mod 30217. From

$$
\begin{array}{ccccc}
 & 12147 & -2381 & 59 & 1 \\
30217 & 4883 & 1161 & 3 & 1
\end{array}
$$

one finds that A_4, A_3, A_2, A_1, A_0 are the divisors of 1, $1-\sqrt{-2}$, $19-20\sqrt{-2}$, $-39+41\sqrt{-2}$, $-97+102\sqrt{-2}$ respectively. The 2 solutions are $\pm(97-102\sqrt{-2})$. Alternatively, $(11,3)$, $(41,11)$ and $(67,20)$ are the divisors of $3-\sqrt{-2}$, $3-4\sqrt{-2}$, and $7+3\sqrt{-2}$ respectively, and the product of these is $97-102\sqrt{-2}$. **4.**

$$
\begin{array}{cccc}
 & 8 & 1 & 1 \\
23 & 3 & 2 & 3
\end{array}
$$

1 is not reached and the divisor is not principal. **5.** A_0 divides $(8-\sqrt{-5})^2=$ $59-16\sqrt{-5}$. Since $(16)^{-1}\equiv -10$ mod 23, $(16)^{-1}\equiv -33$ mod 23^2, A_0 divides $(-33)(59)-\sqrt{-5}\equiv 169-\sqrt{-5}$ mod 23^2. Then

$$
\begin{array}{ccc}
 & 169 & -7 \\
529 & 54 & 1
\end{array}
$$

gives A_0 as the divisor of $22+3\sqrt{-5}$. **6.** p divides $N(x+y\sqrt{-1})$ but not $x+y\sqrt{-1}$ so it is not prime. If p ramifies $p=2=1^2+1^2$. If p splits and (p,u) is principal then $p=a^2+b^2$. It suffices, therefore, to show that all divisors are principal. It was shown in the text that every divisor is equivalent to a divisor with norm at most $2\sqrt{1/3}<2$, i.e. equivalent to I. **7.** If $p|x^2+y^2$ and $p\neq 2$ then $p=u^2+v^2$ and $p\equiv 1$ mod 4. If $4n=p-1$ then $a^{2n}-1$ cannot, by Euler's differencing argument of Sec. 2.4, be 0 mod p for all $a=1,2,\dots,p-1$.

Therefore $p|(a^{2n}+1)$ for some $a=1,2,\ldots,p-1$. **8.** $2\sqrt{2/3}<2.\ 2\sqrt{3/3}=2$, but when $D=-3$ no divisor has norm 2. The last one is proved by Ex. 4. **9.** Not principal because

$$
\begin{array}{ccccccccccc}
8 & 7 & 5 & 2 & 7 & 7 & 2 & 5 & 7 & 8 \\
-1 & 3 & -6 & 7 & -9 & 2 & -9 & 7 & -6 & 3 & -1
\end{array}
$$

10.

$$
\begin{array}{ccccccc}
\frac{9}{2} & \frac{5}{2} & \frac{7}{2} & \frac{3}{2} & \frac{7}{2} & \frac{5}{2} & \frac{9}{2} \\
-1 & 7 & -3 & 5 & -5 & 3 & -7 & 1
\end{array}
$$

$$
\begin{pmatrix} 1 & -1 \\ 1 & 0 \end{pmatrix}
\begin{pmatrix} -2 & -1 \\ 1 & 0 \end{pmatrix}
\begin{pmatrix} 1 & -1 \\ 1 & 0 \end{pmatrix}
\begin{pmatrix} -1 & -1 \\ 1 & 0 \end{pmatrix}
\begin{pmatrix} 2 & -1 \\ 1 & 0 \end{pmatrix}
\begin{pmatrix} -1 & -1 \\ 1 & 0 \end{pmatrix}
\begin{pmatrix} \frac{9}{2}-\frac{1}{2}\sqrt{D} \\ 1 \end{pmatrix}
$$

gives $\frac{1}{2}(-261+25\sqrt{109})$. **11.** One reduction is necessary before $N(r-\frac{1}{2}\sqrt{D})<0$.

$$
\begin{array}{cccc}
-\frac{41}{2} & \frac{1}{2} & \frac{9}{2} \\
79 & 5 & -5 & 1
\end{array}
\qquad
\begin{pmatrix} -4 & -1 \\ 1 & 0 \end{pmatrix}
\begin{pmatrix} -1 & -1 \\ 1 & 0 \end{pmatrix}
\begin{pmatrix} \frac{9}{2}-\frac{1}{2}\sqrt{D} \\ 1 \end{pmatrix}
$$

gives $\frac{1}{2}(35-3\sqrt{101})$. **12.** r_0 can be found from $(1-\sqrt{79})^6\equiv 266+344\sqrt{79}$ mod 3^6, $(-344)^{-1}\equiv 251$, $251(1-\sqrt{79})^6\equiv -302-\sqrt{79}$. Alternatively, solve $r_0^2\equiv 79$ mod 3^j for $j=1,2,\ldots,6$. Then

$$
\begin{array}{cccc}
-302 & 52 & -10 \\
729 & 125 & 21 & 1
\end{array}
\qquad
\begin{pmatrix} -2 & -1 \\ 1 & 0 \end{pmatrix}
\begin{pmatrix} 2 & -1 \\ 1 & 0 \end{pmatrix}
\begin{pmatrix} -10-\sqrt{D} \\ 1 \end{pmatrix}
$$

gives $52+5\sqrt{79}$. **13.** In one step $(\frac{25}{2}-\frac{1}{2}\sqrt{-163})=(197,25)$. **14.** $\pm(5-3\sqrt{-165})$. **15.** The divisor of $r-\sqrt{D}$ (or $r-\frac{1}{2}\sqrt{D}$) is of the form $\amalg(p,u)^{\mu}\cdot\amalg(p,*)$. The divisors which divide this are determined by their norms.

7.5 1.

$$
\begin{array}{ccc}
\frac{7}{2} & \frac{5}{2} & \frac{7}{2} \\
1 & -3 & 3 & -1
\end{array}
$$

$(\frac{7}{2}+\frac{1}{2}\sqrt{61})(\frac{5}{2}+\frac{1}{2}\sqrt{61})(\frac{7}{2}+\frac{1}{2}\sqrt{61})/(-9)=-\frac{1}{2}(39+5\sqrt{61})=\varepsilon$. The computational shortcut of the exercises of Sec. 7.4 can be used to find ε by applying the cyclic method to $A_0=(-1,*)$. The most general unit is $\pm\varepsilon^n$. If $u^2-61v^2=1$ then $u+v\sqrt{61}=\pm\varepsilon^{2n}$ ($N(\varepsilon)=-1$). Clearly ε^2 does not have integer coefficients. $\varepsilon^3=-29718-3805\sqrt{61}$. ε^4 does not have integer coefficients. The smallest solution of Pell's equation is thus $u+v\sqrt{61}=\pm\varepsilon^6$, i.e. $v=2\cdot29718\cdot3805$. **2.** Application of the cyclic method to $A_0=(-1,*)$ gives $\varepsilon=\frac{1}{2}(-261+25\sqrt{109})$. Smallest solution of Pell's equation is ε^6.

7.6 **1.** 2 remains prime, $(-1,*) \sim I \sim (3, \pm 1)$. $\sqrt{D/3} < 5$. All divisors are principal. I is a representative set. **2.** $2, 3, 7$ remain prime. $\sqrt{|D|/3} < 9$. 5 ramifies and $(5,*)$, $(5,*)^2 \sim I$ form a representative set. **3.** The cyclic method applied to $(2,0)$ gives $(2,0) \sim (-1,*)(2,1)^3 \sim (3,-1) \sim (-1,*)(2,0) \sim (2,1)^3 \sim (-1,*)(3,-1)$, from which $(-1,*) \sim I$, $(2,0)^4 \sim I$. Let $A = (2,0)$. Then $A^2 \not\sim I$ by the cyclic method. $\sqrt{D/3} < 7$. $(3,-1) \sim A$. $(5,*) \sim (-1,*)(2,1)(3,1) \sim A^3 A^3 \sim A^2$. Thus $A, A^2, A^3, A^4 \sim I$ is a representative set. **4.** $(-1,*) \sim (2,1)^2 \sim (-1,*)(5,*) \not\sim I$ by the cyclic method. Therefore $(5,*) \sim I$, $(2,1)^4 \sim I$. The cyclic method applied to $(2,1)$ gives a long cycle (8 divisors) and shows $(2,1) \sim (7,2) \not\sim I$. Since $\sqrt{D/3} < 11$, the 4 powers of $(2,1)$ form a representative set. **5.** By the cyclic method $(2,*) \sim (5,-1)(13,-1) \not\sim I$, $(3,*) \sim (43,*)$. Then $(5,-1) \sim (5,-1)(13,-1)(13,1) \sim (2,*)(13,1)$. To find $(5,-1)^2$ set $r = -1+5k$, $-129 \equiv (-1+5k)^2 \bmod 25$, $k \equiv 13 \equiv -2 \bmod 5$, $r \equiv -11$. This gives $(5,-1)^2 \sim (2,*)(5,1) \not\sim I$. $(5,-1)^3 \sim (2,*)$. With $A = (5,-1)$ it follows that $A^2 \sim (2,*)(5,1)$, $A^3 \sim (2,*)$, $A^4 \sim (2,*)(5,-1)$, $A^5 \sim (5,1)$, $A^6 \sim I$. This leaves out $(3,*)$. Let $B = (3,*)$. Then $B^2 \sim I$, $AB \sim (11,-5)$(a short computation), $A^2 B \sim (7,2)$ (another), $A^3 B \sim (2,*)(3,*), A^4 B \sim (7,-2)$, $A^5 B \sim (11,5)$. Since $2\sqrt{D/3} < 14$, this accounts for all possible nonprincipal divisors except prime divisors of 13. However, from $(2,*) \sim (5,-1)(13,-1)$, $A^3 \sim A(13,-1)$, $(13,-1) \sim A^2$, $(13,1) \sim A^4$. Thus $A^i B^j$, $i = 0,1,2,3,4,5; j = 0,1$ is a representative set. **6.** Let $A = (2,*)$, $B = (3,*), C = (5,*)$. Then $I, A, B, C, AB, AC, BC, ABC$ are a representative set. **7.** $(2,*)(3,*)(5,*)$ divides $15 - \sqrt{D}$. **8.** By $r^2 + 163 s^2 =$ prime, neither r nor s is 0. If $u = 0$ then $\sqrt{-163}$ divides $(2x+1)$, which is impossible. If $v = 0$ then u is an integer which divides $\frac{1}{2}(k + \sqrt{D})$ so $u = \pm 1$.

7.7 **1.** See Ex. 15, Sec. 7.4. **2.** Suppose $A \sim B$ and the cyclic method applied to either A or B increases its norm. Let $x + y\sqrt{D}$ have divisor $A\bar{B}$. Let $a = N(A), b = N(B)$. Since $a \leqslant \sqrt{|D|/3}$, $b \leqslant \sqrt{|D|/3}$, one has $x^2 - Dy^2 \leqslant |D|/3$, from which $y = 0$ or $\pm \frac{1}{2}$. If $y = 0$ then x is an integer and the argument of the text gives $A = B$. If $y = \pm \frac{1}{2}$ then, as in the text, if r is the least halfinteger $r \equiv \frac{1}{2}\sqrt{D} \bmod A$, r' the least for which $r' \equiv \frac{1}{2}\sqrt{D} \bmod B$, then $|r| = |r'|$ and $a = b$. If $|r| = \frac{1}{2}a$ then, as in the text, $D = -3$; in this case $a \leqslant \sqrt{|D|/3} = 1$ implies $A = I = B$. Otherwise B is the successor of A. **4.** **Theorem.** *A matrix has the form* (4) *if and only if* $X \geqslant Z > W \geqslant 0$, $X \geqslant Y \geqslant W \geqslant 0$ *and* $XW - YZ = \pm 1$. *The representation is unique.* PROOF. To prove that the 3 conditions are necessary is simple. Let the Euclidean algorithm applied to X, Z give $X = q_0 Z + r_1, Z = q_1 r_1 + r_2, r_1 = q_2 r_2 + r_3, \ldots, r_{m-1} = q_m r_m + 1$ (X and Z are relatively prime by $XW - YZ = \pm 1$). Then the matrix on the right in (4) has X, Z as its first column if $n_k = q_0, n_{k-1} = q_1, \ldots, n_2 = q_m$, $n_1 = r_m$ (and, in particular, $k = m + 2$). The determinants of the two sides are equal if and only if $XW - YZ = (-1)^k$. If this is not the case, add another equation $r_m = (r_m - 1) \cdot 1 + 1$ to the Euclidean algorithm so that $n_k = n_{m+3} = q_0, n_{k-1} = q_1, \ldots, n_3 = q_m, n_2 = r_m - 1, n_1 = 1$. Then the two sides of (4) have equal determinants and equal first columns. Thus $X(W - W') = Z(Y - Y')$ and it is to be shown that $W = W', Y = Y'$. From $Y \equiv Y' \bmod X$ and $X \geqslant Y \geqslant 0, X \geqslant Y' \geqslant 0$ the desired conclusions follow unless Y or Y' is 0. But $Y \geqslant W, Y' \geqslant W'$ shows this is impossible. **5.** There are 2 ways of inserting signs in the right side of (4) to get a product as in the right side of (2), namely, n_1 can be given either sign. Either of these merely puts signs in front of X, Y, Z, W,

393

leaving their absolute values unchanged. Given $|X|$ and $|Z|$, there are two possibilities for (4), one with an odd and one with an even number j of factors. X and Y have like signs if and only if $n_1 > 0$. X and W have like signs if and only if j is odd. The remaining sign in front of (2) makes it possible to achieve each of the 8 possible patterns of signs in just 1 way. **6.** In the proof in the text $x^2 \geqslant r^2 y^2$ because $x = bu, y = bv, u^2 - Dv^2 = 1, D > r^2$. Similarly $ab + Dy^2 > (s - |b|)^2 y^2$.

7.8 **1.** Arguments in the text show "$p|(x^2 - Dy^2)$ for some relatively prime integers x, y" if and only if "p splits in quadratic integers for the determinant D" provided $p \neq 2$. If D is a square then $p|(x^2 - Dy^2), x, y$ relatively prime is always possible. Let $p_0|(x_0^2 - Dt^2 y_0^2)$ for relatively prime x_0, y_0 and let $p_0 \equiv p_1 \bmod 4Dt^2$; it is to be shown that $p_1|(x_1^2 - Dt^2 y_1^2)$ for relatively prime x_1, y_1. The only way x_0 and ty_0 can fail to be relatively prime is for p_0 to divide t, in which case $p_1 = p_0$. If $p_0 \neq p_1$ then there are, by assumption, relatively prime x_2, y_2 with $p_1|(x_2^2 - Dy_2^2)$. Set $d = $ g.c.d. $(t, y_2), t' = t/d, y_2' = y_2/d, x_1 = t' x_2, y_1 = y_2'$. **3.** If $q = 2$ then of the 4 classes $1, 3, 5, 7$ only 1 is a square. Assume $q \neq 2$. If y is relatively prime to q and $x^2 \equiv y^2 \bmod 4q$ then $(xy^{-1})^2 \equiv 1 \bmod 4q$, i.e. $x = ty$ where $t^2 \equiv 1 \bmod 4q$. Then $t \equiv \pm 1 \bmod 4$ and $t \equiv \pm 1 \bmod q$ and by the Chinese remainder theorem there are precisely 4 possibilities for t. (In fact, they are $\pm 1, 2q \pm 1$.) Thus squaring is a 4 to 1 mapping. **4.** When $D = -1$, 1 splits so -1 does not split. When $D = 2$, 1 splits so -1 splits and the remaining classes do not split. When $D = -2$, $3|1^2 + 2 \cdot 1^2$ splits so -1, -3 do not split. **5.** Let $a = \prod p^\mu$ be relatively prime to $4D$. If p splits then p^μ is in a splitting class. If p remains prime then $A = (p)^\nu A'$ where A' is prime to p, $\mu = 2\nu$, and $p^\mu = (p^\nu)^2$ is in a splitting class. Therefore a is in a splitting class. **6.** Let S represent splitting and N nonsplitting. Since multiplication by an S carries S to S, it must, by counting, carry N to N. Thus S times N is always N. Since multiplication by N carries S to N, it must, by counting, carry N to S, as was to be shown. **7.** $\left(\frac{-7}{15}\right) = +1$ means that if p is a *prime* $\equiv 15 \bmod 4 \cdot 7$ then -7 is a square mod p. (For example, $6^2 \equiv -7 \bmod 43$.) **8.** It suffices to consider the case where q is prime. Let $p = p_1 \cdots p_\mu p_1' \cdots p_\nu'$ where the p_i are 1 mod 4 and the p_i' are 3 mod 4. Then

$$\left(\frac{q}{p}\right) = \prod \left(\frac{q}{p_i}\right) \prod \left(\frac{q}{p_i'}\right) = \prod \left(\frac{p_i}{q}\right) \prod \left(\frac{-p_i'}{q}\right) = \left(\frac{(-1)^\nu p}{q}\right)$$

as desired. **9.** Suppose Euler's theorems hold for D'. Each class of integers mod $4D$ relatively prime to $4D$ is contained in a class mod $4D'$ relatively prime to $4D'$. Call a class a splitting class for D if it is contained in a splitting class for D'. The argument of Ex. 1 shows that if p is in a splitting class for D then $p|(x^2 - Dy^2)$ for x, y relatively prime. Euler's theorems are easily deduced.

7.9 **1.** By Ex. 5, Sec. 7.8, if $m = N(A)$ where A is relatively prime to $4D$ then $\left(\frac{D}{m}\right) = +1$. Since $\left(\frac{D}{m}\right) = \left(\frac{k}{m}\right)\left(\frac{m}{p_1}\right)\cdots\left(\frac{m}{p_\nu'}\right)$ where $k = D/p_1 \cdots p_\mu(-p_1')\cdots(-p_\nu') \equiv D \bmod 4$ and in fact $k \equiv D \bmod 8$ when $2|D$, the Jacobi symbol is simply the product of the signs in the genus of A. By Sec. 8.3 every divisor is equivalent to such an A. Given any possible character, use the Chinese remainder theorem to find an integer m such that $\left(\frac{k}{m}\right), \left(\frac{m}{p_1}\right), \cdots, \left(\frac{m}{p_\nu'}\right)$ are the signs specified. By Dirichlet's theorem there is a prime $p \equiv m \bmod 4D$. Since $(\frac{D}{p}) = 1$, p splits and its prime factors have the specified character. **2.** Only (3) is nontrivial. Because $(x/y)^2 \equiv -n \equiv (u/v)^2 \bmod k$, $xv \equiv \pm yu \bmod k$. Change the sign of y if necessary to make $xv \equiv yu$. Then $k^2 = (xu + nyv)^2 + n(xv - yu)^2$ is divisible by k^2 to give $1 = a^2 + nb^2$. If $b = 0$ then $xv = yu, x^2k = x^2u^2 + nx^2v^2 = u^2k, x = \pm u, v = \pm y$ as desired. If $b \neq 0$ then $n = 1$, $a = 0$, $xu = -yv$ and $x^2k = v^2k, x = \pm v, y = \pm u$. If $n \not\equiv 1 \bmod 4$ then the proof is easier using divisor theory. **3.** When $D = -165$, $A \sim \overline{A}$ for all divisors A. $A_1 \neq \overline{A}_1$ because (1) and (2) show that $(2, *), (3, *), (5, *), (11, *)$ do not divide A_1. If $u + v\sqrt{D}$ has divisor $\overline{A}_1 A_2$ and $x + y\sqrt{D}$ has divisor $A_1 A_2$ then $x \pm y\sqrt{D}$ is not a unit times $u + v\sqrt{D}$, contrary to (3). If $x + y\sqrt{D}$ has divisor $(p, u)^n$ then $(p)(p, u)^{n-2}$ is principal, say the divisor of $r + s\sqrt{D}$, and again (3) fails. **4.** If $a = $even, $b = $odd. Assume $a = $odd. ak is the norm of the divisor of $(ax + y\sqrt{D})$ where $D = -ab$. This divisor has the form $(p_1, *) \cdot (p_2, *) \cdots (p_n, *)A$ where $k = N(A)$. If k is prime then (1) and (2) are obvious. As for (3), if $k = au^2 + bv^2$ then $au + v\sqrt{D}$ has divisor $(p_1, *) \cdots (p_n, *)B$ where $k = N(B)$. If k is prime then $B = A$ or $B = \overline{A}$, and $au \pm v\sqrt{D}$ is ± 1 times $ax + y\sqrt{D}$, as desired. Conversely, the argument of Ex. 3 shows that if A is not prime then there is $c + d\sqrt{D}$ with norm ak but not equal to $\pm(ax \pm y\sqrt{D})$; since a must divide c, this contradicts (3). **5.** The exclusions are $165 + 2^2 = 0 \cdot 165 + 13^2$, $165 + 14^2 = 361 = 0 \cdot 165 + 19^2$, $165 + 26^2 = 0 \cdot 165 + 29^2$, $165 + 28^2 = 2^2 \cdot 165 + 17^2$, and $165 + 32^2 = 2^2 \cdot 165 + 23^2$. **6.** $5 \not\equiv 3 \bmod 4$. There are 2 classes and their characters are distinct $(+ +$ and $- -)$. **7.** If $p \equiv 3 \bmod 4$, $p \equiv 3$ or $7 \bmod 10$ then $(\frac{-5}{p}) = (\frac{-1}{p})(\frac{p}{5}) = (-1)(-1) = +1$ and p splits when $D = -5$. The character of its prime divisors is $- -$. The product of a prime divisor of p_1 and one of p_2 is in the principal genus, therefore in the principal class. **8.** $-Dy^2 \equiv -D$ or $0 \bmod 4D$. $p + D = 180 = 5 \cdot 6^2$. By trial and error $5 \equiv 20^2 \bmod 79$. Then $59^2 \equiv 5 \bmod 4 \cdot 79$ and $38^2 \equiv 180 \bmod 4 \cdot 79$ are easily found. Then $733 = 38^2 - 79 \cdot 3^2$ is quickly found. Divide by $3, 5, \ldots, 23$ to prove that 733 is prime. Thus 733 splits and its prime divisors are principal. From $101 \equiv 733 \bmod 4D$ it follows not only that 101 splits but that its prime divisors are in the same genus as those of 733, namely, the principal genus. It is to be shown that they are not in the principal class. By trial and error $79 \equiv 1089 = 33^2 \bmod 101$. The cyclic method gives $(101, 33) \sim (2, *)(5, 2)$ and, using $B = (3, 1)$ as in Sec. 7.6, $(2, *)(5, 2) \sim I \cdot B^2 \not\sim I$.

7.10 **1.** I and $(2, *)(3, *)(5, *)$ are principal; 4 ambiguous classes. **2.** I and $(31, *)$ are principal; 1 ambiguous class. **3.** $I, (2, *), (31, *), (2, *)(31, *)$ are principal; $(-1, *)$ is not principal; 2 ambiguous classes. **4.** $I, (-1, *) \cdot (2, *)(7, *), (3, *)(5, *)$, and $(-1, *)(2, *)(3, *)(5, *)(7, *)$ are principal. From the characters $- + - +$, $- - + +$, $- - - -$, $+ - - +$ of $(-1, *)$, $(2, *), (3, *), (7, *)$ it is clear that all 8 characters with an even number of minuses occur as characters of ambiguous classes. Therefore there are at least 8 ambiguous classes. Therefore there are 8 ambiguous classes. **5.** I, $(-1, *), (61, *)$, and $(-1, *)(61, *)$ are principal; 1 ambiguous class. **6.** $(-1, *)$ is not in the principal genus. I, $(-1, *)(2, *)$, $(-1, *)(59, *)$, and $(2, *)(59, *)$ are principal; 2 ambiguous classes. **7.** For $D > 0$ there are at least 4 principal ambiguous divisors. When D is prime and $D \equiv 1 \bmod 4$ there are just 4 ambiguous divisors. Thus $(-1, *)$ is principal. When D is prime and $D \equiv 3$, mod 4, $(-1,*)$ is not in the principal genus. **8.** There are 8 divisors $(-1,*)^a(p,*)^b(q,*)^c$. The character of $(-1,*)$ is $--$. Thus there are 2 classes, each making a separate genus. The character of $(p,*)(q,*)\sim(-1,*)$ is $--$ so $(p,*)$ and $(q,*)$ cannot both have character $--$ and cannot both have character $++$. **9** The cycle of I contains $(-1,*)$, say $(-1,*)=A_{2j+1}$. Then $N(A_i)=-N(A_{2j+1-i})$ and therefore $N(A_j)=-N(A_{j+1})$. When $D=13$, $j=0$ and $N(\frac{3}{2}-\frac{1}{2}\sqrt{13})=-1$ gives $3^2-13=-2^2$, $13=3^2+2^2$ as desired. When $D=233$, $a_4=-a_5=4$, $r_4=13/2$, from which $13^2+(2\cdot4)^2=D$.

7.11 **1.** $n=1\geqslant s$ gives $s=1,h=sa=a$. That is, every class is ambiguous. **2.** In VI, $(-1,*)$ has character $--$. The second sign of the character of $(p,*)\sim$ $(-1,*)(2,*)$ is $+1$ because $p\equiv3$ mod 8, $D\equiv-2$ mod 8. Therefore the first sign must also be $+1$. **3.** If $g=a$ then $n=s$.

8.1 **1. (a)** The techniques of Chap. 7 show that the only quadratic integers with divisor $(3,1)$ are $\pm\frac{1}{2}(1-\sqrt{-11})$ and these are not in the suborder. **(b)** (2) is the divisor of 2 and (2)(3,1) is the divisor of $\pm(1-\sqrt{-11})$. **(c)** $(3,1)^2$ is the divisor of $\pm(1-\sqrt{-11})^2/4=\pm\frac{1}{2}(5+\sqrt{-11})$ and $(3,1)^3$ the divisor of $\pm(4-\sqrt{-11})$. **(d)** $(5,2)(3,-1)$ is the divisor of $2-\sqrt{-11}$. **(k)** If x,y are odd integers then $y\pm x$ is divisible by 4 for one of the choices of sign and $(x+y\sqrt{-11})(1\pm\sqrt{-11})/4$ has integer coefficients. **2.** If $(2)B\sim I$ then $(2)AB$ is principal if and only if A is. This is false when $A=(3,1)$. **3.** (13,5) divides $5-\sqrt{-1}$. Therefore it divides $30-6\sqrt{-1}$ and $4-\sqrt{-36}$ which has divisor $(2,*)^2(13,5)$. The next step reduces -4 mod 4 to give $-\sqrt{-36}$ with divisor $(2,*)^2(3,1)(3,-1)$. The divisors are thus $(13,5)$, $(2,*)^2$, $(3,1)(3,-1)$, $(2,*)^2$, $(3,1)(3,-1),\ldots$. Both $(2,*)^2$ and $(3,1)(3,-1)$ are principal, so neither is equivalent to $(13,5)$. **5.** In the case $D'\equiv1$ mod 4 let v be the least positive integer such that $u+v\omega$ is in the order. (The order contains at least one element $\pm(x+y\omega)$ in which $y\neq0$.) If $x+y\omega$ is any element of the order let $y=qv+r$ where $0\leqslant r<v$. Since $x+y\omega-q(u+v\omega)=(x-qu)+r\omega$ is in the order, $r=0$. Thus the order is the order of index v. The same method applies when $D'\not\equiv1$ mod 4. **6.** Let A be a given divisor relatively prime to 3. All divisors are principal when $D=2$, so A is the divisor of $x+y\sqrt{2}$ for some x,y. The most general element with divisor A is $\pm(x+y\sqrt{2})\cdot(1-\sqrt{2})^{2n}$. Mod 3 this is $\pm(x+y\sqrt{2})$ or $\pm(-y+x\sqrt{2})$ and it is in the order for some n if and only if x or y is divisible by 3. Thus $(-1,*)$ is not principal. If A is not principal then $A(-1,*)$ is. **7.** $(1-\sqrt{D})^3=1+3D-(3+D)\sqrt{D}\equiv0$ mod 8. If D is negative and $\neq-1$ or -3 then the only units are ±1. Therefore no element of the order has the same divisor as $\frac{1}{2}(1-\sqrt{D})$ except when $D=-3$. When $D=-3$ there is only one divisor class. **10.** $11+1^2$ has none of the 4 forms, so 11 should not be convenient. This was shown in Ex. 7. $13+1=2\cdot7$, $13+4=17$, $13+9=2\cdot11$, $13+16=29$, $13+25=2\cdot19$, $13+36=7^2$, so 13 should be convenient. When $D=-13$, I and $(2,*)$ are a representative set. Of the numbers from 80 to 89 only 85 and 88 satisfy Euler's criterion.

8.2 **2.** The congruence $a^2\equiv-1$ mod 65 or, what is the same, $a^2\equiv-1$ mod 5 and $a^2\equiv-1$ mod 13 has solutions $a\equiv\pm2$ mod 5, $a\equiv\pm5$ mod 13, i.e. $a\equiv\pm8$ or ±18 mod 65. By the cyclic method

$$\begin{array}{cccc} & -18 & -2 & 0 \\ 65 & 5 & 1 & 1 \end{array}$$

Reverse this and set $u=0$, $v=1$ to find

$$\begin{pmatrix} 0 & -1 \\ 1 & -2 \end{pmatrix}\begin{pmatrix} 0 & -1 \\ 1 & -4 \end{pmatrix}\begin{pmatrix} 0 \\ 1 \end{pmatrix}=\begin{pmatrix} 4 \\ 7 \end{pmatrix},$$

$4^2+7^2=65$. In the same way the solution -8 gives $1^2+8^2=65$. **3.** $a^2\equiv-19$ mod 121 gives $a^2\equiv3$ mod 11, $a\equiv\pm5$ mod 11, $a\equiv\pm5+11k$ mod 121, $a\equiv\pm49$ mod 121. The cyclic method shows that the form $121r^2+98rs+20s^2$ is equivalent to $5g^2-2gh+4h^2$. This latter is equivalent to the given form. Reversing the steps and multiplying matrices as in the text gives $g=-5$, $h=-2$ as the representation and therefore $x=2$, $y=-5$ as a solution of the original problem. **4.** $x=8$ is a solution of $x^2\equiv-5$ mod 23. Then

$$\begin{array}{cccccc} 20 & 7 & 1 & 8 \\ 15 & 27 & 2 & 3 & 23 \end{array}$$

gives the solution $x=-11$, $y=8$. **5.** The solution $a=49$ of $a^2\equiv79$ mod $3\cdot43$ does not lead to a representation. The solution $a=-49$ leads to

$$\begin{array}{cccccc} 59 & 22 & 8 & -5 & -49 \\ 42 & 81 & 5 & -3 & 18 & 129 \end{array}$$

and the representation $x=-15$, $y=13$. **6.** The two solutions 19 and 33 of $a^2\equiv-3\bmod182$ gives the two representations $x=-1$, $y=10$ and $x=-6$, $y=11$.

8.3 1. I and $(-1,*)$ is a representative set (Sec. 7.6). The forms x^2-67y^2 and $-x^2+67y^2$ therefore have the desired property. **2.** x^2+165y^2, $2x^2+2xy+83y^2$, $3x^2+55y^2$, $5x^2+33y^2$, $6x^2+6xy+174y^2$, $10x^2+10xy+19y^2$, $11x^2+15y^2$, $13x^2+4xy+13y^2$ and the negatives of all of these. **3.** x^2-79y^2, $3x^2\pm2xy-26y^2$, $5x^2\pm4xy-15y^2$, $-x^2+79y^2$. **4.** x^2+161y^2, $3x^2\pm2xy+54y^2$, $9x^2\pm2xy+18y^2$, $6x^2\pm2xy+27y^2$, $2x^2+2xy+81y^2$, $7x^2+23y^2$, $10x^2\pm6xy+17y^2$, $11x^2\pm4xy+15y^2$, $5x^2\pm4xy+33y^2$, $14x^2+14xy+15y^2$, and the negatives of all of these. **5.** x^2+11y^2, $3x^2\pm2xy+4y^2$ and their negatives. **6.** $\pm(x^2-18y^2)$.

8.4 1. The number of reduced binary quadratic forms for a given determinant is obviously finite. **2.** Let a unit with norm 1 be given, and let M be the 2×2 matrix corresponding to multiplication of $ax+(b-\sqrt{D}\,)y$ by this unit. The proof in the text shows that E_1^nM is a truncation of E_1^N for large N, from which $M=\pm E_1^k$ for integral k. Thus the given unit is $\pm\varepsilon^k$. If $ax^2+2bxy+cy^2$ is reduced then the cyclic method gives $A=QAQ^t$ where

$$A=\begin{pmatrix} a & b \\ b & c \end{pmatrix}, \qquad Q=\begin{pmatrix} n_1 & -1 \\ 1 & 0 \end{pmatrix}\begin{pmatrix} n_2 & -1 \\ 1 & 0 \end{pmatrix}\cdots\begin{pmatrix} n_k & -1 \\ 1 & 0 \end{pmatrix},$$

the n's being the quotients $(r_i+r_{i+1})/a_{i+1}$. The unit $\varepsilon=g+h\sqrt{D}$ can then be found using the formula

$$Q=\begin{pmatrix} g+bh & -ah \\ ch & g-bh \end{pmatrix}$$

$(Q,a,b,c$ known), which follows from formulas in the text. The cyclic method applied to $-ax^2+2bxy-cy^2$ is the same as it is applied to $ax^2+2bxy+cy^2$ except that the signs on the lower row are reversed. Thus one is reduced if and only if the other is reduced and if they are reduced and equivalent the cyclic method gives an explicit equivalence

$$\begin{pmatrix} -a & b \\ b & -c \end{pmatrix} = Q \begin{pmatrix} a & b \\ b & c \end{pmatrix} Q^t, \quad Q = \begin{pmatrix} n_1 & -1 \\ 1 & 0 \end{pmatrix} \cdots \begin{pmatrix} n_j & -1 \\ 1 & 0 \end{pmatrix}.$$

This corresponds to multiplication by $u+v\sqrt{D}$ followed by division by a where $N(u+v\sqrt{D})=-a^2$. Then $(u+v\sqrt{D})/a$ is a quadratic integer with norm -1. Explicitly it is $g+h\sqrt{D}$ where

$$Q = \begin{pmatrix} g+bh & -ah \\ -ch & -g+bh \end{pmatrix}.$$

Conversely, if $g+h\sqrt{D}$ is any quadratic integer with norm -1 this formula gives a proper equivalence between $ax^2+2bxy+cy^2$ and $-ax^2+2bxy-cy^2$. The proof that the above procedure for finding a unit with norm -1 (when there i. one) gives a unit whose square is ε seems a bit tricky. Perhaps the simplest method is to observe that, except for signs,

$$Q = \begin{pmatrix} |n_1| & 1 \\ 1 & 0 \end{pmatrix} \begin{pmatrix} |n_2| & 1 \\ 1 & 0 \end{pmatrix} \cdots \begin{pmatrix} |n_k| & 1 \\ 1 & 0 \end{pmatrix}.$$

3. The cyclic method applied to x^2-67y^2 and the formulas of Sec. 8.2 show that $QAQ^t=A$ where

$$A = \begin{pmatrix} 1 & 8 \\ 8 & -3 \end{pmatrix}, \quad Q = \begin{pmatrix} -5 & -1 \\ 1 & 0 \end{pmatrix} \begin{pmatrix} 2 & -1 \\ 1 & 0 \end{pmatrix} \cdots \begin{pmatrix} 16 & -1 \\ 1 & 0 \end{pmatrix}$$

$$= \begin{pmatrix} -96578 & 5967 \\ 17901 & -1106 \end{pmatrix}.$$

$-ah=5967, h=-5967, g+bh=-96578, g=-48842.$

4. The cyclic method gives

$$\begin{pmatrix} 1 & \frac{7}{2} \\ \frac{7}{2} & -3 \end{pmatrix} = \begin{pmatrix} 37 & 5 \\ -15 & -2 \end{pmatrix} \begin{pmatrix} -1 & \frac{7}{2} \\ \frac{7}{2} & 3 \end{pmatrix} \begin{pmatrix} 37 & -15 \\ 5 & -2 \end{pmatrix}.$$

Then $-ah=5$, $h=5$, $g+bh=37$, $g=39/2$ and the unit is $\frac{1}{2}(39+5\sqrt{61})$. **5.** Since $(5,1)$ divides $1-\sqrt{11}$ it divides $3-\sqrt{99}$ and the cyclic method gives

3		-3	8	6	3	7	8
5	-20	5	-7	9	-10	5	

The divisors $(5,1)$ and $(-1,*)(2,*)(5,-1)$ are equivalent because they correspond to forms in the same period. $(5,1)^2\sim(-1,*)(2,*)$, $(5,1)^4\sim I$, and no power of $(5,1)$ is equivalent to $(-1,*)$. Every form is equivalent to one in which $0 < b < \sqrt{99}$, $b + |a| > \sqrt{99}$, and $b + |c| > \sqrt{99}$. It is easily checked that all such forms are equivalent to forms in the 8 periods $(-1,*)^a(5,1)^b$ that have been found. **6.** $(5,2)^3 \sim (11,*)$ and $(5,2)^j$ for $0 \leqslant j < 6$ is a representative set. **7.** Choose u, v so that $xv - yu = 1$. Then

$$\begin{pmatrix} x & y \\ u & v \end{pmatrix}\begin{pmatrix} a & b \\ b & c \end{pmatrix}\begin{pmatrix} x & u \\ y & v \end{pmatrix} = \begin{pmatrix} m & n \\ n & k \end{pmatrix}$$

where $n = axu + bxv + byu + cyv$ and $k = au^2 + 2buv + cv^2$ gives a form $mX^2 + 2nXY + kY^2$ properly equivalent to the given form whose first coefficient is m. Thus $n^2 - D \equiv 0 \bmod m$. If, in the process of Sec. 8.2, the solution d of $d^2 \equiv D \bmod m$ is $\equiv n \bmod m$ then the forms $mX^2 + 2nXY + kY^2$ and $mX^2 + 2dXY + eY^2$ are properly equivalent by

$$\begin{array}{ccc} n & -n & d \\ m & k & m & e. \end{array}$$

The theorem of this section then proves that the cyclic method reduces $mx^2 + 2dxy + ey^2$ and $ax^2 + 2bxy + cy^2$ to the same period and therefore gives an explicit equivalence. **8.** Find an explicit equivalence of the given form with a reduced form so that the problem is to find all representations by a reduced form. Use the method of Sec. 8.2 to find one representation (if there is one). This can be written as a column matrix and, when $D > 0$, the most general representation is a power of a 2×2 matrix times this column matrix, and the 2×2 matrix can be found explicitly. **9.** $(3,1)$ is not in the principal class. The method of Ex. 11, Sec. 5.6 gives $37^{38} \equiv 43$ as a square root of $37 \bmod 151$. The method of Sec. 8.2 then gives $274^2 - 37 \cdot 45^2 = 151$.

8.5 1. See Ex. 6, Sec. 8.4. **2.** $(2,*) \not\sim I$, $(7,1)^2 \sim (13,*)$. A representative set is $(2,*)^i(7,1)^j$ for $i = 0, 1; j = 0, 1, 2, 3$. **3.** A representative set is $(5,1)^j$, $0 \leqslant j < 18$. The reduced forms corresponding to these powers, in order, are $(1,0,531)$, $(5,3,108)$, $(25,12,27)$, $(17,8,35)$, $(7,-1,76)$, $(20,7,29)$, $(4,-1,133)$, $(20,3,27)$, $(19,-1,28)$, $(9,0,59)$, $(19,1,28)$, ... **4.** $(7,2)^j$ where $j = 0, 1, 2, 3$ is a representative set. **5.** $(2,1) \sim (-1,*)(2,0)$ and $(2,1)^j$ for $j = 0, 1, 2, 3$ is a representative set. **6.** $(5,1)^j$ for $0 \leqslant j \leqslant 8$ is a representative set. **7.** $(5,1)^3 \sim I$. **8.** $(5,1)^6 \sim I$. **11.** All forms with determinant 5 are properly equivalent to $x^2 + 4xy - y^2$.

8.6 1. Quadratic reciprocity shows that -5 is a square mod p if and only if p is $1, 3, 7$, or $9 \bmod 20$ ($p \neq 2$ or 5). Let $d^2 \equiv -5 \bmod p$ and $e = (d^2 + 5)/p$. This form has character $- -$ and is therefore equivalent to $2x^2 + 2xy + 3y^2$. Thus p_1, p_2 both have representations by this form. Then apply formula (2). **2.** Actually the assumption that a and a' be odd is not essential and with $a = 4, b = b' = 0, a' = 9$ one has the composition $(4x^2 + 9y^2)(9u^2 + 4v^2) = 36X^2 + Y^2$ where $X = xu - yv$ and $Y = 4xv + 9yu$. Alternatively, the class of the given form corresponds to the divisor $(13,5)$. $(13,5)^2$ divides $(5 - \sqrt{-1})^2$, $12 - 5\sqrt{-1}$, $34 \cdot 12 - 170\sqrt{-1}$, $-99 - \sqrt{-1}$, $82 - \sqrt{-36}$. Then the cyclic method shows that $(13,5)^2 \sim I$.

9.2 1. For $D = -6$ and -10 see Table 7.8.1. The other cases are $-13, -14$, and -17. For these use Euler's theorems. **2.** No.

9.5 **2.** $c_1 = -i/2 = -c_3$, $c_2 = c_4 = 0$, $L(1,\chi) = -(i/2)\log((1+i)/(1-i)) = -i\log\sqrt{i}$. **3.** Let $k = 1 + \chi(2)2 + \chi(3)3 + \cdots$. Then $k = -m\Sigma + 2\Sigma'$ and $k = -m(\frac{D}{2})\Sigma + 4(\frac{D}{2})\Sigma'$, from which $k = -m(\frac{D}{2})\Sigma[2(\frac{D}{2})-1]^{-1}$. Then $h = L(1,\chi)\sqrt{m}/\pi = |\Sigma|$ when $(\frac{D}{2}) = 1$ and $= \frac{1}{3}|\Sigma|$ when $(\frac{D}{2}) = -1$, except when $D = -3$. **4.** $(\frac{-7}{n}) = (\frac{n}{7})$. $h = |\Sigma| = |1+1-1| = 1$. $(\frac{-11}{n}) = (\frac{n}{11})$. $h = \frac{1}{3}|\Sigma| = \frac{1}{3}|1-1+1 +1+1| = 1$. $(\frac{-15}{7}) = -(\frac{-15}{8}) = -(\frac{-15}{2}) = -1$ and $h = |1+1+0+1+0+0-1| = 2$. $(\frac{-19}{7}) = -(\frac{-19}{12}) = -(\frac{-19}{3}) = +1$ and $h = \frac{1}{3}$ the sum of $+ - - + + + + - + = 1$. For $D = -23, -31, -35, -39, -43, -47$ the class numbers are $h = 3, 3, 2, 4, 1, 5$ respectively. **5.** $(\frac{D}{1}) + (\frac{D}{3})3 + (\frac{D}{5})5 + \cdots = (\frac{D}{1})[1-(2D-1)+(2D+1)-(4D-1)] + (\frac{D}{3})[3-(2D-3)+(2D+3)-(4D-3)] + \cdots = -4D\Sigma$, from which $h = |\Sigma|$. The required class numbers are $2, 2, 2, 2, 4, 4, 4$ respectively. **6.** When $D = -5$ the values of $(\frac{s}{p})$ are $+ - - +$ and $h = 2$. When $D = -13$ they are $+ - + | + - - | - + | + - -$ and $h = 2$. The other class numbers are $4, 4, 6, 4, 2, 8$, and 6, in that order. **7.** When $D = -6$ the value $s = 1$ lies in the 3rd octant $6/8 < s \leqslant 9/8$ and $h = 2$. The other class numbers are 4, 2, 4, and 6. **8.** The required class numbers are $2, 6, 4,$ and 4. **9.** (a) When $\alpha = e^{2\pi i/\lambda}$ Gauss's sum is $1 + \alpha + \alpha^4 + \alpha^9 + \cdots = 1 + 2\theta_0 = \theta_0 - \theta_1$. (b) If m is prime then $D \equiv 1 \bmod 4$, and $|D|$ is prime. Thus the left side of (6) is $\theta_0 - \theta_1$. The right side is $\sqrt{\lambda} = \sqrt{D} = \sqrt{m}$ if $D > 0$ and $i\sqrt{\lambda} = i\sqrt{m}$ if $D < 0$. (c) $i - (-i) = 2i = i\sqrt{4}$. $\sqrt{i} + i\sqrt{i} - (-\sqrt{i}) - (-i\sqrt{i}) = 2\sqrt{i} \cdot (1+i) = 2\sqrt{i}\sqrt{2i} = i\sqrt{m}$. $\sqrt{i} - i\sqrt{i} - (-\sqrt{i}) + (-i\sqrt{i}) = 2\sqrt{i}(1-i) = \sqrt{m}$. (d) $\Sigma\alpha^k(\frac{D}{k}) = \Sigma\alpha^{4a}\alpha^{pb}(\frac{-1}{4a+pb})(\frac{4a+pb}{p}) = (\Sigma\alpha^{pb}(\frac{-1}{pb}))(\Sigma\alpha^{4a}(\frac{a}{p})) = (\frac{-1}{p}) \cdot 2i \cdot \Sigma\alpha^{4a}(\frac{-D}{a})$. If $D = p \equiv -1 \bmod 4$ this is $(-1) \cdot 2i(i\sqrt{p}) = 2\sqrt{p} = \sqrt{m}$. If $D = -p$ where $p \equiv 1 \bmod 4$ it is $2i\sqrt{p} = i\sqrt{m}$. (e) When $D = 2p$, $\Sigma\alpha^k(\frac{D}{k}) = \Sigma\alpha^{8a+pb}.$ $(\frac{2}{8a+pb})(\frac{p}{8a+pb})$. When $p \equiv 1 \bmod 4$ this is $(\frac{2}{p})(\Sigma\alpha^{pb}(\frac{2}{b}))(\Sigma\alpha^{8a}(\frac{8a}{p})) = \sqrt{8}\,\Sigma\alpha^{8a}(\frac{p}{a}) = \sqrt{m}$. When $p \equiv -1 \bmod 4$ it is $(\frac{-2}{p})(\Sigma\alpha^{pb}(\frac{-2}{b}))(\Sigma\alpha^{8a}(\frac{8a}{p})) = (\frac{-1}{p})i\sqrt{8} \cdot i\sqrt{p} = \sqrt{m}$. The proof when $D = -2p$ is analogous. (f) like (e). (g) (6) has been deduced from (5) for prime n. (6) says $m\bar{c}_1 = i\sqrt{m}$ when $D < 0$. Thus $\mu = c_1 = -i\sqrt{m}/m$ determines the sign in (3). This determines the sign in (4). (h) is a straightforward check. (i) If $p \equiv -1 \bmod 4$ set $D = -p$. Then h or $3h$ is $C_1 + C_2 + C_3 + C_4 = $ sum of the first half of the Legendre symbols mod p. If $p \equiv 1 \bmod 4$ and $D = -p$ then $h = 2(C_1 + C_2)$ is twice the sum of the first quarter of the Legendre symbols. (j) The sign of the class number formula determines the sign of the Gaussian sum.

9.6 **1.** See Ex. 5, Sec. 8.4. **2.** Using (3), $h = 1 \cdot \frac{1}{3} \cdot 6 \cdot (1 + \frac{1}{2})(1 - \frac{1}{3}) = 2$. I and $(-1, *)$ constitute a representative set. **3.** Since $t = 2$, h_0 is multiplied by $2 - (\frac{D}{p})$ as was to be shown. The rule also works when $D = -3$.

9.7 **6.** Let a be the product of p_i to the μ_i for $i = 1, 2, \ldots, n$. The real characters mod a are of the form $\chi_1\chi_2 \cdots \chi_n$ where the character χ_i corresponding to p_i is either the principal character mod p_i or the Legendre symbol mod p_i (if $p_i \neq 2$) or one of the 4 real characters mod 8 (if $p_i = 2$). Thus it can be written in the form $\chi(n) = (\frac{\varepsilon}{n})(\frac{r}{n})(\frac{n}{q_1})(\frac{n}{q_2}) \cdots (\frac{n}{q_p})(\frac{n}{q'_1})(\frac{n}{q'_2}) \cdots (\frac{n}{q'_\sigma})$ where the q_i are primes $\equiv 1 \bmod 4$ from the p_i, the q'_i are primes $\equiv -1 \bmod 4$ from the p_i, $\varepsilon = \pm 1$ and $r = 1$ or 2, except that χ is zero whenever n is divisible by one of the p_i whereas the right side is zero only when n is divisible by one of the q_i or q'_i. The right side is $(\frac{D}{n})$ where $D = \varepsilon r q_1 q_2 \cdots q_\rho (-q'_1)(-q'_2) \cdots (-q'_\sigma)$ (see Sec. 7.8). Thus $L(s, \chi) = \prod(1 - \chi(p)p^{-s})^{-1} = (1 - \chi(r_1)r_1^{-s}) \cdots (1 - \chi(r_\tau)r_\tau^{-s})L(s, \chi_D)$ where the r's are the p's that are not q's.

A.1   **1.** If $a = a'd$ and $b = b'd$ then $qa + rb = (qa' + rb')d$. **2.** If $a = ub$ and $b = va$ then $a = uva = 1 \cdot a$ and $uv = 1$. If $u > 1$ then $1 = uv > v \geqslant 1$ which is impossible. Therefore $a = 1 \cdot b = b$. **3.** Let $a = a_0, b = a_1, a_{i+1} = q_i a_i + a_{i-1}$, $a_0 > a_1 > \cdots > a_n = d$. Then $a_{n-2} = q_{n-1} a_{n-1} + d$. If $ua_i = d + va_{i+1}$ then $ua_i + q_i va_i = d + v(q_i a_i + a_{i+1})$, $Ua_i = d + va_{i-1}$ and if $ua_i + d = va_{i+1}$ then $Ua_i + d = va_{i-1}$ where $U = u + q_i v$. If $ax + 1 \equiv 0 \bmod b$ then $a^2 x^2 + 2ax + 1 \equiv 0 \bmod b$, $a^2 x^2 + 2(ax + 1) \equiv 1 \bmod b$, $Ax \equiv 1 \bmod b$ where $A = a^2 x$. If a and b are relatively prime then $y(= ax) \equiv 0 \bmod a$, $y \equiv 1 \bmod b$ has a solution. By symmetry $z \equiv 1 \bmod a$, $z \equiv 0 \bmod b$ has a solution. Then $u = cz + dy$ satisfies $u \equiv c \bmod a$, $u \equiv d \bmod b$, as required. **4.** (a) no solutions. (b) $x = 2 + 5k \equiv 3 \bmod 7, 8 + 20k \equiv 12 \bmod 7, 1 \equiv 5 + k \bmod 7, 3 \equiv k \bmod 7, k = 3 + 7n, x = 17 + 35n$. (c) $x = 2 + 295k, 7 \equiv 2 + 15k \bmod 56, 1 \equiv 3k \bmod 56, 19 \equiv 57k \bmod 56, x = 5607 + 56 \cdot 295n$. **5.** See Ex. 3 above. **6.** Find $b_2, b_3, \ldots, b_n$ such that $b_i \equiv 1 \bmod c_1$, $b_i \equiv 0 \bmod c_i$. Set $d_1 = b_2 b_3 \cdots b_n$. Then $d_1 \equiv 1 \bmod c_1$, $d_1 \equiv 0 \bmod c_i$ for $i = 2, 3, \ldots, n$. In the same way find d_i such that $d_i \equiv 1 \bmod c_i$, $d_i \equiv 0 \bmod c_j$ $(j \neq i)$. The given problem is then solved by $a_1 d_1 + a_2 d_2 + \cdots + a_n d_n$. **7.** If a is negative set $ax = (-a)(-x)$. Let d be the greatest common divisor of a and b. By Ex. 3, $d = ua + vb$ for integers u and v. If c is not divisible by d then clearly there is *no* solution. If $c = qd$ then $x = -qu$, $y = -qv$ is a solution. If X, Y is a particular solution then the most general solution is $x = X + r, y = Y + s$ where $ar + bs = 0$. This is equivalent to $a'r + b's = 0$ where $a = a'd$, $b = b'd$. In any solution of this equation $a'r \equiv 0 \bmod b'$, $r \equiv 0 \bmod b'$ (b' and a' are relatively prime), $r = kb'$, $s = -kb'a'/b' = -ka'$. Thus the most general solution is $x = -qu + kb'$, $y = -qv - ka'$.

A.2   **2.** $\gamma^a \equiv \gamma^b \bmod p$ if and only if $a \equiv b \bmod p - 1$. If k is relatively prime to $p - 1$ then $(\gamma^k)^a \equiv 1 \bmod p$ only if $p - 1$ divides a, and γ^k is a primitive root $\bmod\ p$. Conversely, if γ^k is a primitive root then $\gamma \equiv (\gamma^k)^a$ for some a, $ak \equiv 1 \bmod p - 1$, and k is relatively prime to $p - 1$.

Bibliography

The boldface numbers in the entries below indicate the pages of this book on which the entries are cited.

[A1]   *Comptes Rendus de l'Académie des Sciences*, Paris, vol. 24 (1847), p. 310 et seq. **76**

[B1]   Bell, E. T., *The Last Problem*, Simon and Schuster, New York, 1961. **24**

[B2]   Borevich, Z. I. and Shafarevich, I. R., *Theoria Chisel*, Nauka, Moscow, 1964 (English transl., *Theory of Numbers*, Academic Press, New York, 1966). **233, 244**

[B3]   Bourbaki, N., *Élements d'Histoire des Mathématiques*, 2nd ed., Hermann, Paris, 1969. **250**

[C1]   Cayley, A., Tables des formes quadratiques binaires, *Jour. für Math.* (Crelle), 60 (1862) 357–372 (*Mathematical Papers*, vol. 5, Cambridge, 1892, Johnson Reprint Corp., New York and London, 141–156). **333**

[C2]   Cohn, H., *A Second Course in Number Theory*, Wiley, New York and London, 1962. **333**

[C3]   Colebrooke, H. T., *Algebra with Arithmetic and Mensuration from the Sanskrit of Brahmegupta and Bhaskara*, London, 1817. **27**

[D1]   Dedekind, R., see [D7].

[D2]   Dickson, L. E., *History of the Theory of Numbers*, (3 vols.), Carnegie Institute of Washington, 1919, 1920, and 1923 (reprint, Chelsea Pub. Co., New York, 1971). **10, 11, 13, 27, 37, 45, 70, 74, 312, 375**

[D3]   Diophanti Alexandrini arithmeticorum libri sex, et de numeris multangulis liber unus. Cum commentariis C. G. Bacheti V. C. et observationibus D. P. de Fermat Senatoris Tolosani, Toulouse, 1670. **2**

[D4]   Dirichlet, P. G. L., Mémoire sur l'impossibilité de quelques équations indéterminées du cinquième degré, *Jour. für Math.* (Crelle) 3 (1828) 354–375 (*Werke*, vol. 1, 21–46).

403

[D5] Dirichlet, P. G. L., Démonstration du théorème de Fermat pour le cas des 14ièmes puissances, *Jour. für Math.* (Crelle) 9 (1832) 390–393 (*Werke*, vol. 1, 189–194). **73**

[D6] Dirichlet, P. G. L., Über eine neue Anwendung bestimmter Integrale auf die Summation endlicher oder unendlicher Reihen, *Abh. König. Preuss. Akad. Wiss.*, 1835, 391–407. **360**

[D7] Dirichlet, P. G. L., *Vorlesungen über Zahlentheorie*, herausgegeben und mit Zusätzen versehen von R. Dedekind, Vieweg und Sohn, Braunschweig, 1893 (reprint Chelsea Pub. Co., New York, 1968). **308, 326, 343, 358, 360, 365**

[E1] Edwards, H. M., *Advanced Calculus*, Houghton Mifflin, Boston, 1969. **143, 216**

[E2] Edwards, H. M., *Riemann's Zeta Function*, Academic Press, New York, 1974.

[E3] Edwards, H. M., The background of Kummer's proof of Fermat's Last Theorem for regular primes, *Arch. Hist. Exact Sci.*, 14 (1975) 219–236. **80, 128**

[E4] Edwards, H. M., Postscript to "The background of Kummer's proof of Fermat's Last Theorem for regular primes", (to appear). **80, 95**

[E5] Euclid, *Elements*, T. L. Heath, ed., Cambridge Univ. Press, New York, 1908, 3 vols. (reprint Dover, New York, 1956).

[E6] Euler, L., *Introductio in Analysin Infinitorum*, Bousquet et Socios., Lausanne, 1748 (*Opera* (1), vol. 8). **351**

[E7] Euler, L., Theoremata circa divisores numerorum in hac forma $paa \pm qbb$ contentorum, *Comm. Acad. Sci. Petrop.* 14 (1751) 151–181; *Opera* (1), 2, 194–222. **287**

[E8] Euler, L., De numeris, qui sunt aggregata duorum quadratorum, *Nov. Comm. Acad. Sci. Petrop.* 4 (1758) 3–40; *Opera* (1), 2, 295–327. **46**

[E9] Euler, L., *Vollständige Anleitung zur Algebra*, St. Petersburg, 1770; *Opera* (1), vol. 1. **33, 43**

[E10] Euler, L., Extrait d'une lettre de M. Euler à M. Beguelin en mai 1778, *Nouv. Mém. Acad. Sci. Berlin*, 1776, 1779, 337–339; *Opera* (1), 3, 418–420. **311**

[E11] Euler, L., Observationes circa divisionem quadratorum per numeros primos, *Opuscula Analytica*, 1 (1783) 64–84; *Opera* (1), 3, 497–512. **287**

[E12] Euler, L., *Opera* (1), 4, Vorwort des Herausgebers (III. Grosse Primzahlen) and related papers 708, 715, 718, 719, and 725 of Euler. **311, 312**

[F1] de Fermat, P., *Oeuvres*, 3 vols., Gauthier–Villars, Paris, 1891, 1894, 1896.

[F2] de Fermat, P., *Observations on Diophantus*, originally published in [D3]; also [F1], 1, 291–342; French transl. [F1], 3, 241–274.

[F3] de Fermat, P., Letter to Pascal, 25 Sept. 1654. *Oeuvres de Pascal*, 4, 437–441; also [F1], 2, 310–314. **55**

[F4] de Fermat, P., Letter to Digby, sent by Digby to Wallis on 19 June 1658, published by Wallis in [W3]; republished [W2] and [F1], 2, 402–408; French transl. [F1], 3, 314–319. **48**

[F5] de Fermat, P., Letter to Carcavi, dated August 1659, [F1], 2, 431–436. **24, 56**

[F6] Fuss, P.-H., ed., *Correspondance Mathématique et Physique*, Imp. Acad. Sci., St. Petersburg, 1843, vol. 1 (reprint, Johnson Reprint Corp., New York and London, 1968). **45, 46, 287**

[G1] Gauss, C. F., *Disquisitiones Arithmeticae*, Leipzig, 1801; republished, 1863, as vol. 1 of *Werke*; French transl., *Recherches Arithmétiques*, Paris, 1807; republished, Hermann, Paris, 1910; German transl. in [G2]; English transl., Yale, New Haven and London, 1966.

[G2] Gauss, C. F., *Untersuchungen über Höhere Arithmetik*, (German transl. of [G1] and other works on number theory) H. Maser, transl., Springer-Verlag, Berlin, 1889 (reprint, Chelsea Pub. Co., New York, 1965). **331, 346**

[G3] Gauss, C. F., Theorematis arithmetici demonstratio nova, *Comm. Soc. Reg. Sci. Gott.*, 16 (1808); *Werke*, 2, 3–8; German transl. in [G2]. **287, 302**

[G4] Gauss, C. F., Summatio quarumdam serierum singularium, *Comm. Soc. Reg. Sci. Gott. Rec.* 1 (1811); *Werke* 2, 11–45; German transl. in [G2]. **359, 360**

[G5] Gauss, C. F., Theorematis fundamentalis in doctrina de residuis quadratici demonstrationes et ampliationes novae, *Comm. Soc. Reg. Sci. Gott. Rec.* 4 (1818); *Werke*, 2, 49–64; German transl. in [G2]. **174**

[G6] Gauss, C. F., Theoria residuorum biquadraticorum. Commentatio prima, *Comm. Soc. Reg. Sci. Gott. Rec.* 6 (1828), Commentatio secunda, 7 (1832); *Werke*, vol. 2, 67–148; German transl. in [G2]. **85**

[G7] Gauss, C. F., *De nexu inter multitudinem classium, in quas formae binariae secundi gradus distribuunter, earumque determinantem*, commentatio prior Societati Regiae exhibita, 1834; *Werke*, vol. 2, 269–303; German transl. in [G2].

[G8] Gauss, C. F., Letter to Sophie Germain, *Werke* 10 (part 1), 70–73; also in *Oeuvres Philosophiques de Sophie Germain*, Paris, 1896, p. 275. **61**

[G9] Grube, F., Über einige Eulersche Sätze aus der Theorie der quadratischen Formen, *Zeitschr. Math. Phys.*, 19 (1874) 492–519. **312**

[H1] Heath, T. L., *Diophantus of Alexandria*, Cambridge University Press, 1910 (reprint, Dover, New York, 1964). **5, 12, 15, 26, 27**

[H2] Hecke, E., *Vorlesungen über die Theorie der Algebraischen Zahlen*, Akad. Verlag, Leipzig, 1923 (reprint, Chelsea Pub. Co., New York, 1948). **174**

[H3] Hilbert, D., Die Theorie der algebraischen Zahlkörper (called "the Zahlbericht"), *Jahresber. der Deut. Math. Verein.*, 4 (1897) 175–546; *Gesammelte Abhandlungen*, 1, 63–363. **173**

[I1] Ince, E. L., *Cycles of Reduced Ideals in Quadratic Fields*, Brit. Assn. for the Advancement of Science, Mathematical Tables, vol. 4, Camb. Univ. Press, New York 1966. **333**

[J1] Jacobi, C. G. J., Über die Kreistheilung und ihre Anwendung auf die Zahlentheorie, *Monatsber. Akad. Wiss. Berlin*, 1837, 127–136; also *Jour. für Math.* (Crelle) 30 (1846) 166–182; and *Werke*, 6, 254–274. **58, 290**

[J2] Jacobi, C. G. J., Über die complexen Primzahlen, welche in der theorie der Reste der 5ten, 8ten, und 12ten Potenzen zu betrachten sind, *Monatsber. der Akad. Wiss. Berlin*, 1839, 86–91; also *Jour. für Math.* (Crelle) 19 (1839) 314–318; and *Werke*, 6, 275–280; French transl., *Jour. de Math.*, 8 (1843) 268–272. **79, 89**

[J3] Jacobi, C. G. J., *Canon Arithmeticus*, Academie-Verlag, Berlin, 1956 (originally published by Typis Academicus Berolini, 1839.). **379**

[J4] Johnson, W., Irregular primes and cyclotomic invariants, *Math. of Computation*, 29 (1975) 113–120. **244**

[K1a] Kline, M., *Mathematical Thought from Ancient to Modern Times*, Oxford, New York, 1972. **25**

Bibliography

[K1] Kronecker, L., Zur Geschichte des Reciprocitätsgesetzes, *Monatsber. Akad. Wiss. Berlin*, 1875, 267–274; *Werke*, vol. 2, 1–10. **291**

[K2] Kronecker, L., Ein Fundamentalsatz der allgemeinen Arithmetik, *Jour. für Math.* (Crelle) 100 (1887) 490–510; *Werke*, vol. 3, 211–240. **86**

[K3] Kronecker, L., Über den Zahlbegriff, *Jour. für Math.* (Crelle) 101 (1887) 337–355; *Werke*, vol. 3, 251–274. **86**

[K4] Kummer, E. E., *Collected Papers*, André Weil, ed., vol. 1, *Contributions to Number Theory*, Springer-Verlag, Berlin Heidelberg New York, 1975.

[K5] *Festschrift zur Feier des* 100. *Geburtstages Eduard Kummers*, Abh. Gesch. Math. Wiss., Teubner, Berlin and Leipzig, 1910; reprint, [K4], 31–133. **335**

[K6] Kummer, E. E., De numeris complexis, qui radicibus unitatis et numeris integris realibus constant, Gratulationschrift der Univ. Breslau zur Jubelfeier der Univ. Königsberg; reprint, *Jour. de Math.*, 12 (1847) 185–212, and [K4], 165–192. **78, 102**

[K7] Kummer, E. E., Zur Theorie der Complexen Zahlen, *Monatsber. Akad. Wiss. Berlin*, 1846, 87–96; also *Jour. für Math.* (Crelle) 35 (1847) 319–326 and [K4] 203–210. **79, 142, 152, 320, 335**

[K8] Kummer, E. E., Über die Zerlegung der aus Wurzeln der Einheit gebildeten complexen Zahlen in ihre Primfactoren, *Jour. für Math.* (Crelle) 35 (1847) 327–367; also [K4] 211–251. **79, 128, 141, 142, 153, 250**

[K9] Kummer, E. E., Extrait d'une lettre de M. Kummer à M. Liouville, *Jour. de Math.*, 12 (1847) p. 136; also [K4] p. 298.

[K10] Kummer, E. E., Beweis des Fermat'schen Satzes der Unmöglichkeit von $x^\lambda + y^\lambda = z^\lambda$ für eine unendliche [sic] Anzahl Primzahlen λ, *Monatsber. Akad. Wiss. Berlin*, 1847, 132–141, 305–319; also [K4] 274–297. **182**

[K11] Kummer, E. E., Bestimmung der Anzahl nicht äquivalenter Classen für die aus λten Wurzeln der Einheit gebildeten complexen Zahlen, *Jour. für Math.* (Crelle) 40 (1850) 93–116; also [K4] 299–322. **153, 250**

[K12] Kummer, E. E., Zwei besondere Untersuchungen über die Classen-Anzahl und über die Einheiten der aus λten Wurzeln der Einheit gebildeten complexen Zahlen, *Jour. für Math.* (Crelle) 40 (1850) 117–129; also [K4] 323–335.

[K13] Kummer, E. E., Allgemeiner Beweis des Fermat'schen Satzes, dass die Gleichung $x^\lambda + y^\lambda = z^\lambda$ durch ganze Zahlen unlösbar ist, für alle diejenigen Potenz-Exponenten λ, welche ungerade Primzahlen sind und in den Zählern der ersten $(\lambda - 3)/2$ Bernoulli'schen Zahlen als Factoren nicht vorkommen, *Jour. für Math.* (Crelle) 40 (1850) 130–138; also [K4] 336–344.

[K14] Kummer, E. E., Mémoire sur la théorie des nombres complexes composés de racines de l'unité et de nombres entiers, *Jour. de Math.* 16 (1851) 377–498; also [K4] 363–484. **226, 232**

[K15] Kummer, E. E., Über die Irregularität von Determinanten, *Monatsber. Akad. Wiss. Berlin*, 1853, 194–200; also [K4] 539–545. **226**

[K16] Kummer, E. E., Über die den Gaussischen Perioden der Kreistheilung entsprechenden Congruenzwurzeln, *Jour. für Math.* (Crelle) 53 (1857) 142–148; also [K4] 574–580. **128**

[K17] Kummer, E. E., Über diejenigen Primzahlen λ, für welche die Klassenzahl der aus λten Einheitswurzeln gebildeten complexen Zahlen durch λ teilbar ist, *Monatsber. Akad. Wiss. Berlin*, 1874, 239–248; also [K4] 945–954. **226**

[L1] Lagrange, J. L., *Sur la solution des problèmes indéterminés du second degré*,

Mém. de l'Acad. Roy. Sci. et Belles-Lettres, Berlin, 23 (1769) (*Oeuvres*, 2, 377–535). **180, 339**

[L2] Lagrange, J. L., *Additions* to Euler's *Algebra*. First published 1774, Lyon, in vol. 2 of a French translation of Euler's *Algebra* [E9]; republished in vol. 7 of Lagrange's *Oeuvres*, 5–180, and in vol. (1) 1 of Euler's *Opera*. **298**

[L3] Lagrange, J. L., *Oeuvres*, vol. 2, Paris, 1868, 531–535. **77**

[L4] Lagrange, J. L., *Oeuvres*, vol. 14, Paris, 1892, 298–299. **60**

[L5] Lamé, G., Démonstration général du théorème de Fermat, *Comptes Rendus*, 24 (1847) 310–315.

[L6] Lamé, G., Mémoire sur la résolution, en nombres complexes, de l'équation $A^5 + B^5 + C^5 = 0$, *Jour. de Math.* 12 (1847) 137–184. **97**

[L7] Legendre, A. M., *Sur quelques objets d'analyse indéterminée et particulièrement sur le théorème de Fermat*, Mém. Acad. R. Sc. de l'Institut de France, 6, Paris 1827; also appeared as 2nd Supplement to 1808 edition of [L8]. **62, 70**

[L8] Legendre, A. M., *Théorie des Nombres*, vol. 2, Paris, 1830, pp. 361–368 (reprint, Blanchard, Paris, 1955).

[M1] Mordell, L. J., On a simple summation of the series $\Sigma e^{2s^2\pi i/n}$, *Messenger of Math.*, 48 (1919) 54–56. **360**

[N1] Neugebauer, O., *The Exact Sciences in Antiquity*, Brown University Press, Providence, 1957, Chapter 2. **4**

[N2] Newton, I., *The Correspondence of Isaac Newton*, vol. 2, H. W. Turnbull, ed., Cambridge, 1960, 110–160. **346**

[N3] Nörlund, N. E., *Differenzenrechnung*, Springer-Verlag, Berlin, 1924. **232**

[O1] Ohm, M., Etwas über die Bernoullischen Zahlen, *Jour. für Math.* (Crelle) 20 (1840) 11–12. **231**

[S1] Shanks, D., Five Number-theoretic Algorithms, *Proceedings of the 2nd Manitoba Conference on Numerical Mathematics*, 51–70, Univ. of Manitoba, Winnipeg, 1972. **180**

[S2] Smith, D. E., *A Source Book in Mathematics*, McGraw-Hill, New York, 1929 (reprint, Dover, New York, 1959). **79**

[S3] Smith, H. J. S., *Report on the Theory of Numbers*, originally published in six parts as a Report of the British Assn; reprinted, 1894, in *The Collected Mathematical Papers of H. J. S. Smith*; (reprinted, both separately and as part of the *Mathematical Papers*, Chelsea Pub. Co., New York, 1965). **32, 81, 232, 287, 302, 360**

[S4] Steinig, J., On Euler's idoneal numbers, *Elemente der Math.*, 21 (1966) 73–88. **311**

[T1] Toeplitz, O., *Die Entwicklung der Infinitesimal Rechnung*, Springer-Verlag, Berlin-Göttingen-Heidelberg, 1949; English transl. *The Calculus: A Genetic Approach*, Univ. of Chicago, Chicago, 1963.

[W1] Wagstaff, S., Fermat's Last Theorem is true for any exponent less than 100000 (Abstract), *AMS Notices*, No. 167 (1976) p. A–53. **244**

[W2] Wallis, J., *Opera Mathematica*, 3 vols., Oxford, 1695–1699 (reprint, G. Olms, Hindesheim-New York, 1972).

[W3] Wallis, J., ed., *Commercium Epistolicum*, Oxford, 1658 (reprint, [W2]; French transl., [F1]).

[W4] Weil, A., La cyclotomie jadis et naguère, *L'Enseignment Mathématique* (2) 20 (1974) 247–263. **360**

Index

Graduate Texts in Mathematics

(continued from page ii)